CABI is a trading name of CAB International

CABI Head Office
Nosworthy Way
Wallingford
Oxfordshire OX10 8DE
UK

Tel: +44 (0)1491 832111
Fax: +44 (0)1491 833508
Email: cabi@cabi.org
Web site: www.cabi.org

CABI North American Office
875 Massachusetts Avenue
7th Floor
Cambridge, MA 02139
USA

Tel: +1 617 395 4056
Fax: +1 617 354 6875
Email: cabi-nao@cabi.org

A catalogue record for this book is available from the British Library, London, UK.

ISBN-13: 978-0-85198-758-3
ISBN-10: 0-85198-758-3

First published 1993
Transferred to print on demand 2008

Printed and bound in the UK by CPI Antony Rowe, Eastbourne.

Dedicated to

J. Basil Goodey (1914–1965)

Photograph by courtesy of Rothamsted Experimental Station.

'The classification of any group of organisms is a matter of convenience; it is made by man to enable him to name and catalogue specimens easily and is, therefore, largely artificial. Allied to classification is the relationship between the organisms being classified. The systematist always has this in mind, and is pleased when his classification seems to show inter-relationships, for he can then suppose that his system reflects some natural law. It is important, however to keep classification and inter-relationships separate, for it is all too easy to confuse them and be plunged into immediate difficulties. Only when a group of organisms has been extensively catalogued can inter-relationships be seriously considered; collection and cataloguing of nematodes have not yet progressed far enough for their inter-relationships to be seriously studied.'

J.B. Goodey

'. . . species-making is a nice job for biologists who have an eye for small differences but lack that peculiar quality which makes the good experimenter. Some of them abuse one another with all the zeal of grammarians for breaking various sacred rules which have been laid down for the naming of species. Such rules are, of course, needed to avoid confusion, but a life spent in enforcing them leads one a long way from the scientific spirit.'

J.B.S. Haldane

lowed here are not intended to be controversial (although the recognition of the Aphelenchida might be seen as provocative in some quarters!) and a belief, or otherwise, in their veracity need not be prejudicial to, or compromise, the purpose of the book. Indeed, the systematic arrangement employed here should be regarded as a rather plastic matrix within which the genera are somewhat tentatively embedded, rather than a rigid framework, as the emphasis is placed on the genus as the most accessible, easily definable and important taxon.

Keys to generic level are included. They are unlikely to be perfect (there seems to be an exception to every rule in Nematology), but should be used judiciously and with some allowance for heterogeneity. Each genus is diagnosed, illustrated and provided with a comprehensive list of nominal species and, where appropriate, synonyms. The illustrations have been carefully selected from a variety of sources and can be used either in conjunction with the diagnoses or independently. The synonymy is by no means complete, particularly in genera such as *Aphelenchoides*, where many descriptions are of low standard and type material unavailable. General notes on the bionomics and distribution are apppended and some 22 species, mostly those of economic importance, are individually described in greater detail. In addition, a general introduction to each of the three systematic sections includes an historical review, basic morphology and a summary of the bionomics.

The fourth section, the bibliography, is intended to be an intrinsic component of the book by providing a complete literature source to the authorities for all the specific epithets and their combinations. In addition, a selection of specialized publications on topics which, of necessity, could only be dealt with briefly in the text is also included.

Almost inevitably, such a wide-ranging study is predominantly based on a study of the literature, modulated by personal experience of many of the genera and examination of type and other material of the more contentious taxa. The species lists are largely based on the standard checklists and on information gleaned from the CAB International database and are up to date as of October, 1992.

When I was first asked to write this book, it was billed as a sequel to M.R. Siddiqi's *Tylenchida, Parasites of Plants and Insects.* An awesome thought! However, after some debate, I decided to attempt a text which, although not as taxonomically high powered, would appeal to, and hopefully be more accessible to, a wider audience – yet still be of value to the dedicated taxonomist. Whether this intention has been satisfactorily achieved, or whether the book has fallen ignominiously between two stools, awaits the verdict of that ultimate arbiter, the user . . . .

I enjoyed writing it, anyway.

David Hunt
November, 1992

Preface

In recent years there has been a number of monographic-style works, of varying accessibility to the non-specialist, dealing with the systematics and taxonomy of the Nematoda. Andrássy, with *A Taxonomic Review of the Suborder Rhabditina*, published in 1983, covered one of the major saprophagous groups of Nematoda; Siddiqi's *magnum opus* encompassed the plant and insect parasitic Tylenchida in 1986; Jairajpuri covered the Mononchida and Jairajpuri and Ahmad, in 1992, produced a generic monograph of the Dorylaimida. The raison d'être of this book was to provide a practical, working guide to the taxa of the Aphelenchida, together with the Longidoridae and Trichodoridae, in such a form as to facilitate usage by both specialist taxonomists and pluripotential nematologists.

These three groups, totalling almost 1000 species from some 40 genera, contain a number of species which are of known agricultural importance, both in their own right as parasites of fungi or higher plants and/or as vectors of plant nepo and tobraviruses. The aphelenchs, longidorids and trichodorids are found worldwide with representatives frequenting all edaphic types. In addition, the aphelenchs also colonize the aerial parts of some plants and form a variety of associations with insects ranging from simple phoresy through to complex endoparasitism. Indeed, two remarkable species of aphelench, *Bursaphelenchus xylophilus*, the pine wilt nematode, and *Rhadinaphelenchus cocophilus*, causing red ring disease of coconut palms – both extremely damaging phytoparasites of economically important crops – exploit their association with insects to vector them efficiently from host to host.

In the last decade, nematode systematics has witnessed a considerable turmoil, often leading to confusion and despair for the non-taxonomist as genera and other taxa rose and fell and new combinations proliferated in the literature. That some consolidation was both necessary and long overdue is beyond dispute, although many of the proposals have almost certainly gone too far and some judicious inflationary adjustment would not go amiss. The schemes followed here are not intended to be controversial (although the recognition of the Aphelenchida might be seen as provocative in some quarters!) and a belief, or otherwise, in their veracity need not be prejudicial to, or compromise, the purpose of the book. Indeed, the systematic arrangement employed here should be regarded as a rather plastic matrix within which the genera are somewhat tentatively embedded, rather than a rigid framework, as the emphasis is placed on the genus as the most accessible, easily definable and important taxon.

Keys to generic level are included. They are unlikely to be perfect (there seems to be an exception to every rule in Nematology), but should be used judiciously and with some allowance for heterogeneity. Each genus is diagnosed, illustrated and provided with a comprehensive list of nominal species and, where appropriate, synonyms. The illustrations have been carefully selected from a variety of sources and can be used either in conjunction with the diagnoses or independently. The synonymy is by no means complete, particularly in genera such as *Aphelenchoides*, where many descriptions are of low standard and type material unavailable. General notes on the bionomics and distribution are apppended and some 22 species, mostly those of economic importance, are individually described in greater detail. In addition, a general introduction to each of the three systematic sections includes an historical review, basic morphology and a summary of the bionomics.

The fourth section, the bibliography, is intended to be an intrinsic component of the book by providing a complete literature source to the authorities for all the specific epithets and their combinations. In addition, a selection of specialized publications on topics which, of necessity, could only be dealt with briefly in the text is also included.

Almost inevitably, such a wide-ranging study is predominantly based on a study of the literature, modulated by personal experience of many of the genera and examination of type and other material of the more contentious taxa. The species lists are largely based on the standard checklists and on information gleaned from the CAB International database and are up to date as of October, 1992.

When I was first asked to write this book, it was billed as a sequel to M.R. Siddiqi's *Tylenchida, Parasites of Plants and Insects.* An awesome thought! However, after some debate, I decided to attempt a text which, although not as taxonomically high powered, would appeal to, and hopefully be more accessible to, a wider audience – yet still be of value to the dedicated taxonomist. Whether this intention has been satisfactorily achieved, or whether the book has fallen ignominiously between two stools, awaits the verdict of that ultimate arbiter, the user . . . .

I enjoyed writing it, anyway.

David Hunt
November, 1992

Acknowledgements

I thank M.R. Siddiqi and D.J. Hooper for generously making available their extensive knowledge and experience and for their numerous comments, suggestions and useful discussions during the writing of this book. Especial gratitude goes to Jo Gill for inputting the major portion of the text and bibliography, for tracking down elusive references and for her unlimited *enthusiasm*. I also thank Ann Hall and Gill Kaser for their valuable contributions to the inputting, Richard Tranfield and Janet Rowe for photographic assistance, Chris Small for library assistance, Sue Hines for SEM work, Sarah Cox for German translations and all those who provided type and other material for examination or made available photographs and illustrations.

Glossary of some Nomenclatorial Terms

apud. After; in the sense of

binomen (*pl.* **binomina**). The combination of two names, generic and specific, to form the scientific name of a species

Code. Abbreviation of 'International Code of Zoological Nomenclature'

et. And

genus dubium (*pl.* **genera dubia**). A genus, or genera, of doubtful validity

genus incertae sedis (*pl.* **genera incertae sedis**). A genus, or genera, of uncertain systematic position

genus inquirendum (*pl.* **genera inquirenda**). Genus or genera requiring further study to validate

homonym. The same nomen for a taxon. It may be **junior** or **senior**, depending on date of authorship; **objective** or **subjective** depending on whether it is based on the same taxon or arises from a matter of opinion, or **primary** or **secondary** depending on whether the homonymy arose from an original combination of two names under the same genus or arose from a subsequent combination

incertae sedis. Of uncertain taxonomic position

lapsus calami. Literally a slip of the pen. A printer's error or inadvertent spelling mistake by the author

nec. Not

new combination. The first combination of a generic name and previously published specific name. Usually abbreviated to n.comb.

nomen (*pl.* **nomina**). A name or names

nomen novum (*pl.* **nomina nova**). New or replacement name(s)

nomen nudum (*pl.* **nomina nuda**). A name, or names, without a valid description (literally a naked name)

objective. Factual, not based on opinion (see 'subjective')

partim. In part

sensu lato. In the widest or broad sense

sensu stricto. In the restricted or narrow sense
species dubia (*pl.* **species dubiae**). Species of dubious validity
species incertae sedis. Species of uncertain systematic position
species indeterminata (*pl.* **species indeterminatae**). Species where the description is too poor to allow reidentification
species inquirenda (*pl.* **species inquirendae**). Species requiring further study in order to validate
subjective. Based on opinion (see 'objective')
synonym. A taxon regarded as identical to another. It may be **junior** or **senior**, depending on date of authorship
taxon (*pl.* **taxa**). Taxonomic unit(s) in a systematic scheme
vel. Or

A more extensive glossary can be found in the *International Code of Zoological Nomenclature*, Third Edition, 1985, International Trust for Zoological Nomenclature, 338pp.

I Order Aphelenchida

1 Historical Review

The oldest known reference to an aphelench nematode is that of Gasparrini (1864) when he described *Anguillula caprifici*. The species was not well described and its placement has suffered dispute. The following year H. Charlton Bastian (Fig. 1.1.a), in an illustrated monograph containing descriptions of 100 new species (Bastian, 1865), proposed four species in his new genus *Aphelenchus*, viz. *A. avenae*, from the lower sheaths of oat leaves from Broadmoor, Berkshire (and subsequently designated as the type of *Aphelenchus* by Bastian in 1905, unpublished letter); *A. parietinus*, from a yellow lichen, later transferred to *Aphelenchoides* by Steiner (1932a), and *A. pyri* and *A. villosus*, from decaying pears and moss respectively, both of which are now regarded as *species inquirendae* (Fig. 1.1.b). Bastian's elegantly written monograph provides an intriguing insight into the beginning of nematology as a science and reveals that even in 1865 systematics could be a contentious subject!

> Elberth has also unfortunately attempted to range all the twenty-three new marine species discovered by himself under five of the old genera, with the result of greatly confusing the nomenclature, since I feel convinced, from an examination of the beautiful figures he has given of these forms, that they cannot properly be included under less than from eight to twelve distinct genera.
>
> (Bastian, 1865)

Bütschli (1873) described *Aphelenchus rivalis* (Fig. 1.2) and three years later de Man (1876) described *Aphelenchus modestus* and von Linstow (1876), *A. erraticus*. De Man also described *A. agricola* in 1881 (now regarded as a synonym of *A. avenae*) and *A. tenuicaudatus* in 1895, a species which now resides in *Seinura*. N.A. Cobb (1888, 1891, 1893b) published several species of aphelench of which *Aphelenchus longicaudatus* is now in *Seinura*.

The first plant parasitic aphelench to be named was *Aphelenchus* (now *Aphelenchoides*) *fragariae* by Ritzema Bos (1890). He subsequently proposed

Fig. 1.1.a. Henry Charlton Bastian (1837–1915), who described *Aphelenchus avenae*, the type species of the Aphelenchida in his 1865 monograph. (Photograph courtesy The Royal Society of London.)
Opposite: **Fig. 1.1.b.** Plate X from Bastian's 1865 monograph. Illustrations from the original descriptions of *Aphelenchus avenae* (97, 98), *A. villosus* (99, 100, 101), *A.* [= *Aphelenchoides*] *parietinus* (102, 103) and *A. pyri* (103a,b,c).

Aphelenchus olesistus in 1892, but this is now regarded as a synonym of *Aphelenchoides fragariae*. Other species were named by Noack (1898), Zimmermann (1898) and Maupas (1900) and from then on the pace of new descriptions began to accelerate.

The genus *Aphelenchoides* was proposed by Fischer (1894) with *A. kuehnii* as the type species, but was overlooked until Steiner's paper (1932a). Cobb

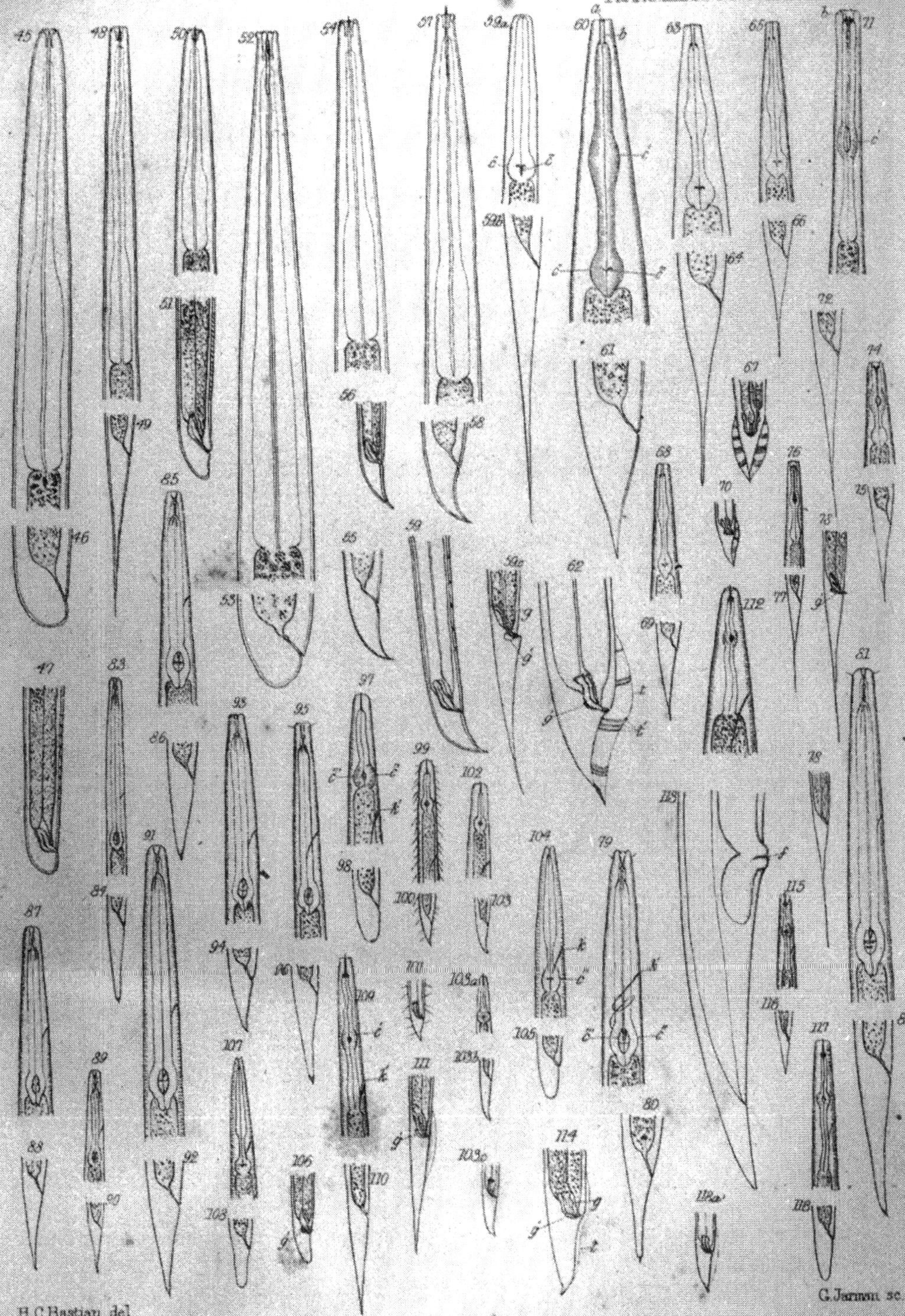

H. C. Bastian del.

G. Jarman sc.

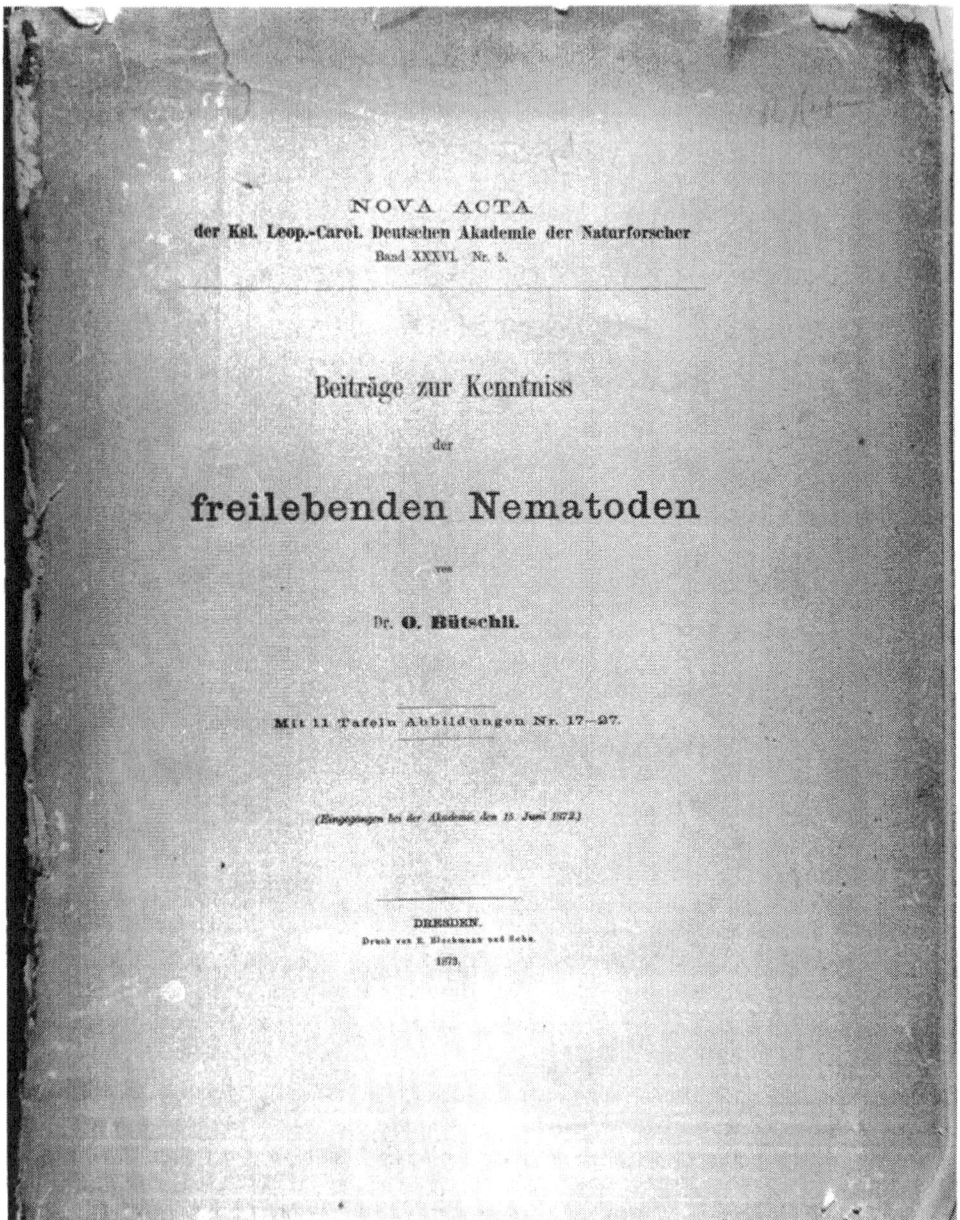

NOVA ACTA
der Ksl. Leop.-Carol. Deutschen Akademie der Naturforscher
Band XXXVI. Nr. 5.

Beiträge zur Kenntniss
der
freilebenden Nematoden
von
Dr. O. Bütschli.

Mit 11 Tafeln Abbildungen Nr. 17–27.

(Eingegangen bei der Akademie den 15. Juni 1873.)

DRESDEN.
Druck von E. Blochmann und Sohn.
1873.

Fig. 1.2. Title page of O. Bütschli's 1873 paper.

(1913) erected the genus *Isonchus*, subsequently to be synonymized under *Aphelenchus*.

After *A. fragariae*, the next important plant parasite to be named was *Aphelenchus ritzemabosi* by Schwartz (1911). Taylor (1917) described the same species as *Tylenchus ribes*, as did Stewart (1921) under the binomen

Aphelenchus phyllophagus. *Aphelenchus ritzemabosi* was placed in *Aphelenchoides* by Steiner (1932a) along with many other species originally described under the genus *Aphelenchus*.

Cobb (1919), working on material submitted for identification by Mr William Nowell, Mycologist to the Imperial Department of Agriculture for the West Indies, described the highly pathogenic nematode from diseased coconut palms on Grenada as *Aphelenchus cocophilus*, now better known as *Rhadinaphelenchus cocophilus*, the red ring nematode. A subsequent series of papers on red ring disease was remarkable more for the dispute between entomologists and nematologists as to which discipline should be ascribed the greater causal role in the disease, than for objective science. Subsequently, Cobb (1923) made another major contribution when he recorded the fact that in aphelenchs all three oesophageal glands opened into the oesophageal lumen within the confines of the median bulb, whereas in tylenchs the dorsal gland debouched far more anteriorly near the stylet base. This seminal discovery resulted in an ever-increasing systematic divide between the two groups which eventually culminated in separate ordinal status (Siddiqi, 1980b).

Other workers on the aphelenchs included Micoletzky (1922, 1925) and Steiner (1931, 1932a), but the next major contribution of lasting significance was made by Anton Gilbert Fuchs, a forest entomologist in Austria, in a sequence of detailed and well illustrated papers (1929, 1931, 1937). The 1931 paper proposed the genus *Seinura* whilst the 1937 treatise involved a thorough systematic re-arrangement with the erection of two new genera, *Bursaphelenchus* and *Laimaphelenchus*, and several subgenera, including *Cryptaphelenchus* and *Parasitaphelenchus*, to contain the unusual forms associated with the bark beetles in which he specialized. Fuchs' efforts, although undeniably of considerable importance, were somewhat tarnished by his penchant for the subspecific rank, a penchant which, incidentally, Walter Rühm (1956) subsequently embraced in his *magnum opus* on bark beetle nematodes, 'Die Nematoden der Ipiden'.

J.R. Christie (1942) described *Aphelenchoides besseyi*, the causal agent of white tip disease of rice, a serious seed borne infection of a major economic crop and Steiner (1943) erected *Metaphelenchus*, an unusual genus related to *Aphelenchus*, but now usually regarded as being synonymous. Skarbilovich (1947) overhauled the systematics and erected the Aphelenchoidinae which was subsequently raised first to family and then superfamily status by Paramonov (1953) and Siddiqi (1980b) respectively.

The next phase of aphelench development started when T. Goodey (1951) published *Soil and Freshwater Nematodes*, a book which sustained its status of a nematologist's *vade mecum* for many years and, thanks to the extensive revision of J.B. Goodey (1963), is still useful today. Allen (1952) thoroughly reviewed the literature and taxonomy of the *Aphelenchoides* species known to be pests of plant foliage and provided good illustrations and a key.

Wachek (1955) working on the nematode parasites of rove beetles (Coleoptera: Staphylinidae) and sexton beetles (Coleoptera: Silphidae) described *Entaphelenchus* and *Peraphelenchus*, two unusual new genera each having three adult forms. Unfortunately, Wachek did not preserve type material of his new genera or species. Rühm (1956), working like his predecessor Fuchs on the nematodes associated with wood boring and bark beetles of the family Scolytidae, described a profusion of new species representing both the Tylenchida and Aphelenchida. He also dealt with the ecology, morphology and phylogeny of the bark beetle nematodes and proposed several systematic adjustments. The value of his work was somewhat compromised by a failure to designate and preserve type material, yet it nevertheless represents a considerable achievement and remains an essential reference on the entomophilic aphelenchs.

The next thorough exposition on the aphelench systematics was 'The classification of the Aphelenchoidea, Fuchs, 1937' by J.B. Goodey (1960b). In this work he considered the published work on the systematics since T. Goodey (1951), i.e. that of Allen (1952), Paramonov (1953), Franklin (1955), Wachek (1955), Rühm (1956), Chitwood (1958) and Hopper and Cairns (1959). The resulting synthesis, with a small number of modifications, was also used in his 1963 revision of *Soil and Freshwater Nematodes*. Included in his scheme was the Sphaerulariidae, a group of obligate insect parasites which was subsequently shown to belong not to the aphelenchs, but to the tylenchs. His 1960 paper included the erection of three new genera (*Megadorus*, *Rhadinaphelenchus* and *Cryptaphelenchoides*) and comprehensive species lists for every genus.

Paramonov (1964) proposed the subfamilies Ektaphelenchinae, Bursaphelenchinae and Rhadinaphelenchinae and Geraert (1966) suggested that the aphelenchs be elevated to subordinal status with his proposal of the Aphelenchina. The recognition that the aphelenchs had long deserved was starting to gain substance.

Nickle (1970a) described two new genera, one, *Sheraphelenchus* being presumed to be mycetophagous and the other, *Roveaphelenchus*, an obligate parasite of staphylinid beetles. The latter genus exhibited three adult forms and was related to *Entaphelenchus* Wachek, 1955 and *Praecocilenchus* Poinar, 1969, an obligate parasite of a palm weevil. Nickle (1970b) also presented a well illustrated review of the genera of the Aphelenchoidea in a publication which remains pertinent to the study of the group over two decades later.

In the meantime Massey, also working in the USA, was describing a number of nematode species and genera associated with bark beetles. His work was summarized in a USDA publication (Massey, 1974) which included two new genera of aphelenchs (*Teragramia* and *Berntsenus*). This publication is essential for anyone studying the nematodes associated with the bark beetles of North America and contains some excellent illustrations.

Andrássy (1976) presented an up-to-date overview of the systematics of the Aphelenchina in his book *Evolution as a Basis for the Systematization of Nematodes*. He did not provide keys to genera or lists of nominal species and his diagnoses suffered from being rather succinct.

Poinar (1977b) erected the genus *Oligoaphelenchoides* to contain fossil aphelenchs found in amber from Mexico. Although structural details could not be discerned, the sharply recurved male tail suggests strong affinities with *Bursaphelenchus*. The early years of the next decade produced a flurry of activity. Baujard (1980b, 1981, 1985) produced useful reviews of *Bursaphelenchus* and *Laimaphelenchus* respectively and proposed a number of new species.

Siddiqi (1980b) published a stimulating and provocative overview of the systematics and phylogeny of the aphelenchs and proposed ordinal status together with associated elevations in the subordinal rankings. His systematic views, although somewhat controversial, were initially accepted by Maggenti (1981, 1982, 1983) and Fortuner (1984), although in recent years they have generated considerable criticism, the tendency being to regard the aphelenchs as being only of subordinal status under the Tylenchida (see, for example, Maggenti, 1991b).

Baranovskaya (1981) proposed the Paraseinurinae to include two unusual genera, namely *Paraseinura* Timm, 1960 and *Aprutides* Scognamiglio, Talame' and s'Jacob, 1970. Her publication provided a useful, illustrated overview of the group.

The genus *Acugutturus* was erected by Hunt (1980a) to contain a remarkable new ectoparasitic aphelench from cockroaches. The genus was in part characterized by the longest recorded stylet in the aphelenchs – some 60 μm in length – but also on its unique biology. All stages, including the unusually large eggs, were found on the external body surfaces of *Periplaneta americana* L. from St Lucia, West Indies and the juveniles and adults appeared to feed on the cockroach haemolymph by penetrating the soft intersegmental membranes with their needle-like stylet. Hunt speculated that the genus was related to the Ektaphelenchinae but, because of the unusual morphology and life cycle, proposed a new subfamily, the Acugutturinae, to accommodate the new genus.

Remillet and Silvain (1988) described an even more remarkable aphelench when they proposed the genus *Noctuidonema*. This genus, ectoparasitic on the abdomen of adult noctuid moths, was characterized by a needle-like stylet up to 185 μm in length! In addition, the body became considerably swollen so that the nematodes moved about on their host like leeches. The male spicule of the type species, *N. guyanense*, was very large and of such unusual form that a separate paper was subsequently written to describe it (Anderson and Laumond, 1990). *N. guyanense* has been recorded from South America, the West Indies and, on migratory moths, from the southern USA. I have also received males of an undescribed species of *Noctuidonema* from Aldabra Island in the Indian Ocean (leg. Vaughan-Spaull), so the genus is probably very much more widespread than is currently suspected. Recent discoveries such as *Acugutturus* and *Noctuidonema* show how relatively incomplete our knowledge of this group is, even at the generic level, and many more weird forms remain to delight the curious worker. This must be particularly so for the insect-associated/parasitic species from the subtropical and tropical areas which have so far received scant attention.

Recent reviews of the group include those of Baujard (1988), advocating the use of Expert System Technology using eight nests of species as an identification aid and Nickle and Hooper (1991), who briefly described and illustrated all the genera they considered valid, but stopped short of an in-depth taxonomic and biological treatment of the group.

Notwithstanding the fact that the first aphelench was described over 125 years ago, the group largely remains a backwater both in taxonomic and biological studies, the only notable exceptions being a handful of species of economic importance on crops, such as *Aphelenchoides besseyi*, *Bursaphelenchus xylophilus* and *Rhadinaphelenchus cocophilus*, or those with possible biocontrol potential, the current doyen being *Noctuidonema guyanense*. As a consequence of this neglect, the Aphelenchida is burdened with an inordinate number of poorly illustrated and inadequately characterized taxa, the attempted reidentification of which would doubtless defeat even their nominators.

The systematic arrangement of genera used in this book is based more on convenience than conviction. Nickle (1970b) wrote that he 'felt [that] the genus is the best defined and most objective taxonomic category in the aphelenchs at this time'. The group has become more varied, more complex and better known and yet, despite the speculative systematic and phylogenetic odysseys of the last 20 years, Nickle's sober view remains pertinent.

2 General Morphology

Aphelenchs are mostly slender, vermiform nematodes usually varying from 0.2 mm to 2 mm in length. Some ectoparasites and endoparasites of insects are swollen, the endoparasites becoming particularly obese and sometimes reaching lengths of 3 mm or more (Fig. 2.1). The tails of both sexes are usually short and conoid with a pointed or rounded terminus, often bearing one or more mucrons (Fig. 2.3, L, M, O). In some genera the tail is elongate and filiform, e.g. *Seinura* or *Ektaphelenchoides* (Fig. 2.3, P), but it may be extremely short as in *Noctuidonema*. In the female the anal aperture takes the form of a transverse, arcuate slit with a prominent, overhanging, anterior lip (Siddiqi, 1980b).

Cuticle and associated structures: The cuticle is thin and comprises a number of layers (see Maggenti, 1979 and 1991a for a more detailed discussion). It is usually annulated with fine transverse striae. The cephalic region may be continuous with the body contour or offset, but is usually low and rounded in shape (Fig. 2.3, D–I) except in some unusual cases such as *Anomyctus* where the head is almost spherical, but with a prominent sclerotized oral disc. The cephalic region is typically hexaradiate, the sectors being of equal size except in *Ektaphelenchus* where the laterals are reportedly narrower. One undescribed species of Entaphelenchidae was shown to have only four head sectors when studied with the scanning electron microscope (Giblin and Kaya, 1983c). The cephalic sclerotization is usually rather weak and is never strongly developed. The oral aperture is circular to oval in shape and is surrounded by a circlet of six papillae, possibly the inner labial papillae, according to Hooper and Clark (1980). External to the oral disc are six bumps which may represent the outer labial papillae and four cephalic papillae in the submedian position (Fig. 2.2). The amphidial apertures are small and pore-like and are located dorsosublaterally on the lateral lip sectors (Hooper and Clark, 1980). The lateral fields are usually distinct with the number of incisures ranging from none to as many as 12 or more, the

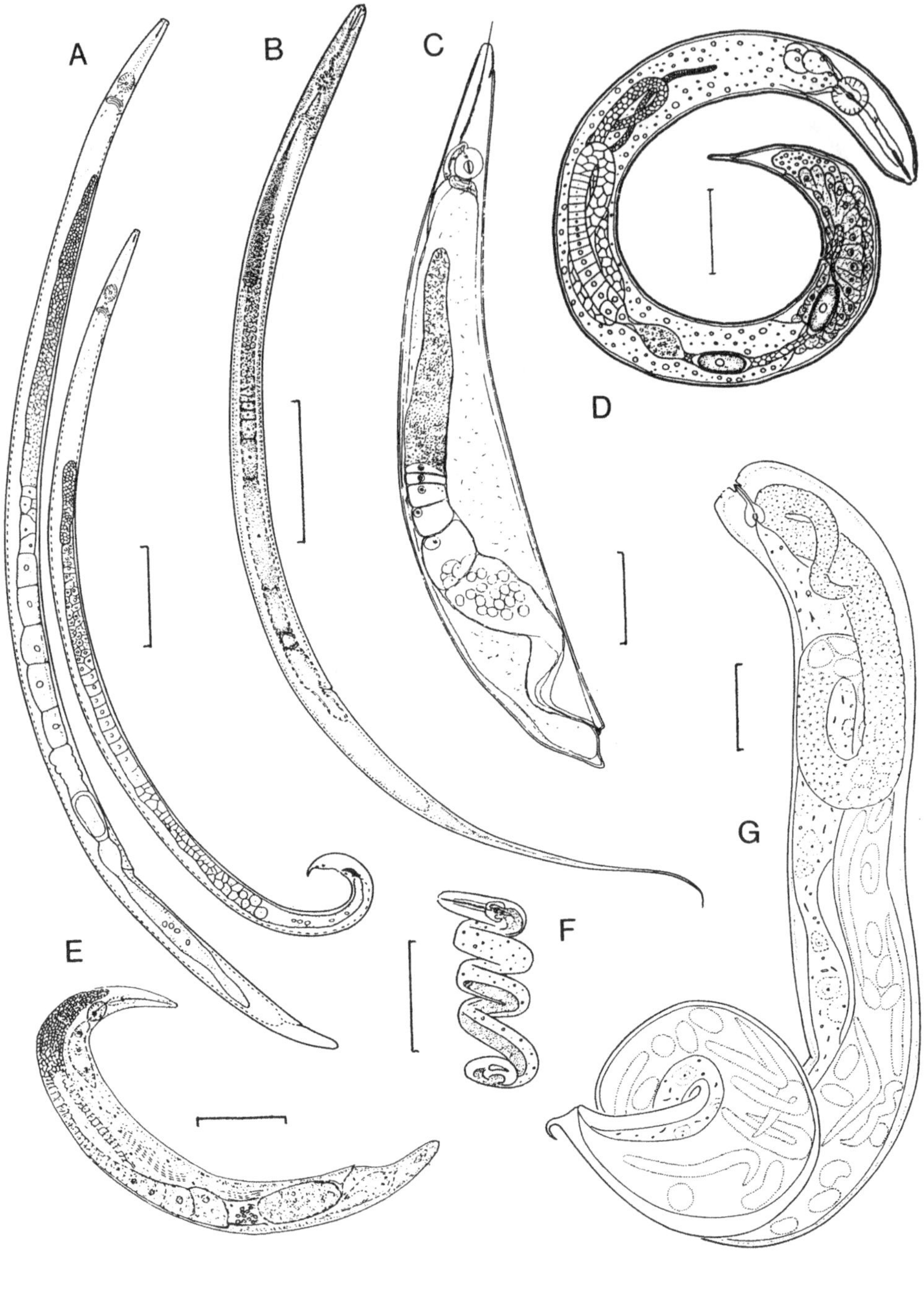

Fig. 2.1. General morphology of Aphelenchida;. Body form: A. *Bursaphelenchus naujaci*, female and male; B. *Ektaphelenchoides attenuata*, female; C. *Noctuidonema guyanense*, female; D. *Entaphelenchus oxyteli*, female; E. *Vampyronema daptria*, female; F. *Peraphelenchus necrophori*, male; G. *Roveaphelenchus jonesi*, female. (After various authors. Scale bar = 100 μm).

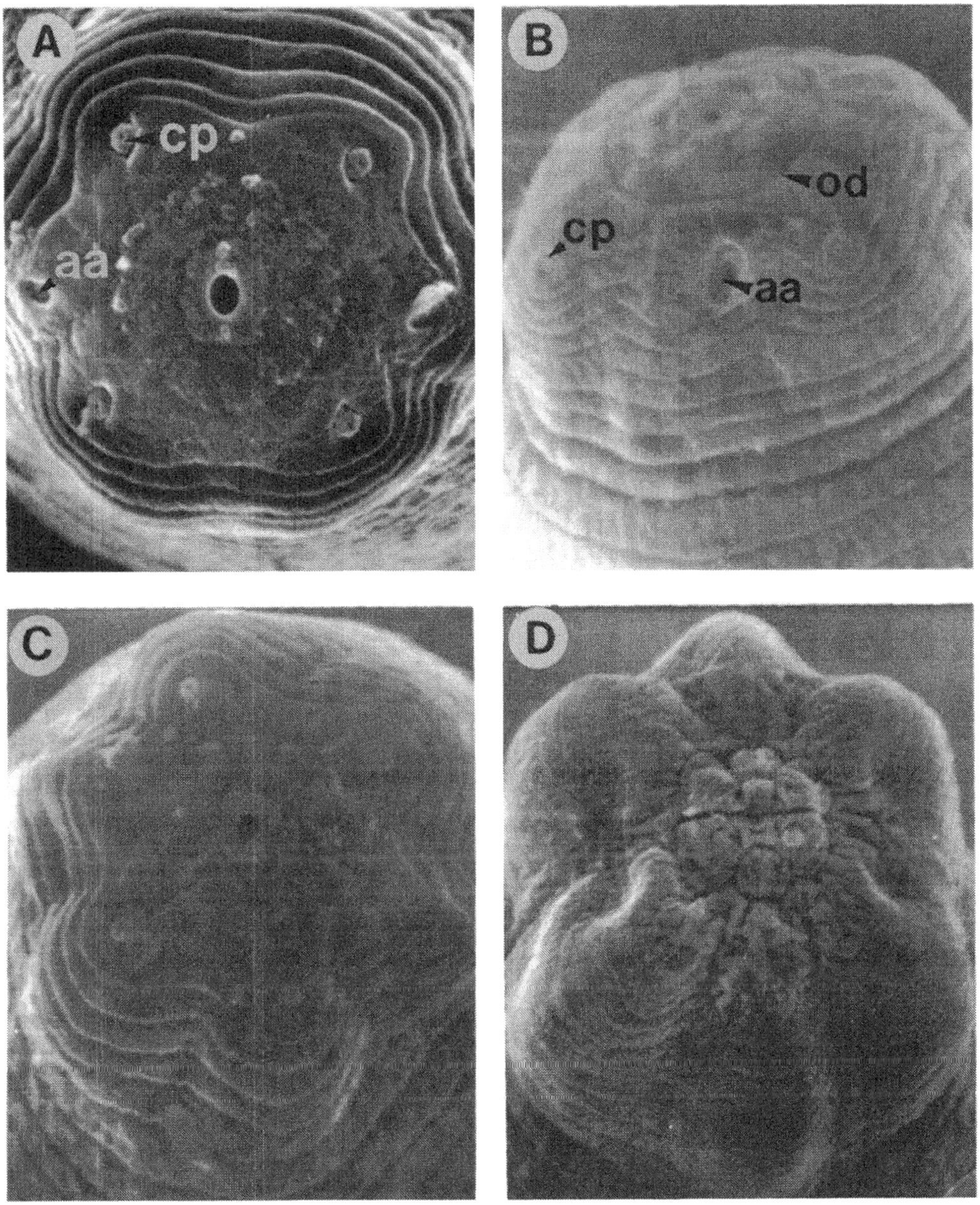

Fig. 2.2. Cephalic region of four aphelenchs;. A. *Aphelenchoides* sp.; B. *A. blastophthorus*; C. *Aphelenchus* sp.; D. *Seinura* sp. aa = amphidial aperture; cp = cephalic papilla; od = oral disc. (Photographs courtesy D.J. Hooper, Rothamsted Experimental Station).

most common complement being four (Fig. 2.3, J, K). The hemizonid is located posterior to the excretory pore which usually opens on the ventromedian line behind the median bulb. However, in *Noctuidonema* the pore opens near the cephalic region. The renette cell may be enormously developed as in *Vampyronema daptria* (see Anderson and Laumond, 1992).

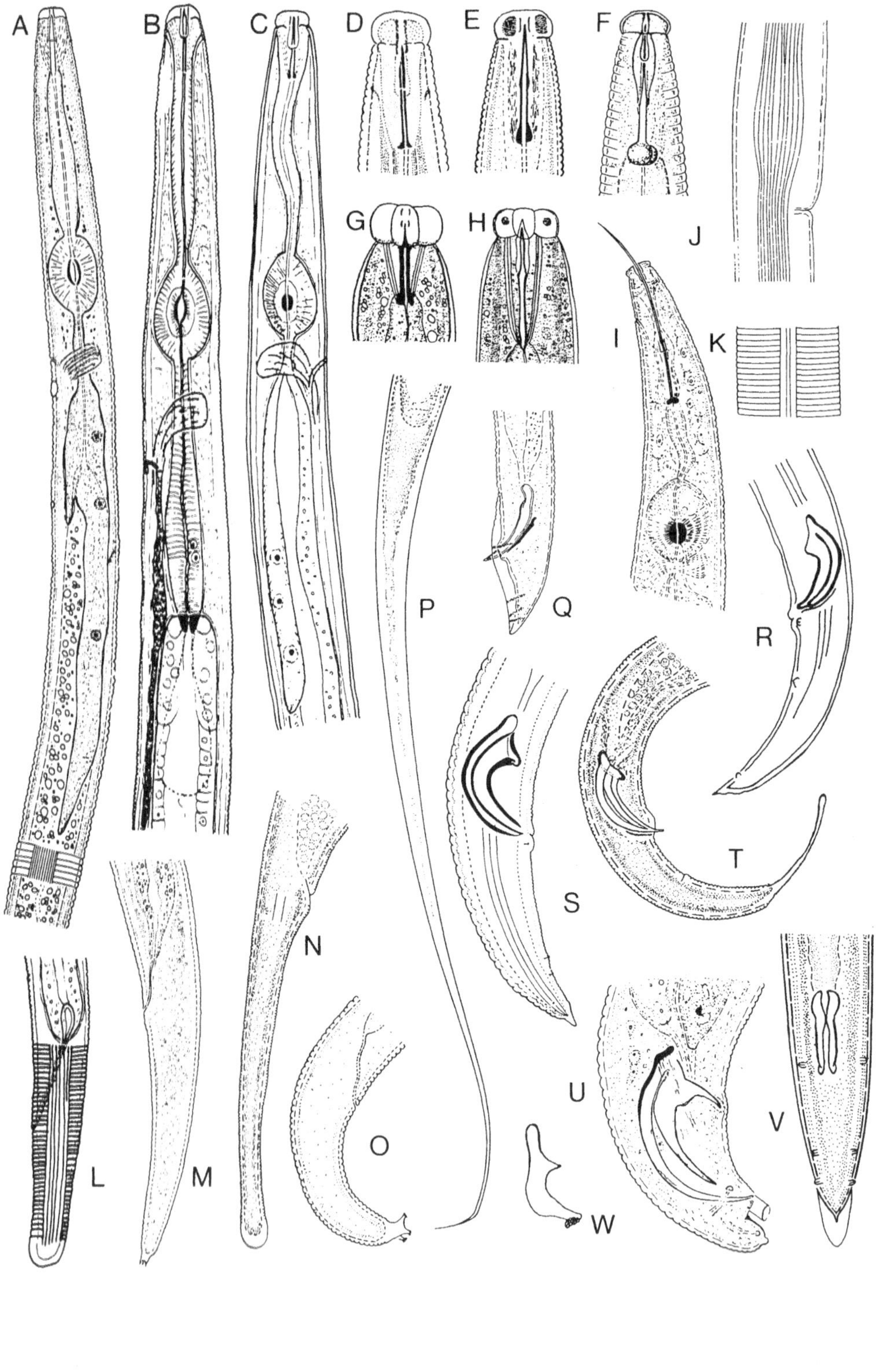
A
B
C
D
E
F
G
H
I
J
K
L
M
N
O
P
Q
R
S
T
U
V
W

Buccal cavity and armature: The stylet is a hollow, axial tube, usually between 10–20 μm long, but exceptionally up to 185 μm in some insect ectoparasites (Fig. 2.3, D–I). The conus is shorter than the shaft (except in the forms with extremely long, attenuated stylets) and has a subterminal ventral aperture. According to Shepherd *et al.* (1980) the conus is composed of six layers and, in *Aphelenchoides blastophthorus*, has a narrow, 0.2 μm diameter, lumen. Anteriorly the conus is surrounded by a guiding-ring formed from the invagination, sometimes referred to as the vestibule extension, of the cheilostome. The shaft is embedded in the body tissue and is composed of five layers, one less than for the conus. The stylet is weakly to moderately sclerotized with a narrow or wide lumen, depending on genus. In *A. blastophthorus* (Shepherd *et al.*, 1980) the main layer of the stylet is traversed by a network of tiny channels which open through pores localized on the outer surface of the shaft just behind the conus/shaft junction, the area being sheathed by a gustatory neuron. Basal swellings or knobs may be present or absent, but only show massive development in *Megadorus* (Fig. 2.3, F). There are six stylet protractor muscles, three being larger than the others and attached further anterior on the knobs. Distally these muscles are not attached to the cheilostome extension, as occurs in tylenchs, with the notable exception of *Vampyronema daptria* which has the protractors uniquely inserted. The cell bodies of these muscle cells are reportedly located in the median bulb (Shepherd *et al.*, 1980). The homology of the aphelench stylet is said to be different from that of the tylenchs (J.B. Goodey, 1963; Calaway and Tarjan, 1973; Siddiqi, 1980b), thus lending additional weight to the ordinal status of the aphelenchs.

Alimentary tract: The oesophagus comprises a cylindrical procorpus of varying length and development which leads to the strongly developed, muscular, median bulb or metacorpus (Fig. 2.3, A–C). The procorpus is uniform in structure, contains no nuclei and has a cuticular lining with five layers, i.e. the same number as the stylet shaft. Immediately before the bulb, the procorpus narrows sharply. The median bulb is markedly offset both anteriorly and posteriorly from the rest of the oesophagus and is typically spheroid to rounded rectangular in shape. The sclerotized valve apparatus may be situated centrally or post-median, the lining of the pump chamber being of more complex form than in tylenchs (Shepherd *et al.*, 1980). The valve apparatus is operated by six powerful, obliquely radial, muscles. The

Fig. 2.3. General morphology of Aphelenchida;. Oesophagus: A. *Aphelenchus avenae*; B. *Paraphelenchus afsi* (= *micoletzkyi*); C. *Aphelenchoides siddiqii*. Cephalic region: D. *Aphelenchoides besseyi*; E. *Laimaphelenchus cocuccii*; F. *Megadorus megadorus*; G. *Tylaphelenchus leichenicola*; H. *Cryptaphelenchoides macrobulbosus*; I. *Vampyronema daptria*. Lateral fields: J. *Aphelenchus avenae*; K. *Aphelenchoides ritzemabosi*. Female tail: L. *Aphelenchus isomerus*; M. *Aphelenchoides ritzemabosi*; N. *Aprutides martuccii*; O. *Laimaphelenchus cocuccii*; P. *Ektaphelenchoides attenuata*. Male tail: Q. *Aphelenchus avenae*; R. *Aphelenchoides eltayebi*; S. *Anomyctus xenurus*; T. *Aprutides martuccii*; U. *Vampyronema daptria*; V. *Bursaphelenchus maxbassiensis* (ventral view showing terminal bursa); W. Spicule and 'gubernaculum' of *Cryptaphelenchoides macrobulbosus*. (After various authors. Not to scale.)

bulb contains all three oesophageal gland orifi, the dorsal opening anterior to the valve plates and the ventrosublaterals posterior. This is in marked contrast to the tylenchs where the dorsal gland orifice is far anterior in the procorpus, usually just behind the stylet base. There is a valve, operated by six radial muscles, situated in the lumen between the dorsal gland orifice and the pump chamber. Anteriorly the bulb has six muscles, the function of which is not clear. Shepherd *et al.* (1980) referred to them as 'constraining muscles' and suggested that they might stabilize the bulb when the powerful pumping muscles were in operation. In the Aphelenchoidea, the genera *Aphelenchus* and *Paraphelenchus* have a well developed isthmus (Fig. 2.3, A, B), but this is not the situation in all the remaining aphelench genera which reside in the Aphelenchoidoidea (Fig. 2.3, C). In *Aphelenchoides blastophthorus* the oesophageal tissue is invaginated into the base of the median bulb to form a very short 'isthmus', a structure which is not normally visible under the light microscope. By way of contrast, some other species do have a rudimentary isthmus which is, however, shorter than the distance from the valve plates to the base of the median bulb. The three oesophageal glands almost invariably form a dorsally overlapping lobe (Fig. 2.3, A, C), the exception being *Paraphelenchus* in which they are small and restrained within an abutting bulb (Fig. 2.3, B). The gland lobe has the dorsal gland posteriormost and in some genera extends for a considerable distance down the body. The oesophago-intestinal junction is usually at the base of the median bulb, but may be slightly posterior. A circumoesophageal (Aphelenchoidea) or circumintestinal (Aphelenchoidoidea) nerve ring is located posterior to the median bulb (Fig. 2.3, A–C).

The intestine is cellular and endodermal in origin. It usually has a distinct lumen lined with microvilli and a sphincter at the junction with the cuticle-lined rectum. A detailed account of its ultrastructure was given by Shepherd *et al.* (1980). Some insect associates lack a functional rectum and anus and in these cases the intestine ends as a blind diverticulum extending partially into the tail region. In addition, some *Seinura* have the anus reduced to a pore-like structure which may be non-functional. In some insect parasites such as *Acugutturus* and *Noctuidonema* the intestine may contain numerous needle-like crystals which may represent stored excretory products.

Female genital system: The female genital system is characterized by a posterior vulva, usually located at about 70–75% of the body length, but varying from 60–98% (Fig. 2.1). The vulval lips may be slightly protuberant or flush, the anterior lip overhanging the vulval opening in *Rhadinaphelenchus cocophilus* and in some species of *Laimaphelenchus* and *Bursaphelenchus*. The vulva is a transverse slit or, very exceptionally, a small, transverse oval pore-like structure, as in *Aphelenchus*. The vagina may penetrate the body at right angles to the body axis or be directed anteriorly. Some species are reported to exhibit a sclerotized ring around the vagina, but this might be circum-vaginal musculature. The genital tract is monoprodelphic and outstretched, but rarely a double flexure is present. It consists of a uterus; axial spermatheca, which may be present or absent; oviduct and telogonic

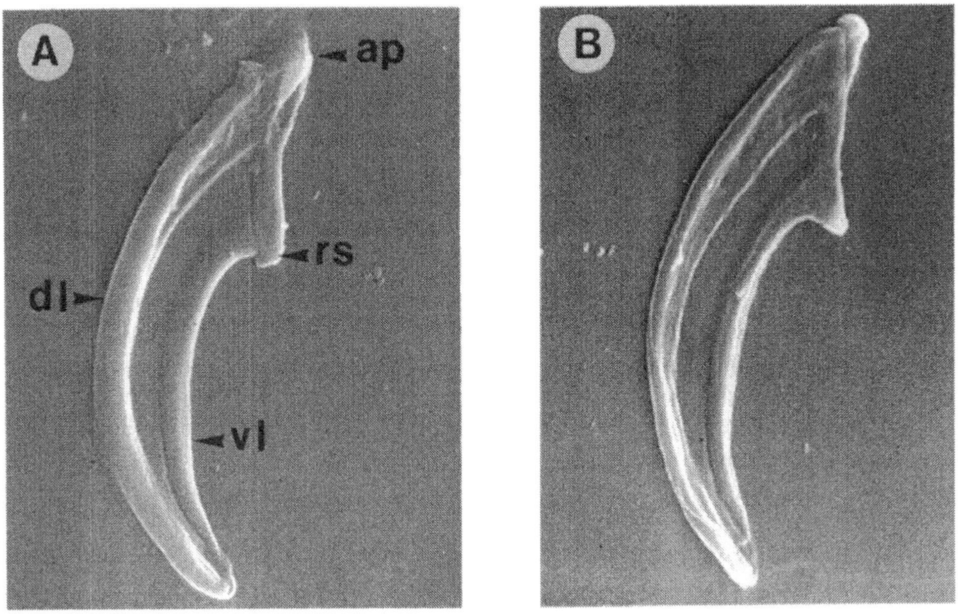

Fig. 2.4. Spicule of *Aphelenchoides*;. A. Outer face; B. Inner face. ap = apex; dl = dorsal limb; vl = ventral limb; rs = rostrum. (Photographs courtesy D.J. Hooper, Rothamsted Experimental Station.)

ovary. Developing oocytes are arranged in one to several rows. A postuterine sac may be absent or present and, if present, short or long. The postuterine sac often acts as a seminal receptacle, even in those species possessing a spermatheca. Some species are ovoviviparous although most are oviparous.

Male genital system: The male genital system is monorchic and outstretched or terminally reflexed. The development is reported to be different to that of the tylenchs (Shepherd and Clark, 1976a). The sperm are typically large and rounded (except in *Schistonchus* where they are flagelliform) with much cytoplasm and lie in one or two rows in the gonoduct. Their ultrastructure is also different to that of tylench sperm (Shepherd and Clark, 1976a) and the sperm nucleus has discrete chromosomes (Triantaphyllou, 1971; Siddiqi, 1980b). The spicules are paired and ventrally curved. In Aphelenchoidoidea they are typically robust and characteristically rosethorn-shaped, or derived therefrom (Fig. 2.3, R–T). In *Noctuidonema* and *Vampyronema* the spicules are exceedingly large (Fig. 2.3, U) and have some unique morphological features (Anderson and Laumond, 1990). Clark and Shepherd (1977) reported on the spicules and caudal region of *Aphelenchoides blastophthorus* and provided some excellent transmission electron micrographs of the spicule ultrastructure. Typically, the shaft of each spicule comprises a curved dorsal and ventral limb, connected by a thin scleroprotein membrane and joined proximally by a prominent transverse bar with a dorsal apex and ventral rostrum of varying shape and development (Fig. 2.4). When protruded the ventral limbs of the spicules abut and interdigitate, although the dorsal limbs remain far apart. In the Aphelenchoidea, on the other hand, the spicules of the genera *Aphelenchus* and *Paraphelenchus* are not rosethorn-shaped but slender, cephalated and tylenchiform in shape (Fig. 2.3, Q). The spicules are usually paired and separate, but may be partially fused as, for example, in *Parasitaphelenchus* and some *Bursaphelenchus*. A gubernaculum is typically absent, but in *Aphelenchus* and *Paraphelenchus* it is well developed, elongate and V-shaped in cross-section (Fig. 2.3, Q). A gubernaculum-like structure has been reported from several other genera, but it is never elongate, existing only as a small, sclerotized area near the distal region of the dorsal limbs of the spicules (Fig. 2.3,W). A true bursa is absent except in *Aphelenchus* (Fig. 2.3, Q) where it is strongly developed with four pairs of papillae or 'ribs' and envelops the tail tip. A vestigial terminal flap of cuticle, usually referred to as a bursa despite its posterior location, is found in *Bursaphelenchus*, *Rhadinaphelenchus* and *Parasitaphelenchus*, but this is often only visible in ventral view (Fig. 2.3, V) and can be easily overlooked. Three pairs of innervated caudal papillae are typically present in members of the Aphelenchoidoidea, but the number varies from two to five pairs, exceptionally more. A single ventromedian preanal papilla may also be present.

More detailed discussion of nematode morphology and ultrastructure can be obtained from the excellent accounts by Bird and Bird (1991) and Maggenti (1991a), the former containing an extensive bibliography.

3 Bionomics

Many species of aphelench reproduce by amphimixis, although a considerable number appear to be parthenogenetic. There are typically four juvenile stages, the J2 emerging from the egg as in tylenchs, although there are some reports, the veracity of which needs to be confirmed, of the eclosion of the J1 in *Bursaphelenchus fungivorus* (Franklin and Hooper, 1962) and *Parasitaphelenchus* (Rühm, 1956; Saunders and Norris, 1961). Some *Seinura* are reported to have only two post-hatch juvenile stages and although two moults may occur within the confines of the egg, some species only have one and thus appear to have deleted a stage from the life cycle (Hechler and Taylor, 1965; Wood, 1975). The life cycle can be extremely rapid and occupy only a few days in certain predacious or mycetophagous species when food is abundant. Other nematodes are closely tied to the life cycle of their insect host and are univoltine.

An ability to enter anhydrobiosis and thus survive desiccation appears to be intrinsic in many genera, particularly those insect associates having a specialized endo- or ectophoretic dauerlarva stage, although in genera such as *Ektaphelenchus* and *Cryptaphelenchoides* the immature female is the resistant stage. However, anhydrobiosis is not limited to such species and has been reported from other genera such as *Aphelenchoides*, *Laimaphelenchus* and, to some extent, in *Paraphelenchus*. *A. besseyi* may survive desiccation for several years on rice seed and *A. ritzemabosi* also exhibits a pronounced anhydrobiotic ability, proving viable after more than three years in dry plant tissue. *A. besseyi* survives best as the preadult juvenile and *A. ritzemabosi* as the preadult and adult stages.

Aphelenchs are, in general, highly active nematodes and boast amongst their number one of the most accomplished swimmers, namely *A. besseyi*, the abilities of which are only equalled by one other nematode, *Ditylenchus angustus*, amongst the plant parasites.

The biology of the majority of the 400 or so nominal species of aphelench is not known in any detail, but there appear to be four, not necessarily mutually exclusive, trophic types, viz. mycetophagous, phytoparasitic, predatory and entomoparasitic. Approximately one third of the described species are associates, in one way or another, of insects, and as most of those were described by Fuchs or Rühm in Europe and Massey in the USA, doubtless many more await discovery, particularly in the tropical and subtropical regions of the world.

Aphelenchs are predominantly free-living and mycetophagous in habit and are found worldwide in soil and decaying plant material, associated with lichens, on tree bark and in the tunnels of wood-boring beetles. *Aphelenchoides* has been recorded from Signy Island in the Antarctic (Maslen, 1979) and some species are even reported to be marine, *A. gynotylurus* being described from a salt water pool on the Bay of Bengal and *A. marinus* from blades of turtle grass (*Thalassia*) in Biscayne Bay, Florida (Timm and Franklin, 1969). The most abundant and widely spread aphelenchs belong to the genera *Aphelenchus* and *Aphelenchoides* and, these being primarily mycetophagous, probably represent the nearest to the ancestral forms. Feeding studies on *Aphelenchus avenae* showed that stylet penetration of the fungal hyphae occurred in 6–7 seconds, the median bulb then pumping rapidly for about 8 seconds as the hyphal contents were ingested. No glandular secretions were injected into the hyphae (Fisher and Evans, 1967). *A. avenae* has also been recorded feeding on plant roots and, occasionally, on other nematodes (e.g. Decker, 1962). Siddiqui and Taylor (1969), studying the feeding of *Aphelenchoides bicaudatus* on fungal hyphae, found it to be a transient feeder with no injection of glandular secretions before ingestion.

Although most *Aphelenchoides* are probably solely mycetophagous, a few species, colloquially known under the sobriquet 'bud and leaf nematodes', parasitize higher plants. Some of these species, such as *A. arachidis*, *A. besseyi* and *A. fragariae*, reproduce readily on fungi and are clearly facultative phytoparasites, but others, such as *A. ritzemabosi*, have usually been regarded as obligate parasites of higher plants, although Hooper and Cowland (1986) showed that reproduction on fungal mycelium was possible. *A. arachidis* is locally important in Nigeria where it attacks the seeds of groundnut whereas *A. besseyi* is widely distributed in rice growing areas where it causes 'white tip' disease. The nematode is largely ectoparasitic and, being highly resistant to desiccation, is able to survive on dry seed, thus increasing the likelihood of dissemination. *A. ritzemabosi*, a phytoparasite of members of the Compositae, enters leaves via the stomata. It can survive in dry tissue for several years. Another phytoparasite, *A. fragariae*, feeds both ectoparasitically and endoparasitically, mainly attacking ferns, Liliaceae, Primulaceae and Ranunculaceae. Klingler (1970) reported entry into the leaf via the stomata in damp conditions, although the epidermis can also be directly penetrated (Strümpel, 1967). The other two species of aphelench implicated in plant diseases of major economic importance are the red ring nematode, *Rhadinaphelenchus cocophilus* and the pine wilt nematode, *Bursaphelenchus xylophilus*. *R. cocophilus* is an obligate phytoparasite vectored by the palm weevil, *Rhynchophorus palmarum*, from infected to healthy palm trees where it proliferates in the stem tissue. *B. xylophilus* is vectored by several species of wood-boring cerambycid beetles to pine trees where it reproduces in the vascular system with an associated blue-stain fungus. Eventually the tree wilts and dies and serves as a breeding ground for the beetle vector, thus completing the cycle.

Predatory nematodes are reported from *Laimaphelenchus* and, more particularly, *Seinura*. Members of the former genus also feed on fungi and lower plants, but *Seinura* is exclusively predatory. On contacting a prey, *Seinura* penetrates the cuticle by rapid thrusts of the stylet and injects immobilizing secretions before ingesting the body contents. A number of other genera, all rare and poorly known, are probably predacious, but their biology has yet to be studied.

Of the remaining genera, the vast majority are known associates or parasites of insects. Many species are commonly found in the tunnels of various bark beetles (Coleoptera: Scolytidae) where they feed on fungi growing on frass, a finely comminuted mixture of beetle faeces and chewed wood fragments, in the humid environment of the galleries. The majority of this rich assemblage of species is known to use the insects as a transport mechanism to get from one tree to the next. Usually this association takes the form of ectophoresy where nematodes, particularly a specialized third stage juvenile known as a 'dauerlarva', swarm over the teneral adult beetles before emergence from the tree. They congregate particularly beneath the elytra and/or between the segmental folds of the tergites and sternites. Dauerlarvae are adapted to resist desiccation and become quiescent until reactivated by increased humidity as the vector insect excavates a new gallery in a fresh site. A single bark beetle may carry several hundred dauerlarvae of divers species. Such ectophoretic nematodes are normally mycetophagous and use the insect purely as an, albeit unwitting, transport device. In the aphelenchs this phenomenon is best exemplified by members of *Bursaphelenchus*, although ectophoresy also occurs in some tylenchs, diplogasterids and rhabditids.

Cryptaphelenchus represents an interesting stage in the association of aphelenchs with insects. At least one juvenile stage is found inside the insect, the usual location being the Malpighian tubules, colleterial glands or genital region. Whether this association is endophoresy or parasitism is not entirely clear, although the latter case is, perhaps, more likely. The adults of *Cryptaphelenchus* are found in the frass of bark beetle tunnels, the females lacking a functional anus and rectum. This feature is found in several genera known to be ectoparasitic on insects, but whether this is indicative of a similar trophic habit in *Cryptaphelenchus* is uncertain.

Ectoparasitism of insects occurs in several genera. At its most primitive it is best exemplified in such genera as *Crytaphelenchoides*, *Ektaphelenchus* and *Ektaphelenchoides*. In *Ektaphelenchus* the immature females are found in cocoon-like masses under the elytra or between the tergites (Massey, 1956; Rühm, 1956) and a similar situation has been reported in *Cryptaphelenchoides scolyti* by Hunt and Hague (1976). An undescribed *Ektaphelenchoides* species has been found with the stylet penetrating the cuticle of certain dipterous larvae, the intestine being filled with a pinkish fluid similar in hue to the insect haemolymph (Hooper, 1991, personal communication). Whether all stages of these genera are ectoparasitic and whether the association is obligate or facultative remains to be elicited.

The most advanced ectoparasitism occurs in the recently described genera *Acugutturus* and *Noctuidonema*. *Acugutturus* is a monotypic, obligate ectoparasite of the American cockroach, *Periplaneta americana*. All stages, including the exceptionally large eggs, are found on the body surface of the host, particularly the abdomen, but also on the wings in adult insects. The nematodes penetrate the host cuticle with a 50–60 μm long stylet and imbibe haemolymph. The intestine contains a yellowish fluid and the rectum and anus appear to be non-functional in the female. All stages remain vermiform and transmission from one cockroach to another probably occurs during communal roosting (Hunt, 1980a). So far this genus has only been recorded from St. Lucia, West Indies, but is almost certainly far more widespread.

Noctuidonema is more specialized than *Acugutturus*. It is ectoparasitic on various noctuid moths in the Neotropical Region, but also occurs on Aldabra in the Indian Ocean, (Hunt, unpublished). All stages are found on the abdomen of the moths, the juveniles and particularly the adults, swelling markedly yet still retaining mobility. The stylet is extremely long and slender and can achieve a maximum length in excess of 170 μm. The nematodes feed on the insect haemolymph (Remillet and Silvain, 1988). Noctuid moths have a holometabolous life cycle and so transmission from one generation to the next is not as facile as with *Acugutturus* and the hemimetabolous cockroach. The procedure is made feasible as the moths that *Noctuidonema* parasitizes have overlapping generations and hence transmission from one generation to another can occur during mating.

Endoparasitism of insects has been recorded for *Schistonchus*, *Parasitaphelenchus* and for all four nominal genera in the Entaphelenchidae. *Schistonchus* is unusual in that the nematodes not only parasitize the haemocoel of a fig wasp, but also feed on plant cells within the *Ficus carica* fruit in which the host/vector insect also breeds (Vovlas *et al.*, 1992). The only other example in the Nematoda of such a relationship is in the tylench genus *Fergusobia* which parasitizes gall wasps on *Eucalyptus*, but also feeds on plant cells in the galls.

In the genus *Parasitaphelenchus*, the adults are mycetophagous in bark beetle galleries (Hunt and Hague, 1974) but the third stage juveniles penetrate the beetle grubs and moult to the fourth stage in the haemocoel. This stage absorbs nourishment from the haemolymph and increases considerably in size. Although an insect may be host to several hundred such juveniles, the detrimental effects are usually small. The nematodes remain in the haemocoel as the insect moults and are thus transported with the adult beetle to a new site. After the beetle has cut a new gallery and laid eggs, it dies, whereupon the *Parasitaphelenchus* juveniles emerge and moult to the adult stages which feed on fungi growing in the frass. Clearly, if the parasites so adversely affected the host as to prevent migration, gallery formation and egg laying, it would be counter-productive to their own dispersal and survival.

The most evolved endoparasites occur in the Entaphelenchidae. The four genera in this family all have three adult forms – male, immature infective female and mature, parasitic female – occuring in the insect haemocoel of various beetles. The mature parasitic females are large (up to 2.5 mm) and swollen and may be oviparous or ovoviviparous. Members of this group appear to be rare and their detailed biology remains a lacuna in our knowledge.

The basic evolutionary trend within the Aphelenchida seems reasonably clear and uncontentious. Paramonov (1970) suggested that the ancestral trophic traits were mycetophagous and predaceous. Siddiqi (1980b, 1983a) argued that both the predatory habit and what he termed the 'active' mode of feeding were ancestral and thereby provided an explanation for the powerful median bulb containing the dorsal gland orifice, thus facilitating not only the storage and subsequent rapid injecton of large volumes of saliva into the prey, but also the efficient ingestion of food. He considered that mycetophagy in aphelenchs developed from predatory ancestors. Whether this is a logical evolutionary trend is a moot point, the development of a true predatory habit from a mycetophagous life style being more plausible than the reverse. In aphelenchs, the predatory habit, utilizing a protrusible stylet and requiring more complex behaviour patterns, is further advanced than mycetophagy and probably evolved from such ancestors. The overwhelming majority of aphelenchs are fungal feeders, the predacious habit being known only from *Seinura* (although the handful of other species in the Seinurinae probably conform with the trophism of the type genus) together with the occasional report from predominantly mycetophagous/phytophagous genera such as *Laimaphelenchus*.

The basic trophic type in the aphelenchs is mycetophagy in soil and decaying organic matter, but also extends to mosses and lichens encrusting tree bark. An intrinsic ability to resist desiccation, probably obtained from the presumed diplogast ancestors, facilitated the development of phytoparasitism on the aerial parts of higher plants, a phytoparasitism that was initially facultative, but later obligate as in *Rhadinaphelenchus cocophilus*. The active type of behaviour developed by aphelenchs for feeding on fungal hyphae, and the possession of a powerful median bulb capable of injecting saliva, allowed the predatory habit to develop in certain genera, but this is almost certainly a specialized sideline. Although Siddiqi (1980b) argued that a powerful median bulb containing the dorsal gland opening was evolved to allow rapid injection of large volumes of immobilizing saliva into prey, this is almost certainly a secondary adaptation of an existing ancestral structure evolved for other reasons. Those mycetophagus species that have been studied do not inject saliva into the host cell, but immediately ingest the contents, aided to some extent by the turgor pressure of the hyphae, with powerful pulsations of the median bulb. If mycetophagy was the primitive trophic state there would be no need to separate the dorsal gland opening from those of the ventrosublateral glands as occurs in the tylenchs (*sensu stricto*) which regularly inject saliva into plant cells – all glandular secretions could be added to the food as it passed rapidly through the median bulb to the intestine. As the powerful median bulb is found in all aphelenchs, regardless of trophic type or life style, it is presumably an ancestral, rather than a derived character. The unusually strong development of the median bulb in aphelenchs is, however, something of a puzzle. A powerful pump is not necessary for a nematode to feed on fungal hyphae – *Hexatylus*, for example, has a non-muscular oesophagus (Shepherd and Clark, 1976b) and relies on the turgor pressure of the cytoplasm within the hyphae, coupled with its own low hydrostatic pressure and the suction caused by defaecation (the so-called rectal pump) to facilitate a comparatively passive mode of feeding (Doncaster and Seymour, 1974). The assertion by Roggen (1975, 1979) that a spherical bulb develops less suction, but more pressure than an equivalent cylinder may be of relevance in explaining the development of the median bulb. Roggen (1979) argued that the logical position of such a pressure pump would be at the point where the pressure is required, i.e. at the junction of oesophagus and intestine, so that the ingested food could be injected efficiently into the intestine against the counter pressure of the surrounding hydrostatic fluid which would otherwise tend to keep the lumen collapsed. Perhaps the most elegant theory for the existence of the well developed bulb is that it needs to be powerful, not to inject saliva or ingest hyphal contents, but to pump food from the oesophagus into the intestine! The fact that the majority of aphelenchs, including the predatory *Seinura*, have a median bulb shape closely approaching a sphere or ovoid, could support the somewhat speculative contention that the aphelench oesophagus is primarily designed as an oesophago/intestinal pressure pump, rather than a device for injecting saliva and that, therefore, the mycetophagous, rather than the predacious trophic habit, is likely to be ancestral.

The intimate association of aphelenchs with decaying organic matter, and particularly tree bark, would allow contact with a wide range of insects colonizing the same niches. Entry into bark beetle galleries in trees initially provided a new and protected fungal pabulum in the humid environment of the frass-filled tunnels. Transport of nematodes on adult beetles from the old tree to a fresh site would initially have been a risky, haphazard process, but the development of the ectophoretic habit with a desiccation-resistant dauerlarva stage, would be a natural advance and, indeed, one for which the group appears to be preadapted. Having established a closer association with certain insects, the development of endophoresy, ectoparasitism and endoparasitism are logical evolutionary trends. The development of a sophisticated ectoparasitism on cockroaches and, particularly, on noctuid moths, is more difficult to explain, but our knowledge of aphelench diversity is still rudimentary, particularly where insect associations are concerned, and intermediate states may await discovery.

4 Systematics and Taxonomy

The majority of classification schemes for the aphelenchs regard them as belonging to the suborder Aphelenchina Geraert under the order Tylenchida (Andrássy, 1976; Luc *et al.*, 1987; Maggenti *et al.*, 1987; Nickle and Hooper, 1991). This is in contrast to the proposal of Siddiqi (1980b) who argued that the aphelenchs were significantly different enough from the tylenchs to warrant ordinal status. His conclusions were based on a combination of morphological, biological, and phylogenetic grounds, the latter coming in for some scathing criticism from Maggenti (1991b).

The scheme presented in this book retains the order Aphelenchida as I feel that the group is significantly distinct in both morphology and certain biological aspects to justify the rank. Morphological differences include: the position of the dorsal oesophageal gland orifice; the strongly developed, offset median bulb; the crescentic anus with overhung anterior lip in the female; the structure and development of the female and male genital tracts; ultrastructure of the large, rounded, sperm; spicule shape and in possessing caudal papillae or rays, etc. I do not, however, accept the phylogenetic arguments of Siddiqi (1980b), partly because all such arguments tend to be, in the absence of fossil data, highly speculative flights of fancy and, if taken too seriously risk obfuscating, rather than elucidating, the issues.

The subordinal systematic scheme adopted here is independent of whether the aphelenchs are regarded as having ordinal status or not, and hence such differences of opinion on rank can be easily accommodated without compromise.

For comparative purposes, I include the outline systematic schemes proposed by Andrássy (1976), Siddiqi (1980b) and Nickle and Hooper (1991). The authority for ordinal rank is ascribed to the proposer. For other alterations in rank, the proposing authority is placed in parentheses after the original author.

Outline Classification of Andrássy (1976)

TYLENCHIDA Thorne, 1949

Aphelenchina Geraert, 1966

Aphelenchoidea Fuchs, 1937 (Thorne, 1949)

Aphelenchidae Fuchs, 1937 (Steiner, 1949)
Aphelenchus Bastian, 1865

Paraphelenchidae T. Goodey, 1951 (J.B. Goodey, 1960)
Paraphelenchus Micoletzky, 1922 (Micoletzky, 1925)

Aphelenchoididae Skarbilovich, 1947 (Paramonov, 1953)
Aphelenchoidinae Skarbilovich, 1947
Aphelenchoides Fischer, 1894
Aprutides Scognamiglio, Talame' and s'Jacob, 1970
Laimaphelenchus Fuchs, 1937
Megadorus J.B. Goodey, 1960
Papuaphelenchus Andrássy, 1976
Paraseinura Timm, 1960
Ruehmaphelenchus J.B. Goodey, 1963
Schistonchus Cobb, 1927 (Fuchs, 1937)
Seinura Fuchs, 1931
Sheraphelenchus Nickle, 1970
Tylaphelenchus Rühm, 1956

Bursaphelenchinae Paramonov, 1964
Bursaphelenchus Fuchs, 1937
Huntaphelenchoides Nickle, 1970
Omemeea Massey, 1971
Parasitaphelenchus Fuchs, 1929

Rhadinaphelenchinae Paramonov, 1962
Rhadinaphelenchus J.B. Goodey, 1960

Ektaphelenchinae Paramonov, 1964
Cryptaphelenchus Fuchs, 1937 (Rühm, 1954)
Ektaphelenchus Fuchs, 1937 (Skrjabin, Shikhobalova, Sobolev, Paramonov and Sudarikov, 1954)
Peraphelenchus Wachek, 1955

Anomyctinae J.B. Goodey, 1960 (Decker, 1972)
Anomyctus Allen, 1940

Entaphelenchidae Nickle, 1970
Entaphelenchus Wachek, 1955
Praecocilenchus Poinar, 1969
Roveaphelenchus Nickle, 1970

Outline Classification of Siddiqi (1980)

APHELENCHIDA Siddiqi, 1980

Aphelenchina Geraert, 1966

Aphelenchoidea Fuchs, 1937 (Thorne, 1949)

Aphelenchidae Fuchs, 1937 (Steiner, 1949)
Aphelenchinae Fuchs, 1937 (Schuurmans Stekhoven and Teunissen, 1938)

Paraphelenchidae T. Goodey, 1951 (J.B. Goodey, 1960)
Paraphelenchinae T. Goodey, 1951

Aphelenchoidoidea Skarbilovich, 1947 (Siddiqi, 1980)

Aphelenchoididae Skarbilovich, 1947 (Paramonov, 1953)

Aphelenchoidinae Skarbilovich, 1947
Acugutturinae Hunt, 1980
Anomyctinae J.B. Goodey, 1960
Ektaphelenchinae Paramonov, 1964
Seinurinae Husain and Khan, 1967

Entaphelenchidae Nickle, 1970
Entaphelenchinae Nickle, 1970 (Siddiqi, 1980)

Parasitaphelenchidae Rühm, 1956 (Siddiqi, 1980)
Parasitaphelenchinae Rühm, 1956
Bursaphelenchinae Paramonov, 1964
Rhadinaphelenchinae Paramonov, 1962

NOTE: Genera were not included in Siddiqi's scheme.

Outline Classification of Nickle and Hooper (1991)

TYLENCHIDA Thorne, 1949

Aphelenchina Geraert, 1966

Aphelenchidae Fuchs, 1937 (Steiner, 1949)
Aphelenchus Bastian, 1865

Paraphelenchidae T. Goodey, 1951 (J.B. Goodey, 1960)
Paraphelenchus Micoletzky, 1922 (Micoletzky, 1925)

Aphelenchoididae Skarbilovich, 1947 (Paramonov, 1953)

Aphelenchoidinae Skarbilovich, 1947
Aphelenchoides Fischer, 1894
Anomyctus Allen, 1940
Laimaphelenchus Fuchs, 1937
Megadorus J.B. Goodey, 1960

Ruehmaphelenchus J.B. Goodey, 1963
Schistonchus Cobb, 1927 (Fuchs, 1937)
Sheraphelenchus Nickle, 1970
Tylaphelenchus Rühm, 1956

Ektaphelenchinae Paramonov, 1964
Ektaphelenchus Fuchs, 1937 (Skrjabin, Shikhobalova, Sobolev, Paramonov and Sudarikov, 1954)
Berntsenus Massey, 1974
Ektaphelenchoides Baujard, 1984
Ipsaphelenchus Lieutier and Laumond, 1978
Cryptaphelenchus Fuchs, 1937 (Rühm, 1954)
Cryptaphelenchoides J.B. Goodey, 1960

Bursaphelenchinae Paramonov, 1964
Bursaphelenchus Fuchs, 1937
Huntaphelenchoides Nickle, 1970
Omemeea Massey, 1971
Parasitaphelenchus Fuchs, 1929
Teragramia Massey, 1974

Rhadinaphelenchinae Paramonov, 1962
Rhadinaphelenchus J.B. Goodey, 1960

Seinuridae Husain and Khan, 1967 (Baranovskaya, 1981)
Seinurinae Husain and Khan, 1967
Seinura Fuchs, 1931
Aprutides Scognamiglio, Talame' and s'Jacob, 1970
Papuaphelenchus Andrássy, 1973
Paraseinura Timm, 1960

Entaphelenchidae Nickle, 1970

Entaphelenchinae Nickle, 1970
Entaphelenchus Wachek, 1955
Peraphelenchus Wachek, 1955
Praecocilenchus Poinar, 1969
Roveaphelenchus Nickle, 1970

Acugutturinae Hunt, 1980
Acugutturus Hunt, 1980
Noctuidonema Remillet and Silvain, 1988

Genera inquirenda:

Caballeroides Chaturvedi and Khera, 1977
Devibursaphelenchus Kakulia, 1967
Paraphelenchoides Haque, 1967

Proposed Outline Classification of the Aphelenchida

APHELENCHIDA Siddiqi, 1980

Aphelenchina Geraert, 1966
Aphelenchoidea Fuchs, 1937 (Thorne, 1949)

Aphelenchidae Fuchs, 1937 (Steiner, 1949)
Aphelenchinae Fuchs, 1937
Aphelenchus Bastian, 1865

Paraphelenchidae T. Goodey, 1951 (J.B. Goodey, 1960)
Paraphelenchinae T. Goodey, 1951
Paraphelenchus Micoletzky, 1922 (Micoletzky, 1925)

Aphelenchoidoidea Skarbilovich, 1947 (Siddiqi, 1980)

Aphelenchoididae Skarbilovich, 1947 (Paramonov, 1953)

Aphelenchoidinae Skarbilovich, 1947
Aphelenchoides Fischer, 1894
Laimaphelenchus Fuchs, 1937
Megadorus J.B. Goodey, 1963
Ruehmaphelenchus J.B. Goodey, 1963
Schistonchus Cobb, 1927 (Fuchs, 1937)
Sheraphelenchus Nickle, 1970
Tylaphelenchus Rühm, 1956

Anomyctinae J.B. Goodey, 1960
Anomyctus Allen, 1940

Seinuridae Husain and Khan, 1967 (Baranovskaya, 1981)
Seinurinae Husain and Khan, 1967
Seinura Fuchs, 1931
Aprutides Scognamiglio, Talame' and s'Jacob, 1970
Papuaphelenchus Andrássy, 1973
Paraseinura Timm, 1960

Ektaphelenchidae Paramonov, 1964

Ektaphelenchinae Paramonov, 1964
Ektaphelenchus Fuchs, 1937 (Skrjabin, Shikhobalova, Sobolev, Paramonov and Sudarikov, 1954)
Cryptaphelenchus Fuchs, 1937 (Rühm, 1954)
Cryptaphelenchoides J.B. Goodey, 1960
Ektaphelenchoides Baujard, 1984

Genus incertae sedis:

Berntsenus Massey, 1974

Acugutturidae Hunt, 1980, n.rank

Acugutturinae Hunt, 1980
Acugutturus Hunt, 1980

Noctuidonematinae n.subfam.
Noctuidonema Remillet and Silvain, 1988

Vampyronema n.gen.

Parasitaphelenchidae Rühm, 1956 (Siddiqi, 1980)

Parasitaphelenchinae Rühm, 1956
Parasitaphelenchus Fuchs, 1929

Bursaphelenchinae Paramonov, 1964
Bursaphelenchus Fuchs, 1937
Rhadinaphelenchus J.B. Goodey, 1960

Entaphelenchidae Nickle, 1970

Entaphelenchinae Nickle, 1970
Entaphelenchus Wachek, 1955
Peraphelenchus Wachek, 1955
Praecocilenchus Poinar, 1969
Roveaphelenchus Nickle, 1970

Genus incertae sedis:

Oligoaphelenchoides Poinar, 1977

A fossil genus impossible to place reliably in the above scheme. However, on the character of the extreme curvature of the male tail, it probably belongs to the Bursaphelenchinae and may well be synonymous with *Bursaphelenchus*.

Genera inquirenda:

Devibursaphelenchus Kakulia, 1967
Paraphelenchoides Haque, 1967 (an ill-defined genus best regarded as a synonym of *Aphelenchoides*)

ORDER APHELENCHIDA SIDDIQI, 1980

Diagnosis: (modified after Siddiqi, 1980b): Soil dwelling or insect associates; trophic habit mycetophagous, phytoparasitic, predacious or entomoparasitic. Body small to long (0.2–2.5 mm), vermiform, rarely obese except in some insect parasites. Cuticle thin, usually finely annulated and with lateral fields marked by 0 to more than 12 incisures. Cephalic region low, rounded, continuous or offset and with weak or moderate sclerotization. Four submedian cephalic papillae and six readily visible labial papillae (plus possibly six, more obscure, labial papillae) around the oral aperture. **Amphidial apertures oval, pore-like, dorsosublateral on labial region**. Stylet always present, length usually 10–20 μm, but exceptionally up to 185 μm; conus usually shorter than the shaft, but much longer in certain insect ectoparasitic forms. **Basal swellings or knobs usually weakly developed or entirely absent. Oesophagus comprising a narrow, cylindrical procorpus, a strongly-developed, offset, ovoid to rounded rectangular median bulb with crescentic valve plates and well developed oesophageal glands forming a dorsally overlapping lobe** in all genera apart from *Paraphelenchus* where the glands are small and enclosed in a non-overlapping basal bulb. **All three gland orifi (i.e. including the dorsal gland orifice) located within the median bulb**. Isthmus usually short or absent. Nerve ring circumoesophageal or circumintestinal. Intestine cellular with distinct lumen. Rectum usually distinct except in some insect associates. **Anus a broad transverse slit with an overhanging anterior lip**, but absent or degenerate in some insect parasites or associates. **Vulva posterior at 60–98%**, usually in the form of a transverse slit or, exceptionally, an oval pore (*Aphelenchus*). **Genital tract monoprodelphic, usually outstretched**, but occasionally with a double flexure. Spermatheca axial, if present. Post-uterine sac usually present and may act as a spermatheca. Male genital system monorchic, outstretched. **Sperm large, rounded**, arranged in one or two rows in the gonoduct. **Spicules typically rosethorn-shaped with prominent apex and rostrum,** or derived therefrom, **but elongate and cephalated in *Aphelenchus* and *Paraphelenchus*.** Gubernaculum usually absent, but well developed and elongate in *Aphelenchus* and *Paraphelenchus*. **Bursa usually absent**, but a true peloderan bursa with supporting ribs is present in *Aphelenchus* only, although some genera may have a terminal flap of cuticle (= bursa). **Usually three pairs of caudal papillae present**, but may number as few as two or as many as five pairs.

TYPE GENUS: *Aphelenchus* Bastian, 1865

SUBORDER APHELENCHINA GERAERT, 1966

DIAGNOSIS: Aphelenchida. With the characters of the order.

TYPE GENUS: *Aphelenchus* Bastian, 1865

TYPE SUPERFAMILY: Aphelenchoidea Fuchs, 1937 (Thorne, 1949)

OTHER SUPERFAMILY: Aphelenchoidoidea Skarbilovich, 1947 (Siddiqi, 1980)

KEY TO SUPERFAMILIES

1. Spicules slender, cephalated. Gubernaculum well developed, elongate; V-shaped in cross-section. Lateral fields with six or more incisures. Oesophagus with distinct isthmus and nerve ring circumoesophageal. . . . . . Aphelenchoidea

– Spicules robust, rosethorn-shaped or derived therefrom; typically with a dorsal and ventral limb joined by a transverse bar. Gubernaculum absent or, if present, consisting of a small structure near the distal tip of the dorsal limb of the spicule; never elongate or V-shaped in cross-section. Lateral fields usually with four or fewer incisures, exceptionally six. Oesophagus lacking a distinct isthmus which, if visible at all, is a short stump less than the distance from the valve plates to the base of the bulb in length; nerve ring circumintestinal. . . . . Aphelenchoidoidea

Superfamily Aphelenchoidea Fuchs, 1937 (Thorne, 1949)

DIAGNOSIS: Aphelenchina. Cephalic region low, flattened, continuous with body contour. **Lateral fields with six or more incisures. Oesophagus with a distinct isthmus**; glands either in a dorsally overlapping lobe (Aphelenchidae) or restrained within a non-overlapping basal bulb (Paraphelenchidae). **Nerve ring circum-oesophageal**. Vulva in the form of either a transverse oval pore (Aphelenchidae) or a transverse slit (Paraphelenchidae). Female tail short, sub-cylindroid to conoid and with a broadly rounded terminus which may be mucronate. **Spicules slender, ventrally arcuate; cephalated. Gubernaculum well developed, elongate; V-shaped in cross-section. Bursa** present or absent; **if present then well developed, peloderan and supported by four pairs of ribs** (only three pairs reported by Thorne (1961) in *A. eremitus*).

TYPE GENUS: *Aphelenchus* Bastian, 1865

TYPE FAMILY: Aphelenchidae Fuchs, 1937 (Steiner, 1949)

OTHER FAMILY: Paraphelenchidae T. Goodey, 1951 (J.B. Goodey, 1960)

KEY TO FAMILIES

1. Male with prominent peloderan bursa supported by four pairs of ribs. Female vulval aperture in the form of an oval pore. Oesophageal glands forming a long dorsally overlapping lobe .... Aphelenchidae

– Male lacking a bursa. Vulva in the form of a transverse slit. Oesophageal glands small, retained within a non-overlapping basal bulb .... Paraphelenchidae

Family Aphelenchidae Fuchs, 1937 (Steiner, 1949)

DIAGNOSIS: Aphelenchoidea. Soil dwelling or found in decaying plant material. More than six lateral lines (usually 10 to 12). Vulval aperture in the form of a transverse oval pore. Oesophageal glands free, forming a dorsally overlapping lobe. Female tail short, cylindroid with a rounded tip. Male bursa well developed, peloderan in form and with four pairs of supporting ribs, one pair adanal and the other three postanal.

TYPE GENUS: *Aphelenchus* Bastian, 1865

TYPE SUBFAMILY: Aphelenchinae Fuchs, 1937

OTHER SUBFAMILIES: Monotypic.

Subfamily Aphelenchinae Fuchs, 1937

DIAGNOSIS: Aphelenchidae. With the characters of the family.

TYPE GENUS: *Aphelenchus* Bastian, 1865

OTHER GENERA: Monotypic.

GENUS *APHELENCHUS* BASTIAN, 1865 (Fig. 4.1)

syn. *Isonchus* Cobb, 1913
Metaphelenchus Steiner, 1943
A. (Anaphelenchus) Anderson and Hooper, 1980

DIAGNOSIS: Aphelenchinae. Medium sized to fairly long nematodes (0.5–1.2 mm). Heat relaxed form more or less straight to ventrally arcuate. Cephalic region low, rounded and slightly offset. Cuticle finely annulated. **Lateral fields broad with numerous incisures, usually in excess of ten**, exceptionally fewer (*A. sparsus* only has four to six, for example). Stylet slender with slight basal swellings. Procorpus cylindrical, narrowing just before joining the **large, ovoid, median bulb with centrally placed, crescentic valve plates. Oesophageal glands in a dorsal lobe.** Nerve ring just behind the median bulb and more or less opposite the excretory pore. Vulva posterior at about 70–80% of the body length. Vulval lips slightly protuberant and body often narrowing sharply just behind the vulva. Vagina sloping anteriorly. Genital tract monoprodelphic, outstretched. Developing oocytes mostly in a single row. Post-uterine sac extending for up to half the vulva/anus distance. **Tail short, cylindroid, sometimes slightly ventrally concave, ending in a broadly rounded terminus. Male**, when present, **with paired, slender spicules** which are ventrally arcuate and slightly cephalated proximally. **Gubernaculum linear. Bursa well developed**, extending from the proximal region of the spicules to the tail tip **and supported by four** (exceptionally three) **pairs of ribs**; one pair preanal and the other three in a subterminal group. **Tail short, conoid, tapering to a narrowly rounded point.**

TYPE SPECIES: *Aphelenchus avenae* Bastian, 1865
syn. *Aphelenchus agricola* de Man, 1881
Paraphelenchus agricola (de Man, 1881) Filipjev, 1934
Aphelenchus avenae bicaudata Adilova, 1973
Aphelenchus avenae tricaudata Krall, 1959
Aphelenchus cylindricaudatus (Cobb in Steiner, 1926) Steiner, 1931
Tylenchus cylindricaudatus Cobb in Steiner, 1926
Metaphelenchus goldeni Chaturvedi and Khera, 1979
Aphelenchus macrobolbus Steiner, 1942
Aphelenchus (Paraphelenchus) maupasi Micoletzky, 1922

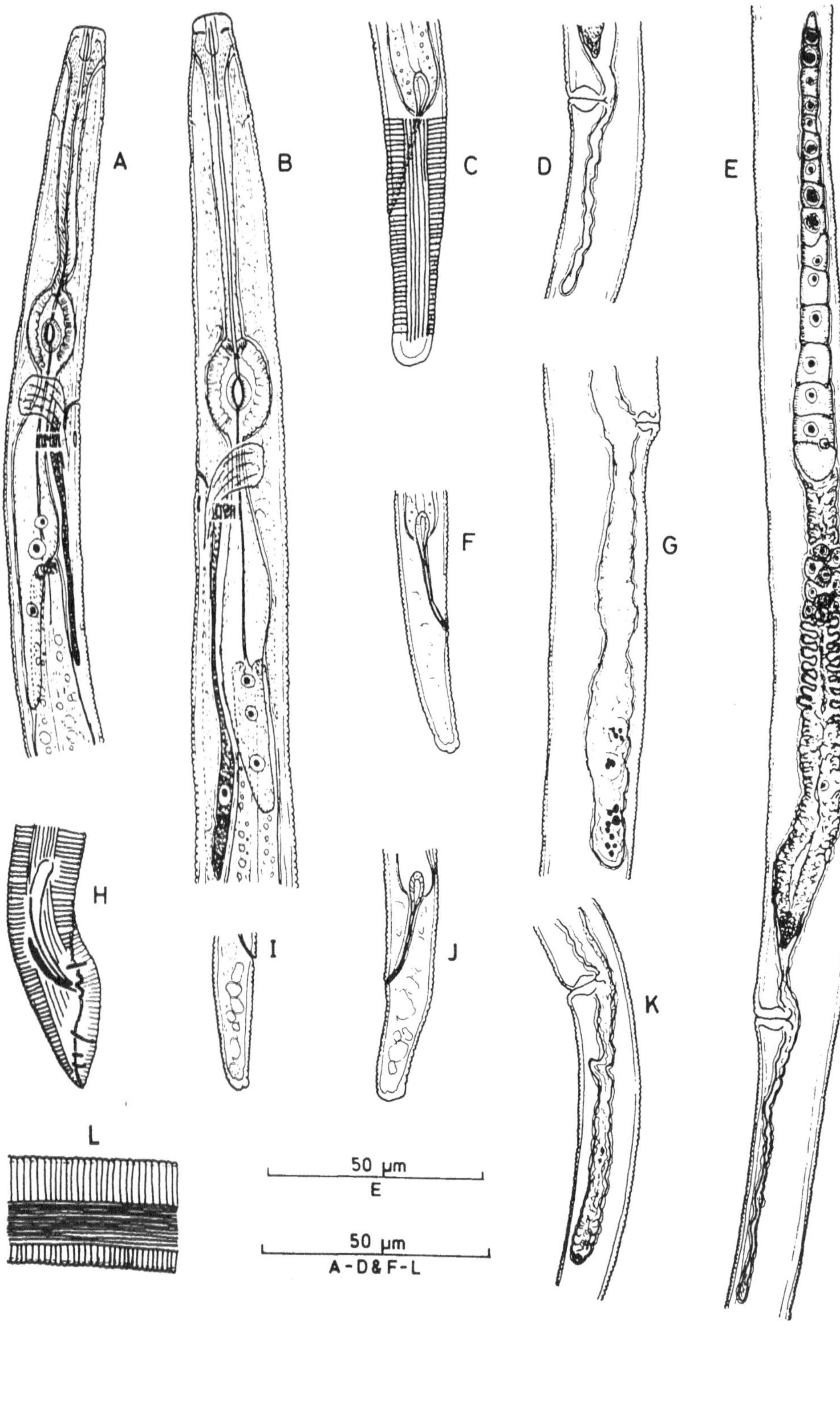
A
B
C
D
E
F
G
H
I
J
K
L
50 μm
E
50 μm
A-D&F-L

Aphelenchus maximus Das, 1960
Paraphelenchus maupasi Micoletzky, 1922 (Bally and Reydon, 1931)
Paraphelenchus micoletzkyi Steiner, 1941
Metaphelenchus micoletzkyi (Steiner, 1941) Steiner, 1943
Aphelenchus radicicolus (Cobb, 1913) Steiner, 1931
Isonchus radicicolus Cobb, 1913
Metaphelenchus rhopalocercus Steiner, 1943
Aphelenchus solani (Steiner, 1935) T. Goodey, 1951
Aphelenchoides solani Steiner, 1935

Other species:
A. amorphophallusi Singh and Agrawal, 1976
A. bastiani Shavrov, 1967
A. deckeri Zeidan and Geraert, 1992
A. eremitus Thorne, 1961
A. isomerus Anderson and Hooper, 1980 (Ebsary, 1991)
syn. *A. (Anaphelenchus) isomerus* Anderson and Hooper, 1980
A. kralli Samibaeva, 1966
A. mirzai Das, 1960
syn. *A. allahabadi* Edward and Misra, 1969
A. paramonovi Nesterov and Lisetskaya, 1965
A. siddiqii Kumar, 1982
A. sparsus Thorne and Malek, 1968
A. tumidicaudatus Shavrov, 1967

Species inquirendae, dubiae vel incertae sedis:
A. aderholdi Schwartz, 1911
A. erraticus von Linstow, 1876
A. gracilis (Cobb, 1888) Cobb, 1891
syn. *Tylenchus gracilis* Cobb, 1888
T. cobbi de Man, 1907
A. kongoensis Allgén, 1933
A. littoralis Hofmänner, 1915 in Hofmänner and Menzel, 1915
A. microlaimus Cobb, 1891
A. modestus de Man, 1876
A. mycogenes Schwartz, 1911
A. pyri Cobbold, 1865
A. rivalis Bütschli, 1873
A. saccharae Akhtar, 1962
A. striatus Steiner, 1914
A. villosus Bastian, 1865

Fig. 4.1. *Aphelenchus isomerus*;. A, B. Male and female anterior ends; C, F, J, I. Female tails; D, G, K. Variation in post-vulval uterine sac and vagina shapes; E. Female reproductive system; H. Male tail; L. Lateral field. (After Zeidan and Geraert, 1992. Courtesy *Nematologica*.)

NOTES:

1. Bastian (1865) described *Aphelenchus pyri* as 'n.sp.' , but Cobbold (1865), in an earlier publication, gave sufficient information to justify the nomen (see Tarjan and Hopper, 1974).
2. Zeidan and Geraert (1992) described *Aphelenchus deckeri* and noted that although it had the non-overlapping gland lobe of *Paraphelenchus*, the female genital tract was more typical of *Aphelenchus*. Males of this species, characters of which should prove decisive in generic placement, are not currently known.

REMARKS: The genus contains about 30 nominal species of which very few are adequately characterized to allow reliable identification. Anderson and Hooper (1980), having studied various populations, erected the subgenus *Anaphelenchus* on the basis of differences in the shape or proportions of the vagina. *Aphelenchus* (*Aphelenchus*) was characterized by having a vagina of unequal proportions (= anisomorphic) whereas *Aphelenchus* (*Anaphelenchus*) has a vagina with equal development (= isomorphic). However, this proposal has not met with general acceptance and Ebsary (1991) synonymized the subgenus.

BIONOMICS: Widespread in temperate and tropical soils. Males are only present in some populations – usually those from warmer soils. The nematodes are mycetophagous, but are often found in damaged or decaying plant material where they presumably feed on fungal hyphae.

DISTRIBUTION: Worldwide in all soil types and often in decaying plant material.

ETYMOLOGY: Derived from the Greek *apheles* = smooth and *enchus* = spear.

SPECIES DESCRIPTION

APHELENCHUS AVENAE BASTIAN, 1865 (Fig. 4.2)

syn. (see above)

MEASUREMENTS: (After Goodey and Hooper, 1965)
Neotype female: L = 0.71 mm; a = 30; b = 5.6; b' = 8.1; c = 35; V = $^{42}78^{10}$; stylet = 15 μm.
Female topotypes (n = 17) : L = 0.67 mm ± 0.75 (0.55–0.86); a = 33 (25–39); b = 5.9 (4.9–6.9); b' = 8.3 (7.2–9.2); c = 31 (27–35); V = $^{38}76.5^{11}$ ($^{29\text{–}46}74\text{–}78^{9\text{–}14}$).
Male (n = 7) [from other English populations]: L = 0.74 mm ± 0.53 (0.63–0.81); a = 29 (27–33); b = 5.7 (5.5–5.9); b' = 8.5 (8.1–9.6); c = 27 (24–29); T = 60 (48–67).

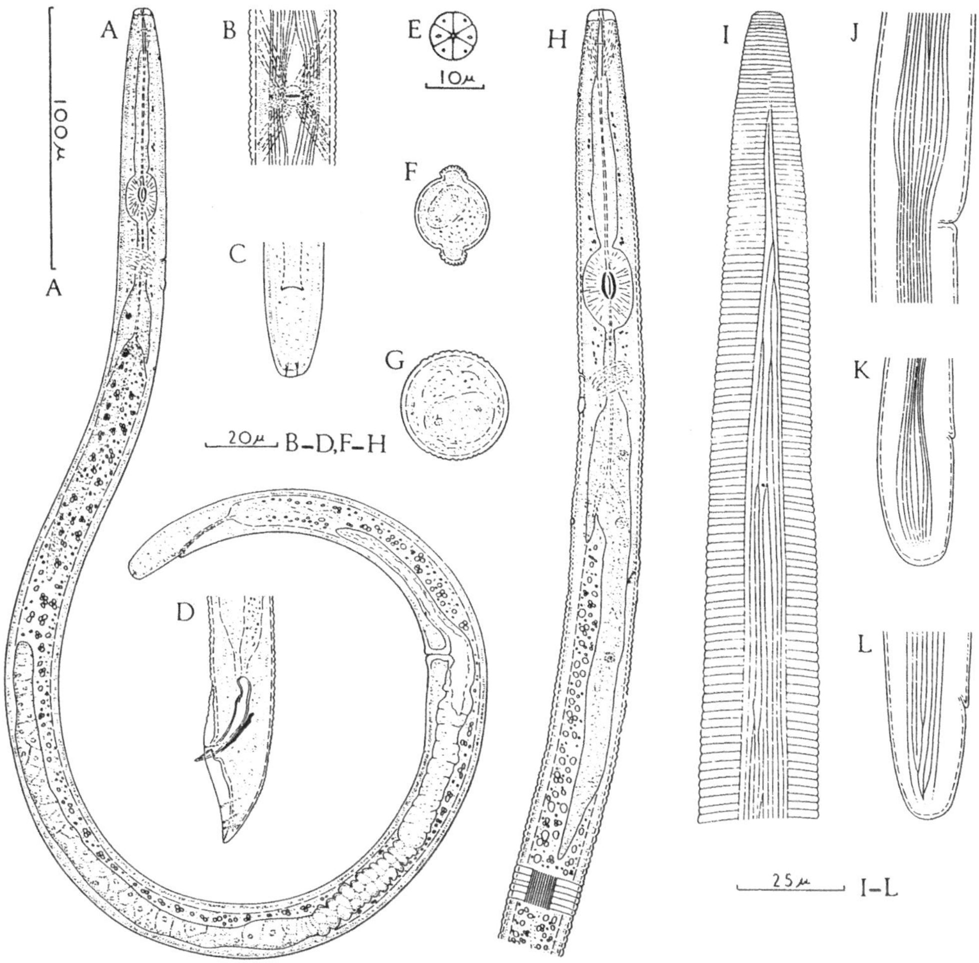

Fig. 4.2. *Aphelenchus avenae.* A. Female; B. Vulva, ventral view; C. Female tail, ventral view; D. Male tail, lateral view; E. Head, *en face*; F. TS through young female; G. TS through mature female; H, I. Anterior parts of female body, lateral view, J. Vulval region, lateral view; K, L. Female tails, lateral view. (After Goodey and Hooper, 1965. Courtesy *Nematologica*.)

NOTE: Considerable morphometric variation occurs in *A. avenae* due to geographical location and/or various environmental factors.

DESCRIPTION: (modified from Hooper, 1974)
Female: Body cylindrical, usually with slight ventral curvature when killed by heat: widest just in front of the vulva and gradually tapering towards the head. Posteriorly the body narrows, sometimes abruptly, just behind the vulva. Cephalic region bluntly rounded to flattened, not offset from body, with three or four very faint head annules barely visible with transmitted light. Body transversely striated, annulations between one and two microns wide. **Lateral fields wide in mid-body region with 12 (10–14) incisures**, narrowing anteriorly

and decreasing in number of incisures; posteriorly the field often widens just in front of the vulva and narrows just posterior to it, the incisures decreasing to about eight at about anal level although the field often expands again in the tail region. Anterior conical part of stylet slightly shorter than the less refractive posterior part **which is slightly thickened but not knobbed at its base**. Procorpus of oesophagus cylindrical; **median bulb widening abruptly from procorpus, well developed, rectangular to oval in shape**, with refractive crescentic valve plates. Nerve ring surrounding oesophageal isthmus half a bulb-length behind the median bulb; oesophageal lumen widens into intestine about two bulb-lengths behind the median bulb and the oesophageal glands partly surround this junction. **Dorsal lobe of oesophageal glands usually well developed, but it can vary in length from very short to up to 100 μm long**; it usually overlaps the intestine on the left subdorsal side. Vulva a short transverse oval, the lips of which tend to protrude a little. Genital tract monoprodelphic, outstretched, **oocytes in a partly single, partly double row**. A central convoluted 'oviduct' region is connected with the uterus. There is also a **postuterine sac reaching about half way from the vulva to the anus**; which is often flattened dorso-ventrally and is sometimes difficult to see. Usually only one egg at a time in the uterus, but more sometimes present in old females. Eggs about three times as long as wide (60–88 × 20–30 μm). Anterior lip of anus protrudes slightly; **tail usually bluntly rounded about 1.75 anal body widths (abw) long**, but may vary between 1.5 and 2.5 abw long; a pair of subterminal phasmids usually visible.

Male: Similar to female except for sexual dimorphism. **Spicules slender, 28–30 μm long**, paired, separate, cephalated, ventrally curved and gradually tapering toward the pointed distal ends. **Gubernaculum present**, 14–16 μm long with a narrow proximal end in lateral view that abruptly widens into the main part which gradually tapers to the distal end. Tail conical and **enveloped by a bursa arising opposite the head of the spicules; four pairs of bursal papillae (ribs) present**, one pair adanal and the other three towards the tail end. **In many populations males are absent or extremely rare**.

Juveniles: Oesophageal region, tail and overall shape of second to fourth stages similar to the female. The number of lateral incisures increases from four in the second stage to eight in the fourth stage.

Bionomics: *Aphelenchus avenae* is primarily mycetophagous and occurs commonly in soil, but also in leaf sheaths, plant crowns and the cortex of some roots, especially if they are infected with fungi. Mankau and Mankau (1963) showed that several phytopathogenic fungi including *Fusarium solani*, *F. lateritum*, *F. oxysporum*, *Periconia* sp., *Rhizoctonia solani*, *Pyrenochaeta* sp., *Sclerotium* sp., *Armillaria mellea* and *Thielaviopsis basicola* were the preferred food of *Aphelenchus avenae* and Ishibashi and Choi (1991) showed the suppressive effect that *A. avenae* could have on such fungi. Other authors have listed a wide range of fungi that are readily fed upon (see

Hooper, 1974). Records of pathogenicity on higher plants are relatively few. Steiner (1936b) observed a female and eggs in phlox root cortex. Decker (1962) found *A. avenae* in the roots of various plants, especially when grown in sandy soil, but it was more frequent in roots also infested with *Pratylenchus* spp. Growth of winter wheat in pots was possibly reduced by this nematode; also it occasionally fed upon other nematodes. Barker (1963) found that *A. avenae* readily reproduced on tobacco callus and, when added in large numbers, they multiplied on the roots and decreased the growth of *Poa pratensis*. Other callus tissues, including lucerne, have also been found to be good hosts. Chin and Estey (1966) showed that *A. avenae* fed and reproduced on the roots of several plants; Jairajpuri (1968) cultured *A. avenae* in *Allium cepa* bulbs; Lordello (1958) reported reproduction in roots of *Cucumis melo* and Lall and Das (1959) found infestations in roots of *Capsicum annuum*.

Males are extremely rare in many populations and reproduction is generally considered to be parthenogenetic. Evans and Fisher (1970) noted that males ranged from very rare to very common between four Australian isolates and that, when cultured on fungus, raising the temperature to 30°C increased their incidence, although they were only necessary for reproduction in one isolate. In an axenic culture of a parthenogenetic population, many males occurred at 30°C (Hansen *et al.*, 1970). Generation time varies from 11–12 days at 20°C to 6–8 days at 30–36°C (Goodey and Hooper, 1965; Jairajpuri, 1968). Fisher and Evans (1967) reported in detail on the feeding of *A. avenae*. Penetration of fungal hyphae was usually accomplished in 6–7 seconds, the median bulb then pulsating for about 8 seconds during which time granular material passed from the oesophagus into the intestine. There was no indication of saliva being injected into host tissues and the cytoplasm was apparently ingested without alteration.

Distribution: *Aphelenchus avenae* is an abundant and cosmopolitan species occurring in most soils. Populations vary considerably in morphometrics and, to some extent, morphology and '*A. avenae*' may represent a species complex.

Family Paraphelenchidae T. Goodey, 1951 (J.B. Goodey, 1960)

Diagnosis: Aphelenchoidea. Soil dwelling. Usually six to eight lateral lines. Oesophageal glands small, enclosed in an abutting basal bulb which is amalgamated with the isthmus. Vulva in the form of a transverse slit. Female tail short, subcylindroid and usually with a mucronate terminus. Male bursa absent, four to five pairs of caudal papillae

Type genus: *Paraphelenchus* Micoletzky, 1922 (Micoletzky, 1925)

Type subfamily: *Paraphelenchinae* T. Goodey, 1951

Other subfamilies: Monotypic

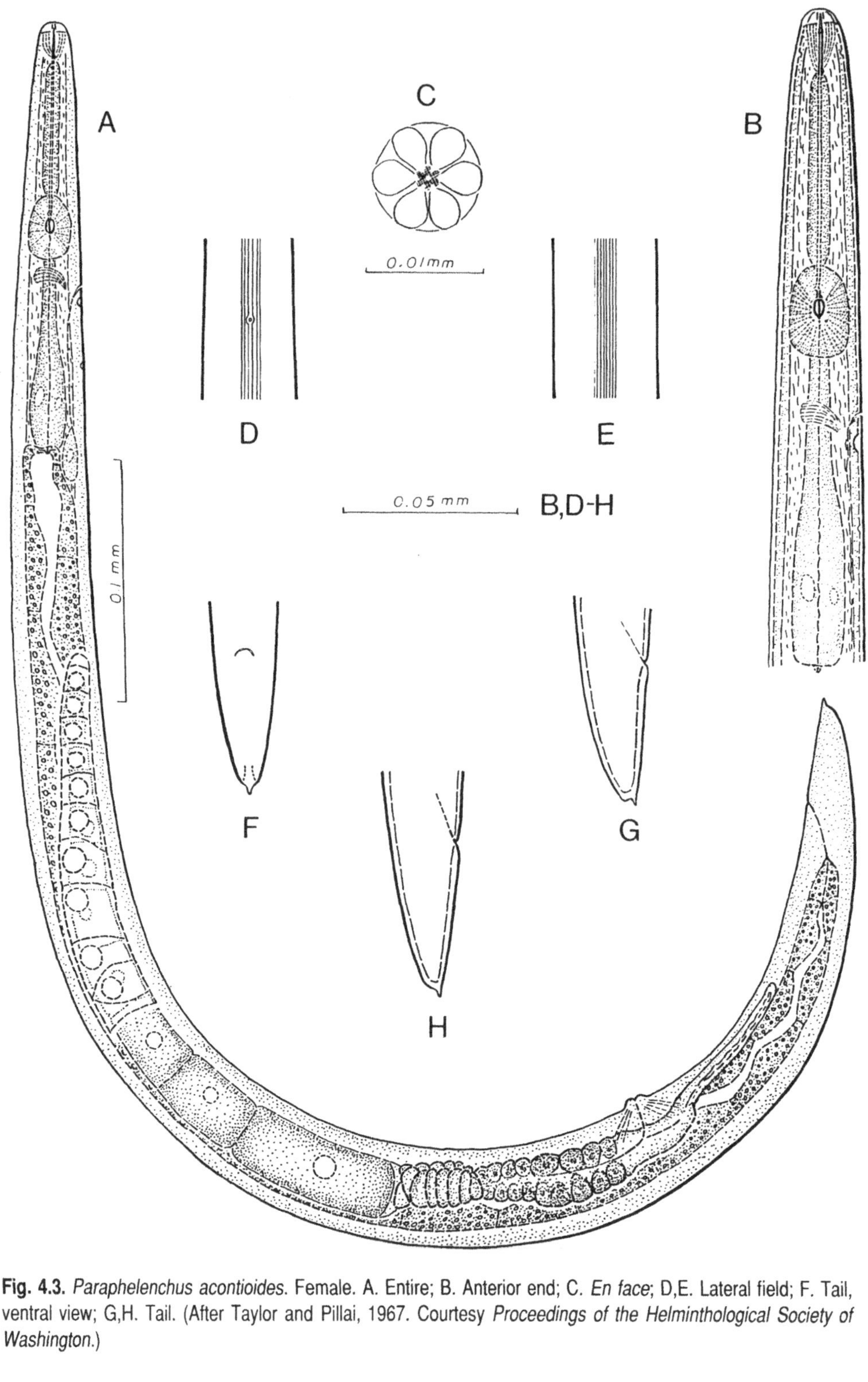

Fig. 4.3. *Paraphelenchus acontioides*. Female. A. Entire; B. Anterior end; C. *En face*; D,E. Lateral field; F. Tail, ventral view; G,H. Tail. (After Taylor and Pillai, 1967. Courtesy *Proceedings of the Helminthological Society of Washington*.)

Subfamily Paraphelenchinae T. Goodey, 1951

DIAGNOSIS: Paraphelenchidae. With the characters of the family

TYPE GENUS: *Paraphelenchus* Micoletzky, 1922 (Micoletzky, 1925)

OTHER GENERA: Monotypic

GENUS *PARAPHELENCHUS* MICOLETZKY, 1922 (MICOLETZKY, 1925) (Fig. 4.3)

syn. *Aphelenchus (Paraphelenchus)* Micoletzky, 1922

DIAGNOSIS: Paraphelenchinae. Medium length nematodes about 0.4–1 mm long. Cephalic region low, rounded, not offset and divided into six lips. Cuticle annulated, the mid-body annules being about 0.8 μm apart. **Lateral fields** occupy about one-third of the body width and are usually **marked by six or eight incisures**. Stylet about 11 μm long with or without basal swellings. Procorpus long, narrow, cylindrical, leading to the ovoid **median bulb containing large crescentic valve plates. Oesophageal glands contained within a spathulate bulb** which abuts, and does not overlap, the intestine. Nerve ring just behind the median bulb with the excretory pore slightly further posterior. Vulva posterior at about 70–80% of the body length and with slightly protuberant lips. Genital tract monoprodelphic, outstretched. **Post-uterine sac long**, extending for several body widths. Tail conoid, tapering to a rounded terminus, usually with a mucron. **Male with slender, paired spicules**, slightly ventrally arcuate and with a slightly expanded capitulum. **Gubernaculum V-shaped, linear. Caudal papillae typically consisting of five pairs**. The first pair is preanal and may be absent or represented by a single midventral papilla; the second pair is adanal, the third about halfway down the tail, the fourth almost terminal and the fifth subterminal but in a sub-dorsal position. **Bursa absent**. Tail short, conoid, with a bluntly rounded terminus.

TYPE SPECIES:
Paraphelenchus pseudoparietinus Micoletzky, 1922 (Micoletzky, 1925)
syn. *Aphelenchus (Paraphelenchus) pseudoparietinus* Micoletzky, 1922

OTHER SPECIES:
P. acontioides Taylor and Pillai, 1967
P. afsi nom. nov.
syn. *P. micoletzkyi* Ali, Farooqui and Suryawanshi, 1970 (= junior homonym) nec *P. micoletzkyi* Steiner, 1941

P. amblyurus Steiner, 1934
P. basili Das, 1960
P. batavicus Filipjev, 1934
syn. *Aphelenchus agricola* of de Man, 1921 nec de Man, 1881
P. arenaceus T. Goodey, 1943
P. crenatus Das and Singh, 1968
P. fidicaudatus Eroshenko, 1966
P. goodeyi Tandon and Singh, 1970
P. heterolineatus Haque, 1967
P. intermedius Thorne and Malek, 1968
P. myceliophthorus J.B. Goodey, 1958
P. obscurus Muchina, 1988
P. octolineatus Shavrov, 1968
P. orientalis Muchina, 1988
P. paramonovi Haque, 1967
P. porrectus Eroshenko, 1966
P. sacchari Husain and Khan, 1967
P. tritici Baranovskaya, 1958
P. ussuriensis Eroshenko, 1966
P. zeae Romanico, 1968
P. zicsii Andrássy, 1989

Remarks: Baranovskaya (1984) produced an illustrated monograph of the genus and a key to species. However, many species remain poorly characterized.

Bionomics: It is probable that all the species are mycetophagous. *P. myceliophthorus* is known to cause damage to the mycelium of cultivated mushroom, *Agaricus bisporus* (Lange), but also feeds on a variety of other fungi. *P. pseudoparietinus* has been recorded from soil, from diseased peanut tissue, in swellings at the stem bases of *Xanthium* spp. and as damaging mushroom mycelium. *P. amblyurus* was recorded from rotting yam tubers by Steiner (1934a). It is likely that the nematode was feeding on fungi, although Steiner (1934a) regarded it as parasitic.

Distribution: Widely distributed and recorded from North America, South America, Europe, Africa and Asia.

Etymology: Derived from the Greek *para* = beside, near to, and *aphelenchus*.

Species description

***Paraphelenchus myceliophthorus* J.B. Goodey, 1958** (Fig. 4.4)

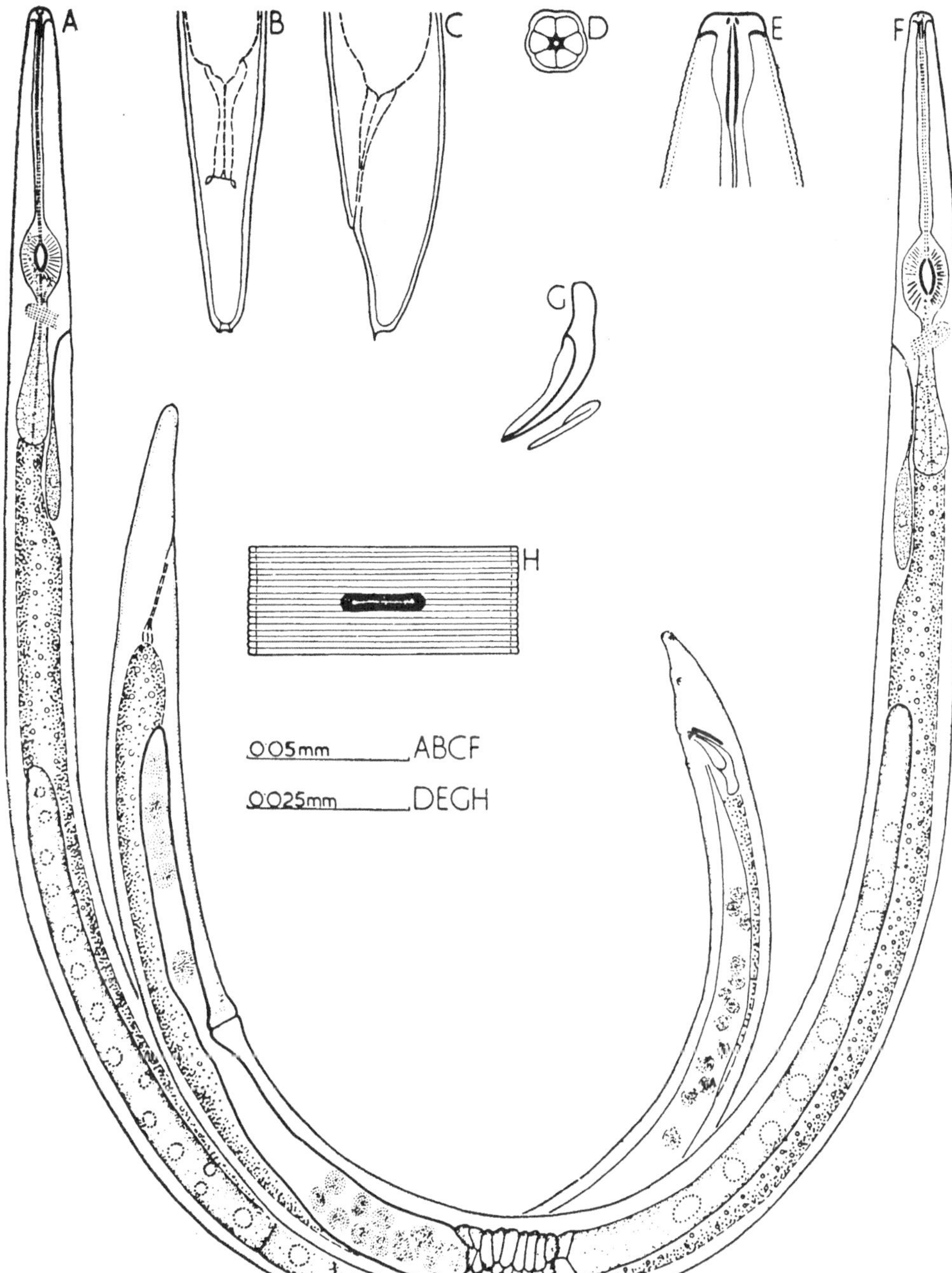

Fig. 4.4. *Paraphelenchus myceliophthorus*. A. Male, entire; B. Female tail, ventral view; C. Juvenile tail, lateral view; D. *En face*; E. Head; F. Female, entire; G. Spicule and gubernaculum; H. Vulva, ventral view. (After J.B. Goodey, 1958. Courtesy *Nematologica*.)

Measurements: (Ex mushroom *Agaricus bisporus* (Lange) Sing, from England; after J.B. Goodey, 1958).
Female (n = 50): L = 0.58–0.82 mm (0.70 ± 0.058); a = 22–34; b = 4.1–6.6; c = 13–24; V = $^{26\text{–}55}71\text{–}78^{9\text{–}15}$.
Male (n = 50): L = 0.56–0.82 mm (0.70 ± 0.060); a = 23–32; b = 3.9–5.9; c = 14–30; T = 46–47.
Holotype female: L = 0.79 mm; a = 26; b = 4.8; c = 18; V = $^{36}75^{10.6}$. (Ex *Botrytis cinerea* Pers. ex Fr. after Naeem, 1967).
Female (n = 20): L = 0.60–0.72 mm (0.64); a = 22–28 (25); b = 4.3–4.9 (4.6); c = 16.3–18.8 (17.8); V = 38–48 (42.7); stylet = 6–19 μm (14.4).
Male (n = 20): L = 0.60–0.75 mm (0.66); a = 23–31 (27); b = 4.2–4.9 (4.6); c = 21.0–25.4 (23.0); stylet = 9.9–23.1 μm (14.9); spicule = 16.9–23.1 μm (20.1); gubernaculum = 9.9–16.5 μm (12.3).
(Ex *Corticium solani* Berkly and Certis after Naeem, 1967).
Female (n = 20): L = 0.72–1.13 mm (1.00); a = 22–32 (24); b = 4.8–6.2 (5.6); c = 18.7–23.7 (20.4); V = 44–67 (63); stylet = 9.9–18.1 μm (16).
Male (n = 20): L = 0.53–1.13 mm (0.88); a = 22–32 (29); b = 4.0–5.6 (5.3); c = 23.3–29.7 (22.5); stylet = 9.9–19.8 μm (16); spicule = 23.1–24.7 μm (19.8); gubernaculum = 6.6–19.8 μm (16.5).
The fungal host affects measurements, particularly body length and vulval position.

Description: (modified after Hooper and Clark, 1985)
Female: Body cuticle with fine transverse striations about 0.8 μm apart; lateral fields about a third of the midbody width wide with **six distinct incisures**. Anteriorly the incisures decrease to four at the base of the oesophagus and two near the stylet base; posteriorly there are six incisures in the vulval region decreasing to four at the end of the postuterine sac and four near the tail tip; the field sometimes expands in the vulval and tail regions. **Cephalic region continuous with the body contour** and appears unstriated when viewed with a light microscope; some six fine transverse striations are visible with the scanning electron microscope (SEM). The head skeleton is rather weak and forms a hexaradiate pattern when viewed *en face*. When viewed with the SEM, the lip region pattern is typical of Aphelenchida (Hooper and Clark, 1980): there are six inner labial papillae around the stomatal aperture, widely spaced amphidial apertures on the lateral lip sectors and a prominent cephalic papilla on the outer edge of each of the other four lip sectors. **Stylet** about 16 μm long, the anterior conus being about 5 μm long and the shaft, which **lacks basal knobs**, 11 μm long. **The oesophageal procorpus is long and cylindrical and the median bulb well developed, roundly elliptical, with distinct crescentic thickenings**. The oesophageal glands form a spathulate basal bulb which is shorter than the procorpus and abuts the intestine. The nerve ring surrounds the oesophageal isthmus just behind the median bulb. **The junction of the oesophagus and intestine is displaced dorsally by a single elongate cell joined to the excretory duct**, the opening of which is about opposite the nerve ring and four to five annules anterior to the hemizonid. A prominent deirid lies in each lateral field opposite the hemizonid. The rectum is about two anal body widths long and the anus is a

short transverse slit with the anterior lip slightly overhanging the aperture. **Tail conical, about three anal body widths long, with a bluntly rounded terminus ending in a pair of short latero-terminal papillae.** There is a short cuticular terminal or ventro-terminal mucro between the terminal caudal papillae. Genital tract monoprodelphic with the ovary separated from the uterus by some 18 closely packed cells surrounding the oviduct. Sperm are often present in the uterus and **postuterine sac** which **extends for two thirds of the vulva to anus distance**; the vulval opening is a distinct transverse slit.

Male: Generally similar to female in appearance apart from sexual dimorphism. Testis single, outstretched. **Spicules elongate with slight ventral curvature**; the ventral flange from the rostrum to the distal tip is rather weak and not easily seen with a light microscope. **Gubernaculum about 14 μm long and in median section when viewed laterally has a folded appearance at the proximal end**; it wraps around the dorsal margins of the spicules, corresponding with the V-shape in transverse section noted by Nickle (1970b) as a feature of *Paraphelenchus*. Also associated with the gubernaculum are two cuticular processes, one on either side of the spicules, which project slightly from the cloacal opening when the spicules are protruded. **Tail somewhat blunter than female, without a bursa, but with caudal papillae**: a single (occasionally a pair) preanal; one pair ventrosublateral about halfway along the tail and two pairs at the tail terminus, one pair being subdorsal with one papilla on the edge of each lateral field and the other pair ventrosublateral. Sometimes a short, terminal cuticular mucro is visible between the ventrosublateral terminal papillae.

Bionomics: The type population reproduced on mushroom mycelium and also on agar plate cultures of various other fungi such as *Alternaria tenuis* Nees, *Botrytis cinerea* Pers. ex Fr., *Fusarium oxysporum* Schlechtendahl, *Gaeumannomyces graminis* (Sacc.) Arx and Oliver (syn. *Ophiobolus graminis*), etc. No detailed studies on the feeding or development of stages have been recorded, but Naeem (1967) found that of 10, 15, 20 and 25°C tested, 20°C was the optimum for reproduction on various fungi and that there was no reproduction at 25°C. *P. myceliophthorus* when dried slowly did not revive when remoistened, unlike *P. pseudoparietinus* which did revive under the same conditions (Naeem, 1967). J.B. Goodey (1960c) found that when *P. myceliophthorus* was added to pots of mushroom compost (8 nematodes per 100 g) the mushroom mycelium gradually disappeared and after 8 weeks there was an average of 443 700 nematodes per 100 g of compost.

Distribution: Recorded from mushroom compost from south east England. Also recorded associated with jute in India and tobacco in Bulgaria.

Control: No control methods specific for *P. myceliophthorus* are recorded, but the general control methods for mushroom nematodes would presumably apply.

Superfamily Aphelenchoidoidea Skarbilovich, 1947 (Siddiqi, 1980)

DIAGNOSIS: Aphelenchina. Cephalic region usually high and offset from body contour. **Lateral fields with four or fewer incisures** (very exceptionally six). Stylet often with basal knobs or swellings. **Oesophagus with isthmus rudimentary or absent. Nerve ring circumoesophageal.** Oesophageal glands in a dorsally overlapping lobe. Vulva in the form of a transverse slit. Female tail conoid to a pointed or narrowly rounded terminus which may be mucronate or otherwise adorned. **Spicules robust, rosethorn-shaped or derived therefrom; usually with a prominent apex and rostrum. Gubernaculum absent or indistinct; if present** it is a small structure located at the distal tip of the dorsal limb of the spicules and is **never elongate, linear or V-shaped in cross-section.** Caudal papillae usually numbering two or three pairs, occasionally as many as five pairs. **Bursa absent around the cloaca**, but a small terminal flap of cuticle is present in the Parasitaphelenchidae.

TYPE GENUS: *Aphelenchoides* Fischer, 1894

TYPE FAMILY: Aphelenchoididae Skarbilovich, 1947 (Paramonov, 1953)

OTHER FAMILIES:
Acugutturidae Hunt, 1980 n.rank
Ektaphelenchidae Paramonov, 1964
Entaphelenchidae Nickle, 1970
Parasitaphelenchidae Rühm, 1956 (Siddiqi, 1980)
Seinuridae Husain and Khan, 1967 (Baranovskaya, 1981)

KEY TO FAMILIES

1. Stylet of both sexes very long (50–180 μm), attenuated; the conus constituting the majority of the stylet length. Ectoparasites of insects . . . . Acugutturidae

– Stylet usually about 10–20 μm long, never over 35 μm long and never attenuate with the conus constituting the majority of the stylet length 2

2. Mature females with swollen body, endoparasitic in the haemocoel of beetles. Three adult forms: male; immature female; mature parasitic female . . . . . . Entaphelenchidae

– Mature females not swollen or endoparasitic. Only two adult forms in the life cycle 3

3. Females with functional anus and elongate tails more than four anal body widths long, often becoming attenuate or filiform, but may be more cylindroid with a rounded or spathulate terminus. Male tail elongate conoid with a spicate terminus. . . . . Seinuridae

– Females with short or medium conoid tails usually less than four anal body widths long, but if longer and with a filiform or spicate terminus then the female anus is non-functional. 4

4. Males with small bursa-like flap of cuticle at tail tip . . . . Parasitaphelenchidae

– Males lacking such a bursa 5

5. Females lacking a functional anus and rectum, the intestine ending as a blind diverticulum in the tail region. Stylet typically with a wide lumen . . . . Ektaphelenchidae

– Females with a functional anus and rectum, intestine not ending as a blind diverticulum. Stylet usually robust, with a narrow lumen and basal knobs or swellings . . . . Aphelenchoididae

Family Aphelenchoididae Skarbilovich, 1947 (Paramonov, 1953)

Diagnosis: Aphelenchoidoidea. Stylet slender, with narrow lumen and usually with small basal knobs or swellings. Post-uterine sac usually present. Female tail of medium length, conoid, with pointed or rounded, often mucronate, terminus. Spicules paired, separate, rosethorn-shaped or derived therefrom. Gubernaculum absent. Bursa absent.

Type subfamily: Aphelenchoidinae Skarbilovich, 1947

Other subfamily: Anomyctinae J.B. Goodey, 1960

Type genus: *Aphelenchoides* Fischer, 1894

Key to subfamilies

1. Cephalic region low, rounded, lacking an obvious oral disc . . . . Aphelenchoidinae

– Cephalic region high, almost spherical, and with a prominent sclerotized oral disc . . . . Anomyctinae

Subfamily Aphelenchoidinae Skarbilovich, 1947

Diagnosis: Aphelenchoididae.

Type genus: *Aphelenchoides* Fischer, 1894

Other genera:

Laimaphelenchus Fuchs, 1937
Megadorus J.B. Goodey, 1963
Ruehmaphelenchus J.B. Goodey, 1963
Schistonchus Cobb, 1927 (Fuchs, 1937)
Sheraphelenchus Nickle, 1970
Tylaphelenchus Rühm, 1956

Key to genera

1. Tail tip bearing four pedunculate tubercles with fringed margins. A vulval flap, formed by the posterior extension of the anterior lip, may be present . . . . *Laimaphelenchus*

– Tail tip not as above, but may be variously mucronate. Vulval flap never present. 2

2. Vulva well posterior, at about 78–93% with the mean value in excess of 80%. Male tail initially short conoid and then narrowing rapidly to a digitiform or short filiform extension 3

– Vulva more anterior at about 60–75% with the mean value less than 80% and usually between 65–70%. Male tail conoid, evenly tapering 4

3. Post-uterine sac absent, female tail subconoid, tapering evenly to a pointed terminus. Apex of spicules greatly elongated and continuing the line of the dorsal limb. Associates of nitidulid beetles. . . . .*Sheraphelenchus*

– Post-uterine sac present, well developed. Female tail dome-shaped with a terminal spike. Apex of spicule small, knob-like. Associates of scolytid beetles . . . .*Ruehmaphelenchus*

4. Stylet very robust, about 17 μm long and with massive, rounded basal knobs. Cuticular annules coarse (1.7 μm) . . . .*Megadorus*

– Stylet not with the above combination of characters, although it may be robust and as long as 21–24 μm. Cuticular annules fine 5

5. Stylet robust, 21–24 μm long with well developed knobs. Found associated with fig wasps or inside figs. . . . .*Schistonchus*

– Stylet and other characters not as above. 6

6. Lips higher than wide, stylet short, robust and with rounded knobs. Conus distinctly shorter than the shaft. Spicules strongly curved and with the apex and rostrum well developed . . . .*Tylaphelenchus*

– Not with the above combination of characters, spicules rosethorn-shaped, apex and rostrum low and rounded or absent . . . .*Aphelenchoides*

NOTE: *Tylaphelenchus* is very close to *Aphelenchoides* and not all of the nominal species are easy to place.

GENUS *APHELENCHOIDES* FISCHER, 1894 (Fig. 4.5)

syn. *Asteroaphelenchoides* Drozdovsky, 1967
Chitinoaphelenchus Micoletzky, 1922
Paraphelenchoides Haque, 1967
Pathoaphelenchus Cobb, 1927
Pseudaphelenchoides Drozdovsky, 1967

DIAGNOSIS: Aphelenchoidinae. Small to long nematodes, usually between 0.4 to 1.2 mm in length. **Heat relaxed** females die straight to ventrally arcuate whereas the **males assume a 'walking-stick' shape** with the tail region sharply curled ventrally. Cuticle finely annulated. Lateral fields often with four incisures but may be two or three. Cephalic region usually rounded in form and slightly offset. There are six equally sized lips and the cephalic skeleton is weak. **Stylet slender**, usually with basal knobs or swellings, **often about 10–12 μm long** and usually less than 20 μm long. Procorpus cylindrical, leading to a well developed ovoid or spherical median bulb with central valve plates. Oesophageal gland lobe well developed and lying dorsal to the intestine. Nerve ring and excretory pore posterior to the median bulb, although the excretory pore may be anterior or posterior to the nerve ring. **Vulva postmedian, usually at between 60–75% of the body length**, only very exceptionally more posterior. Genital tract monoprodelphic, typically outstretched, but may reflex. Developing oocytes in one or more rows. **Post-uterine sac usually present** and often containing spermatozoa, but may be absent. **Tail conoid with a variable terminus which may be bluntly or finely rounded, digitate or bifurcate or with a ventral projection. One or more mucrons of various shapes may be present. Male tail strongly hooked ventrally to form the characteristic 'walking-stick' form**, conoid in shape and tapering to a variable terminus. **Spicules thorn-shaped**, paired and separate. The rostrum and apex are usually well developed, but may be almost absent. Typically there are three pairs of caudal papillae, one pair adanal, one pair subterminal and the other in between. **Bursa absent.**

TYPE SPECIES:
Aphelenchoides kuehnii Fischer, 1894
syn. *A.* (*Aphelenchoides*) *kuehnii* Fischer, 1894 (Filipjev, 1934)

OTHER SPECIES:
A. absari Husain and Khan, 1967
A. abyssinicus (Filipjev, 1931) Filipjev, 1934
syn. *Aphelenchus abyssinicus* Filipjev, 1931
A. africanus Dassonville and Heyns, 1984

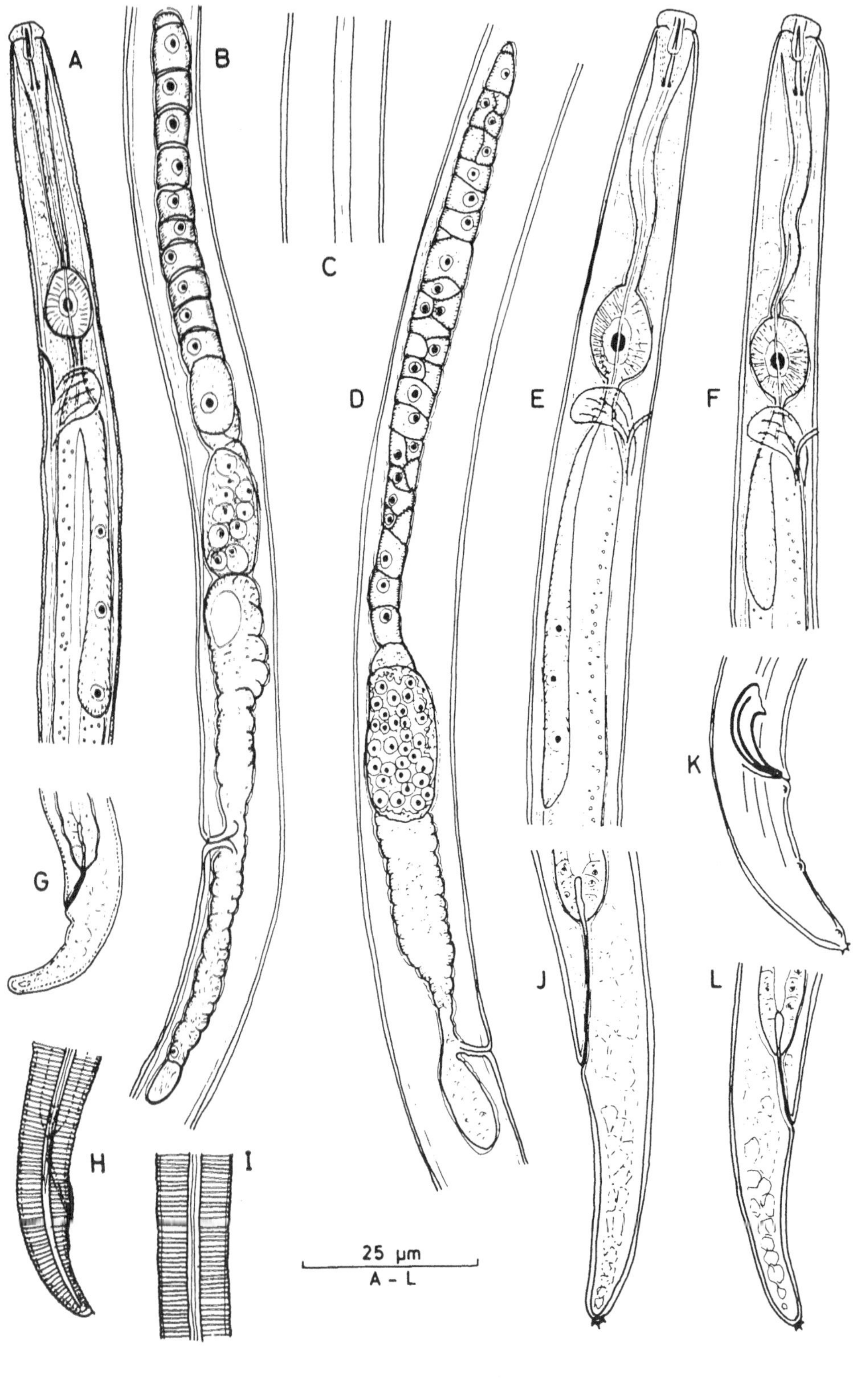
A
B
C
D
E
F
G
H
I
J
K
L
25 µm
A – L

A. agarici Seth and Sharma, 1986
A. aligarhiensis Siddiqi, Husain and Khan, 1967
A. andrassyi Husain and Khan, 1967
A. angusticaudatus Eroshenko, 1968
A. appendurus Singh, 1967
A. arachidis Bos, 1977
A. arcticus Sanwal, 1965
A. asterocaudatus Das, 1960
A. asteromucronatus Eroshenko, 1967
A. besseyi Christie, 1942
    syn. *Aphelenchoides oryzae* Yokoo, 1948
        Asteroaphelenchoides besseyi (Christie, 1942) Drozdovsky, 1967
A. bicaudatus (Imamura, 1931) Filipjev and Schuurmans Stekhoven, 1941
    syn. *Aphelenchus bicaudatus* Imamura, 1931
A. bimucronatus Nesterov, 1985
A. blastophthorus Franklin, 1952
A. brassicae Edward and Misra, 1969
A. brevicaudatus Das, 1960
A. brevionchus Das, 1960
A. brevistylus Jain and Singh, 1984
A. breviuteralis Eroshenko, 1968
A. centralis Thorne and Malek, 1968
A. chalonus Chawla and Khan, 1979
    syn. *Aphelenchoides teres* apud Chawla, Bhamburkar, Khan and Prasad, 1968 nec *A. teres* (Schneider, 1927) Filipjev, 1934
A. chamelocephalus (Steiner, 1926) Filipjev, 1934
    syn. *Aphelenchus chamelocephalus* Steiner, 1926
A. chauhani Tandon and Singh, 1974
A. chinensis Husain and Khan, 1967
A. cibolensis Riffle, 1970
A. clarolineatus Baranovskaya, 1958
A. clarus Thorne and Malek, 1968
A. composticola Franklin, 1957
A. confusus Thorne and Malek, 1968
A. conimucronatus Bessarabova, 1966
A. conophthori Massey, 1974
A. curiolis Gritsenko, 1971
A. cyrtus Paesler, 1959
A. dactylocercus Hooper, 1958
A. daubichaensis Eroshenko, 1968
A. delhiensis Chawla, Bhamburkar, Khan and Prasad, 1968
A. dubitus Ebsary, 1991

Fig. 4.5. *Aphelenchoides* spp. *A. tuzeti* (A, B, G–I). A. Female anterior end; B. Female reproductive system; G, H. Female tails; I. Lateral field. *A. siddiqii* (C–F, J–L). C. Lateral field; D. Female reproductive system; E, F. Female and male anterior ends; J, L, K. Female and male tails. (After Zeidan and Geraert, 1992. Courtesy *Nematologica*.)

syn. *A. dubius* Wasilewska, 1969 (nec Fuchs, 1930)
A. echinocaudatus Haque, 1968
A. editocaputis Shavrov, 1967
A. eltayebi Zeidan and Geraert, 1992
A. emiliae Romaniko, 1966
A. eradicitus Eroshenko, 1968
A. fluviatilis Andrássy, 1960
A. fragariae (Ritzema Bos, 1890) Christie, 1932
syn. *Aphelenchus fragariae* Ritzema Bos, 1890
Aphelenchoides (A.) fragariae (Ritzema Bos, 1890) Christie, 1932
A. (Chitinoaphelenchus) fragariae (Ritzema Bos, 1890) Christie, 1932
Aphelenchus olesistus Ritzema Bos, 1892
Aphelenchoides olesistus (Ritzema Bos, 1892) Steiner, 1932
Aphelenchoides (A.) olesistus (Ritzema Bos, 1892) Steiner, 1932
Aphelenchoides (Chitinoaphelenchus) olesistus (Ritzema Bos, 1892) Steiner, 1932
Aphelenchus olesistus var. *longicollis* Schwartz, 1911
Aphelenchoides olesistus var. *longicollis* (Schwartz, 1911) T. Goodey, 1933
Aphelenchoides (A.) longicollis (Schwartz, 1911) T. Goodey, 1933
Aphelenchus pseudolesistus T. Goodey, 1928
Aphelenchoides pseudolesistus (T. Goodey, 1928) T. Goodey, 1933
Aphelenchoides (A.) pseudolesistus (T. Goodey, 1928) T. Goodey, 1933
Aphelenchus ormerodis of Jegen, 1920 nec Ritzema Bos, 1891
A. franklini Singh, 1969
A. goeldii (Steiner, 1914) Filipjev, 1934
syn. *Aphelenchus goeldii* Steiner, 1914
Aphelenchoides (A.) goeldii (Steiner, 1914) Filipjev, 1934
A. goldeni Suryawanshi, 1971
A. goodeyi Siddiqi and Franklin, 1967
A. graminis Baranovskaya and Haque, 1968
A. gynotylurus Timm and Franklin, 1969
A. haguei Maslen, 1979
A. hainanensis (Rahm, 1938) T. Goodey, 1951
syn. *Aphelenchus hainanensis* Rahm, 1938
A. hamatus Thorne and Malek, 1968
A. helicosoma Maslen, 1979
A. helicus Heyns, 1964
A. helophilus (de Man, 1880) T. Goodey, 1933
syn. *Aphelenchus helophilus* de Man, 1880
A. parietinus helophilus de Man, 1880
Aphelenchoides (A.) helophilus (de Man, 1880) T. Goodey, 1933
Aphelenchus elegans Micoletzky, 1913
A. hylurgi Massey, 1974
A. indicus Chawla, Bhamburkar, Khan and Prasad, 1968
A. involutus Minegawa, 1992
A. jacobi Husain and Khan, 1967

A. *jodhpurensis* Tikyani, Khera and Bhatnagar, 1970
A. *jonesi* Singh, 1977
A. *kungradensis* Karimova, 1957
A. *lagenoferrus* Baranovskaya, 1963
A. *lanceolatus* Tandon and Singh, 1974
A. *lichenicola* Siddiqi and Hawksworth, 1982
A. *lilium* Yokoo, 1964
A. *limberi* Steiner, 1936
  syn. *Paraphelenchoides limberi* (Steiner, 1936) Haque, 1967
A. *longiurus* Das, 1960
A. *longiuteralis* Eroshenko, 1967
A. *loofi* Kumar, 1982
A. *lucknowensis* Tandon and Singh, 1973
A. *macromucrons* Slankis, 1967
A. *macronucleatus* Baranovskaya, 1963
A. *marinus* Timm and Franklin, 1969
A. *menthae* Lisetskaya, 1971
A. *minoris* Ebsary, 1991
  syn. *A. minor* Seth and Sharma, 1986 nec Cobb, 1893
A. *montanus* Singh, 1967
A. *myceliophagus* Seth and Sharma, 1986
A. *neocomposticola* Seth and Sharma, 1986
A. *nonveilleri* Andrássy, 1959
A. *obtusicaudatus* Eroshenko, 1967
A. *obtusus* Thorne and Malek, 1968
A. *orientalis* Eroshenko, 1968
A. *panaxi* Skarbilovich and Potekhina, 1959
A. *parabicaudatus* Shavrov, 1967
A. *parasaprophilus* Sanwal, 1965
A. *parascalacaudatus* Chawla, Bhamburkar, Khan and Prasad, 1968
A. *parasexlineatus* Kulinich, 1984
A. *parasubtenuis* Shavrov, 1967
A. *parietinus* (Bastian, 1865) Steiner, 1932
  syn. *Aphelenchus parietinus* Bastian, 1865
    Pathoaphelenchus parietinus (Bastian, 1865) Steiner, 1931
    Aphelenchoides (A.) parietinus (Bastian, 1865) Steiner, 1932
    Aphelenchus striatus aquaticus Micoletzky, 1913
    A. aquaticus Micoletzky, 1913
A. *petersi* Tandon and Singh, 1970
A. *pityokteini* Massey, 1974
A. *platycephalus* Eroshenko, 1968
A. *polygraphi* Massey, 1974
A. *pusillus* (Thorne, 1929) Filipjev, 1934
  syn. *Aphelenchus pusillus* Thorne, 1929
    Aphelenchoides (A.) pusillus (Thorne, 1929) Filipjev, 1934
A. *rarus* Eroshenko, 1968
A. *rhytium* Massey, 1971

A. richardsoni Grewal, Siddiqi and Atkey, 1992
A. ritzemabosi (Schwartz, 1911) Steiner and Buhrer, 1932
syn. *Aphelenchus ritzemabosi* Schwartz, 1911
Pathoaphelenchus ritzemabosi (Schwartz, 1911) Steiner, 1932
Aphelenchoides (*A.*) *ritzemabosi* (Schwartz, 1911) Steiner and Buhrer, 1932
Aphelenchoides (*Chitinoaphelenchus*) *ritzemabosi* (Schwartz, 1911) Fuchs, 1937
Pseudaphelenchoides ritzemabosi (Schwartz, 1911) Drozdovsky, 1967
Tylenchus ribes Taylor, 1917
Aphelenchus ribes (Taylor, 1917) T. Goodey, 1923
Aphelenchoides (*A.*) *ribes* (Taylor, 1917) T. Goodey, 1933
Aphelenchoides (*Chitinoaphelenchus*) *ribes* (Taylor, 1917) T. Goodey, 1933
Aphelenchus phyllophagus Stewart, 1921
Aphelenchus ormerodis apud Ritzema Bos, 1892 (partim)
A. rosei Dmitrenko, 1966
A. rutgersi Hooper and Myers, 1971
A. sacchari Hooper, 1958
A. sanwali Chaturvedi, Singh and Khera, 1979 in Chaturvedi and Khera, 1979
A. saprophilus Franklin, 1957
A. scalacaudatus Sudakova, 1958
A. seiachicus Nesterov, 1973
A. sexlineatus Eroshenko, 1967
A. shamimi Khera, 1970
A. siddiqii Fortuner, 1970
A. silvester Andrássy, 1968
A. singhi Das, 1960
A. sinodendroni Rühm, 1957
syn. *A.* (*Aphelenchoides*) *sinodendroni* Rühm, 1957
A. spasskii Eroshenko, 1968
A. sphaerocephalus T. Goodey, 1953
A. spicomucronatus Truskova, 1973
A. spinocaudatus Skarbilovich, 1957
A. spinosus Paesler, 1957
A. stammeri Körner, 1954
A. steineri Rühm, 1956
A. submersus Truskova, 1973
A. subparietinus Sanwal, 1961
A. subtenuis (Cobb, 1926) Steiner and Buhrer, 1932
syn. *Aphelenchus subtenuis* Cobb, 1926
Aphelenchoides (*A.*) *subtenuis* (Cobb, 1926) Steiner and Buhrer, 1932
A. (*Chitinoaphelenchus*) *subtenuis* (Cobb, 1926) Steiner and Buhrer, 1932
Aphelenchoides hodsoni T. Goodey, 1935
A. suipingensis Feng and Li, 1986

A. swarupi Seth and Sharma, 1986
A. tagetae Steiner, 1941
A. taraii Edward and Misra, 1969
A. trivialis Franklin and Siddiqi, 1963
A. tumulicaudatus Truskova, 1973
A. tuzeti B'Chir, 1979
A. unisexus Jain and Singh, 1984
A. vaughani Maslen, 1979
A. vigor Thorne and Malek, 1968
A. wallacei Singh, 1977
A. zeravschanicus Tulaganov, 1949

Species inquirendae vel incertae sedis:
A. coffeae (Noack, 1898) Filipjev, 1934
  syn. *Aphelenchus coffeae* Noack, 1898
A. coffeae (Zimmermann, 1898) Filipjev, 1934
  syn. *Aphelenchus coffeae* Zimmermann, 1898
A. dubius (Fuchs, 1930) Filipjev, 1934.
  syn. *Parasitaphelenchus dubius* Fuchs, 1930
A. elongatus Schuurmans Stekhoven, 1951
A. ferrandini Meyl, 1954
A. hessei (Rahm, 1925) Filipjev, 1934
  syn. *Aphelenchus hessei* Rahm, 1925
A. hyderabadensis Das, 1960 (*sp. inq.* Hechler and Taylor, 1965)
A. minimus Meyl, 1953
A. minor (Cobb, 1893) Steiner and Buhrer, 1933
  syn. *Aphelenchus minor* Cobb, 1893
A. mucronatus Paesler, 1946 (regarded as belonging to *Seinura* by Hechler and Taylor, 1965)
A. naticochensis (Steiner, 1920) Filipjev, 1934
  syn. *Aphelenchus naticochensis* Steiner, 1920
A. ormerodis (Ritzema Bos, 1891) Steiner, 1932
  syn. *Aphelenchus ormerodis* Ritzema Bos, 1891
A. retusus (Cobb, 1927) T. Goodey, 1951
  syn. *Aphelenchus retusus* Cobb, 1927
A. richtersi (Steiner, 1914) Filipjev, 1934
  syn. *Aphelenchus richtersi* Steiner, 1914
A. speciosus Andrássy, 1958 (regarded as belonging to *Seinura* by Hechler and Taylor, 1965)
A. teres (Schneider, 1927) Filipjev, 1934
  syn. *Aphelenchus teres* Schneider, 1927

Remarks: A large, reasonably distinct genus of worldwide distribution. Many nominal species are inadequately characterized for reliable recognition and the genus is in urgent need of a major revision. The above species list is based on one by Hooper (unpublished) and contains many species which are of dubious standing by current criteria.

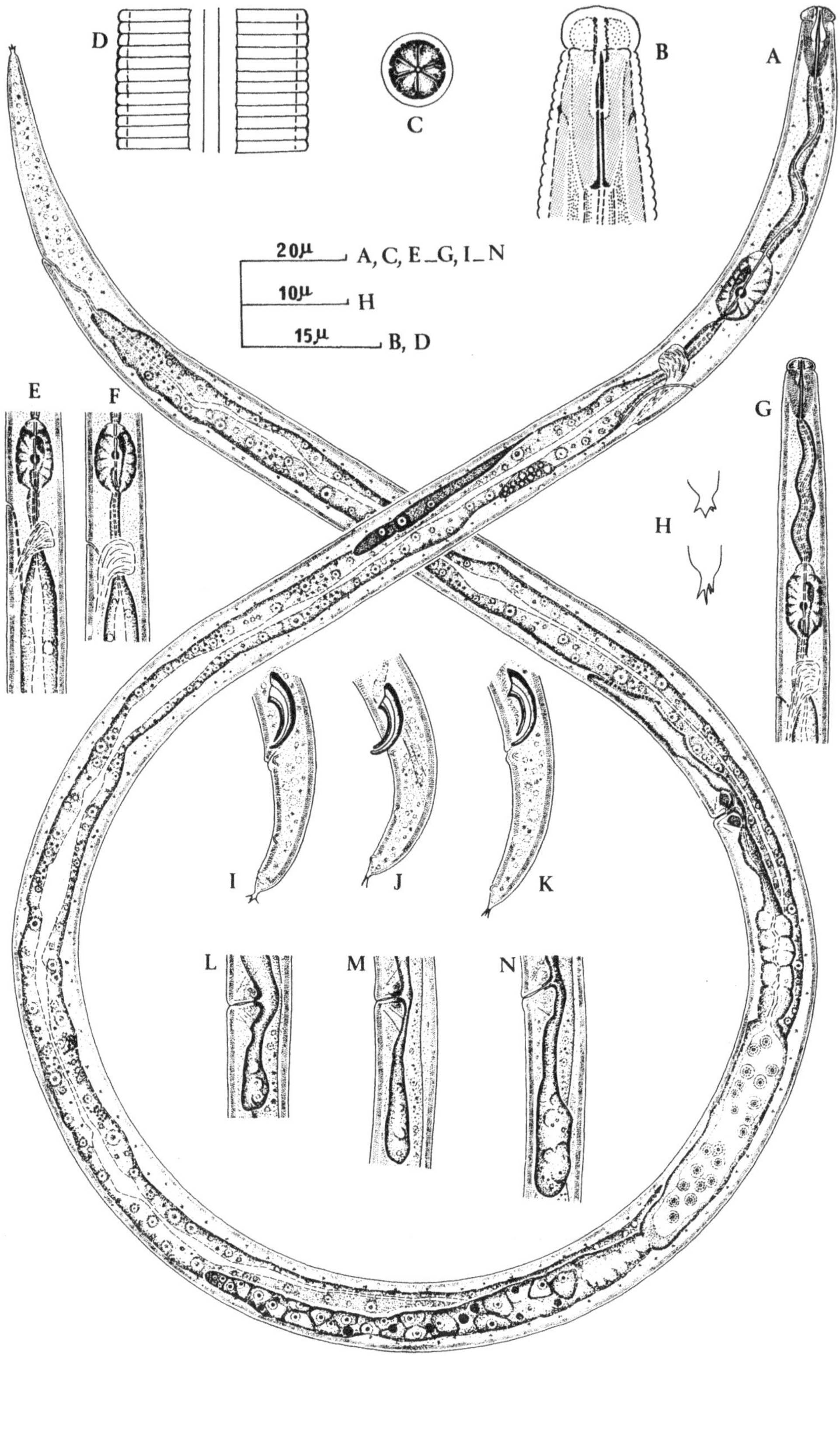

A
B
C
D
E
F
G
H
I
J
K
L
M
N
20μ A, C, E_G, I_N
10μ H
15μ B, D

Bionomics: The majority of species appear to be free-living and are found world-wide in soil, decaying plant material, galleries of wood-boring beetles etc., where they probably lead a mycetophagous lifestyle. However, the genus does contain a few species which are of greater or lesser importance as ecto and endoparasites of plants. Some species have been shown to parasitize higher plants and yet feed and reproduce on a variety of fungi. The major plant-parasitic species include *A. besseyi* Christie, 1942 which causes 'white-tip' disease of rice, *A. fragariae* (Ritzema Bos, 1890) Christie, 1932 and *A. ritzemabosi* (Schwartz, 1911) Steiner, 1932 which attack a wide variety of plants and *A. arachidis* Bos, 1977, a species of restricted distribution, attacking groundnut. Uniquely amongst the Aphelenchida, two species have been reported from the marine littoral habitat (Timm and Franklin, 1969), tolerating wide differences in salinity.

Distribution: Worldwide.

Etymology: Derived from *aphelenchus* and the Greek *oides* = likeness of form.

Species descriptions

Aphelenchoides besseyi Christie, 1942 (Fig. 4.6)

syn. *Aphelenchoides oryzae* Yokoo, 1948
Asteroaphelenchoides besseyi (Christie, 1942) Drozdovsky, 1967

Measurements: (From type host and locality, after Christie, 1942):
Female (n = 10): L = 0.66–0.75 mm; a = 32–42; b = 10.2–11.4; c = 17–21; V = 68–70.
Male (n = 10): L = 0.54–0.62 mm; a = 36–39; b – 8.6–8.8; c = 15–17.
Females: (After Allen, 1952): L = 0.62–0.88 mm; a = 38–58; b = 9–12; c = 15–20; V = $^{43-33}66\text{–}72^{4-8}$.
Neotype female: (from strawberry, Florida, USA): L = 0.7 mm; a = 50; b = 10; c = 19; V = $^{33}70^{6}$; stylet = 10 μm.
Males: L = 0.44–0.72 mm; a = 36–47; b = 9–11; c = 14–19; T = 50–65.
Female: (n = 20) (From rice seeds from Séfa, Senegal; after Fortuner, 1970): L = 0.57–0.84 (0.68) mm; a = 39–53 (47.7); b = 9.2–13.1 (11.46); b′ = 4.06–5.77 (4.85); c = 13.8–20.4 (17.7); V = $^{39.1-19.9}68.7\text{–}73.6\ (71.2)^{4.1-6.2}$; stylet = 10.0–12.5 (11.9) μm.
Male (n = 9): L = 0.53–0.61 (0.57) mm; a = 40.7–46.9 (44.4); b = 8.87–10.70

Fig. 4.6. *Aphelenchoides besseyi.* A. Female; B. Female head end; C. Female *en face* view; D. Lateral field; E, F. Variation in female median oesophageal bulb and position of excretory pore with respect to nerve ring; G. Male anterior end; H. Female tail termini showing variation in shape of mucro; I–K. Male tail ends; L–N. Variation in post uterine sac. (B, D after Siddiqi; the rest after Fortuner, 1970.)

(9.52); b′ = 3.57–4.91 (4.09); c = 16–20 (17.97); T = 28–52 (40.59); stylet = 10.0–12.5 (11.4) μm; spicules (dorsal limb) = 18–21 (19.2) μm.

Description: (modified from Franklin and Siddiqi, 1972)
Female: **Body slender**, straight to slightly arcuate ventrally when relaxed; annules fine, indistinct, about 0.9 μm wide near mid-body. Cephalic region rounded, unstriated, slightly offset and wider than body at lip base; labial framework hexaradiate, lightly sclerotized. **Lateral fields about one-fourth as wide as body, with four incisures**. Anterior part of stylet sharply pointed, about 45% of total stylet length, posterior part with slight **basal swellings** which are **1.75 μm across**. Median oesophageal bulb oval, with a distinct valvular apparatus slightly behind its centre. Oesophageal glands extending dorsally and subdorsally for four to eight body widths over intestine. Nerve ring about one body width behind median oesophageal bulb. **Excretory pore usually near anterior edge of nerve ring** (anterior to nerve ring in neotype). Hemizonid 11–15 μm behind excretory pore. Vulva transverse, with slightly raised lips. **Spermatheca** elongate oval (up to eight times as long as wide when fully distended), **usually packed with sperms. Ovary relatively short and not extending to oesophageal glands, with oocytes in two to four rows. Post-uterine sac narrow, inconspicuous, not containing sperms, 2.5–3.5 times anal body width long but less than one-third distance from vulva to anus. Tail** conoid 3.5–5 anal body widths long; **terminus bearing a mucro of diverse shape with three to four pointed processes**.

Male: **About as numerous as females.** Posterior end of body curved by about 180° in relaxed specimens. Cephalic region, stylet and oesophagus as described for female; **tail** conoid, **with terminal mucro with two to four pointed processes**. First pair of ventrosubmedian papillae adanal, second slightly behind middle of tail and third subterminal. **Spicules typical of the genus except that the proximal end lacks an apex** and has only a moderately developed rostrum. Testis single, outstretched.

Bionomics: *Aphelenchoides besseyi* is best known as the causal agent of 'white tip' disease of rice. It is a facultative ecto- and endoparasite of the leaves and young tissues of rice, the characteristic symptoms including whitening of the top several centimetres of the leaves, necrosis, distortion and crinkling of the flag leaf and reduction in panicle development. The nematode is seed borne and can survive in a state of anhydrobiosis for several years on stored grain, but much less under field conditions. *A. besseyi* also attacks strawberry, causing 'summer crimp' and a number of other plants. It also feeds readily on various fungi. Reproduction is usually amphimictic, the life cycle taking 8 days at 23°C. No development occurs below 13°C, a factor which influences its northerly and southerly distribution. A good general review of this nematode was given by Bridge *et al.* (1990).

Distribution: Widely distributed in the major rice growing areas of the world (Africa, Asia, eastern Europe, North, Central and South America and the Pacific region).

***Aphelenchoides bicaudatus*, (Imamura, 1931) Filipjev and Schuurmans Stekhoven, 1941** (Fig. 4.7)

Measurements:

Female (n = 18) (After Imamura, 1931): L = 0.38–0.47 (0.43) mm; a = 31.3–31.7 (31.5); b= 6.8–8.4 (7.4); c = 9.4–12.6 (10.6); V = 61.7–90.2 (70.4).
Female (n = 50) (After Siddiqui and Taylor, 1967): L = 0.41–0.55 (0.46) mm; a = 25–31 (28.0); b = 7.3–9.6 (8.2); c = 9.8–13.7 (11.4); V = 65–70 (67.5); stylet = 10–12 (11.2) μm.
Male (After Siddiqui and Taylor, 1967): L = 0.385 mm; a = 22.6; b = 7.5; c = 11.4; stylet = 10 μm.

Description: (modified from Siddiqui, 1976)

Female: Body small, slightly attenuated anteriorly, narrowing posterior to anus, lying straight with a slight curvature of the tail region when heat relaxed. Cuticle finely annulated, annules 0.8 μm wide. **Lateral fields** occupy 16–18% of body diameter at the mid-body region; there are **two distinct longitudinal lines between which lies a narrow refractive band**. The two longitudinal lines merge into a single line about one body diameter anterior to the median bulb, the line gradually disappearing approximately one-half body diameter anterior to the merging point; posteriorly the two lines disappear about three body diameters behind anus. Cephalic region distinctly set off from body, appearing smooth and with six equal lips. Refractive cheilorhabdions seen near the oral aperture slightly anterior to the stylet tip. Stylet 10–12 μm long with distinct basal knobs. Cylindrical portion of the stylet slightly longer than the anterior conical section. Procorpus wider anteriorly, gradually narrowing posteriorly, and then widening at the median bulb. Median bulb pyriform to spherical in shape and occupying four-fifths of body width. Oesophageal glands form a lobe overlapping the intestine dorsally, 50–75 μm in length. Nerve ring about one-half body width behind median bulb, encircling anterior end of the oesphageal glands and intestine. **Excretory pore usually opposite anterior margin of nerve ring, but sometimes as far anterior as the basal portion of median bulb**. Hemizonid four annules wide, located 16–18 μm posterior to excretory pore. Deirids and phasmids not seen. Vulva a transverse slit, two-fifths body width long and located approximately two-thirds body length from anterior end. **Genital tract** monoprodelphic, outstretched but reflexed at the apex in well fed specimens. It usually **extends up to the oesophageal gland lobe** but occasionally even beyond the median bulb. **Oocytes arranged in a single row. Post-uterine sac 2.5–3 anal body widths long**, with its terminus directed ventrally. **Tail** 4.5–5 anal body widths long, tapering gradually beyond anus, and **terminating in a bifurcate tip of which the ventral prong is acute and always longer than the dorsal prong**.

Male: **Extremely rare**, slightly smaller than female. Stylet 10.5 μm long. Anterior portion of body similar to female. Single testis extending anteriorly to about three body widths behind the posterior margin of the oesophageal gland

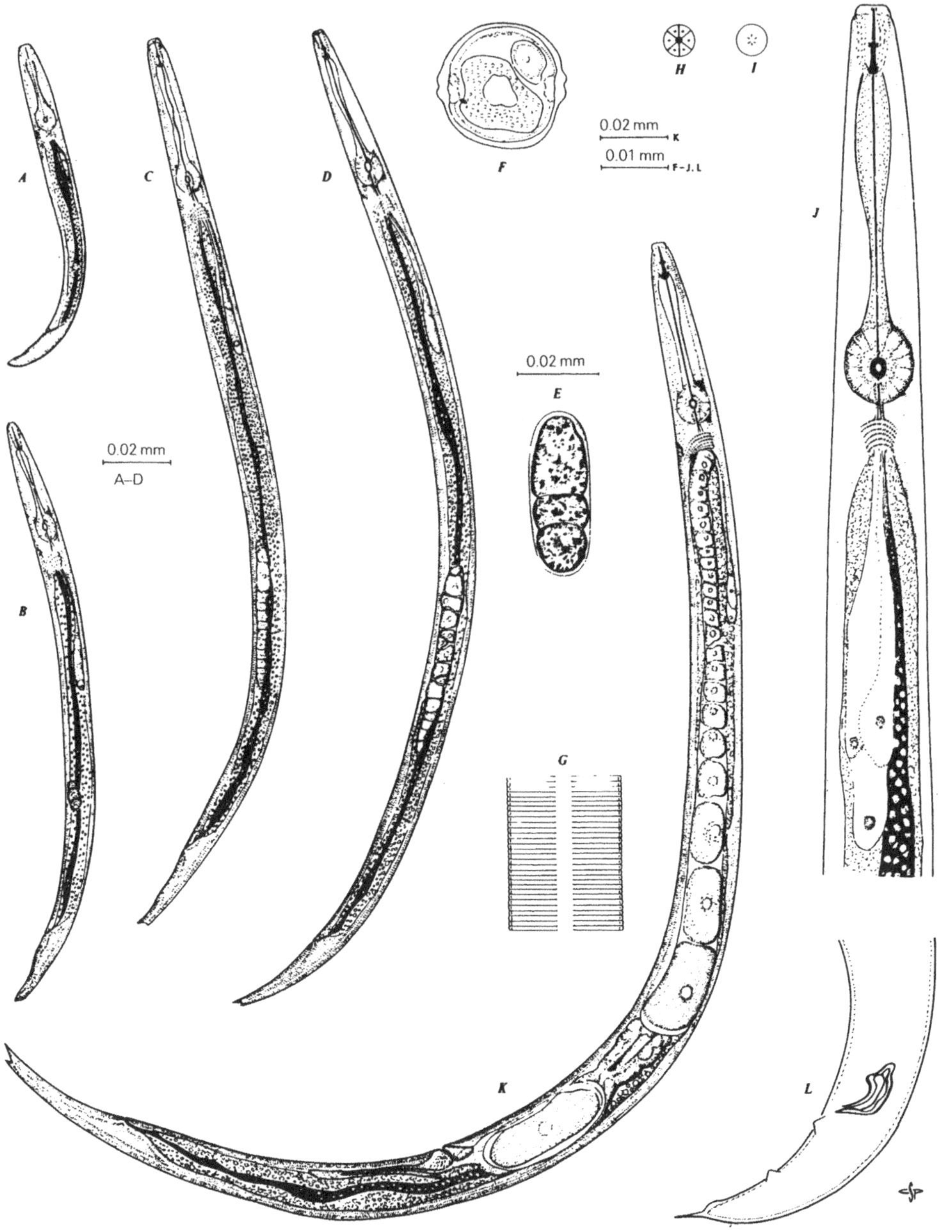

Fig. 4.7. *Aphelenchoides bicaudatus.* A–D. Juveniles, first to fourth stages; E. Egg; F. Cross section of female at mid-body; G. Lateral field; H. *En face* view; I. Framework around oral opening; J. Oesophageal region in dorsal view; K. Entire female; L. Male tail. (After Siddiqui, 1976.)

lobe. Spicules equal in size and typically aphelenchoid in shape. **Tail without bifurcation**, measuring 2.5 body widths long. In the two specimens observed, it narrows gradually and then abruptly to form a centrally located mucro approximately 3.5 μm long. Two pairs of ventrosublateral caudal papillae observed, one pair adanal and the other pair midway along the tail.

Bionomics: *Aphelenchoides bicaudatus* is mycetophagous and is commonly found in the rhizosphere of many crops and other plants in the warmer regions of the world. It is known to feed on mushroom mycelium, but does not appear to be a serious pest. Reproduction is parthenogenetic and the life cycle lasts 5–6 days at 28°C (Siddiqui, 1976).

Distribution: Recorded from most of the tropical and subtropical regions of the world and also from some warmer temperate areas.

Aphelenchoides fragariae **(Ritzema Bos, 1890) Christie, 1932** (Fig. 4.8)

syn. *Aphelenchus fragariae* Ritzema Bos, 1890
Aphelenchus olesistus Ritzema Bos, 1892
Aphelenchoides olesistus (Ritzema Bos, 1892) Steiner, 1932
Aphelenchus olesistus var. *longicollis* Schwartz, 1911
Aphelenchoides olesistus var. *longicollis* (Schwartz, 1911) T. Goodey,1933
Aphelenchus pseudolesistus T. Goodey, 1928
Aphelenchoides pseudolesistus (T. Goodey, 1928) T. Goodey, 1933
Aphelenchus ormerodis of Jegen, 1920 (nec Ritzema Bos, 1891)

Measurements:
Female (After Allen, 1952): L = 0.45–0.80 mm; a = 45–60; b = 8–15; c = 12–20; V = $^{41-22}64\text{–}71^{11-18}$.
Neotype female: L = 0.73 mm; a = 45; b = 11; c = 13; V = $^{41}66^{18}$; stylet = 10 μm.
Male: L = 0.48–0.65 mm; a = 46–63; b = 9–11; c = 16–19; T = 44–61.
Female (after Franklin, 1950): L = 0.552–0.886 (0.796); a = 36–63 (53). Male: L = 0.573–0.864 (0.732); a = 40–63 (51).

Description: (modified from Siddiqi, 1975)
Female: **Body slender (a = 45–63)**, straight or arcuate when relaxed; striae about 0.9 μm apart at the midbody region. **Lateral field a narrow band, with two incisures. Cephalic region appears smooth, elevated, anteriorly flattened with sides straight to rounded, almost continuous with body contour**; framework weakly sclerotized, with outer margins incurved, extending one body annule posteriorly from basal plate; inner margins forming a frail guiding tube four to five body annules long. Stylet slender about 10 to 11 μm long; ante-

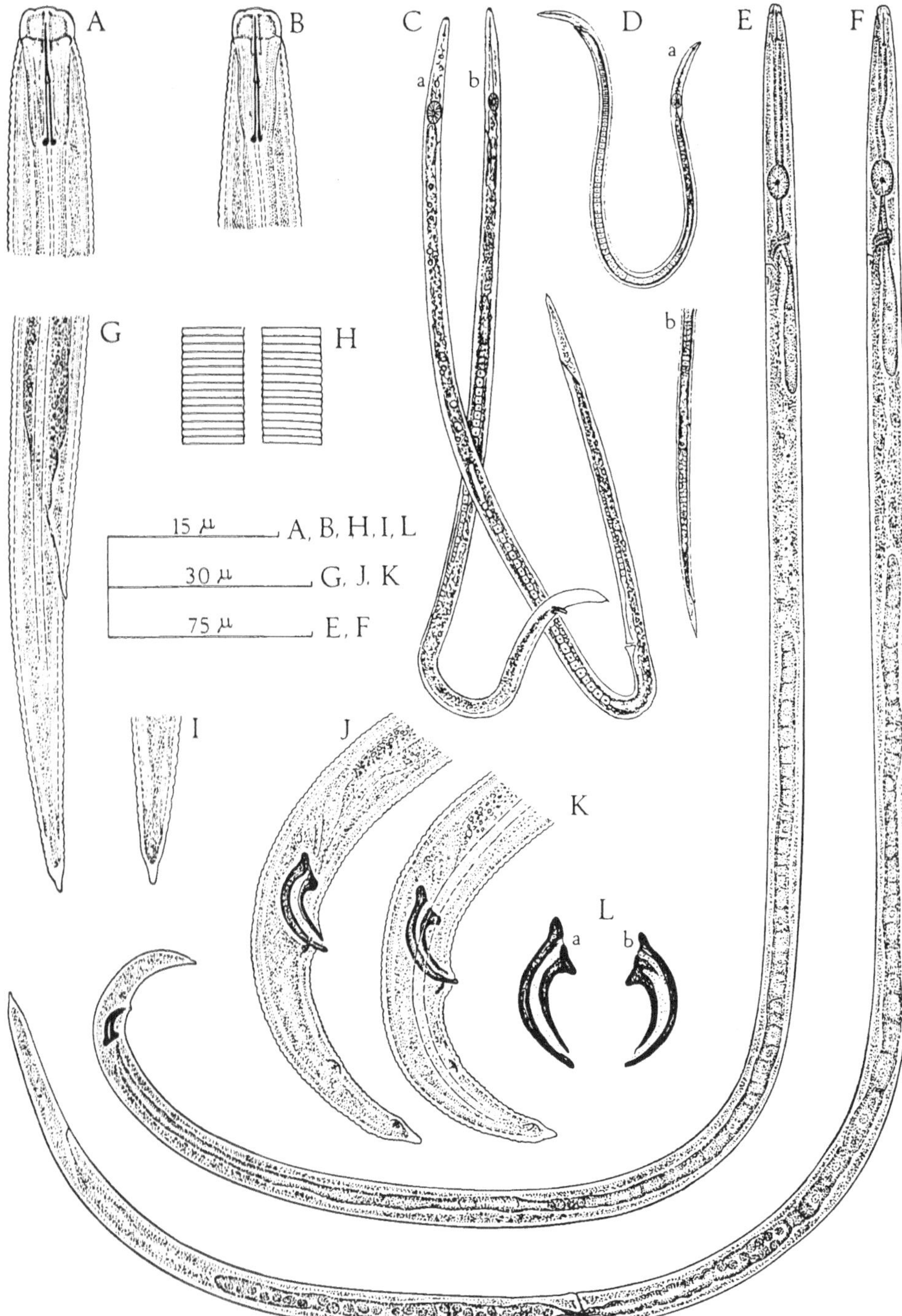

Fig. 4.8. *Aphelenchoides fragariae*. A. Female head end; B. Male head end; C. a, Female; b, Male, of *A. olesistus* Ritzema Bos, 1893 [= *A. fragariae*]; D. a, Male; b, Posterior portion of female, of *Aphelenchus fragariae* Ritzema Bos, 1891; E. Male; F. Female; G. Female tail; H. Lateral field; I. Female tail tip; J, K. Male tails; L. Spicules. (A, B, E-La after Siddiqi, 1975; C, D, after Ritzema Bos, 1893 and 1891, respectively.)

rior portion solid and needle-like for most of its length; basal knobs minute but distinct. Oesophagus typical of genus. Nerve ring encircling isthmus near its base, about one body width behind median bulb which is oval and highly muscular with central valve plates. **Excretory pore level with or close behind nerve ring**. Oesophageal glands elongate, extending over intestine dorsally. Vulva transverse, lips slightly protruding. Genital tract monoprodelphic, outstretched, with oocytes in a single file. Spermatheca elongate-oval. **Post-uterine sac extending for more than half vulva-anus distance**, often with sperms. **Tail** elongate-conoid **ending in a simple blunt spike devoid of any processes**. Hooper (personal communication) has found that in cultures of *A. fragariae* four lines may be present in the lateral field and that the tail tip can be tufted in appearance.

Male: **Abundant**. Essentially similar to female. **Tail arcuate through 45 to 90° when relaxed**, not sharply curved like a hook, with a simple blunt terminal spine. Testis outstretched, with spermatocytes in a row. Three pairs of ventrosubmedian caudal papillae; spicules rosethorn-shaped **with moderately developed apex and rostrum, dorsal limb 14–17 μm long**.

Bionomics: *Aphelenchoides fragariae*, as its specific epithet suggests, is commonly associated with strawberries, on which crop it causes 'strawberry crimp' disease. It is also economically important on ornamental ferns, attacking over 100 species. The host range is extensive, the plants mainly belonging to the Liliaceae, Primulaceae, Pteridophyta and Ranunculaceae [see Siddiqi (1975) for more details]. *A. fragariae* is an ecto- and endoparasite of the above-ground parts of plants, its feeding resulting in distortion, leaf crinkling, leaf blotching and discoloration and browning of tissues, but it can also be mycetophagous. Reproduction is amphimictic, the life cycle lasting 10–11 days at 18°C.

Distribution: Widespread in CIS (formerly USSR), Europe, Japan, and North America.

Aphelenchoides ritzemabosi **(Schwartz, 1911) Steiner and Buhrer, 1932** (Fig. 4.9)

syn. *Aphelenchus ritzemabosi* Schwartz, 1911
Pathoaphelenchus ritzemabosi (Schwartz, 1911) Steiner, 1932
Aphelenchoides (Chitinophelenchus) ritzemabosi (Schwartz, 1911) Fuchs, 1937
Pseudaphelenchoides ritzemabosi (Schwartz, 1911) Drozdovski, 1967
Tylenchus ribes Taylor, 1917
Aphelenchus ribes (Taylor, 1917) T. Goodey, 1923
Aphelenchoides ribes (Taylor, 1917) T. Goodey, 1933
Aphelenchus phyllophagus Stewart, 1921

Measurements:

Female (After Allen, 1952): L = 0.77–1.20 mm; a = 40–45; b = 10–13; c = 18–24; V = $^{48-33}66–75^{14-17}$.

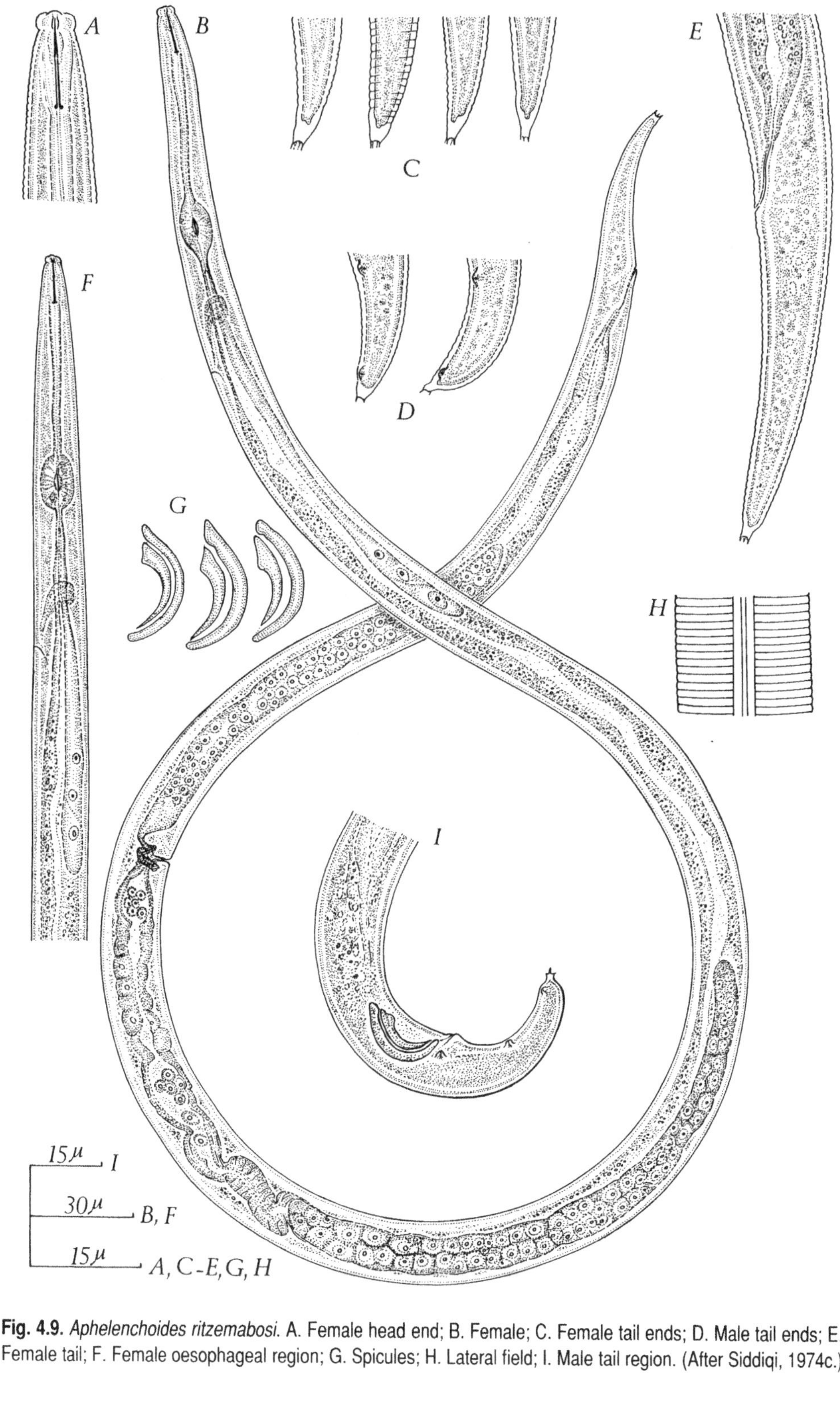

Fig. 4.9. *Aphelenchoides ritzemabosi.* A. Female head end; B. Female; C. Female tail ends; D. Male tail ends; E. Female tail; F. Female oesophageal region; G. Spicules; H. Lateral field; I. Male tail region. (After Siddiqi, 1974c.)

Neotype female: L = 0.85 mm; a = 42; b = 12; c = 18; V = $^{35}68^{17}$; stylet = 12 μm.
Male: L = 0.70–0.93 mm; a = 31–50; b = 10–14; c = 16–30; T = 35–64.

DESCRIPTION: (modified from Siddiqi, 1974c)
Female: **Body** slender **0.77–1.20 mm** long; annules 0.9–1.0 μm wide, distinct; **lateral fields** 1.6–1.5 as wide as body, **with four incisures**. Cephalic region hemispherical, offset by a constriction, slightly wider than adjacent body, no annulations visible with light microscope; framework hexaradiate, weakly sclerotized. **Stylet about 12 μm long, with minute but distinct basal knobs** and sharply pointed anterior part. Procorpus slender; median oesophageal bulb large, somewhat oval in shape, highly muscular, with prominent internal cuticular thickening and orifices of dorsal and ventrosublateral oesophageal glands. Nerve ring in neotype 1.5 body widths behind bulb. **Excretory pore 0.5–2 body widths posterior to nerve ring**. Three oesophageal glands forming a lobe extending about four body widths over intestine dorsally. Oesophago-intestinal junction about 8 μm behind median oesophageal bulb, indistinct and lacking a valve. Intestine with small spherical granules and a distinct lumen throughout. Vulva in the form of a slightly raised, transverse slit. **Post-uterine sac extending for more than half the vulva-anus distance, often containing sperms**. Genital tract monoprodelphic, outstretched; **oocytes in multiple rows. Tail elongate-conoid, bearing a terminal peg which has two to four minute processes pointing posteriorly, thus giving it a paintbrush-like appearance.**

Male: **Common. Posterior end of body usually curved through 180° upon relaxation**. Cephalic region, stylet and oesophagus as in female. Testis single, outstretched. Three pairs of ventro-submedian caudal papillae – first pair adanal, second midway on tail, third near tail end. **Spicules** smoothly curved, rosethorn-shaped, **lacking a dorsal or ventral process at proximal end; dorsal limb 20–22 μm long. Tail peg with two to four processes**, of variable shape.

BIONOMICS: *Aphelenchoides ritzemabosi* is colloquially known as the 'chrysanthemum leaf nematode', on which host it is a major pest. It is a polyphagous ecto- and endoparasite of many plants, mostly from the Compositae, but has also been recorded as reproducing on fungal cultures (Hooper and Cowland, 1986). Together with *A. fragariae*, it also attacks strawberry and can cause extensive damage. Its feeding activities result in growth retardation, distortion and leaf blotches which turn brown and then black. The nematode exhibits considerable cryptobiotic ability and can survive in a dry state for at least three years, mostly as the later stage juveniles and adults. Reproduction is amphimictic, the life cycle taking between 10–14 days, depending on temperature. *A. ritzemabosi* interacts with *Corynebacterium fasciens* to form 'cauliflower disease' of strawberry. For more details of this nematode see Siddiqi (1974c).

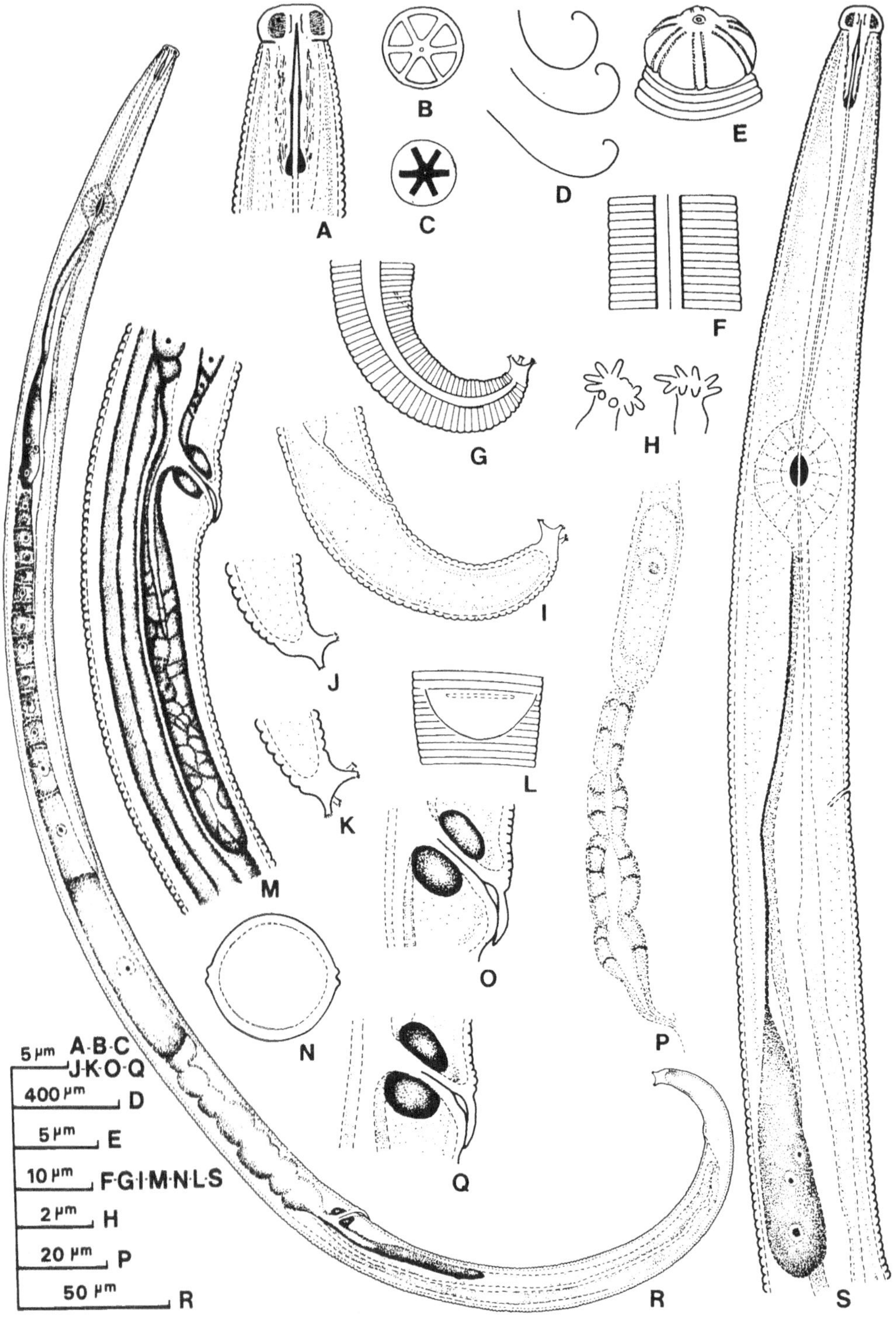
A
B
C
D
E
F
G
H
I
J
K
L
M
N
O
P
Q
R
S
5 μm A·B·C J·K·O·Q
400 μm D
5 μm E
10 μm F·G·I·M·N·L·S
2 μm H
20 μm P
50 μm R

Distribution: Widely distributed in the cooler regions of the world (Australasia, CIS, Europe, North America, South Africa, etc.).

Control of *Aphelenchoides*: Various techniques have been recommended for controling *Aphelenchoides* of which the most usual involve cultural hygiene and/or hot water treatment (HWT). Cultural hygiene may be achieved by destruction of infected plants and by the removal of weeds acting as alternative hosts. *A. arachidis* on peanut can be controlled by thorough sun-drying of the pods in areas where the ambient humidity is low, by planting shelled seeds rather than whole pods or by immersing seeds in four times their volume of water, previously heated to 60°C and then removed from the heat source, for a period of 5 minutes (see Bridge and Hunt, 1985 for a summary and also Minton and Baujard, 1990). The rice nematode, *A. besseyi*, another seed borne species, can also be controlled by HWT or by the use of resistant or tolerant varieties of rice combined with cultural practices (see Bridge *et al.* 1990 for an up-to-date account). *A. fragariae* and *A. ritzemabosi* are effectively controlled by HWT, with various time and temperature combinations depending on host, and systemic nematicides, either as foliar sprays or granular formulations, have also been employed with success. *A. composticola* in mushroom houses has been controlled by treating the compost before spawning, either by heating up to 60°C or with chemicals such as methyl bromide.

GENUS *LAIMAPHELENCHUS* FUCHS, 1937 (Fig. 4.10)

syn. *Ruidosaphelenchus* Laumond and Carle, 1971

Diagnosis: Aphelenchoidinae. Medium sized nematodes **dying strongly ventrally arcuate to C-shaped on heat relaxation**. Body fairly stout and distinctly annulated. Lateral field with three or four incisures. Cephalic region low, rounded, offset. Stylet fairly well developed with distinct basal swellings. Procorpus cylindrical leading to a well developed rounded to rounded-rectangular median bulb with strong valve plates. Oesophageal gland lobe well developed. Nerve ring and excretory pore several body widths posterior to the bulb. Vulva post-median, typically at about 60–70% of the body length. Anterior vulval lip forms a flap over the genital opening

Fig. 4.10. *Laimaphelenchus cocuccii.* Female. A. Anterior region; B. Lip region (*en face* view); C. Cross section through the basal plate; D. Habitus; E. Anterior region (lateral view, based upon SEM pictures); F. Cuticle (lateral field); G,I. Tails (lateral view); H. Tail tubercles; J-K. Tail tips (lateral view); L. Vulva region (ventral view); M. Post-uterine sac; N. Cross section at middle of body; O,Q. Vulval region (lateral view); P. Terminal region of anterior genital tract; R. Entire; S. Oesophageal region. (After Doucet, 1992. Courtesy *Fundamental and Applied Nematology*).

in the type and several other species. **Vagina** usually directed anteriorly and **with a cuticular annulus as it joins the uterus**. Genital tract monoprodelphic, outstretched. Developing oocytes in a single row. Post-uterine sac well developed and extending about halfway to the anus. **Tail conoid, tapering to a distinctive, offset terminus, bearing four pedunculate tubercles with fringed margins**.

Male spicules rather elongate, curved. Apex broadly rounded and rostrum triangular and pointed. Three pairs of caudal papillae: one pair situated pre-anally sometimes as anterior as the spicule head; one pair adanal and the other about mid-way to the tail tip. **Tail similar in shape to that of the female and also bearing the fringed tubercles**.

Type species:

Laimaphelenchus penardi (Steiner, 1914) Filipjev and Schuurmans Stekhoven, 1941
    syn. *Aphelenchus penardi* Steiner, 1914
        Aphelenchoides penardi (Steiner, 1914) Filipjev, 1934
        L. moro Fuchs, 1937
        L. ulmi Khan, 1960
        L. montanus Hirling, 1982
        L. silvaticus Hirling, 1982
        L. praepenardi Hirling, 1986
        Aphelenchoides praepenardi (Hirling, 1986) Ebsary, 1991

Other species:

L. cocuccii Doucet, 1992
L. deconincki Elmiligy and Geraert, 1972
**L. pannocaudus* Massey, 1966
    syn. *Aphelenchoides pannocaudus* (Massey, 1966) Hirling, 1986
L. pensobrinus Massey, 1966
    syn. *Aphelenchoides pensobrinus* (Massey, 1966) Ebsary, 1991
        Ruidosaphelenchus pensobrinus (Massey, 1966) Laumond and Carle, 1971
        R. janasii Laumond and Carle, 1971
**L. phloesini* Massey, 1974
    syn. *Aphelenchoides phloesini* (Massey, 1974) Hirling, 1986
**L. pini* Baujard, 1981
    syn. *Aphelenchoides pini* (Baujard, 1981) Hirling, 1986
        **L. parvus* Hirling, 1982

Species inquirendae:

L. corticilis Truskova and Eroshenko, 1977
    syn. *Aphelenchoides corticilis* (Truskova and Eroshenko, 1977) Ebsary, 1991
L. exilis Truskova and Eroshenko, 1977
    syn. *Aphelenchoides exilis* (Truskova and Eroshenko, 1977) Ebsary, 1991
L. sapinus Truskova and Eroshenko, 1977

syn. *Aphelenchoides sapinus* (Truskova and Eroshenko, 1977) Ebsary, 1991

**L. tenarius* Truskova and Eroshenko, 1977

syn. *Aphelenchoides tenarius* (Truskova and Eroshenko, 1977) Hirling, 1986

L. vescus Truskova and Eroshenko, 1977

syn. *Aphelenchoides vescus* (Truskova and Eroshenko, 1977) Hirling, 1986

Nomen nudum: *L. lichni* Hirling, 1979 (= *L. penardi*)

NOTES:

1. * species thus marked lack a vulval flap.
2. Ebsary (1991) formally proposed a number of 'new' combinations under *Aphelenchoides* that had already been proposed by Hirling (1986).
3. Hirling (1986) consistently misspells *silvaticum* as *sylvaticum*. This is an incorrect subsequent spelling and does not enter into synonymy.

Remarks: *Ruidosaphelenchus* was erected by Laumond and Carle (1971) to contain *R. janasii* Laumond and Carle, 1971 and the type, *R. pensobrinus* (Massey, 1966). The genus was based on spicule shape and was synonymized with *Laimaphelenchus* by Baujard (1985) whose action is accepted here. *Laimaphelenchus* contains two groups of species – those with a vulval flap (including the type) and those without. Hirling (1986) transferred those species without such a flap to *Aphelenchoides* and Ebsary (1991) went further and regarded *Laimaphelenchus* as a synonym of *Aphelenchoides*. Presence or absence of a vulval flap also occurs in *Bursaphelenchus* and for the moment I prefer to retain within *Laimaphelenchus* all those species with fringed tubercles on the tail tip, regardless of the vulval flap character.

Bionomics: Usually recorded from moss, algae and lichens on trees, particularly conifers, or in the tunnels of wood-boring beetles. Steiner (1914a) found *L. penardi* associated with *Chlorella* algae and McBeth (1937) reported that *L. penardi* was predatory on other nematodes in bark beetle tunnels. Laumond and Carle (1971) described *Ruidosaphelenchus janasii* (= *L. janasii*) associated with *Blastophagus destruens* (Coleoptera: Scolytidae), the third stage juveniles clustering around the metathoracic spiracles and using the beetles, 63–90% of which were infested, as a transport mechanism. *L. deconincki* was described from sandy soil in France by Elmiligy and Geraert (1972) and Massey (1974) reported several species as associates of scolytid beetles (Coleoptera: Scolytidae). Hooper and Stone (1975) cultured *Laimaphelenchus penardi* on cultures of *Botrytis cinerea*, the progeny neither attacking nor feeding upon *Aphelenchoides blastophthorus* or *Caenorhabditis* spp. Baujard (1981) found *L. penardi* on *Pinus pinaster* and *P. insignis* without any apparent scolytid associates and also recorded it in tufts of the saxicolous bryophyte *Grimmia pulvinata* (Hedw.), the intestine being coloured green, presumably from the ingestion of chlorophyll. The majority of species are probably phytophagous on lower plants or mycetophagous in habit, predatory behaviour being the exception.

Distribution: Recorded from Europe, Argentina, North America and India.

Etymology: Derived from the Greek *laimos* = throat and *aphelenchus*.

GENUS *MEGADORUS* J.B. GOODEY, 1960 (Fig. 4.11)

Diagnosis: Aphelenchoidinae. Small nematodes, about 0.5 mm in length and slender. **Cuticle coarsely annulated**, the mid-body annules being about 1.7 μm apart. **Lateral field** about one fifth of body width and **marked by three incisures. Cephalic region markedly offset**, about twice as wide as high and in the form of a low dome. Labial region divided into six equal sectors when seen *en face*. **Cephalic framework heavily sclerotized. Stylet robust,** about 17 μm long and **with massive, rounded, basal knobs. Conus** slender and **surrounded by a sclerotized guiding sheath. Procorpus broad near the basal knobs and then attenuating to a narrow cylinder** before joining the prominent, oval, median bulb. Valve plates well developed and situated slightly posterior to the mid-point of the bulb. Oesophageal glands forming a dorsal lobe about three to five body widths long. Nerve ring about a body width behind the bulb with the excretory pore at the same level. Vulva with slightly protuberant lips, posterior (about 70%). Vagina forwardly directed. Genital tract monoprodelphic, outstretched. Developing oocytes in a single file. Post-uterine sac between two to three body widths long. Tail short, conoid; dorsally convex, ventrally concave. Tail tip rounded and lacking a mucron. Male not known and no spermatozoa seen in the genital tract of gravid females.

Type species:
Megadorus megadorus (Allen, 1941) J.B. Goodey, 1960
syn. *Aphelenchoides megadorus* Allen, 1941

Other species: Monotypic.

Remarks: The genus was erected by J.B. Goodey (1960) on the basis of the strongly developed stylet and cephalic region. Subsequent workers have accepted the genus as valid.

Bionomics: The type species was found in two out of fifteen desert soil samples taken from around the roots of shadscale (*Atriplex confertifolia* (Torr. and Frem.)) some two miles west of Utah Lake, Utah, USA (Allen, 1941). Nothing is known concerning its feeding habits although Allen (1941) speculated that the strongly developed stylet, guide sheath and cephalic skeleton were developed for feeding on the woody tissues of desert plants.

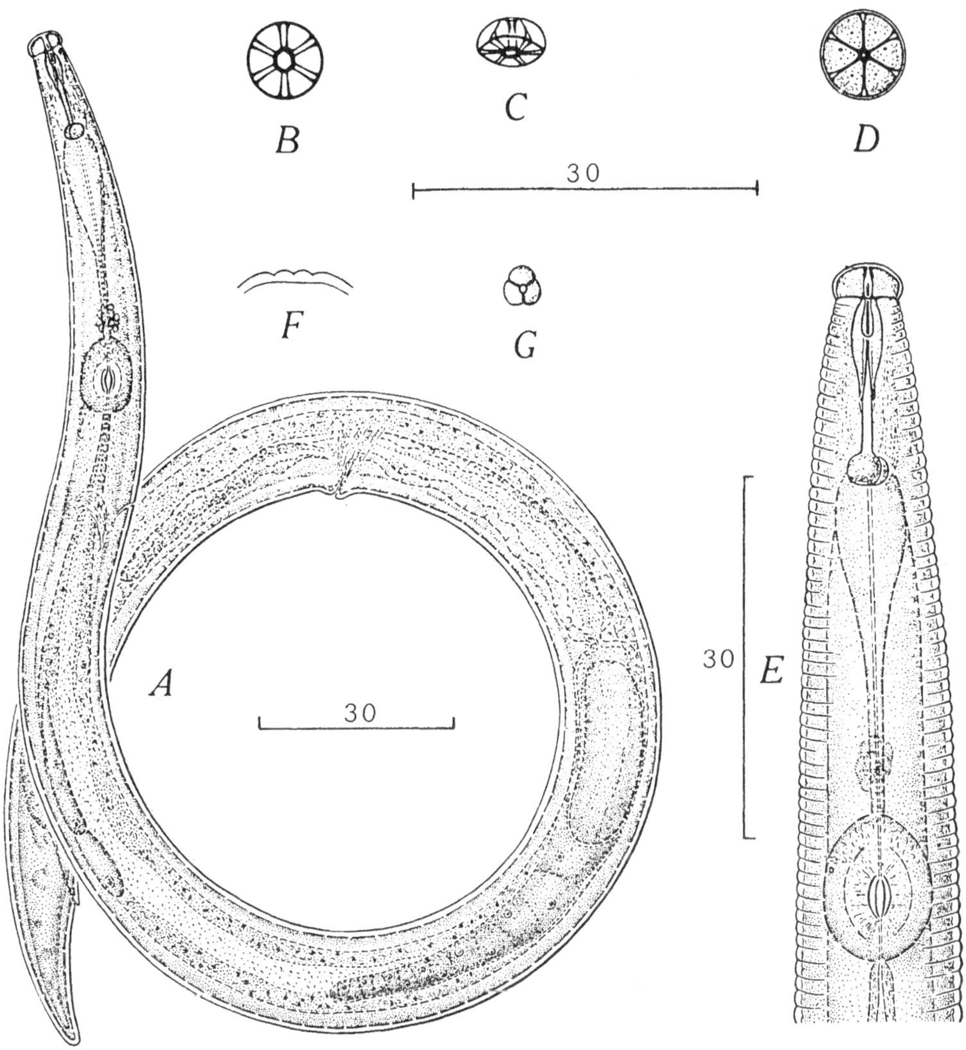

Fig. 4.11. *Megadorus megadorus*. A. Female; B. Basal plate of cephalic framework; C. Cephalic framework; D. *En face* view of head; E. Neck region; F. Lateral field in cross section; G. Knobs of stylet in cross section. (After Allen, 1941. Courtesy *Proceedings of the Helminthological Society of Washington*.)

Distribution: Utah, USA.

Etymology: Derived from the Greek *mega* = large and *dorys* = spear.

GENUS *RUEHMAPHELENCHUS* J.B. GOODEY, 1963 (Fig. 4.12)

Diagnosis: Aphelenchoidinae. Medium sized to fairly long nematodes (0.6–1.1 mm). **Cephalic region** offset with six rounded lips each about as high as

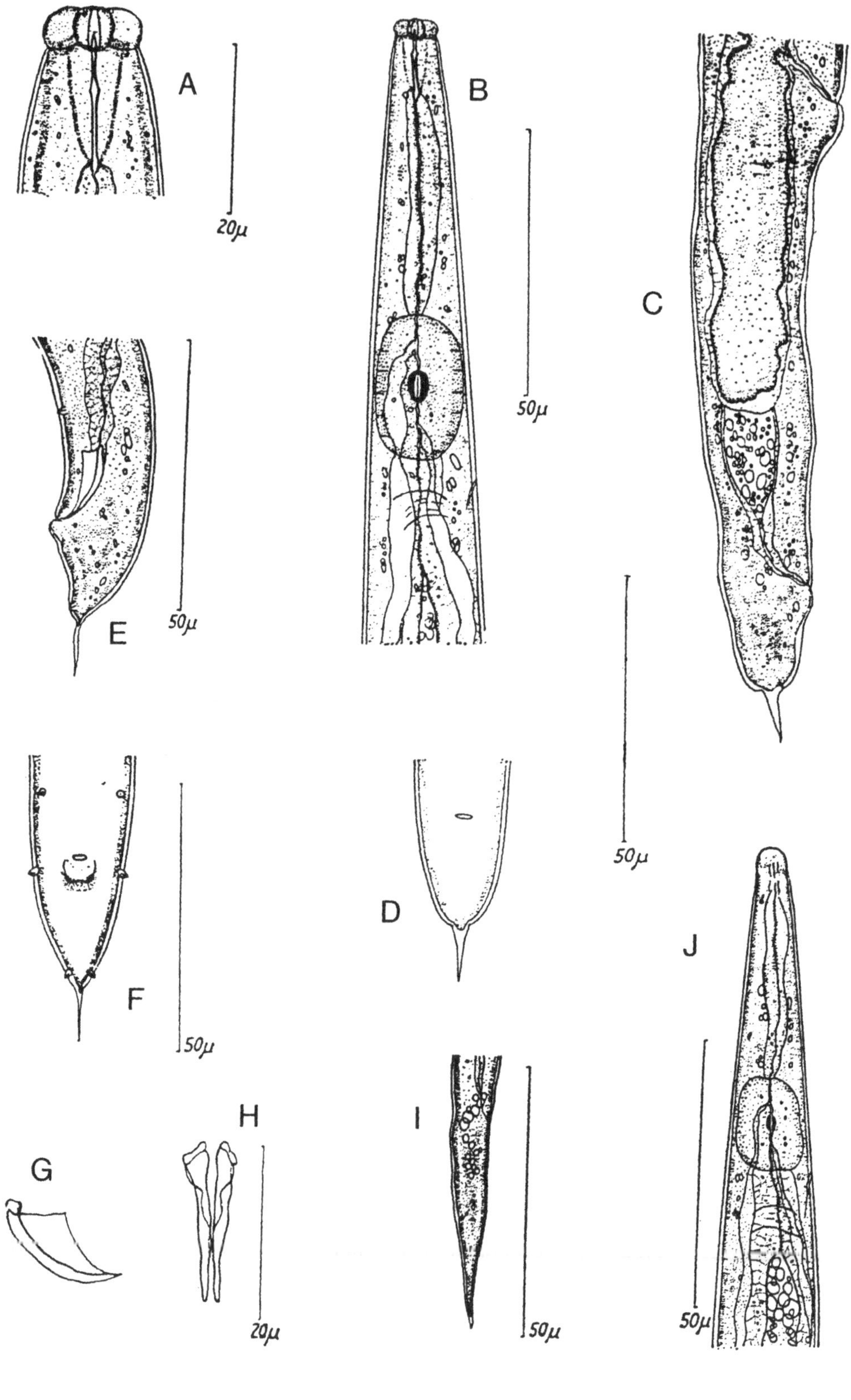

A
20μ
B
50μ
C
50μ
E
50μ
F
50μ
D
J
50μ
G
H
20μ
I
50μ

wide. **Stylet slender and lacking basal knobs** (but the illustration does show slight basal swellings). Procorpus cylindrical leading to an oval median bulb with central valve plates. Excretory pore and nerve ring just behind the bulb. **Vulva well posterior at 83–84%**. Genital tract monoprodelphic, outstretched. Developing oocytes in two or three rows. **Post-uterine sac well developed**, about two to three body widths long. Rectum and anus visible. **Female tail** about an anal body width long, rounded conoid in shape with a terminal spike-like process about half an anal body width long. **Male tail similar to that of the female, but more conical. Terminal spike present. Spicules paired, separate**, rather slender and curved with a ventral flange. Apex and rostrum weakly developed. Three pairs of ventrosublateral papillae: one pair about an anal body width preanal; one pair adanal; one pair subterminal. **Bursa absent**.

TYPE SPECIES:
Ruehmaphelenchus martinii (Rühm, 1955) J.B. Goodey, 1963
syn. *Aphelenchoides martinii* Rühm, 1955

OTHER SPECIES: Monotypic.

REMARKS: On erecting the genus, J.B. Goodey (1963), remarked upon its similarity to the genus *Parasitaphelenchus*, but noted that the male tail was different in structure, i.e. separate spicules and no bursa. The biology is also different in that *R. martinii* is reported to be ectophoretic. The systematic position of this genus involves some uncertainty, but the Aphelenchoidinae appears to be the most appropriate recipient for the moment.

BIONOMICS: Found associated with a wood boring beetle, *Anisandrus dispar* Ferr. (Coleoptera: Scolytidae). The dauerlarvae were found under the elytra and between the segments of the beetle.

DISTRIBUTION: Germany.

ETYMOLOGY: A combination of W. *Rühm*, a nematologist's name and *aphelenchus*.

GENUS *SCHISTONCHUS* COBB, 1927 (FUCHS, 1937) (Figs 4.13, 4.14)

syn. *Aphelenchus* (*Schistonchus*) Cobb, 1927

DIAGNOSIS: Aphelenchoidinae. Medium sized nematodes about 0.6–0.8 mm long. Dying ventrally arcuate to C-shaped on heat relaxation. Cuticle finely

Fig. 4.12. *Ruehmaphelenchus martinii*. Female. A. Head; B. Anterior end; C. Posterior end; D. Ventral view of tail. Male. E. Tail region; F. Ventral view of tail; G. Spicule in lateral view; H. Spicules in ventral view. Dauerlarva. I. Tail; J. Anterior end. (After Rühm, 1955. Courtesy *Zoologischer Anzeiger*.)

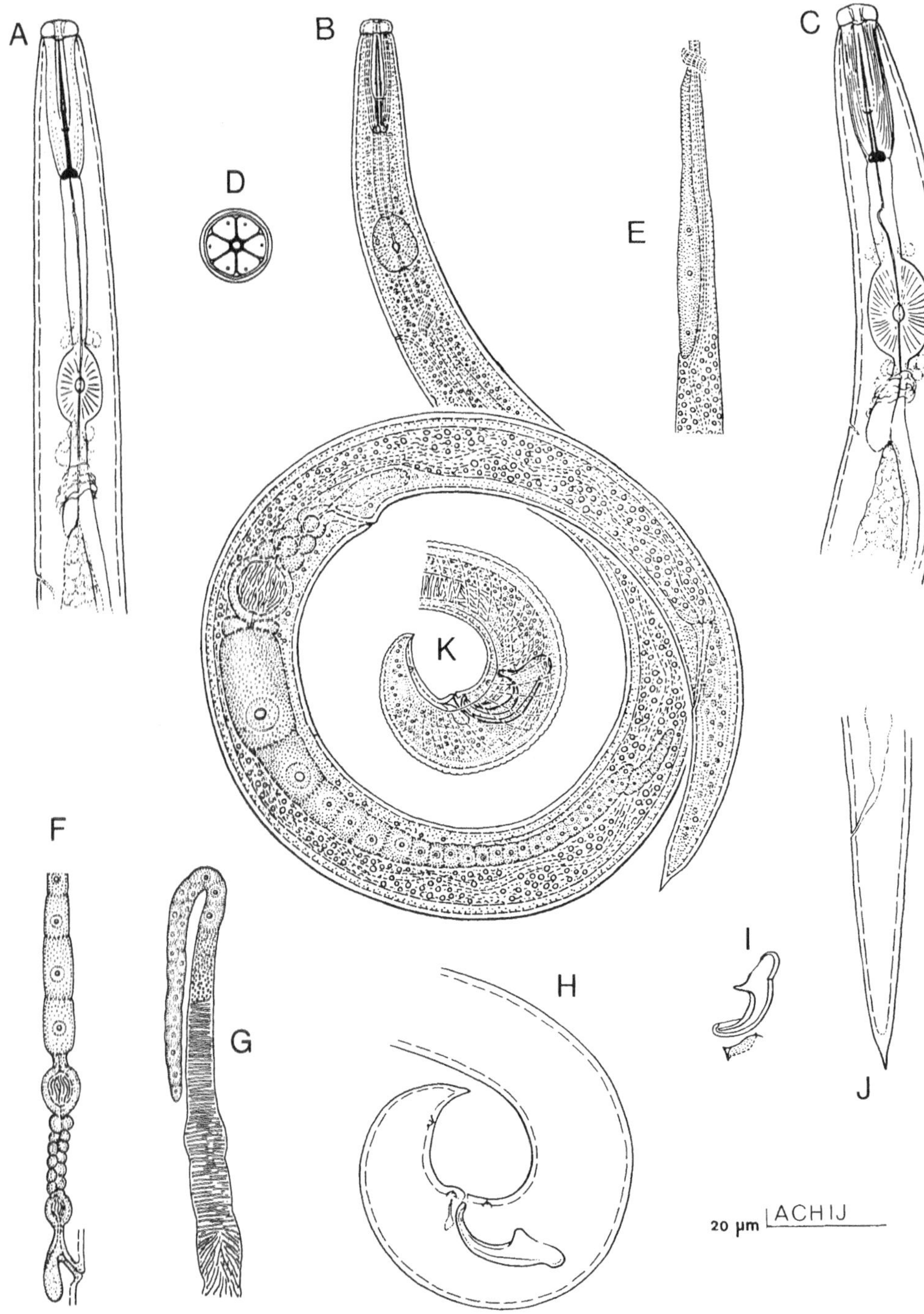

Fig. 4.13. *Schistonchus caprifici.* A,C. Oesophageal region; B. Entire female; D. *En face*; E. Gland lobe; F. Female genital tract; G. Part of male genital tract; H,K. Male tail; I. Spicule; J. Female tail. (A,C,H–J, after Vovlas *et al.*, 1992; remainder after Thorne, 1961. Courtesy *Nematologica* and McGraw Hill, respectively.)

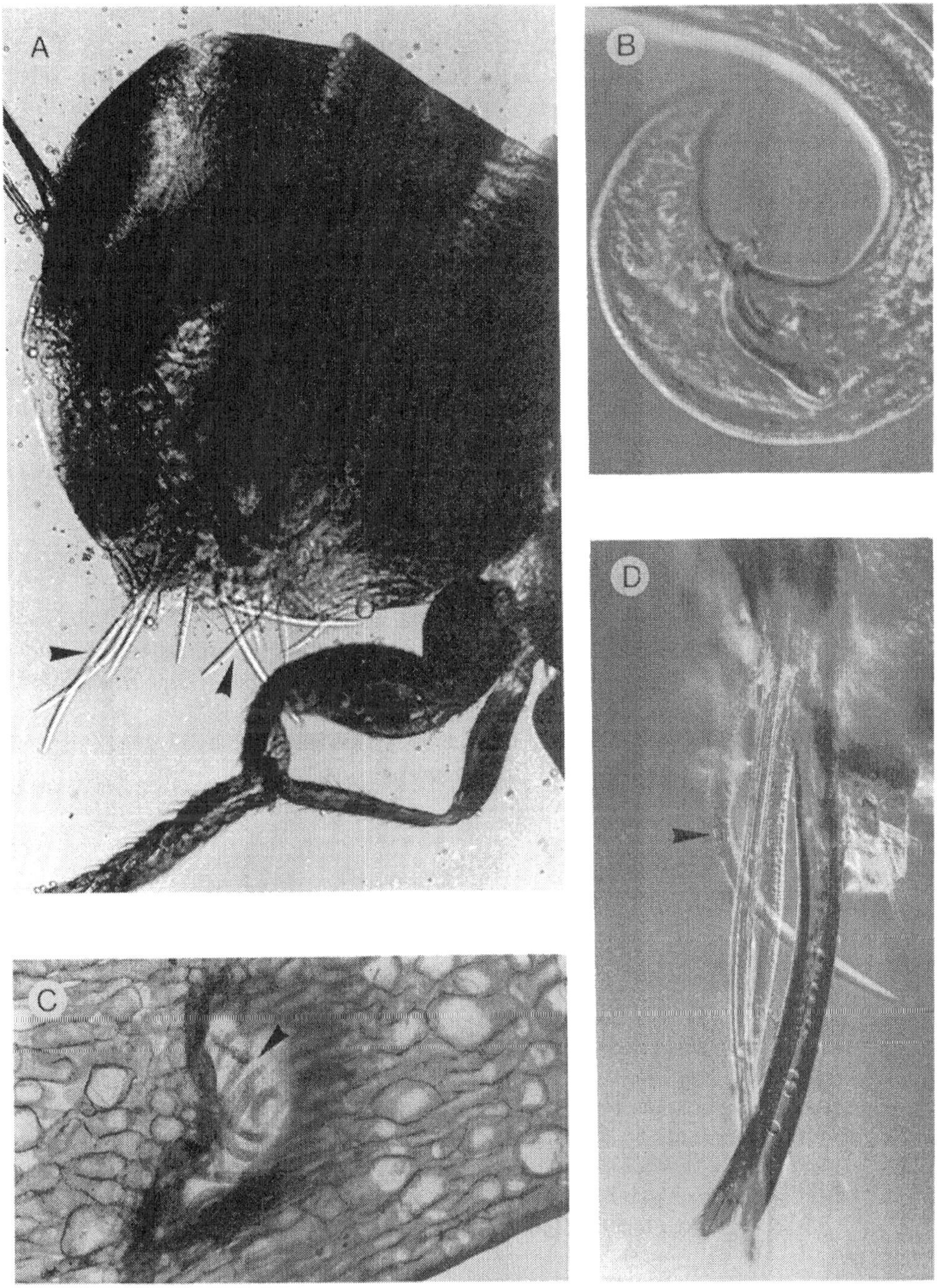

Fig. 4.14. *Schistonchus caprifici.* A. Nematodes (arrowed) from haemocoel of the fig wasp *Blastophaga psenes*; B. Male tail; C. Nematodes (arrowed) in cortical cavity in the fig fruit; d. Nematodes (arrowed) associated with the fig wasp ovipositor. (A,C: courtesy N. Vovlas; B,D: original, specimens courtesy N. Vovlas.)

annulated. Cepahlic region rounded, offset with six equal lips. **Stylet robust, 21–24 μm long and with strong basal knobs**. Procorpus fairly short, cylindrical, leading to a strong, ovoid, median bulb with post-median valve plates. Dorsal oesophageal gland lobe well developed. Vulva posteriorly situated at about 65–70% of the body and with slightly protuberant lips. Vagina directed anteriorly. Genital tract monoprodelphic, outstretched or with a double flexure. Developing oocytes mostly in one or two rows. Post-uterine sac less than a body width long. **Tail of medium length, conoid, tapering to an offset, mucronate tip**. Male tail region sharply recurved. Spicules rosethorn-shaped with a strong, pointed, rostrum and a broadly rounded apex continuing the line of the shaft. **Distally the tip of the dorsal limb is ventrally hooked. A gubernaculum-like structure is present. Spermatozoa flagellate**. Three pairs of caudal papillae: one preanal near the rostrum; one adanal and the other nearer the tail tip. **Tail short, conoid, with a mucronate tip**. The cuticle of the lateral field is slightly thickened ventrally between the last pair of papillae and the tail tip, but this is not sufficient to be classed as a terminal bursa.

TYPE SPECIES:

Schistonchus caprifici (Gasparrini, 1864) Cobb, 1927

syn. *Anguillula caprifici* Gasparrini, 1864

Aphelenchus (Schistonchus) caprifici (Gasparrini, 1864) Cobb, 1927

Pathoaphelenchus (Schistonchus) caprifici (Gasparrini, 1864) Steiner, 1931

Aphelenchoides (Schistonchus) caprifici (Gasparrini, 1864) Filipjev, 1934

OTHER SPECIES:

S. hispida Kumari and Reddy, 1984

S. racemosa Reddy and Rao, 1986

REMARKS: The thickening of the lateral field towards the male tail tip was mentioned and illustrated by Thorne (1961) for *S. caprifici*. Material of *S. caprifici* from Italy, kindly supplied by Dr N. Vovlas, shows this thickening quite clearly in lateral view as a darker line, giving the impression of a possible bursa-like development. However, in ventral view the elevation of this thickening is very slight, being only about 1–1.3 μm, and a terminal bursa-like flap of cuticle is not developed. Studies using the SEM (unpublished) show that the ridges of the lateral field fuse in the vicinity of the last pair of papillae, thus producing the thickening which is visible as a refractive line under the light microscope.

BIONOMICS: *S. caprifici* has been recorded as an associate of *Blastophaga psenes* (L.), a hymenopterous wasp pollinating the fig *Ficus carica* L., by Gasparrini (1864), Cobb (1927), T. Goodey (1928) and Vovlas *et al.* (1992). The biology of the nematode and its relationship with the wasp and the fig has only recently been elucidated (Vovlas *et al.*, 1992).

According to Vovlas *et al.* (1992) the nematode is an obligate phytoparasite in the fig inflorescence, but also parasitizes the haemocoel of the fig wasp (Fig. 4.14). The nematodes are carried by the alate female wasp from an infected to a healthy inflorescence and are deposited as the wasp oviposits in short-style pistillate florets. Chemicals released by the wasp cause the infected florets to gall and the developing wasp larvae and nematodes feed on the tissues. Nematodes migrate to adjacent pistillate and staminate florets and cause necrosis and cavitation in the cortical parenchyma. Nematodes penetrate the haemocoel of the wasp larvae and persist through the metamorphosis to the adult wasp. However, nematodes are only found in the haemocoel of the alate females and not in the apterous males. The nematodes grow and reproduce in the insect and are therefore regarded as parasitic, although no adverse effects were noted. The number of *S. caprifici* per insect ranged from one to 30 in the larvae or pupae and 200 to 400 in the adult females with a female to male sex ratio of 3 : 1. About 40% of larvae, 60% of pupae and 90% of adult females were infected. Mean populations of nematodes within the fig fruit were 550 g^{-1} of fresh floret tissue in the commercial *F. carica* and 12 000 g^{-1} in the wild caprifig, *F. carica sylvestris* L.

Of the other species, *S. hispida* was found associated with a fig wasp in figs of *Ficus hispida* L. and *S. racemosa* was reportedly carried as second stage juveniles in the abdominal folds of *Ceratosolen* sp., a fig wasp pollinating *Ficus racemosa* L.

Distribution: Algeria, Turkey, Italy, India.

Etymology: Derived from the Greek *schistos* = divided or cleft, and *onkos* = tooth

GENUS *SHERAPHELENCHUS* NICKLE, 1970 (Fig. 4.15)

Diagnosis: Aphelenchoidinae. Medium to long nematodes (0.48–1.19 mm). Heat relaxed form slightly ventrally arcuate. Cephalic region continuous with body contour and about half as high as wide. Cuticle finely annulated. Stylet well developed, about 12 μm long and **lacking basal knobs or swellings. Procorpus short, cylindrical,** leading to a powerful, ovoid, median bulb with central valve plates. Oesophageal gland lobe about two to three body widths long. Excretory pore situated near base of median bulb or up to a body width more posterior. Nerve ring just over a body width behind the median bulb. **Vulva posterior** (V = 78–88% in type; 91–93% in other species) with protuberant lips. Genital tract monoprodelphic, outstretched. Developing oocytes in about three rows. **Two or more embryonated eggs often present in the uterus. Post-uterine sac absent.** Tail short or medium conoid, tapering smoothly to a point. **Male spicules partially fused proximally, rosethorn-shaped with the apex elongate and extending the line of the dorsal limb and the rostrum small and pointed.** Three pairs of caudal papillae: one at cloacal level and the

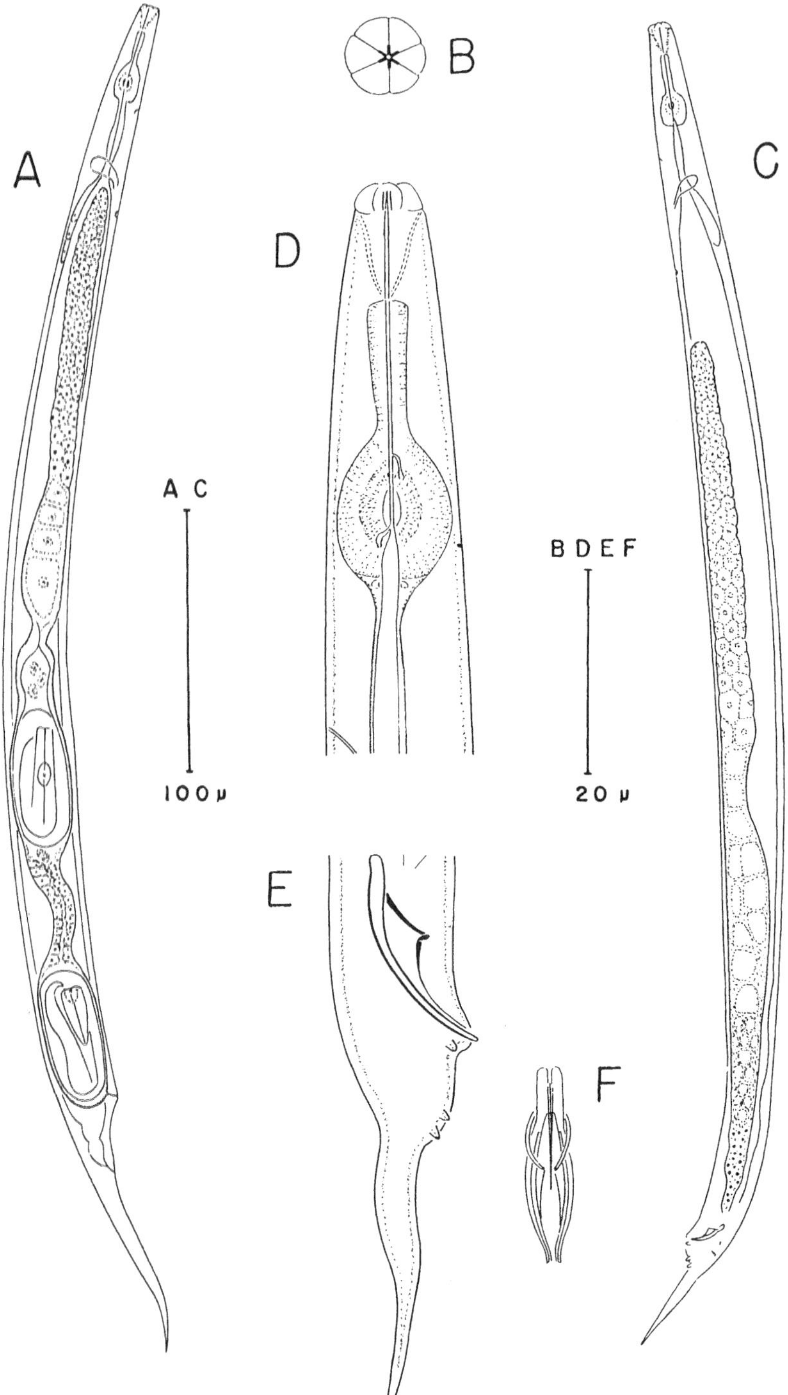

Fig. 4.15. *Sheraphelenchus entomophagus.* A. Female, entire; B. *En face*; C. Male, entire; D. Female, anterior end; E. Male, tail; F. Spicules, ventral view. (After Nickle, 1970. Courtesy *Proceedings of the Helminthological Society of Washington.*)

other two postanal. **Tail broadly conoid at first and then narrowing abruptly just posterior to the last papillae to form a digitiform spike with a finely pointed terminus.**

TYPE SPECIES: *Sheraphelenchus entomophagus* Nickle, 1970

OTHER SPECIES:
S. brevigulonis (Massey and Hinds, 1970) n.comb.
syn. *Entaphelenchus brevigulonis* Massey and Hinds, 1970

BIONOMICS: The type population was collected from decaying oranges in Hawaii. Nickle (1970a) also received material from a fig culture of a nitidulid beetle, *Haptonchus luteolus* (Erichson). This population originated from the nitidulids *Carpophilus humeralis* (Fabricius) and *C. mutilatus* Erichson. Giudice and Lanza (1973) reported *S. entomophagus* associated with *Carpophilus hemipterus* and the fungus *Geotrichum candidum* in split and decaying oranges in Italy. The nematode has a phoretic relationship with nitidulid beetles, Giblin *et al.* (1984a) reporting up to 160 dauerlarvae of *S. entomophagus* beneath the elytra of the pineapple beetle, *Urophorus humeralis*, with all stages being mycetophagous and breeding readily on monoxenic cultures of *Penicillium* sp. and *Monilinia fructicola*. *S. brevigulonis* n.comb. was described from various species of *Epurea* (Coleoptera: Nitidulidae) by Massey and Hinds (1970). They assumed that the available material was the free-living generation of an insect parasite and, as a result, placed the species in *Entaphelenchus*. However, the species is typical of *Sheraphelenchus* in the morphology of the free-living stages and is quite unlike the free-living stages of *Entaphelenchus*.

DISTRIBUTION: The type species was identified from Pahoa, Hawaii and Brazo County, Texas, USA and has also been recorded from California, USA and Italy. *S. brevigulonis* came from Colorado and New Mexico, USA.

ETYMOLOGY: Derived from S. A. Sher, a nematologist's name and *aphelenchus*.

GENUS *TYLAPHELENCHUS* RÜHM, 1956 (Fig. 4.16)

DIAGNOSIS: Aphelenchoidinae. Small nematodes less than 0.5 mm long. Heat relaxed form ventrally arcuate with the tail, particularly that of the male, more sharply curved. Cuticle finely annulated. **Cephalic region offset and divided into six equal sectors, each of which is distinctly higher than wide. Stylet** about 10 μm long, **typically robust with a short conus and with** basal swellings or, more typically, **large, rounded, knobs**. Procorpus cylindrical, leading to a rounded, oval to **barrel-shaped median bulb**. Oesophageal gland lobe dorsal and three to four body widths long. Nerve ring about a body width behind the bulb. Excretory pore usually in the region of the nerve ring but

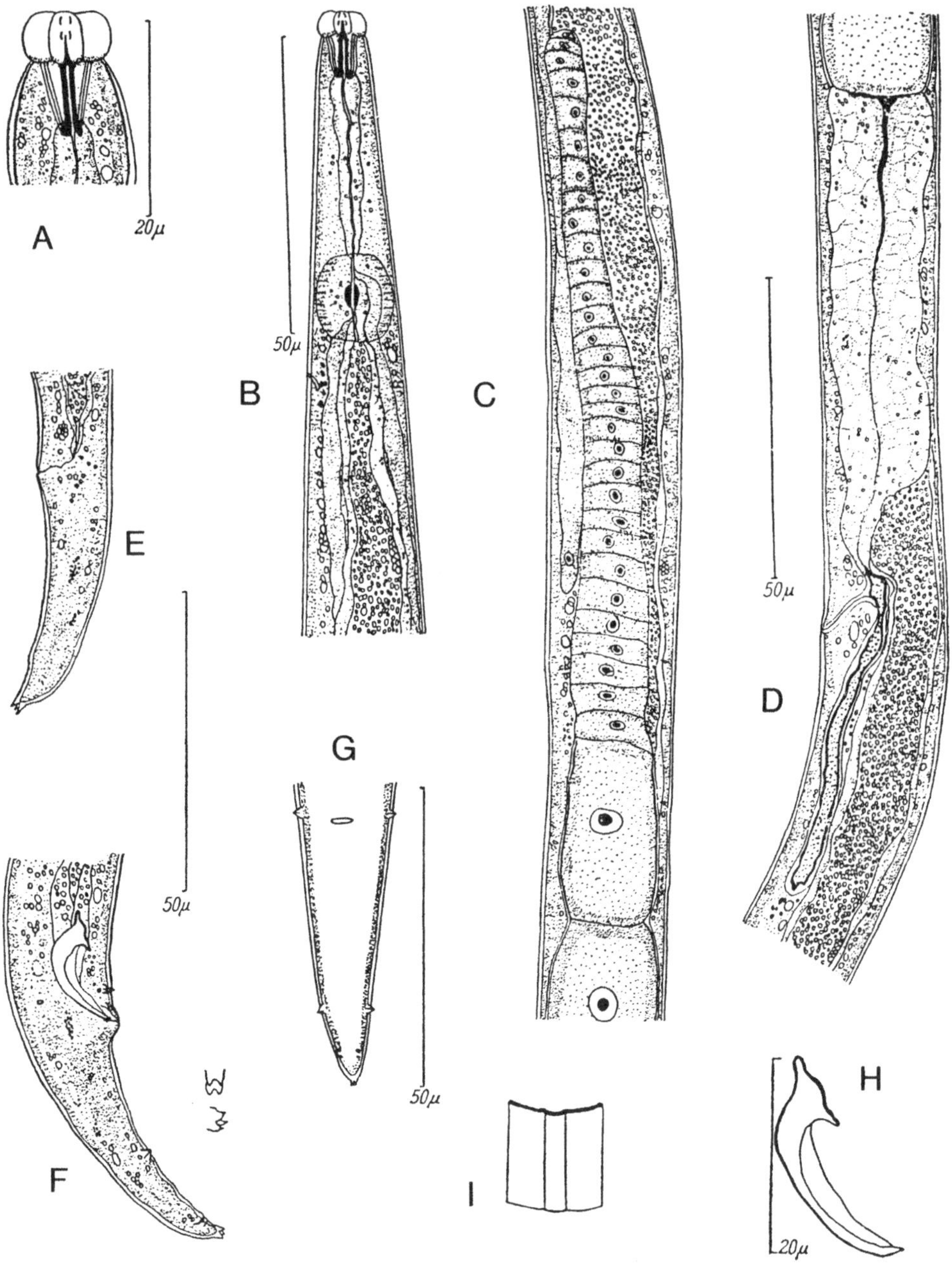

Fig. 4.16. *Tylaphelenchus leichenicola*. Female. A. Cephalic region; B. Oesophagus; C,D. Genital tract; E. Tail. Male. F. Tail, lateral; G. Tail, ventral; H. spicule; I. Lateral field. (After Rühm, 1956. Courtesy *Parasitologische Schriftenreihe*.)

may be more anterior (e.g. *T. christinae*). Vulva posterior (65–70%). Genital tract monoprodelphic, outstretched. Developing oocytes in a single row. Post-uterine sac well developed, about two to four body widths long. **Tail conoid and ending in three (one in *T. christinae*) mucrons. Male spicules separate and strongly curved with a small apex and rostrum**. Two pairs of caudal papillae, one pre-cloacal and the other nearer the tail tip. **Tail conoid with three** (one in *T. christinae*) **terminal mucrons.**

TYPE SPECIES: *Tylaphelenchus leichenicola* Rühm, 1956.

OTHER SPECIES:
T. christinae Lieutier and Laumond, 1978
T. georgiensis Devdariani, 1970
T. grossmannae Rühm, 1965
T. ipidicola (Rühm, 1956) J.B. Goodey, 1960
    syn. *Aphelenchoides ipidicola* Rühm, 1956
T. paramonovi Kakulia, 1963
T. yamani Raski and Valenzuela, 1988

NOTES:
1. *T. christinae* may not belong to this genus as the stylet is more slender and lacks the well developed, rounded, basal knobs of the other species and the tail has only one instead of three mucrons.
2. *T. yamani*, described from Tierra del Fuego, Argentina, has more affinities with *Aphelenchoides* than with *Tylaphelenchus*, particularly with regard to the stylet which is rather long and slender. Indeed, Raski and Valenzuela (1988) note the similarity of *T. yamani* with *A. helicosoma*, described from Signy Island, Antarctica by Maslen (1979).

BIONOMICS: The type species, *T. leichenicola*, was found in the frass of various wood boring insects (bark beetles, weevils and cerambycids) as well as in moss and lichen growing on the bark (Rühm, 1956). It can enter anhydrobiosis in adverse conditions. Most of the other species are also associated with various bark beetles, e.g. *Ips, Dryocoetes*. They are likely to be fungal feeders and may be ectophoretic associates of insects.

DISTRIBUTION: Europe, CIS (formerly USSR), North and South America.

ETYMOLOGY: Derived from the Greek *tylos* = knot or knob, and *aphelenchus*.

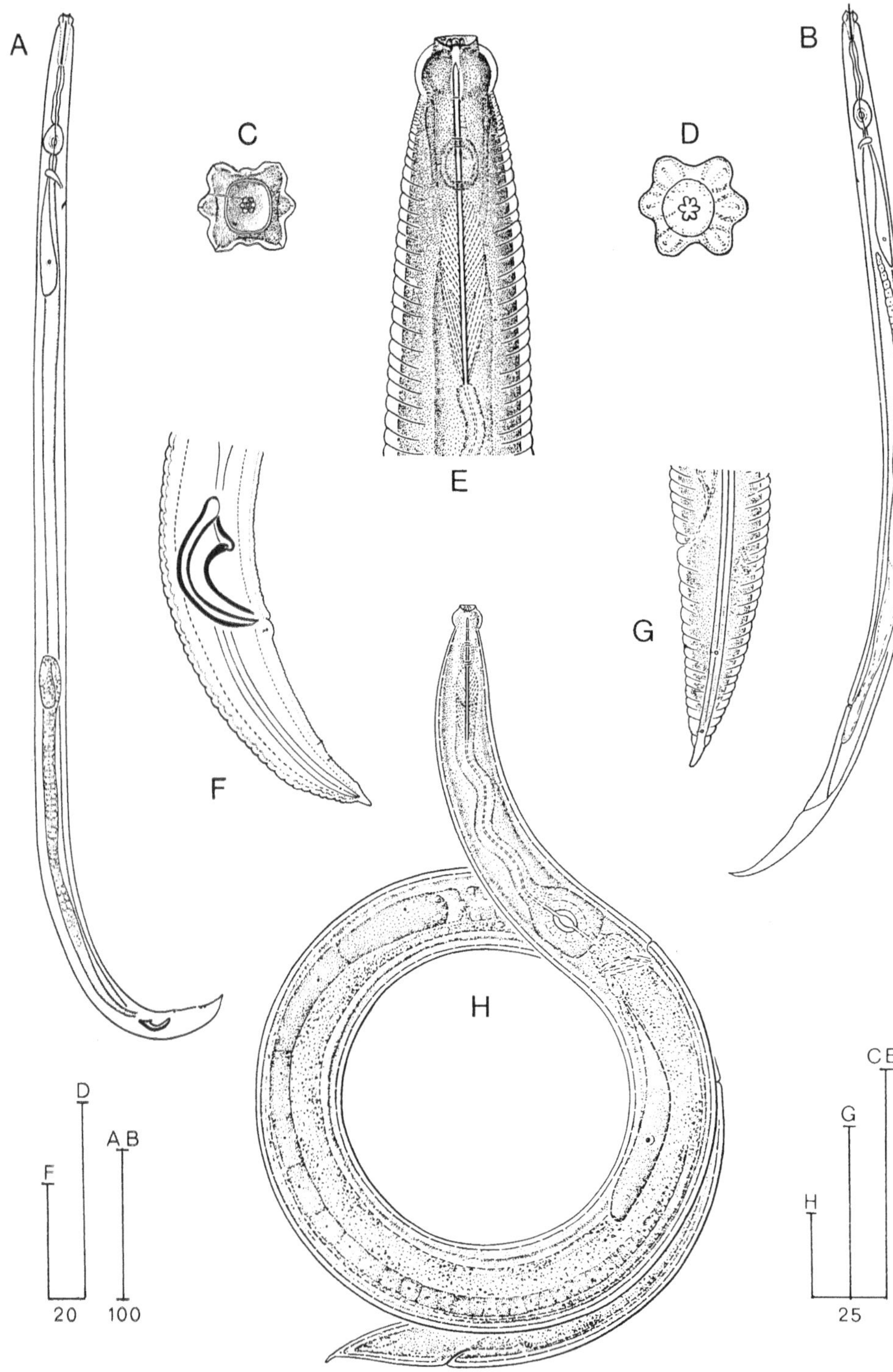

Fig. 4.17. *Anomyctus xenurus.* A. Entire male; B,H. Entire female; C,D. *En face*; E. Stylet; F. Male tail. G. Female tail. (A,B,F, after Nickle, 1971; D. after Hooper and Cooke, 1967; remainder after Allen, 1940. Courtesy *Nematologica* and *Proceedings of the Helminthological Society of Washington.*)

Subfamily Anomyctinae J.B. Goodey, 1960 (Andrássy, 1976)

DIAGNOSIS: Aphelenchoididae. Rare, soil dwelling nematodes. Head rounded, almost spherical, but with a sclerotized, saucer-like frontal disc. Stylet long, slender. Procorpus unusually long.

TYPE GENUS: *Anomyctus* Allen, 1940

OTHER GENERA: Monotypic.

GENUS *ANOMYCTUS* ALLEN, 1940 (Figs 4.17, 4.18)

DIAGNOSIS: Anomyctinae. Medium sized nematodes about 0.7–1 mm long. **Cuticle coarsely annulated**, the mid-body annules being about 1.8 μm apart. Allen (1940) reports that the annules immediately behind the cephalic region are much finer and only 0.3 μm apart. Lateral field marked by three incisures. **Cephalic region highly distinctive** in that it is **strongly offset by a constriction and smoothly rounded with a sclerotized oral disc** in the form of a shallow bowl. Allen (1940) describes and illustrates the oral opening as circular whereas Hooper and Cooke (1967), working with an apparently conspecific population from England, describe it as hexaradiate. Both agree that the opening is surrounded by six prominent liplets. There are six lips, the lateral being narrower than the other four. **The 'stylet' is about 35 μm long and consists of an anterior section about 10 μm long** (the stylet proper) **and a posterior extension between 20 and 27 μm long**. There is a guiding ring near the base of the stylet proper and a second slightly posterior, the two being joined by a circular structure which Hooper and Cooke (1967) described as a local expansion of the stoma-lining. Procorpus cylindrical leading to a strongly developed median bulb with central valve plate. Oesophageal gland lobe extending dorsally for about four body widths. Nerve ring just posterior to the median bulb with the excretory pore situated at the same level or up to a body width posterior. Vulva lacking conspicuous lips, posterior (about 76%). Genital tract monoprodelphic, outstretched; reaching to the region of the oesophageal gland lobe. Developing oocytes in single file. Post-uterine sac about one to two body widths long. **Tail conoid with a short, terminal, digitate projection.** Phasmids situated at about half the tail length and, in addition, Allen (1940) reported a subterminal pair of caudal pores which could sometimes be seen with 'thread-like strings of secretions'.

Allen (1940) found no males and the gravid females on which he based the type description contained no spermatozoa. However, Nickle (1971) described a single male specimen which he attributed to the type species, *Anomyctus xenurus*. Male similar to female in general respects. **Spicules large separate well developed; rosethorn-shaped. Tail conoid with a short terminal process** and three pairs of caudal papillae: one pair just postanal and two pairs close together in the posterior third of the tail. No bursa or gubernaculum.

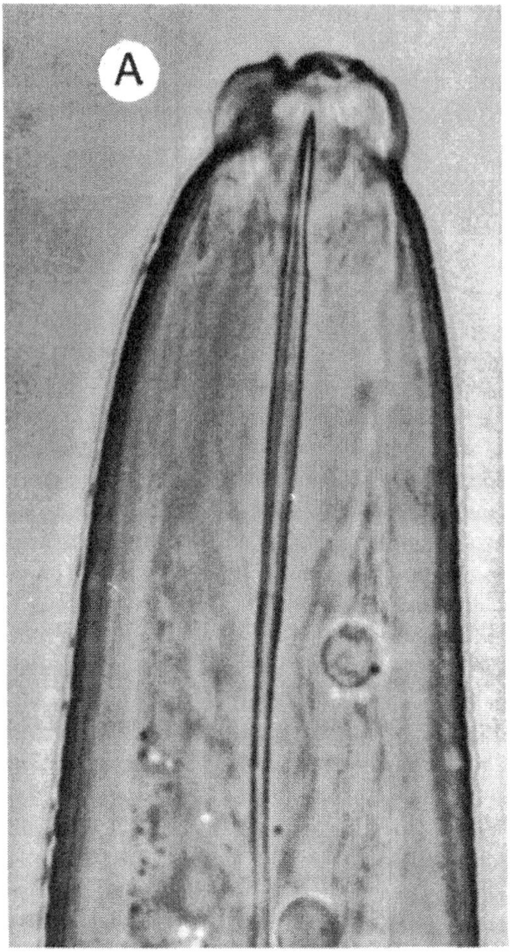

Fig. 4.18. *Anomyctus xenurus.* A. Stylet and anterior region. (After Hooper and Cooke, 1967. Courtesy *Nematologica.*)

Type species: *Anomyctus xenurus* Allen, 1940

Other species: Monotypic.

Remarks: J.B. Goodey (1960) erected the monotypic family Anomyctidae. This was accepted for a time, but when Nickle (1971) described the male, the characteristics of the tail and spicules showed that *Anomyctus* belonged near the *Aphelenchoides* group. Nickle (1971) stated that family status was inappropriate for the genus and in his earlier published paper (Nickle, 1970b) he placed *Anomyctus* in the Aphelenchoididae.

Bionomics: The type description was based on about 25 females recovered from soil about the roots of shadscale (*Atriplex confertifolia* (Torr. and Frem.) S. Wats.) from the desert two miles west of Utah Lake, Utah, USA.

Subsequently *Anomyctus xenurus* (or *Anomyctus* sp.) has been reported from England (Hooper and Cooke, 1967), California, USA (Nickle, 1971) and Australia (Sauer, 1970). All these populations came from desert or sandy soils. Hooper and Cooke (1967) placed five females on each of two agar plates colonized by *Botrytis cinerea* and *Aphelenchus avenae* respectively, but no *Anomyctus* were recovered when the plates were sampled four weeks later.

DISTRIBUTION: Reported from USA, England and Australia.

ETYMOLOGY: Derived from the Greek *ano* = upward or on high, and *mykter* = nose or snout.

Family Seinuridae Husain and Khan, 1967 (Baranovskaya, 1981)

DIAGNOSIS: Aphelenchoidoidea. Most species are known to be, or suspected of being, predacious. Stylet long and slender with a wide lumen. Median bulb exceptionally prominent, elongate oval or rounded-rectangular with the valve plates situated in the posterior half. Spicules with a prominent apex and rostrum. Tail in female long, elongate, with a usually filiform, but exceptionally rounded, terminus.

TYPE SUBFAMILY: Seinurinae Husain and Khan, 1967

OTHER SUBFAMILY: Monotypic.

Subfamily Seinurinae Husain and Khan, 1967

DIAGNOSIS: Seinuridae. With the characters of the family.

TYPE GENUS: *Seinura* Fuchs, 1931

OTHER GENERA:
Aprutides Scognamiglio, Talame' and s'Jacob, 1970
Papuaphelenchus Andrássy, 1973
Paraseinura Timm, 1960

KEY TO GENERA

1. Tail initially conoid and then filiform to a finely pointed terminus 2

– Tail more elongate conoid, tapering to a rounded or clavate terminus; not filiform 3

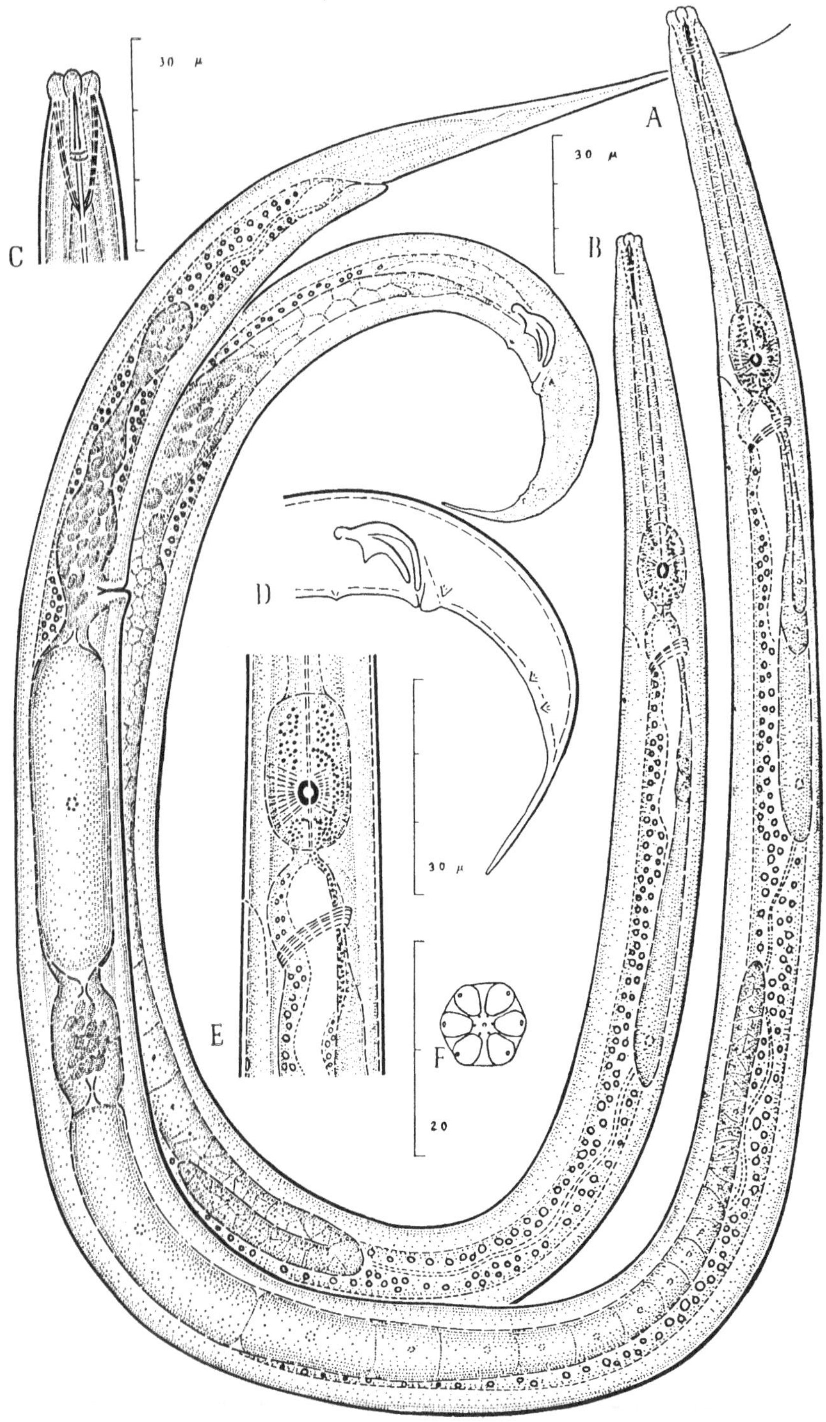

Fig. 4.19. *Seinura tenuicaudata.* A. Female; B. Male; C. Female head; D. Male tail variation; E. Median bulb; F. Female *en face.* (After Hechler and Taylor, 1965. Courtesy *Proceedings of the Helminthological Society of Washington.*)

2. Stylet with the conus apparently jointed . . . . *Paraseinura*

– Stylet of normal form . . . . *Seinura*

3. Tail tip broadly rounded and clavate . . . . *Aprutides*

– Tail tip finely rounded, not clavate . . . . *Papuaphelenchus*

GENUS *SEINURA* FUCHS, 1931 (Fig. 4.19)

Diagnosis: Seinurinae. Small to long nematodes (0.3–1.1 mm). **Cephalic region high, offset. Stylet 10–27 μm long and slender, usually lacking basal knobs** or swellings. Procorpus long and cylindrical. Median bulb oblong or ovoid with prominent valve plates situated in a post-median position. On fixation the central region of the median bulb may contract so that the bulb appears waisted or guitar-shaped. Dorsal oesophageal gland lobe long and well developed. Nerve ring and excretory pore situated within about a body width of the median bulb base. Vulva posterior at 70–80% of the body length. Genital tract monoprodelphic, outstretched; often reaching to the level of the oesophageal glands. Post-uterine sac may be totally absent or well developed. **Tail medium to medium-long, tapering to a fine point**. Male tail sharply recurved on heat relaxation. **Spicules** paired, ventrally curved **with a prominent, pointed rostrum and a strongly developed, elongate apex**. Caudal papillae variable in number, but typically a preanal pair at the level of the transverse bar, an adanal pair and either one or two pairs postanal. Bursa absent. **Tail conoid** at first **and then narrowing rapidly to a finely rounded terminus**.

Type species:
Seinura mali Fuchs, 1931
    syn. *Aphelenchoides mali* (Fuchs, 1931) T. Goodey, 1951

Other species:
S. arguta Kazachenko, 1980
S. arizonensis Massey, 1974
S. aurangabadensis Suryawanshi, 1971
S. celeris Hechler in Hechler and Taylor, 1965
S. chertkovi Dmitrenko, 1966
S. christiei J.B. Goodey, 1960
    syn. *Aphelenchoides tenuicaudatus* apud Christie, 1939 nec de Man, 1895
S. citri (Andrássy, 1957) J.B. Goodey, 1960
    syn. *Aphelenchoides citri* Andrássy, 1957
S. clavata Bajaj and Bhatti, 1982

S. demani (T. Goodey, 1928) J.B. Goodey, 1960
syn. *Aphelenchus demani* T. Goodey, 1928
Aphelenchoides demani (T. Goodey, 1928) T. Goodey, 1933
A. (*Aphelenchoides*) *demani* (T. Goodey, 1928) T. Goodey, 1933
S. diversa (Paesler, 1957) J.B. Goodey, 1960
syn. *Aphelenchoides winchesi diversus* Paesler, 1957
Aphelenchoides diversus (Paesler, 1957) Paesler, 1957
S. elmiraensis (van de Linde, 1938) J.B. Goodey, 1960
syn. *Aphelenchoides elmiraensis* van de Linde, 1938
S. filicaudata (Christie, 1939) J.B. Goodey, 1960
syn. *Aphelenchoides winchesi filicaudatus* Christie, 1939
S. fuchusi Edward and Misra, 1969
S. hechlerae Chaturvedi and Khera, 1979
S. hyderabadensis (Das, 1960) J.B. Goodey, 1960
syn. *Aphelenchoides hyderabadensis* Das, 1960
S. indica Suryawanshi, 1971
S. kherai Singh, 1977
S. linfordi (Christie, 1939) J.B. Goodey, 1960
syn. *Aphelenchoides linfordi* Christie, 1939
S. longicaudata (Cobb, 1893) J.B. Goodey, 1960
syn. *Aphelenchus longicaudatus* Cobb, 1893
Aphelenchoides longicaudatus (Cobb, 1893) T. Goodey, 1933
A. (*Aphelenchoides*) *longicaudatus* (Cobb, 1893) T. Goodey, 1933
S. nagini Husain and Khan, 1965
S. nodosa Andrássy, 1966
S. oahueensis (Christie, 1939) J.B. Goodey, 1960
syn. *Aphelenchoides oahueensis* Christie, 1939
S. obscura Grewal, Siddiqi and Atkey, 1992
S. oliveirae (Christie, 1939) J.B. Goodey, 1960
syn. *Aphelenchoides oliveirae* Christie, 1939
S. oostenbrinki Husain and Khan, 1967
S. oswegoensis (van der Linde, 1938) J.B. Goodey, 1960
syn. *Aphelenchoides oswegoensis* van der Linde, 1938
S. oxurua (Paesler, 1957) J.B. Goodey, 1960
syn. *Aphelenchoides oxurus* Paesler, 1957
S. paratenuicaudata Geraert, 1962
S. paynei Grewal, Siddiqi and Atkey, 1992
S. propora Siddiqi, Husain and Khan, 1967
S. prospera Kazachenko, 1980
S. steineri Hechler in Hechler and Taylor, 1965
syn. *Aphelenchus tenuicaudatus* of Steiner (1927) and T. Goodey (1928)
S. sutura Massey, 1971
S. tandoni Singh, 1977
S. tenuicaudata (de Man, 1895) J.B. Goodey, 1960
syn. *Aphelenchus tenuicaudatus* de Man, 1895
Aphelenchoides tenuicaudatus (de Man, 1895) Christie and Arndt, 1933

A. (*Aphelenchoides*) *tenuicaudatus* (de Man, 1895) Christie and Arndt, 1933

S. tritica Bajaj and Bhatti, 1982

S. variobulbosa Haque, 1966

S. winchesi (T. Goodey, 1927) J.B. Goodey, 1960
syn. *Aphelenchus winchesi* T. Goodey, 1927
Aphelenchoides (*Aphelenchoides*) *winchesi* (T. Goodey, 1927) Filipjev, 1934

Species inquirendae:

S. *mucronatus* (Paesler, 1946) J.B. Goodey, 1960
syn. *Aphelenchoides mucronatus* Paesler, 1946

S. speciosus (Andrássy, 1958) J.B. Goodey, 1960
syn. *Aphelenchoides speciosus* Andrássy, 1958

Note: Thorne (1961) consistently spelled the genus as *Seineura*. This is an incorrect subsequent spelling and does not enter into synomymy.

Bionomics: Many *Seinura* species are known to be predatory on other nematodes, Linford (1937), Linford and Oliveira (1937) and Hechler (1963) recording *Seinura* feeding on plant parasitic and fungal feeding nematodes. Wood (1975) extended the list of prey nematodes to include bacterial feeders when he recorded *S. demani* feeding on *Acrobeloides*, *Chiloplacus*, *Eucephalobus*, *Plectus* and *Wilsonema*, in addition to *Aphelenchoides*, *Aphelenchus*, *Tylenchus* and juvenile *Tylencholaimus*. Feeding appeared to be less successful on fast moving or thick-cuticled nematodes such as adult *Tylencholaimus*.

On contacting potential prey *Seinura* attempts to penetrate the cuticle with rapid stylet thrusts. If successful, an oesophageal gland secretion is injected (Hechler, 1963; Wood, 1975) and the prey is immobilized. Ingestion of body contents then takes place. Cannibalism has also been reported (Wood, 1975).

The life cycle can be extremely rapid. Wood (1975) recorded an egg to egg time for *S. demani* of 4.5–5 days at 20°C. One or two moults can take place within the egg and one stage may be deleted from the life cycle (Hechler and Taylor, 1965; Wood, 1975). Reproduction is either amphimictic or parthenogenetic.

Distribution: Widespread throughout the world.

Etymology: Derived from the Greek *seios* = move to and fro, and *oura* = tail.

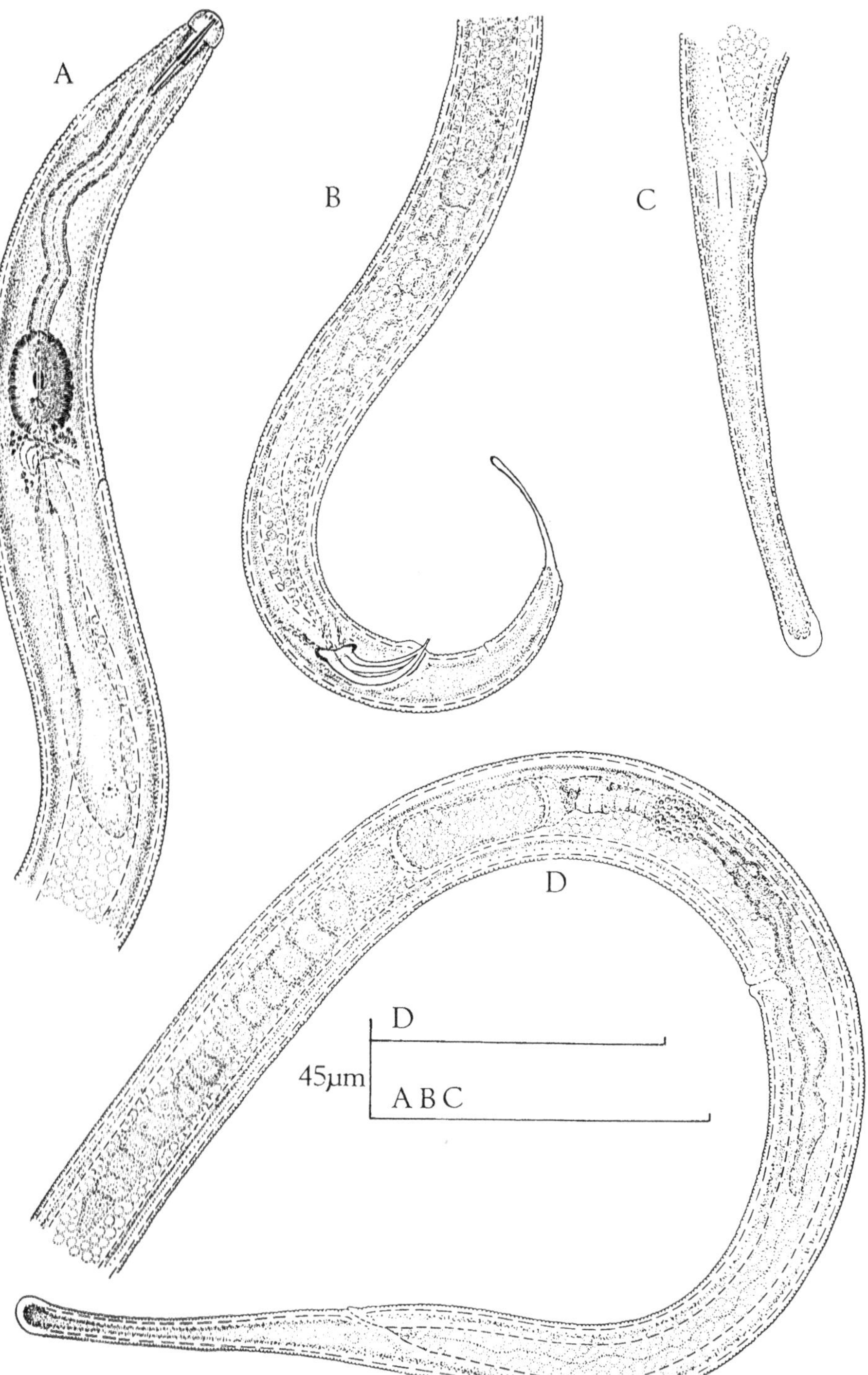

Fig. 4.20. *Aprutides martuccii.* A. Oesophagus; B. Male posterior region; C. Female tail; D. Female posterior region. (After Scognamiglio *et al.*, 1970.)

GENUS *APRUTIDES* SCOGNAMIGLIO, TALAME' AND S'JACOB, 1970 (Fig. 4.20)

DIAGNOSIS: Seinurinae. Small nematodes (0.3–0.5 mm long), ventrally curved to C-shaped on heat relaxation. Cuticle finely annulated, the mid-body annules being about 1 μm apart. Lateral field about one fifth of body width and marked by two incisures in the mid-body region. **Cephalic region slightly offset**, rounded, nearly as high as wide and **divided into two annules**, the posterior one being about twice as wide as the anterior. There is an oral disc in *A. guidettii*. In *en face* view the cephalic region in divided into six equal sectors. Cephalic framework lightly sclerotized. **Stylet about 13–16 μm long, lightly sclerotized and with sclerotized attachment points for the protractor muscles; distinct basal swellings present or absent.** Procorpus cylindrical leading to the well developed oval to rounded rectangular median bulb. Valve plates situated more or less centrally. Oesophageal glands forming a mostly dorsal lobe about three to six body widths long. Nerve ring immediately behind the median bulb with the excretory pore at the same level or slightly posterior. Vulva with slightly protuberant lips, post-median (62–70%). Vagina at right angles to the body axis. Genital tract monoprodelphic, outstretched. Developing oocytes mostly in single file. Spermatozoa may be present in a spermatheca situated between the quadricolumella and the oviduct. Post-uterine sac about two to three body widths long. **Tail elongate**, about 34–65 μm long, **with a clavate tip**. Male testis reaching almost to the oesophageal gland lobe. **Spicules rosethorn-shaped with well developed rostrum and apex**. A weak gubernaculum is present. **Tail medium conoid at first and then narrowing abruptly on the dorsal surface to form an elongate, ventrally curved process with a clavate tip**. One pair of caudal papillae reported about midway along the broader section of the tail.

TYPE SPECIES: *Aprutides martuccii* Scognamiglio, Talame' and s'Jacob, 1970

OTHER SPECIES: *A. guidettii* Scognamiglio, 1974

REMARKS: Scognamiglio *et al.* (1970) assigned the genus to the Aphelenchoididae and commented on similarities with *Seinura* and *Paraseinura*. Andrássy (1976) agreed and placed *Aprutides* in the Aphelenchoidinae along with *Seinura* and *Paraseinura*. In Fortuner's scheme (1984) *Aprutides* is accepted as belonging in the Paraseinurinae along with *Paraseinura* as proposed by Baranovskaya (1981). In the present scheme all genera are placed in the Seinurinae under the Seinuridae.

BIONOMICS: The type species was found in low numbers in the rhizosphere of *Vitis vinifera* L. and *Ficus carica* L. near the village of Miglianico in the Abruzzi region of Italy. The soil type is described as clay over sand. Only a few males were found and no observations on the biology were made. A second species, *A. guidettii*, was described from Italy (Scognamiglio, 1974) and additional data provided by Saltukoglu *et al.* (1976) on two populations of the species from grassland near Istanbul, Turkey.

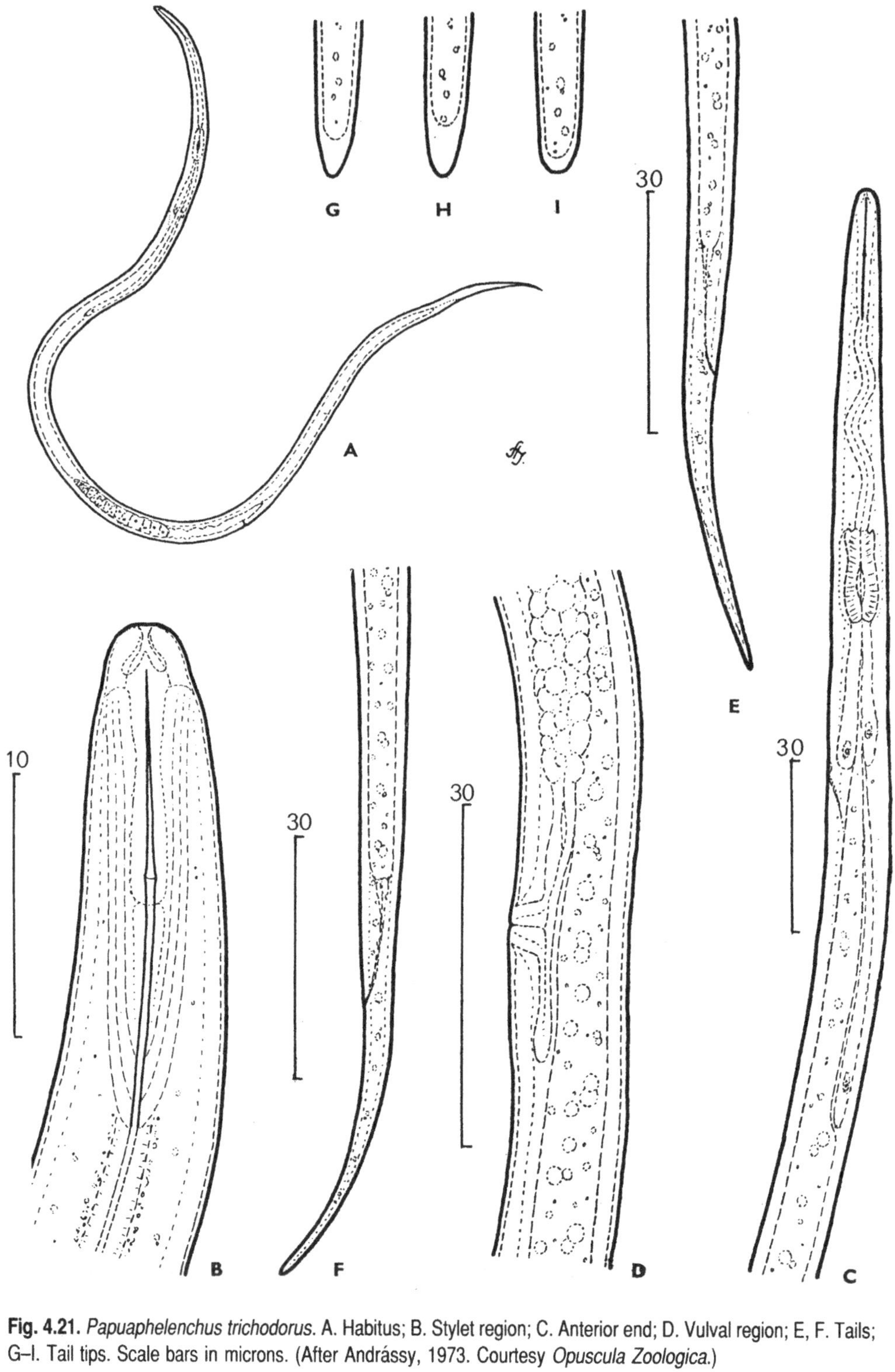

Fig. 4.21. *Papuaphelenchus trichodorus*. A. Habitus; B. Stylet region; C. Anterior end; D. Vulval region; E, F. Tails; G–I. Tail tips. Scale bars in microns. (After Andrássy, 1973. Courtesy *Opuscula Zoologica*.)

Distribution: Recorded from Italy (both species) and Turkey (*A. guidettii*).

Etymology: Derived from *Aprutium*, the Latin name for Abruzzi, the region where the species was found.

GENUS *PAPUAPHELENCHUS* ANDRÁSSY, 1973 (Fig. 4.21)

Diagnosis: Seinurinae. Medium sized nematodes about 0.65 mm long. **Body slender** (a = 55–58) with a smooth cuticle. Lateral fields about one fifth to one sixth of the mid-body width. **Amphids prominent. Stylet 17–19 μm long** (about five times the head width), **slender and lacking basal knobs or swellings**. Procorpus elongate, cylindrical, leading to a well developed, **rounded rectangular, median bulb** with strong valve plates situated slightly post-median. **Oesophageal glands free**, the ventrosublateral glands extending for 24–28 μm and the dorsal gland extending for about 85–94 μm (i.e. six body widths). Excretory pore about two body widths behind the bulb. Vulva posterior at about 67–69% of the body length. Vagina thick walled. **Genital tract short**, monoprodelphic, outstretched. Developing oocytes in a single row. **Post-uterine sac less than a body width long. Tail elongate, attenuating evenly to a finely rounded terminus** (the juvenile tails are stated to be comparatively plump). Males not known.

Type species: *Papuaphelenchus trichodorus* Andrássy, 1973

Other species: Monotypic.

Bionomics: Unknown, but probably predaceous. The type species was found in soil at 1500 m on Mt Kaindi, Papua New Guinea.

Distribution: Papua New Guinea.

Etymology: From *Papua*, the locality and *aphelenchus*.

GENUS *PARASEINURA* TIMM, 1960 (Fig. 4.22)

Diagnosis: Seinurinae. Small to medium sized nematodes (0.5–0.85 mm). Cuticle very finely annulated, the striae being less than 0.5 μm apart. Lateral field indistinct. Cephalic region continuous with body contour and anteriorly flattened. Cephalic framework lightly sclerotized. **Stylet fairly long (22–25 μm in the female) and consisting of a shaft with basal swellings and a conus, the anterior section of which appears to be jointed and with sclerotized thickenings at the junction with the posterior conus.** Procorpus cylindrical. Median bulb well

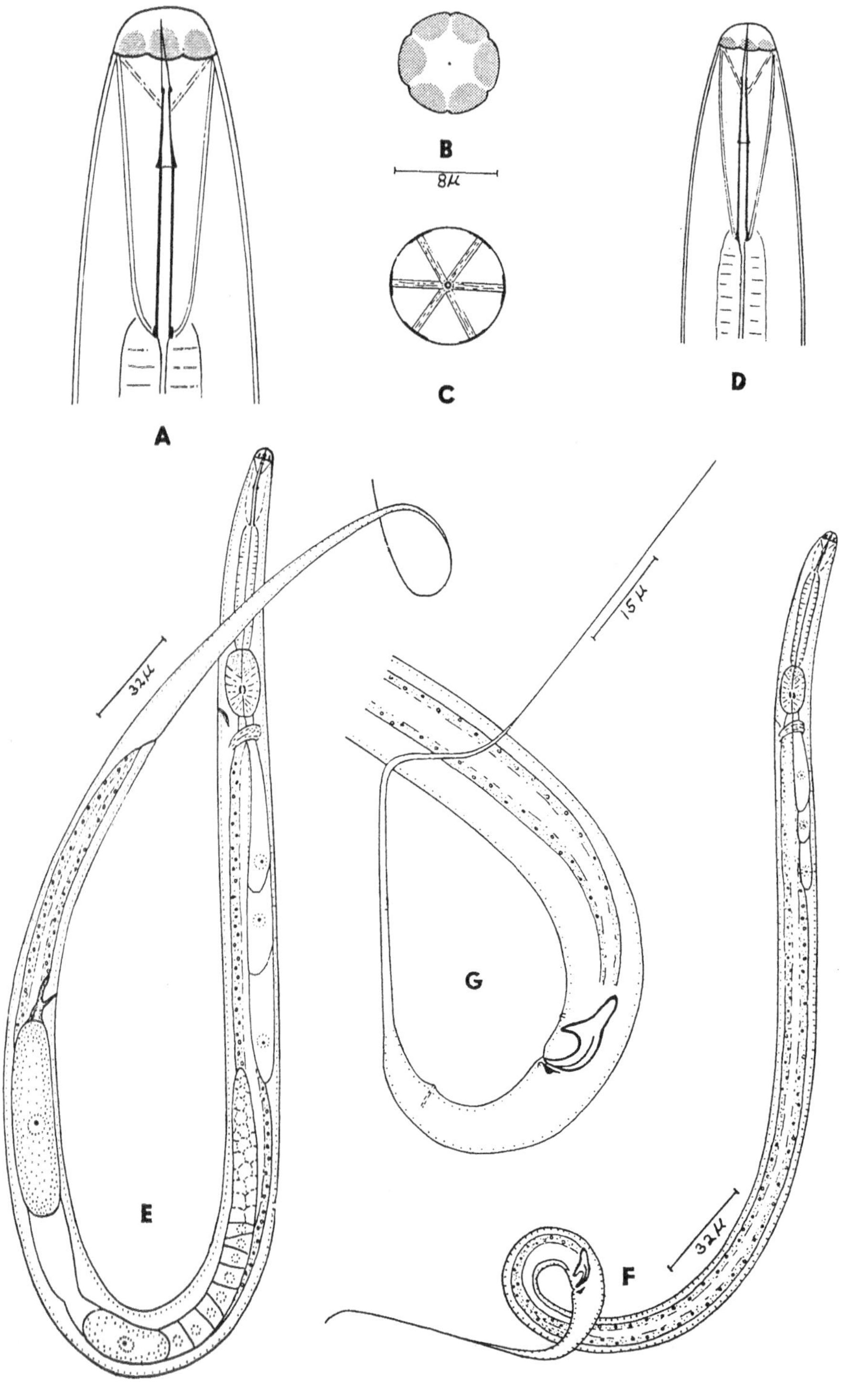

Fig. 4.22. *Paraseinura musicola.* A. Female head; B. *En face* view of female head; C. Optical section at base of lips; D. Male head; E. Entire female; F. Entire male; G. Male tail. (After Timm, 1960. Courtesy *Nematologica*.)

developed, rounded-rectangular in shape with the valve plates situated at two thirds of its length. The anterior third of the bulb is granular in appearance with a few vacuoles whilst the remainder is muscular. Oesophageal glands forming a long, dorsal lobe about four to five body widths in extent. Excretory pore situated at the base of the median bulb with the nerve ring slightly posterior. **Vulva median to post-median (53–63%) in position as a result of the exceedingly long tail**. Genital tract monoprodelphic, outstretched; **postuterine sac about half a body width long**. Developing oocytes initially in two rows. Vagina sloping anteriorly and sclerotized. **Tail long, conoid at first and then attenuating to a filiform tip; about 20 anal body widths long**. Male similar to female but the stylet is shorter (16–17 μm) and the tripartite structure is less evident. **Spicules** paired, **rosethorn-shaped, with a prominent, rounded apex and a pointed rostrum. A small, gubernaculum-like structure is present** near the distal tip of the spicules. Three pairs of caudal papillae, one pair at the level of the proximal end of the spicules, one pair adanal and a much more prominent pair about two anal body widths posterior to the cloaca. **Tail conoid at first and then contracting sharply to a long, attenuated, section with a filiform terminus**.

TYPE SPECIES: *Paraseinura musicola* Timm, 1960

OTHER SPECIES: Monotypic.

REMARKS: A separate subfamily, the Paraseinurinae, has been proposed for this genus (Baranovskaya, 1981), but this is not justified.

BIONOMICS: The type species was recovered from the rotting inflorescence of a banana at Nôtre Dame College, Dacca, Bangladesh. Females were reported to be about four times as common as males. No further specimens were found from nearby soil and root samples. Information on the biology is lacking, but the nematode is probably predatory in habit like the closely related *Seinura*.

DISTRIBUTION: Dacca, Bangladesh.

ETYMOLOGY: Derived from the Greek *para* = beside or near to, and *seinura*.

Family Ektaphelenchidae Paramonov, 1964

DIAGNOSIS: Aphelenchoidoidea. Associates of insects, particularly bark beetles (Coleoptera: Scolytidae). Stylet lumen wide in three genera, but narrow in *Cryptaphelenchus*. Rectum and anus absent, or indistinct and nonfunctional. Intestine extends into tail as a blind diverticulum.

TYPE GENUS: *Ektaphelenchus* Fuchs, 1937 (Skrjabin, Shikhobalova, Sobolev, Paramonov and Sudarikov, 1954)

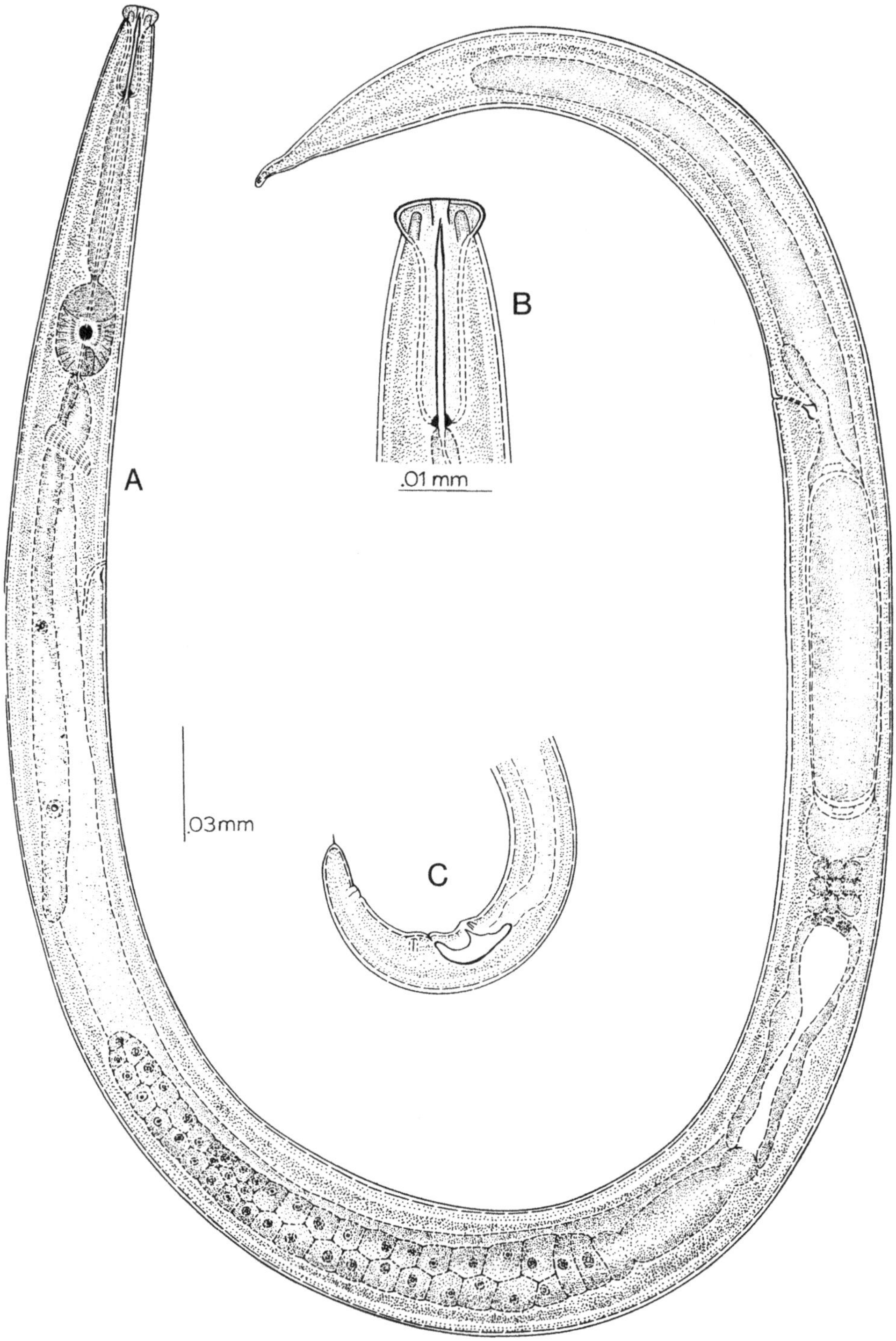

Fig. 4.23. *Ektaphelenchus josephi.* A. Female; B. Head; C. Male tail. (After Massey, 1974. Courtesy USDA.)

Other genera:

Cryptaphelenchoides J.B. Goodey, 1960
Cryptaphelenchus Fuchs, 1937 (Skrjabin, Shikhobalova, Sobolev, Paramonov and Sudarikov, 1954)
Ektaphelenchoides Baujard, 1984

Key to genera

1. Body length < 0.5 mm and usually even shorter. Stylet with narrow lumen . . . . *Cryptaphelenchus*

– Body length > 0.5 mm, stylet with wide lumen 2

2. Tail long, filiform in female, spicate in male . . . . *Ektaphelenchoides*

– Tail short with bluntly, or finely rounded terminus in both sexes. 3

3. Lips cap-like with the laterals narrower than the others . . *Ektaphelenchus*

– Lips not cap-like, equal in size . . . . *Cryptaphelenchoides*

GENUS *EKTAPHELENCHUS* FUCHS, 1937 (SKRJABIN, SHIKHOBALOVA, SOBOLEV, PARAMONOV AND SUDARIKOV, 1954) (Fig. 4.23)

syn. *Parasitaphelenchus* (*Ektaphelenchus*) Fuchs, 1937
Ectaphelenchus Rühm, 1956 (an unjustified emendation, Article 33c, iii)
Caballeroides Chaturvedi and Khera, 1977

Diagnosis: Ektaphelenchidae. Medium sized to long nematodes (0.45–1.2 mm); slender. **Cephalic region distinctly offset and cap-like. The two lateral lips are narrower than the other four** when seen *en face*. **Stylet** 15–23 μm long **with a wide lumen** and distinct basal swellings or knobs in the majority of species. Procorpus cylindrical, often tapering before joining the prominent, rounded-rectangular median bulb. Valve plates post-median in position. Oesophageal gland lobe well developed and extending for a considerable distance down the body. Nerve ring and excretory pore situated close behind the median bulb. **Rectum and anus in female very difficult to discern and almost certainly non-functional. Intestine forming a blind diverticulum** which extends for most of the tail length. Vulva posterior, usually at about 80% of the body length, but exceptionally as anterior as 65%. Vulval lips not protuberant. **Vagina strongly sclerotized in many species**. Genital tract monoprodelphic, outstretched. Developing oocytes usually in two or three rows. Post-uterine sac of variable length and development. Tail conoid, tapering to a pointed or rounded tip. Male spicules paired, separate. The shape varies but they are typically strongly curved ventrally with a prominent rostrum and well developed apex continuing the line of the shaft. Caudal papillae consist of one preanal pair and either one or two postanal pairs. A median preanal papilla may also be present. A gubernaculum-like structure is illustrated for some species lying near the distal tip of the dorsal limb. Bursa absent.

Type species:

E. hylastophilus (Fuchs, 1930) Skrjabin, Shikhobalova, Sobolev, Paramonov and Sudarikov, 1954
syn. *Parasitaphelenchus hylastophilus* forma *ateri* Fuchs, 1930
P. (*Ektaphelenchus*) *hylastophilus* forma *ateri* Fuchs, 1930 (Fuchs, 1937)
P. (*E.*) *hylastophilus* forma *cunicularii* Fuchs, 1930 (Fuchs, 1937)
Aphelenchus hylastophilus (Fuchs, 1930) Filipjev, 1934
Aphelenchus ateri (Fuchs, 1930) Filipjev, 1934
E. cunicularii (Fuchs, 1930) Rühm, 1956

Other species:

E. amitini (Fuchs, 1937) Rühm, 1956
syn. *Parasitaphelenchus* (*Cryptaphelenchus*) *amitini* Fuchs, 1937
Aphelenchoides amitini (Fuchs, 1937) T. Goodey, 1951
E. betulae Rühm, 1956
E. dendroctoni Rühm, 1956
E. goffarti Rühm, 1956
E. josephi Massey, 1974
E. larici Lazarevskaya, 1963
E. olitorius (Chaturvedi and Khera, 1977) Baujard, 1984
syn. *Caballeroides olitorius* Chaturvedi and Khera, 1977
E. piniperdae Kakuliya and Lazarevskaya, 1966
E. prolobos Massey, 1964
syn. *E. lepidus* Massey, 1971
E. smaelus Massey, 1974
E. sandiaensis Massey, 1964
E. skrjabini Lazarevskaya, 1961
E. stammeri (Körner, 1954) J.B. Goodey, 1960
syn. *Aphelenchoides stammeri* Körner, 1954
E. tenuidens (Thorne, 1935) Rühm, 1956
syn. *Aphelenchoides tenuidens* Thorne, 1935
E. tuerkorum Rühm, 1956
E. typographi (Fuchs, 1930) Fuchs, 1937
syn. *Parasitaphelenchus typographi* Fuchs, 1930
Aphelenchus typographi (Fuchs, 1930) Filipjev, 1934
Parasitaphelenchus (*Ektaphelenchus*) *typographi* (Fuchs, 1930) Fuchs, 1937
E. zwoelferi Rühm, 1957

Species incertae sedis:

E. alni (Steiner, 1942) Rühm, 1956
syn. *Aphelenchoides alni* Steiner, 1942
E. terebranus Massey, 1974
E. obtusus Massey, 1956

NOTES:

1. Baujard (1984) synonymized *E. terebranus* Massey under *E. macrobulbosus* (Rühm) (herein regarded as belonging to *Cryptaphelenchoides*) and regarded *E. obtusus* Massey as also embracing *E. macrostylus* Khan (= *C. macrostylus* herein). I do not accept Baujard's arguments for these synonymies, in part because the proposals encompass a wide range in the morphometrics. However, although both *E. terebranus* and *E. obtusus* share some characters with *Cryptaphelenchoides*, information on the labial conformation is not conclusive and so I prefer to leave both species in *Ektaphelenchus* for the moment. In addition, Baujard (1984) also synonymized *E. tenuidens* (Thorne, 1935) under *Berntsenus brachycephalus* (Thorne, 1935) on somewhat tenuous grounds, a proposal which I have not accepted.
2. *Aphelenchoides alni* Steiner, 1942 was transferred to *Ektaphelenchus* by Rühm (1956). However, its position is doubtful as Steiner described and illustrated a prominent rectum and anus. It also has affinities with *Bursaphelenchus*.

BIONOMICS: According to Rühm (1956) the immature inseminated females form cocoon-like structures under the elytra and between the tergites of adult scolytid beetles (Coleoptera: Scolytidae), particularly the conifer-breeding genera. The juveniles are associated with the beetle larvae and pupae and grow and moult. Whether nourishment is gained by penetration of the insect with the stylet is not known for certain, but assumed (Rühm, 1956). The cocoons are initially bright yellow before darkening to a cocoa-brown colour. They are of wrinkled appearance with minute pores visible under the microscope. Rühm (1956) speculated that the cocoon-like structures might be formed from secretions of the vaginal glands or by a stimulated response from the beetle host. Cocoon development is most pronounced on bark beetles of the genus *Hylastes*; *E. goffarti*, found on *Dryocoetes*, only develop very delicate cocoons whilst those species found on *Scolytus* have no protective cocoons at all (Rühm, 1956). After the beetle has cut the maternal gallery and laid eggs, the female *Ektaphelenchus* rehydrate before leaving the cocoons and laying their eggs in the frass. Rühm (1956) implied that no further development occurred in the absence of beetle larvae. Thorne (1935a) reported *E. tenuidens* within cocoon-like material under the elytra of *Dendroctonus monticolae* in Utah and Massey (1974) recorded up to 75 females and a solitary male of *E. obtusus* in small leathery cocoons under the elytra of *D. rufipennis*. *Ektaphelenchus* has also been recorded from the haemocoel of bark beetles (Thorne, 1935a; Massey, 1956, 1974), but this appears to be a rare phenomenon.

DISTRIBUTION: Europe, CIS (formerly USSR), North America and probably throughout the Palaearctic region.

ETYMOLOGY: Derived from the Greek *ektos* = outside and *aphelenchus*.

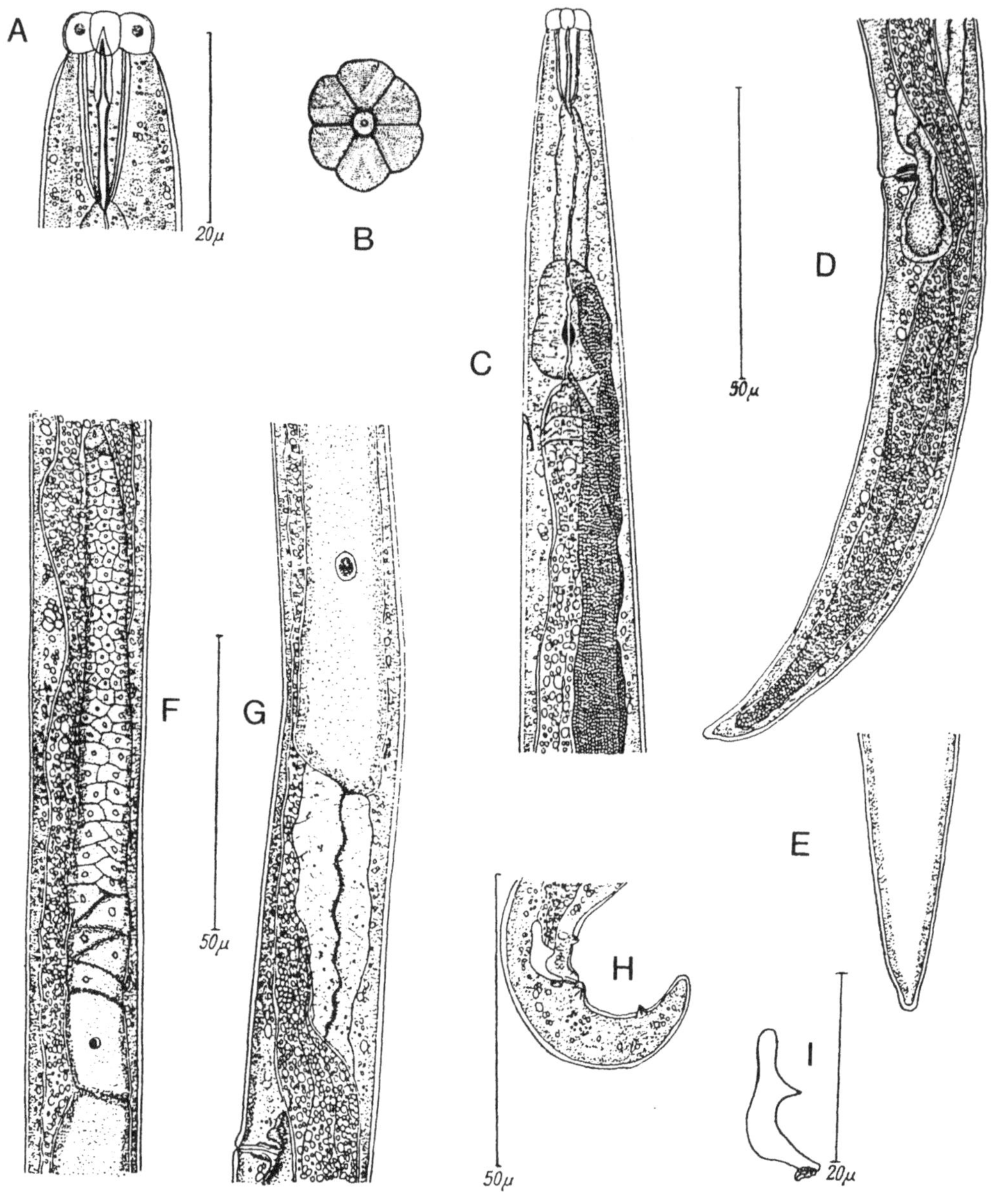

Fig. 4.24. *Cryptaphelenchoides macrobulbosus.* Female: A. Head; B. *En face*; C. Anterior end; D. Posterior end; E. Tail, ventral view; F. Anterior part of genital tract; G. Posterior part of genital tract. Male: H. Tail region; I. Spicule. (After Rühm, 1956. Courtesy *Parasitologische Schriftenreihe.*)

GENUS *CRYPTAPHELENCHOIDES* J.B. GOODEY, 1960 (Fig. 4.24)

DIAGNOSIS: Ektaphelenchidae. Small to medium sized nematodes (0.4–0.6 mm) dying ventrally arcuate on heat relaxation with the male tail more sharply recurved. **Cephalic region offset with six rounded lips of equal size.** Stylet slender, about 15 μm long and with a wide lumen and small basal swellings. Procorpus short, cylindrical, leading to a rounded-oblong median bulb with post-median valve plates. Oesophageal gland lobe lying dorsal to the **intestine and well developed. Nerve ring just posterior to median bulb with the excretory pore at the same level. Although an anus and rectum can be discerned, the intestine ends in a blind diverticulum which extends almost to the tail tip in the female. Vulva posterior**, situated at about 80% of the body length. Vagina at right angles to the body axis and strongly sclerotized in the type species. Genital tract monoprodelphic, outstretched. Developing oocytes mostly in two or three rows. Post-uterine sac short (less than a body width) in the type species, but longer in *C. macrostylus*. Tail conoid, tapering to a rounded terminus. **Male spicules mitten-shaped** with the rostrum pointed and well developed and the bluntly rounded apex forming a dorsally directed curve to the shaft. **Distally the spicules have a small, dorsally situated, apophysis**. There are three pairs of caudal papillae: one preanal; one adanal and one postanal. Bursa absent.

TYPE SPECIES:
Cryptaphelenchoides macrobulbosus (Rühm, 1956) J.B. Goodey, 1960
    syn. *Aphelenchoides macrobulbosus* Rühm, 1956
        Ektaphelenchus macrobulbosus (Rühm, 1956) Massey, 1974

OTHER SPECIES:
C. macrostylus (Khan, 1960) Nickle, 1970
    syn. *Ektaphelenchus macrostylus* Khan, 1960
C. propora Yang, 1985
    syn. *Ektaphelenchus propora* (Yang, 1985) Ebsary, 1991
C. riograndensis (Massey, 1964) Hunt and Hague, 1976
    syn. *Ektaphelenchus riograndensis* Massey, 1964
C. scolyti (Rühm, 1956) Hunt and Hague, 1976
    syn. *Ektaphelenchus scolyti* Rühm, 1956

BIONOMICS: The type species was found under bark of a dead tree, *Pinus sylvestris*, in Germany (Rühm, 1956). It had no apparent insect associate although such is likely to be the case. The other species are associated with bark beetles. Hunt and Hague (1976) gave an account of the biology of *C. scolyti*, an associate of *Scolytus scolytus*. Inseminated immature females were found on the cuticle of the fourth and fifth stage larvae and beneath the elytra of the pupae and adult beetles. Male nematodes were found in the frass and on the larvae and pupae, but never on the adult insects. Female nematodes were attached beneath the anterior inner angles of the elytra by means of a sheath-like structure about the neck region of each nematode. This sheath or invagination appeared to be formed by the nematode pushing itself into the soft cuticle of the teneral adult and was partially tanned in the region of the nematode head (Fig. 4.25). The females remained in an anabiotic state until gradually rehydrated as the beetle excavated a new gallery system. Eggs were laid in the frass and the subsequent stages fed on fungal hyphae until the adults were produced some five weeks later. Whether the nematodes are ectoparasitic is not known. Rühm (1956) suggested that this might be the case. If so, such ectoparasitism would be more likely to occur on the pupae and teneral adults where the cuticle is soft and not tanned.

DISTRIBUTION: Europe, CIS (formerly USSR), North America.

ETYMOLOGY: A combination of *cryptaphelenchus* and the Greek *oides* = likeness of form.

GENUS *CRYPTAPHELENCHUS* FUCHS, 1937 (SKRJABIN, SHIKHOBALOVA, SOBOLEV, PARAMONOV AND SUDARIKOV, 1954) (Fig. 4.26)

syn. *Parasitaphelenchus* (*Cryptaphelenchus*) Fuchs, 1937
Parasitaphelenchus (*Steineria*) Fuchs, 1937
(nec. *Steineria* Micoletzky, 1922)

DIAGNOSIS: Ektaphelenchidae. **Small nematodes** (usually less than 0.5 mm). Cuticle finely to moderately coarsely annulated. Lateral field indistinct with up to four incisures. **Cephalic region usually continuous with the body contour, the lips being rounded and forming a low, flattened, cap.** Labial region divided into six sectors when seen *en face*. Cephalic framework moderately sclerotized. **Stylet short, usually about 8–10 μm long. Conus delicate, shaft more robust** and with distinct basal knobs in most species, but reportedly lacking even basal swellings in others. Procorpus tubular, **median bulb spherical** with central valve plates. Gland lobe dorsal and usually about three body widths long. Nerve ring more or less opposite the excretory pore and situated about a body width behind the bulb. **Anus and rectum absent in the female, the gut ending in a blind sac. Vulva posterior at about 75–85% of the body length** and with indistinct lips. Vagina sloping forwards. Some species are reported as having strongly sclerotized vaginal walls. Genital tract monoprodelphic, outstretched. Developing oocytes arranged in two or three rows. **Post-uterine sac absent**. Tail ventrally arcuate, conoid; terminus pointed. **Male spicules** paired, **mitten-shaped**, usually ending in a fine point distally. **Rostrum prominent, pointed. Apex well developed, with a bluntly rounded tip. Posterior to the spicule tips there is often a sclerotized structure** which is reportedly not gubernacular. According to Massey (1974) it is the heavily sclerotized, posterior cloacal lip. **Tail conical, ventrally curved and ending in a fine point**. Two pairs of caudal papillae; one pair pre-cloacal and the other near the tail tip.

TYPE SPECIES:
Cryptaphelenchus minutus (Fuchs, 1930) Fuchs, 1937
syn. *Parasitaphelenchus minutus* Fuchs, 1930
P. (*C.*) *minutus* Fuchs, 1930 (Fuchs, 1937)
Aphelenchoides (*Schistonchus*) *minutus* (Fuchs, 1930) Filipjev, 1934

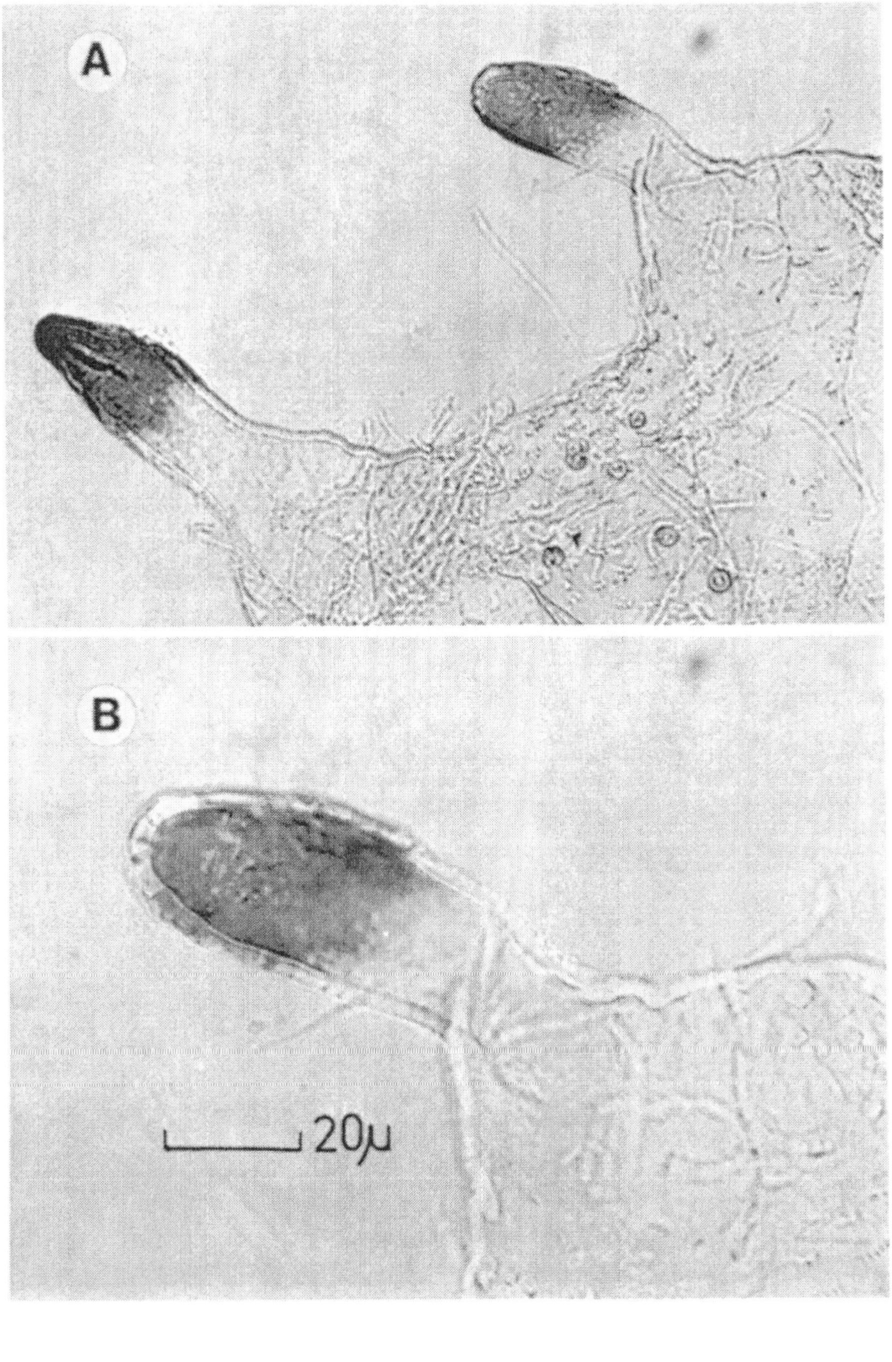

Fig. 4.25. *Cryptaphelenchoides scolyti.* A, B. Cuticular sheaths from beneath elytron of *Scolytus scolytus.* The sheaths are invaginations of the insect cuticle, each being formed by the action of an immature female of *C. scolyti.* The invaginations enclose the anterior oesophageal region, the insect cuticle tanning in the region between the lips and the median bulb of the *in situ* nematode. (After Hunt and Hague, 1976. Courtesy *Nematologica.*)

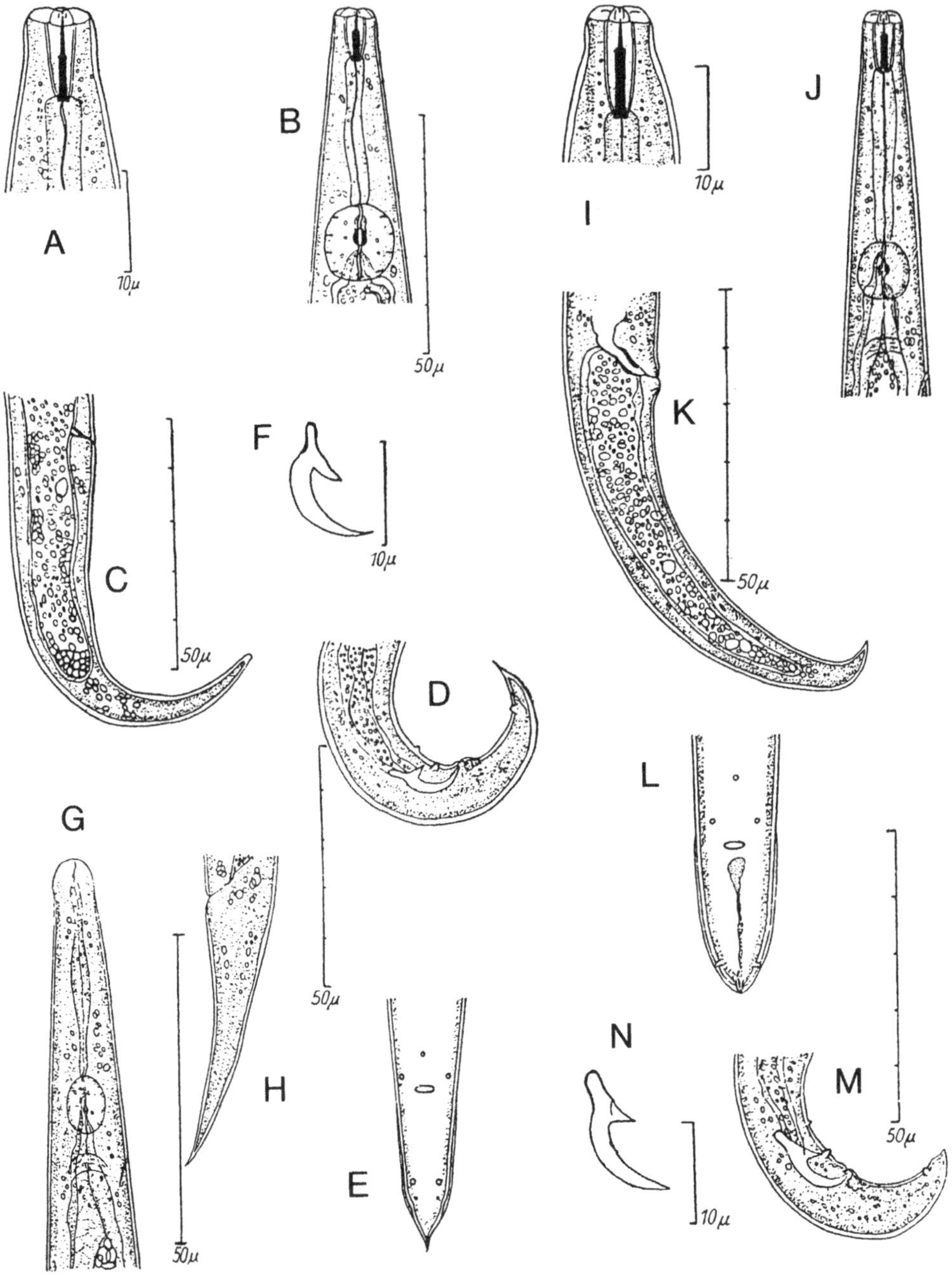

Fig. 4.26. *Cryptaphelenchus* spp. *C. koerneri.* Female. A. Cephalic region; B. Anterior end; C. Posterior end. Male. D. Tail region; E. Tail, ventral; F. Spicule. Dauerlarva. G. Anterior end; H. Tail. *C. macrogaster.* Female. I. Cephalic region; J. Anterior end; K. Posterior end. Male. L. Tail, ventral; M. Tail; N. Spicule. (After Rühm, 1956. Courtesy *Parasitologische Schriftenreihe.*)

Other species:

C. aedili Lazarevskaya, 1961
C. bicoloris Devdariani, 1971
C. cirrus Massey, 1974
C. diversispicularis Korenchenko, 1987
C. hectographi Rühm in Rühm and Chararas, 1957
C. ipinius Massey, 1974
C. koerneri Rühm, 1956
C. latus (Thorne, 1935) Rühm, 1956
  syn. *Aphelenchoides latus* Thorne, 1935
C. leptocaudus Rühm, 1956
C. macrogaster macrogaster (Fuchs, 1915) Fuchs, 1937
  syn. *Tylenchus macrogaster* Fuchs, 1915
    Parasitaphelenchus macrogaster (Fuchs, 1915) Fuchs, 1930
    P. (C.) macrogaster (Fuchs, 1915) Fuchs, 1930 (Fuchs, 1937)
    Aphelenchoides (Schistonchus) macrogaster (Fuchs, 1915) Filipjev, 1934
C. macrogasteroides macrogasteroides Rühm, 1956
C. malpighius bidentati Rühm, 1956
C. malpighius malpighius Fuchs, 1937 (Rühm, 1956)
  syn. *Parasitaphelenchus (C.) malpighius* Fuchs, 1937
    Aphelenchoides malpighius (Fuchs, 1937) T. Goodey, 1951
C. piceae Rühm, 1956
C. poligraphi Rühm, 1956
C. quadridens Kakuliya, 1963
C. sutoricus Devdariani, 1974
C. typographi Kakuliya, 1966
C. viktoris (Fuchs, 1937) Rühm, 1956
  syn. *Parasitaphelenchus (Steineria) viktoris* Fuchs, 1937
    Aphelenchoides viktoris (Fuchs, 1937) T. Goodey, 1951

Species inquirendae:

C. cryptus Rühm, 1956
C. macrogaster acuminati Rühm, 1956
  syn. *C. acuminati* Rühm, 1956 (Ebsary, 1991)
C. macrogasteroides villosi Rühm, 1956
  syn. *C. villosi* Rühm, 1956 (Ebsary, 1991)
C. piceoides Rühm, 1956
C. pygmaeus (Fuchs, 1930) Fuchs, 1937
  syn. *Parasitaphelenchus pygmaeus* Fuchs, 1930
    P. (C.) pygmaeus (Fuchs, 1930) Fuchs, 1937
    Aphelenchoides (Schistonchus) pygmaeus (Fuchs, 1930) Filipjev, 1934

Nomina nuda:

C. ligniperdae Kurashvili, Kakuliya and Devdariani, 1980
C. monochasmus Kakuliya, Devdariani and Maglakelidze, 1980

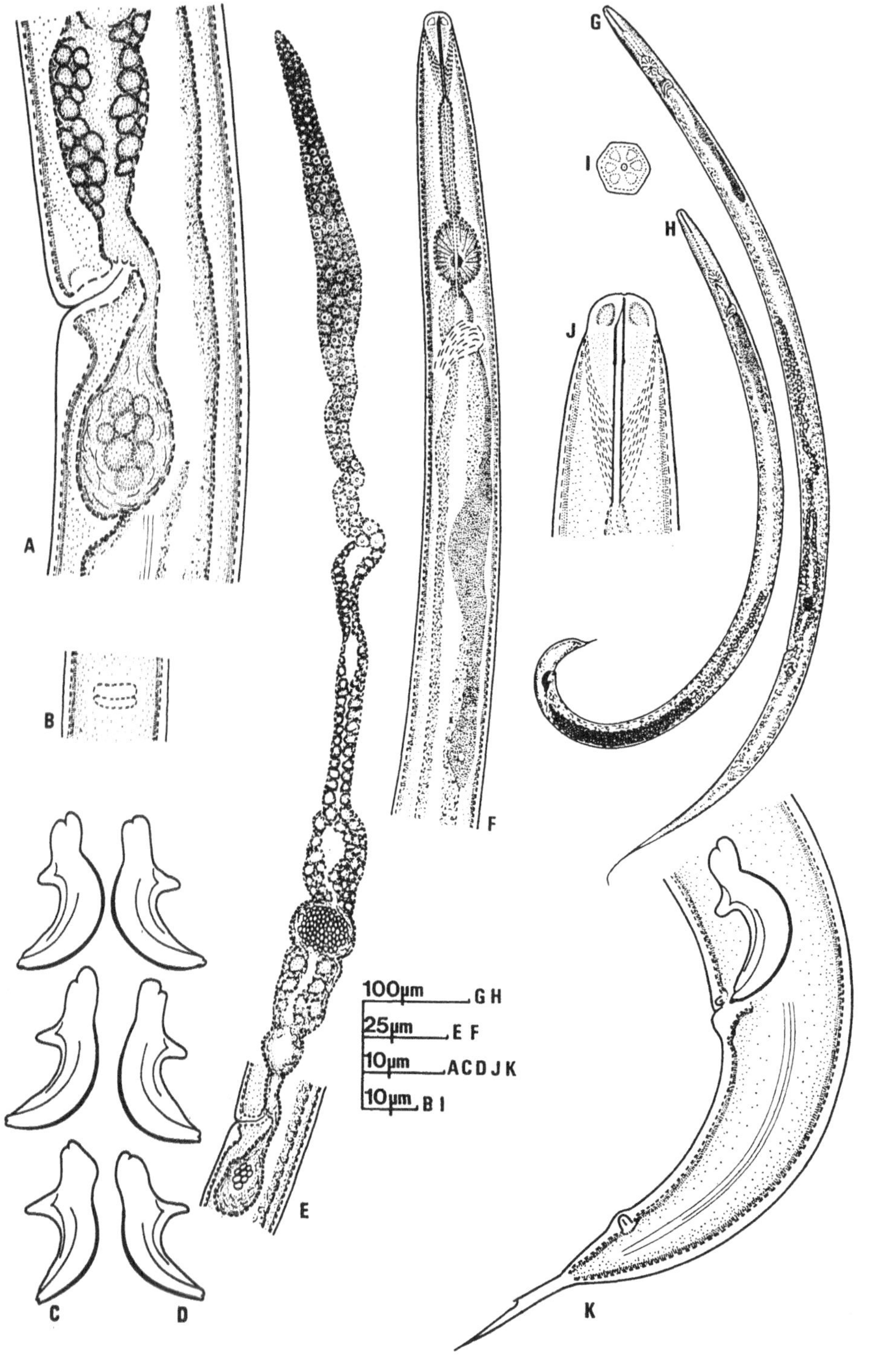
A
B
C
D
E
F
G
H
I
J
K
100µm G H
25µm E F
10µm A C D J K
10µm B I

C. ryjikovi Kurashvili, Kakuliya and Devdariani, 1980
C. sutor Kakuliya, Devdariani and Maglakelidze, 1980
C. vorontzovi Kurashvili, Kakuliya and Devdariani, 1980

Bionomics: Virtually all the nominal species are found associated with bark beetles (Coleoptera: Scolytidae). The adults are found living in the frass of the beetle tunnels and are presumably mycetophagous. At least one of the juvenile stages (the J3?) is found within the adult beetle in the Malpighian tubules, the colleterial gland or the genital region (Rühm, 1956). This stage is more probably endophoretic rather than parasitic and uses the beetle as a transport mechanism in a similar way to the ectophoretic species which are found under the elytra and intersegmentally. All the described species are from Europe, the USSR, and North America, although their distribution is probably throughout the Palaearctic region wherever bark beetles of the family Scolytidae occur. Several authors (e.g. Devdariani, 1974; Korenchenko, 1987) also report *Cryptaphelenchus* spp. as associates of *Monochamus*, a cerambycid wood-boring beetle. Korenchenko (1987) reported that the inseminated, immature females of *C. diversispicularis* were located in the anterior coxal depressions of teneral beetles, about 10–50 nematodes being found on each insect. The nematodes were also reared on fungal cultures for about a year.

Distribution: Widespread in North America and the Palaearctic region.

Etymology: Derived from the Greek *kryptos* = hidden and *aphelenchus*.

GENUS *EKTAPHELENCHOIDES* BAUJARD, 1984 (Fig. 4.27)

Diagnosis: Ektaphelenchidae. Medium sized nematodes, 0.5–0.9 mm long; heat relaxed form ventrally arcuate with the male tail more sharply recurved. Cuticle finely annulated. Lateral fields with up to three incisures. **Cephalic region high, rounded, not offset** from body and comprising six equal lips. **Stylet slender,** about 17–26 μm long **with a wide lumen. Sclerotized thickenings mark the junction of conus and shaft**, but the latter lacks basal knobs or thickenings. **Procorpus** cylindrical, **narrowing just before it joins the large rounded rectangular median bulb** whose **valve plates** are situated **in the posterior half** of the bulb. Oesophageal gland lobe lying dorsal to the intestine and well developed. Nerve ring about a body width behind the bulb and posterior to the oesophago-intestinal junction. Excretory pore posterior to the nerve ring in two species, but level or anterior in the remaining species. **Rectum and anus absent, the**

Fig. 4.27. *Ektaphelenchoides pini.* Female. A. Vulval region, lateral view; B. Vulval region, ventral view; E. Genital tract; F. Oesophageal region; G. Entire; I. Anterior part, *en face*; J. Anterior part (lateral view); Male. C. Left spicules; D. Right spicules; H. Entire; K. Posterior part. (After Baujard, 1984. Courtesy *Revue de Nématologie*.)

intestine forming a blind diverticulum in the female. Vulva posterior at about 60–77% of the body length. Vagina directed anteriorly, usually with perivaginal sclerotization. Genital tract monoprodelphic, outstretched. Developing oocytes in two or three rows. Post-uterine sac about a body width long, or less, in three species, but about three body widths long in the fourth. **Tail long, conoid**, tapering and attenuate in three species, but much shorter in *E. compsi*. **Male spicules** paired, ventrally curved **with a prominent rounded apex** and a pointed rostrum. Caudal papillae comprising two pairs in the type species; one pair adanal and the other near the tail tip. In *E. compsi* there is also a pre-cloacal ventral papilla and another near the tail tip. **Tail conoid, tapering to a strongly developed spicate process.**

TYPE SPECIES:
Ektaphelenchoides pini (Massey, 1966) Baujard, 1984
syn. *Seinura pini* Massey, 1966

OTHER SPECIES:
E. attenuata (Massey, 1974) Baujard, 1984
syn. *Seinura attenuata* Massey, 1974
E. musae Baujard, 1984
E. compsi Baujard, 1984

BIONOMICS: *Ektaphelenchoides pini* was recorded associated with *Dendroctonus terebrans* (Coleoptera: Scolytidae) in loblolly pine and *E. attenuata* with *D. adjunctus* in ponderosa pine by Massey (1974). Baujard (1984) also recorded *E. pini*, but from xylem of *Pinus laricio* in France. *E. musae* was found in banana corm tissue from the Ivory Coast and *E. compsi* from xylem of *Pinus laricio*, France (Baujard, 1984). Hooper (1991, personal communication) identified an undescribed species from oak xylem in England. The nematodes were apparently ectoparasitic on dipterous larvae of the genus *Xylodiplosis* (Diptera: Cecidomyidae), and were quite firmly attached by means of the stylet which was inserted into the insect's integument (Fig. 4.28). The nematode intestine was coloured pinkish, the same hue as the insect haemolymph.

DISTRIBUTION: England, France, Ivory Coast, USA.

ETYMOLOGY: A combination of *ektaphelenchus* and the Greek *oides* = likeness of form.

Genus Incertae Sedis

GENUS *BERNTSENUS* MASSEY, 1974 (Fig. 4.29)

DIAGNOSIS: ? Ektaphelenchidae. Medium sized nematodes ranging from 0.7 to

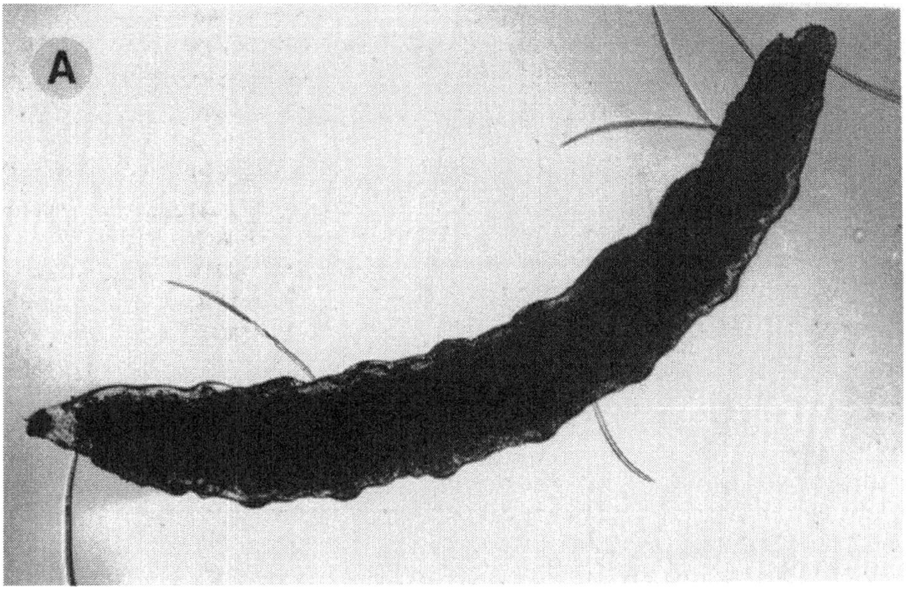

Fig. 4.28. *Ektaphelenchoides* sp. parasitizing maggot of *Xylodiplosis* sp. The nematodes have the stylet piercing the insect integument. (Photograph courtesy of D.J. Hooper, Rothamsted Experimental Station.)

1.0 mm in length. Cuticle finely annulated, lateral incisures not discernible. **Cephalic region expanded, about three times wider than high. Stylet** of unusual form, the **conus** region being **heavily sclerotized with a wide lumen and prominent aperture,** but **leading to a lightly sclerotized shaft** with or without basal knobs or swellings. Stylet 17–18 μm long in the type species. Procorpus short, cylindrical but tapering at either end. Median bulb massive, round to rounded oblong in shape and with well developed valve plates located centrally. Oesophageal gland lobe dorsal, about four body widths long and stout. Nerve ring within a body width of the posterior margin of the median bulb. The excretory pore is either just anterior to, or just posterior to, the nerve ring. Vulva posterior at about 70–75% with non-protuberant or slightly protuberant lips. Genital tract monoprodelphic, outstretched. Developing oocytes mostly arranged in two rows. Post-uterine sac well developed, two to five body widths long. Tail conoid to a subacute or mucronate terminus. **Anus and rectum conspicuous.** Male spicules rosethorn-shaped with a low rounded apex hardly distinguishable from the dorsal limb and a very weak, pointed rostrum. One pair of ventrosubmedian papillae located two-thirds of the distance to the tail tip. Tail dorsally convex-conoid to a finely pointed terminus. Bursa absent.

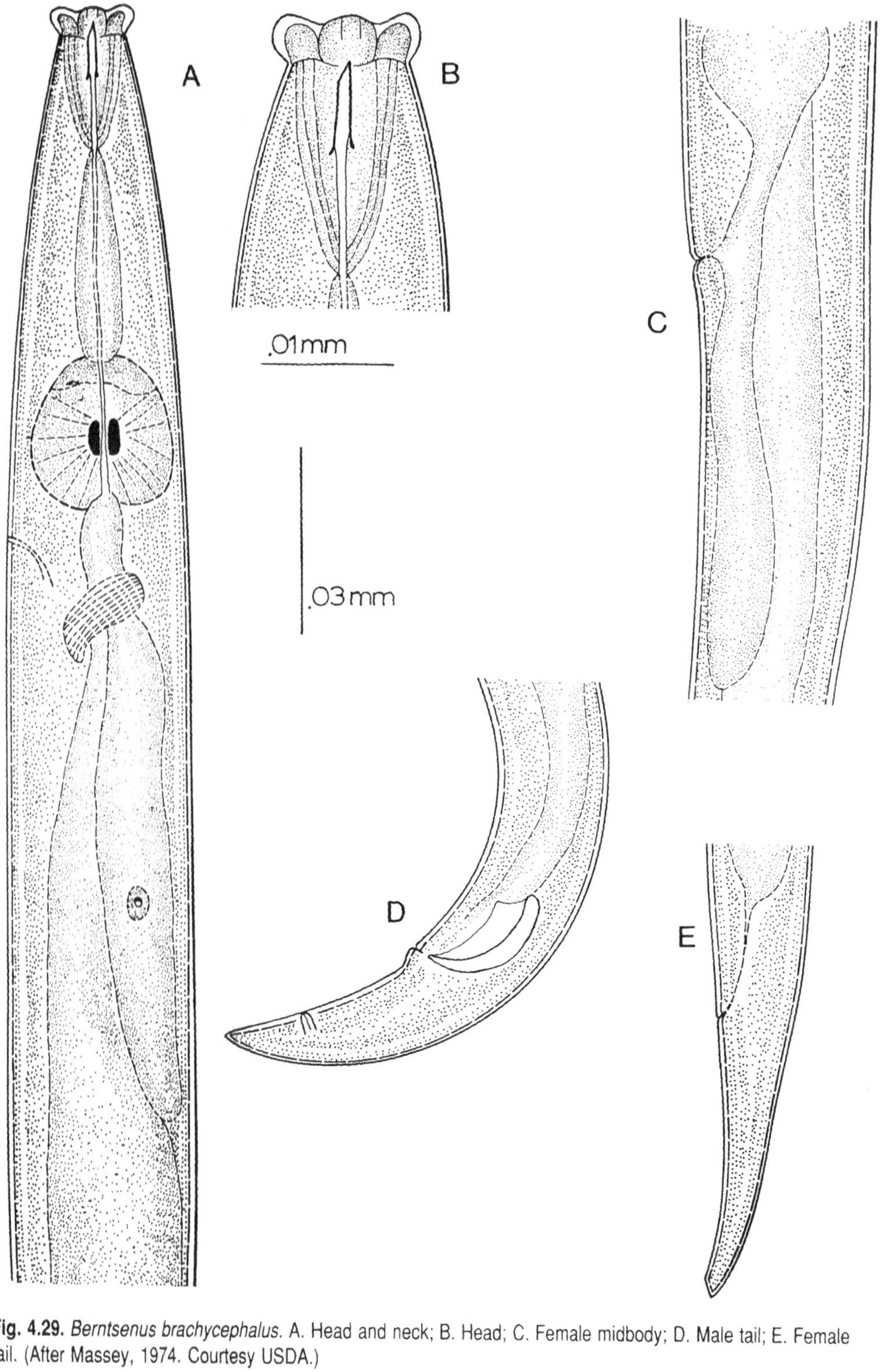

Fig. 4.29. *Berntsenus brachycephalus*. A. Head and neck; B. Head; C. Female midbody; D. Male tail; E. Female tail. (After Massey, 1974. Courtesy USDA.)

Type species:
Berntsenus brachycephalus (Thorne, 1935) Massey, 1974
syn. *Aphelenchoides brachycephalus* Thorne, 1935
Ektaphelenchus brachycephalus (Thorne, 1935) J.B. Goodey, 1960
Laimaphelenchus brachycephalus (Thorne, 1935) Massey, 1956

Other species: *B. labiosus* Massey, 1974

note: This genus was proposed by Massey (1974) with A. *brachycephalus* Thorne, 1935 as type. Massey (1974) regarded *Berntsenus* as being related to *Ektaphelenchus*, presumably because of the expanded cephalic region and the wide lumen of the stylet, but noted differences such as the prominent anus and rectum. Andrássy (1976) overlooked *Berntsenus*, but Fortuner (1984) regarded it as being closer to *Bursaphelenchus* and placed it in the Bursaphelenchinae. Baujard (1984) accepted *Berntsenus* as belonging to the Ektaphelenchinae and proposed *A. tenuidens* Thorne 1935 (= *Ektaphelenchus tenuidens* (Thorne, 1935) Rühm, 1956) as a synonym of *B.* brachycephalus. Such a proposal is unacceptable as the descriptions and illustrations of the two species differ in a number of important respects. Nickle and Hooper (1991) retained *Berntsenus* in the Ektaphelenchinae without further comment.

Berntsenus does not fit comfortably within the Ektaphelenchidae because of the presence of a prominent rectum and apparently functional anus, features which conflict with the other genera in the family. In addition, the spicules of *B. brachycephalus* (males are not known for *B. labiosus*) are atypical of the ektaphelenchs in lacking a prominent apex and well developed rostrum – in fact they more closely resemble some *Aphelenchoides* species. The definitive placement of such a genus must await the description of additional species supplemented by biological data.

Bionomics: Associated with the bark beetles (Coleoptera: Scolytidae) *Dendroctonus ponderosae* in lodgepole pine and *Pityokteines elegans* in white fir. No other details are known.

Distribution: Utah and New Mexico, USA.

Etymology: Derived from Dr Carl *Berntsen*.

Family Acugutturidae Hunt, 1980, n.rank

Diagnosis: Aphelenchoidoidea. Ectoparasites of cockroaches or noctuid moths, all stages being found on the insect. Stylet elongate, 50–185 μm long, attenuated. Anus and rectum difficult to discern and probably nonfunctional.

Type subfamily: Acugutturinae Hunt, 1980

Other subfamily: Noctuidonematinae n.subfam.

Key to subfamilies

1. Adults and juveniles not markedly swollen, spicules of normal size (< 30 μm) and rostrum not tubular, thorn-shaped. Parasites of cockroaches .... Acugutturinae

– Adults and juveniles markedly swollen, spicules very large (38–109 μm) with a tubular rostrum, massive. Parasites of moths....... Noctuidonematinae

Subfamily Acugutturinae Hunt, 1980

Diagnosis: Acugutturidae. Ectoparasites of cockroaches. Body not markedly swollen. Excretory pore posterior to median bulb. Spicules rosethorn-shaped and of normal size and development.

Type genus: *Acugutturus* Hunt, 1980

Other genera: Monotypic.

GENUS *ACUGUTTURUS* HUNT, 1980 (Fig. 4.30)

Diagnosis: Acugutturinae. Medium sized nematodes (0.6–0.9 mm). Heat relaxed form ventrally arcuate with the tail region more sharply curved, particularly in the male. Cuticle finely annulated, the annules being about 0.7 μm apart in the mid-body region. Lateral field marked by a single incisure. **Cephalic region offset, rounded and rather knob-like. Stylet very long (50–60 μm), slender and needle-like. The conus forms about 75% of the length and the shaft lacks basal knobs or swellings**. Protractor muscles are attached to the proximal region of the shaft and extend to the cephalic skeleton. The guiding sheath or vestibule extension extends to beyond the conus/shaft junction. **Procorpus, slender, cylindrical and reflexed to allow for the protrusion of the long stylet**. Median bulb ovoid, strongly developed and with central valve plates. Oesophageal gland lobe dorsal, well developed, extending for about four body widths. Nerve ring just posterior to the median bulb and slightly anterior to the excretory pore. **Intestine filled with a pale yellowish liquid**, presumed to be the haemolymph of the cockroach host. **Rectum and anus very indistinct or apparently absent in the females and presumably non-functional**. Vulva posterior (73–78%) with slightly protuberant lips. Vagina sloping anteriorly and sclerotized. Genital tract monoprodelphic, outstretched. Developing oocytes in two or three rows. **Post-uterine sac absent**. Deposited eggs measure about

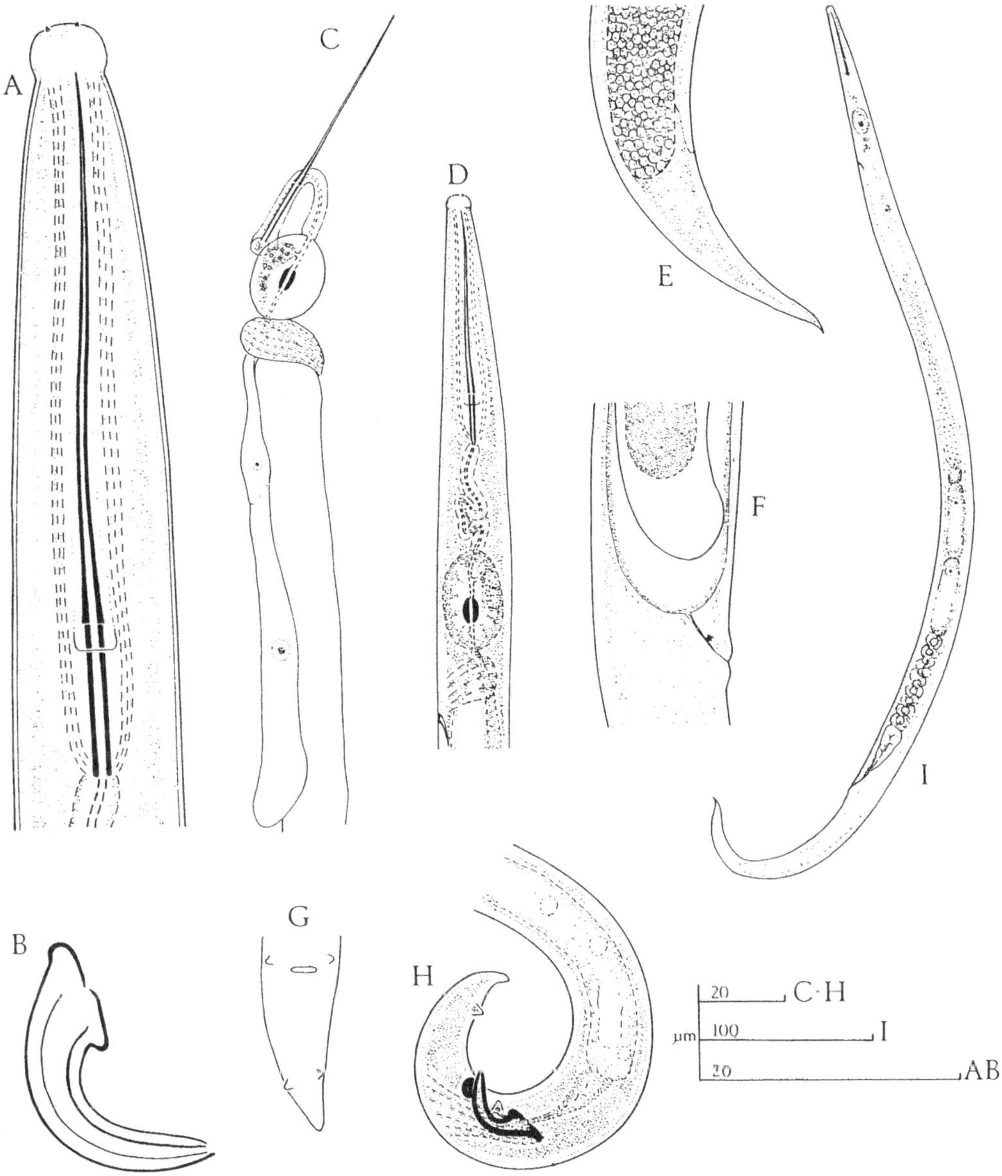

Fig. 4.30. *Acugutturus parasiticus.* A. Stylet; B. Spicules; C,D. Oesophageal region; E. Female tail; F. Vulval region with part of uterine egg; G. Ventral view of male tail; H. Male tail region; I. Entire female. (After Hunt, 1980. Courtesy *Systematic Parasitology.*)

126 × 27μm. Tail conoid, ventrally arcuate and terminating in a fine point. Male spicules separate, rosethorn-shaped with prominent apex and rostrum. **A sclerotized, gubernaculum-like structure, is situated ventral to the spicule tips.** Tail conoid, strongly curved ventrally with a rounded tip. Two pairs of caudal papillae, one pre-cloacal and the other near the tail tip.

Type species: *Acugutturus parasiticus* Hunt, 1980.

Other species: Monotypic.

Remarks: In view of the unique characteristics of this nematode, Hunt (1980a) proposed the subfamily Acugutturinae to contain the new genus. He concluded, on the basis of morphological and biological criteria, that the nearest existing subfamily was the Ektaphelenchinae. Subsequently, *Noctuidonema* Remillet and Silvain, 1988, an even more unusual insect-parasitic aphelench was described and placed in the Acugutturinae. With the description of yet another bizarre form (Anderson and Laumond, 1992) I consider that family rank is now appropriate to recognize the morphological and biological characteristics of this unique group of aphelenchs.

Bionomics: *Acugutturus parasiticus* was recovered from the external surfaces of the cockroach, *Periplaneta americana* L., collected in St. Lucia, West Indies. All stages of the nematode were washed from the cockroaches, the majority being under the wings, on the abdomen and between the limbs. Mature females were sluggish, and when on the wings, were usually situated alongside the veins accompanied by a number of large eggs. Immature females, males and juveniles were far more active and moved freely about the cockroach cuticle, thus facilitating dispersal as cockroaches are gregarious insects and roost in confined spaces. Infestations were usually between 20–30 nematodes per cockroach but one specimen yielded about 2500 nematodes. Hunt (1980a) presumed that the species was an obligate ectoparasite as all attempts to culture the nematode on a variety of fungi failed and nematodes were only recovered from the insects. The intimate association of all stages of *A. parasiticus* with the cockroach, the long needle-like stylet and the pale yellow colour (identical in hue to cockroach haemolymph) of the intestinal contents strongly support the contention that a parasitic relationship exists.

Distribution: St. Lucia, West Indies.

Etymology: Derived from the Latin *acus* = needle and *guttur* = throat.

Sufamily Noctuidonematinae n.subfam.

Diagnosis: Acugutturidae. Ectoparasites of various noctuid moths. Body of all stages swollen, particularly posteriorly. Excretory pore located far anterior to median bulb near the cephalic region. Spicules exceptionally long (38–109 μm) and massive with a tubular rostrum.

Type genus: *Noctuidonema* Remillet and Silvain, 1988

Other genera: *Vampyronema* n.gen.

KEY TO GENERA

1. Stylet over 100 μm long (106–185 μm), lacking prominent basal knobs. Spicules 74–109 μm long with prominent, bluntly rounded apex and complex sheaths and sclerotizations on the blade .... *Noctuidonema*

– Stylet under 80 μm long (53–69 μm), with prominent basal knobs. Spicules 38–49 μm long, lacking a prominent rounded apex and with simple blades .... *Vampyronema*

GENUS *NOCTUIDONEMA* REMILLET AND SILVAIN, 1988 (Figs 4.31, 4.32)

DIAGNOSIS: Noctuidonematinae. Nematodes of medium length (0.5–0.8 mm), obese and dying slightly ventrally arcuate on heat relaxation. **Body slender anteriorly but, in the female progressively swelling for about 75% of the body length** before narrowing slightly towards the tail. Cephalic region rounded, continuous with body contour. Cuticle with annules about 2.2 μm apart. **Stylet extremely long (106–185 μm), consisting of an attenuated, needle-like conus comprising about 84% of the total length, and a short shaft with basal swellings** which are more easily seen before fixation. **Procorpus long, cylindrical and reflexed. Median bulb strongly developed spheroid** with the well developed valve plates situated slightly posterior to the mid-point. Oesophageal gland lobe dorsal, but rather obscure. Nerve ring immediately posterior to the median bulb. **Excretory pore prominent, very anterior in position** (within 12–23 μm of the cephalic extremity in the type species). **Intestine almost filling the body cavity and filled with a yellow-coloured fluid** and small, needle-like, crystals. Rectum and anus visible before fixation. **Vulva extremely posterior** (V = 88.5–98.4% in the type) and in the form of a broad, transverse slit. **Anterior lip strongly developed and overhanging the vulval opening**. Genital tract monoprodelphic, outstretched. **Developing oocytes in multiple rows**. Post-uterine sac absent. **Tail very short, conoid**, with a finely rounded tip. Male similar to female in general respects, but not as swollen, spicules paired, separate and massive (74–109 μm long in the type species. In shape they represent an extreme modification of the typical rosethorn shape. **The rostrum is 22–29 μm long, tubular and pointed and appears to fit into a cuticular 'pocket'** attached to the ventral side of the body. The spicule appears to rotate about this fixed point during exsertion. **The apex is ventrally directed, well developed and rounded whilst the shaft of the spicule is broad** with a complicated arrangement of sheaths and sclerotized parts. Essentially, the dorsal limb is extended to form a tubular, sigmoid structure 34–41 μm long, dorsal to which is a spongy matrix and a sclerotized, distally expanded, sheath (a detailed account of the structure of this highly unusual spicule is presented by Anderson and Laumond, 1990). In addition, a long, dorsally located, flexible midrib is associated with the spicules and deforms as the spicules are exserted. There are two pairs of papillae, one pre- and one postanal. **A small terminal flap of cuticle (= 'bursa') present**. Tail very short. Juveniles more

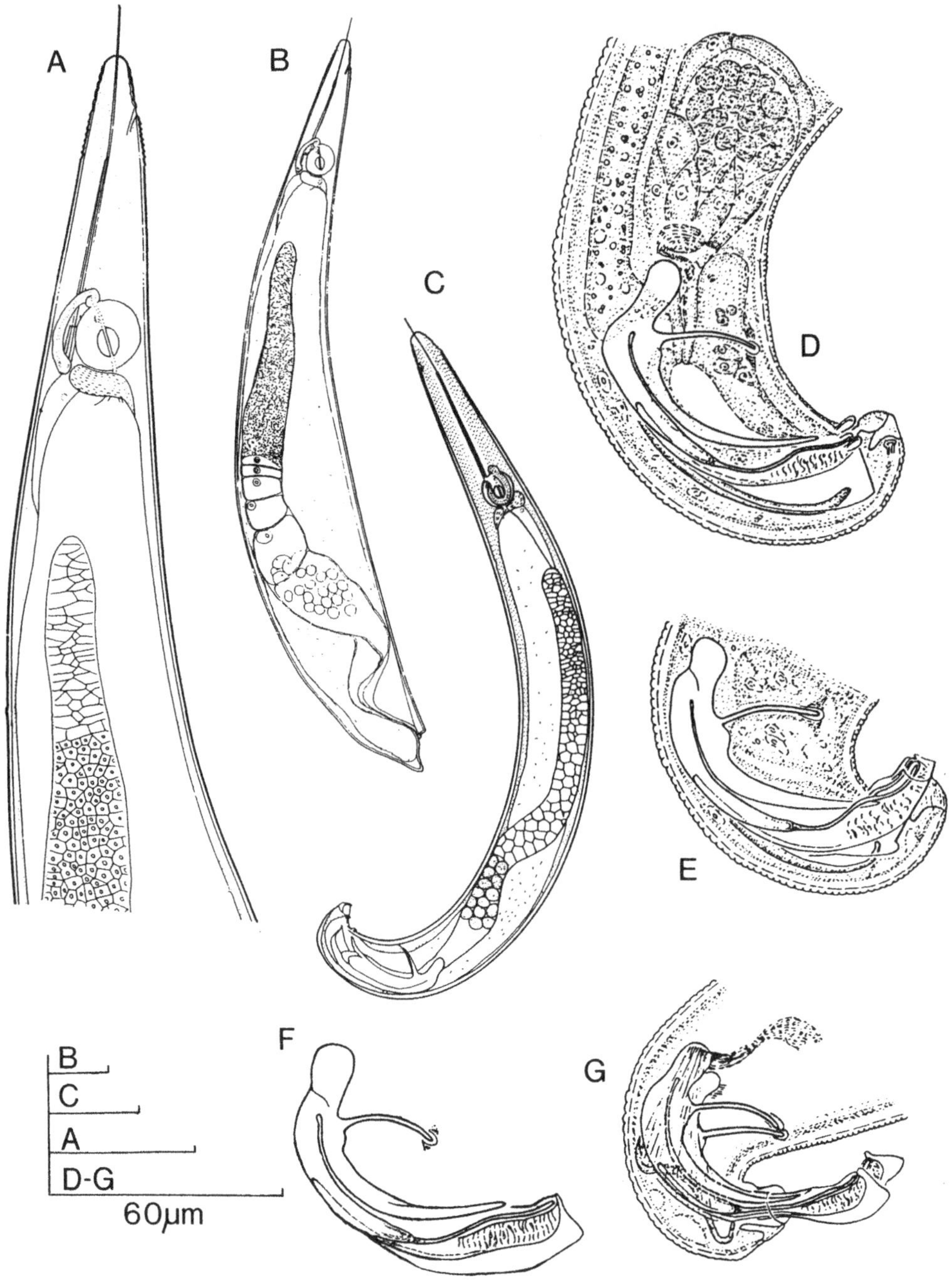

Fig. 4.31. *Noctuidonema guyanense.* A. anterior region; B. Entire female; C. Entire male; D. Spicules; E,F,H. Male tail. (A–C after Remillet and Silvain, 1988; remainder after Anderson and Laumond, 1990. Courtesy *Revue de Nématologie.*)

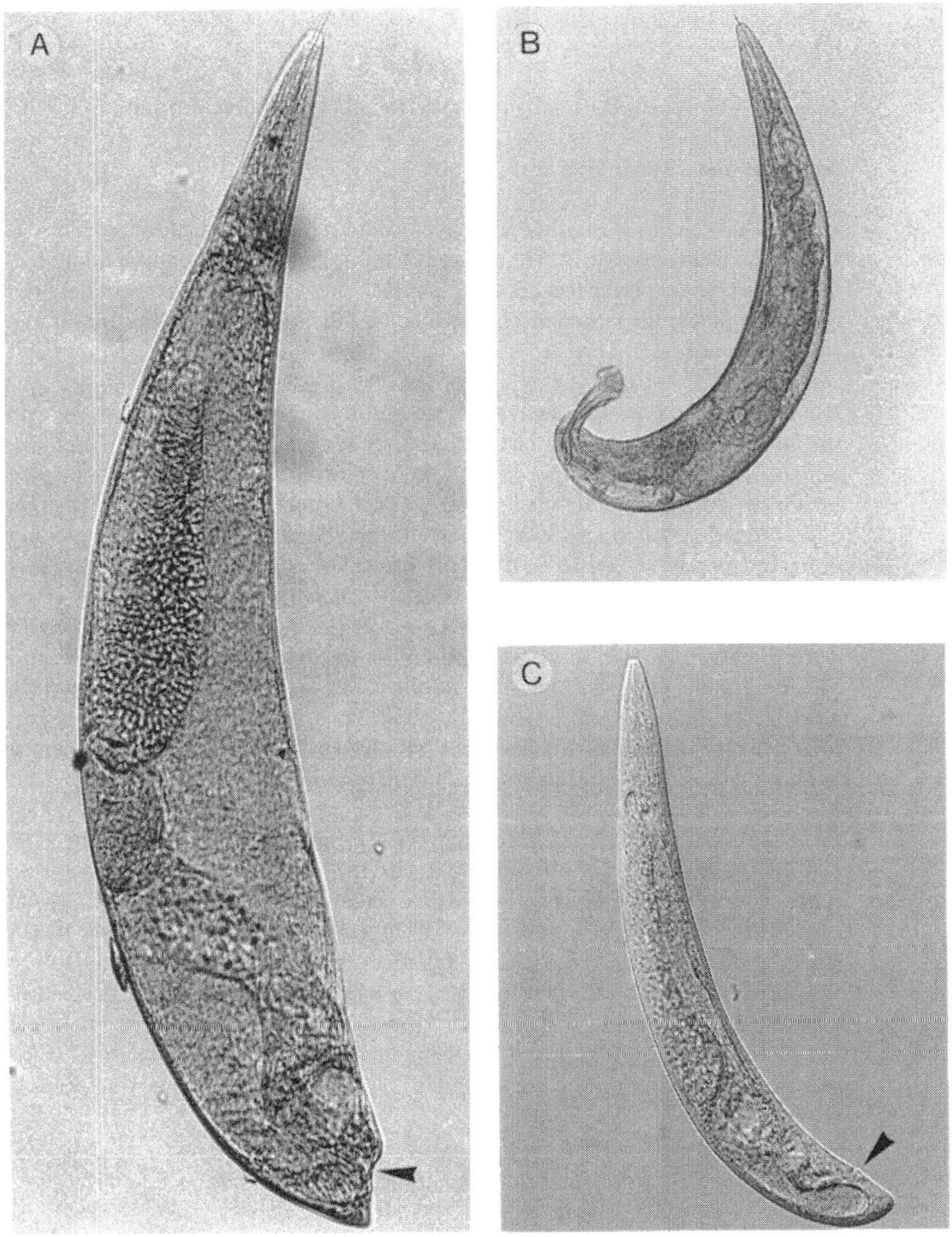

Fig. 4.32. *Noctuidonema guyanense*. A. Mature female (vulva is arrowed); B. Mature male with spicules exserted; C. Immature female (vulva is arrowed.) (A,B: types; C: ex. Florida. Original.)

cylindrical, but possessing the long stylet. Eggs large (108 × 54 μm in the type).

TYPE SPECIES: *Noctuidonema guyanense* Remillet and Silvain, 1988

OTHER SPECIES: Monotypic

NOTE: I have males of another species of *Noctuidonema* from Aldabra Island in the Indian Ocean (leg. V.W. Spaull) and Remillet (personal communication) has mentioned other species from South America. The Aldabra specimens share the complex spicule structure of *N. guyanense*, but differ in the position of the excretory pore.

BIONOMICS: The type species was found on the abdominal region, particularly the last two segments, but also the thorax of adult noctuid moths (*Spodoptera androgea* (Cram.), *S. frugiperda* (J. E. Smith), *S. latifascia* (Walk.), *S. marima* (Schaus.), *Anicla infecta* (Och.) and *Leucania* spp.) from French Guyana, South America. All stages were found on the abdomen. They are presumably obligate ectoparasites and feed on the insect haemolymph (hence the yellow colour of the gut contents). Quite how the nematodes are transmitted from one generation of adult moths to the next is unknown, but probably occurs during mating as the adult generations of the hosts overlap. Remillet and Silvain (1988) observe that the swollen, gravid females may either be attached to the insect cuticle with the stylet penetrating the host integument, or freely creeping about the abdomen like leeches – 'on libres se déplaçant à la façon des sangsues'. Rogers *et al.* (1990a) gave further data on the host range of *N. guyanense*. In addition to being found on Noctuidae they also recorded it less commonly from other lepidopteran families (Lasiocampidae, Notodontidae, Pyralidae and Sphingidae). The number of nematodes found on each host varied from five to 189, the usual average being between 20–30 and the incidence of infection ranged from 4–51% (Rogers *et al.*, 1990). Marti *et al.* (1990) discussed the pathological effects of *N. guyanense* on *Spodoptera frugiperda* moths and reported minor pathological changes in the insect cuticle. Additional data on dispersal, prevalence, host records and the effect of temperature and humidity on development and population dynamics of *N. guyanense* were given by Rogers *et al.* (1990b) and Simmons and Rogers (1990a,b; 1991).

DISTRIBUTION: Central America, West Indies, southern USA.

ETYMOLOGY: A combination of *Noctuididae*, the host family and the Greek *nema* = thread.

GENUS *VAMPYRONEMA* N.GEN. (Figs 4.33, 4.34, 4.35)

DIAGNOSIS: Noctuidonematinae. **Nematodes of small to medium length (0.23–**

0.77 mm), plump, with a tendency to become obese in the middle third or so of the body, particularly in the mature females. Dying ventrally arcuate to varying degrees on heat relaxation. **Cephalic region sexually dimorphic; low, rounded and lacking annules in the male, hemispherical, offset and with five or six annules in the female**. Cuticle with annules about 1.5–1.9 μm apart. **Stylet long (53–69 μm)**, and consisting of an attenuated, needle-like conus and a short shaft **with well developed basal knobs** 3–6 μm wide. **Protracter muscles** attached proximally to the basal knobs and extending anteriorly before **diverging to points of attachment on the body wall below the cephalic basal plate. A second set runs from the basal knobs to the base of the vestibule extension**. Procorpus short, subcylindroid and reflexed when the stylet is retracted. Median bulb powerful, ovoid and with strong valve plates located post-median. Oesophageal gland lobe extending dorsally for about two or three body widths. Nerve ring immediately behind base of median bulb. **Excretory pore very anterior**, within two head widths of the anterior extremity and with the duct heavily sclerotized, particularly in the female. **Renette cell enormously developed in the female, extending posteriorly for about two thirds of the body length**. Intestinal lumen with numerous needle-like crystals arranged lamella fashion. **Rectum apparently absent and anus not observed in the females. Vulva posterior at 82–88%** and with a protuberant anterior lip. Genital tract monoprodelphic, outstretched with the developing oocytes arranged in multiple rows and, in mature females, extending anterior to the median bulb. **Spermatheca present. Post-uterine sac absent. A bilateral uterine gland complex is present, comprising one pair of lobes anterior to, and two bilobed pairs posterior to, the vagina. Postvulval region subconoid, tapering slightly to a broadly rounded terminus with a short, more or less axial, mucron**. Male testis usually outstretched, but may reflex. **Spicules paired, separate and 38–49 μm long. The dorsal and ventral linbs are long, curved and well separated; they lack the complex modifications typical of *Noctuidonema***. The apex is heavily sclerotized and continues the line of the shaft whilst **the rostrum is elongate, tubular and with the tip inserted into a cuticular pocket-like invagination of the body wall. A short circumcloacal cylinder of cuticle protrudes from the cloaca**. There are two pairs of caudal papillae, one slightly preanal and the other subterminal. **Tail short with a broadly rounded terminus and an axial mucron.**

TYPE SPECIES:
Vampyronema daptria (Anderson and Laumond, 1992) n.comb.
syn. *Noctuidonema daptria* Anderson and Laumond, 1992

OTHER SPECIES: Monotypic.

RELATIONSHIPS: Anderson and Laumond (1992), in describing the species, commented on the unique combination of characters which made generic placement difficult, before provisionally placing it in *Noctuidonema*. Having examined type material of both *N. guyanense* and *N. daptria* and also material of an undescribed species of *Noctuidonema* from Aldabra Island (leg. V. W. Spaull), I consider that a new genus is appropriate to contain *N.*

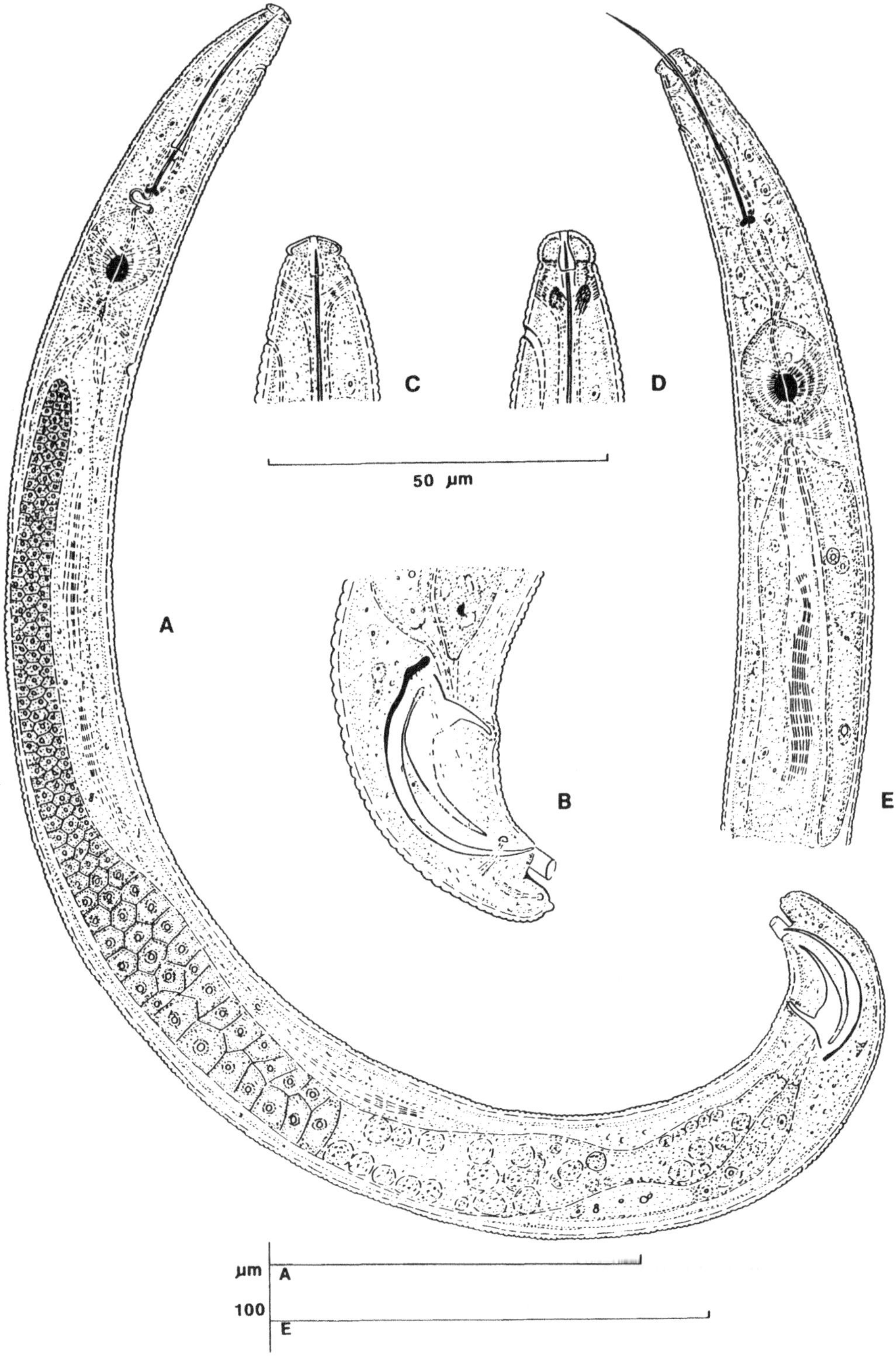
A
B
C
D
E
50 µm
µm
100
A
E

daptria as its morphological features are at variance with those of *Noctuidonema, sensu stricto*. The differential characters of *Vampyronema* n.gen. include: the sexual dimorphism of the cephalic region; the much shorter stylet with strong basal knobs; the anterior insertion of the protracter muscles; the renette cell development in the female; the bilateral uterine gland complex; the shorter and less complex spicules.

Bionomics: All stages were found on the abdomen of adult moths, *Lesmone porcia* Stoll (Lepidoptera: Noctuidae). They appear to be obligate ectoparasites and presumably share a similar biology to species of *Noctuidonema*.

Distribution: Guadeloupe, West Indies.

Etymology: A combination of *vampira*, a Serbian word for a nocturnal blood sucking demon, and the Greek *nema* = thread.

Family Parasitaphelenchidae Rühm, 1956 (Siddiqi, 1980)

Diagnosis: Aphelenchoidoidea. Mostly associates or endoparasites of insects, particularly bark beetles (Coleoptera: Scolytidae). Two insect associates are known to be implicated in serious diseases of palms or pines. Male tail tip enveloped by small, bursa-like flap of cuticle. Spicules stout, may be partially fused and with a prominent rostrum.

Type subfamily: Parasitaphelenchinae Rühm, 1956

Other subfamily: Bursaphelenchinae Paramonov, 1964

Key to subfamilies

1. Fourth stage juvenile endoparasitic in the insect haemocoel. Vulva very posterior (85–90%), spicules partially fused, male tail not strongly recurved. . . . . Parasitaphelenchinae

– Third stage dauerlarvae ectophoretic, exceptionally endophoretic. Vulva more anterior at 70–80%. Spicules usually separate, but reported to be partially fused in some species, male tail strongly recurved . . . . Bursaphelenchinae

Fig. 4.33. *Vampyronema daptria.* Male, A,B,C,E. Female, D. A. Adult in typical relaxed state; B. Tail region; C. Head end; D. Female head; E. Oesophageal region. (After Anderson and Laumond, 1992. Courtesy *Fundamental and Applied Nematology*.)

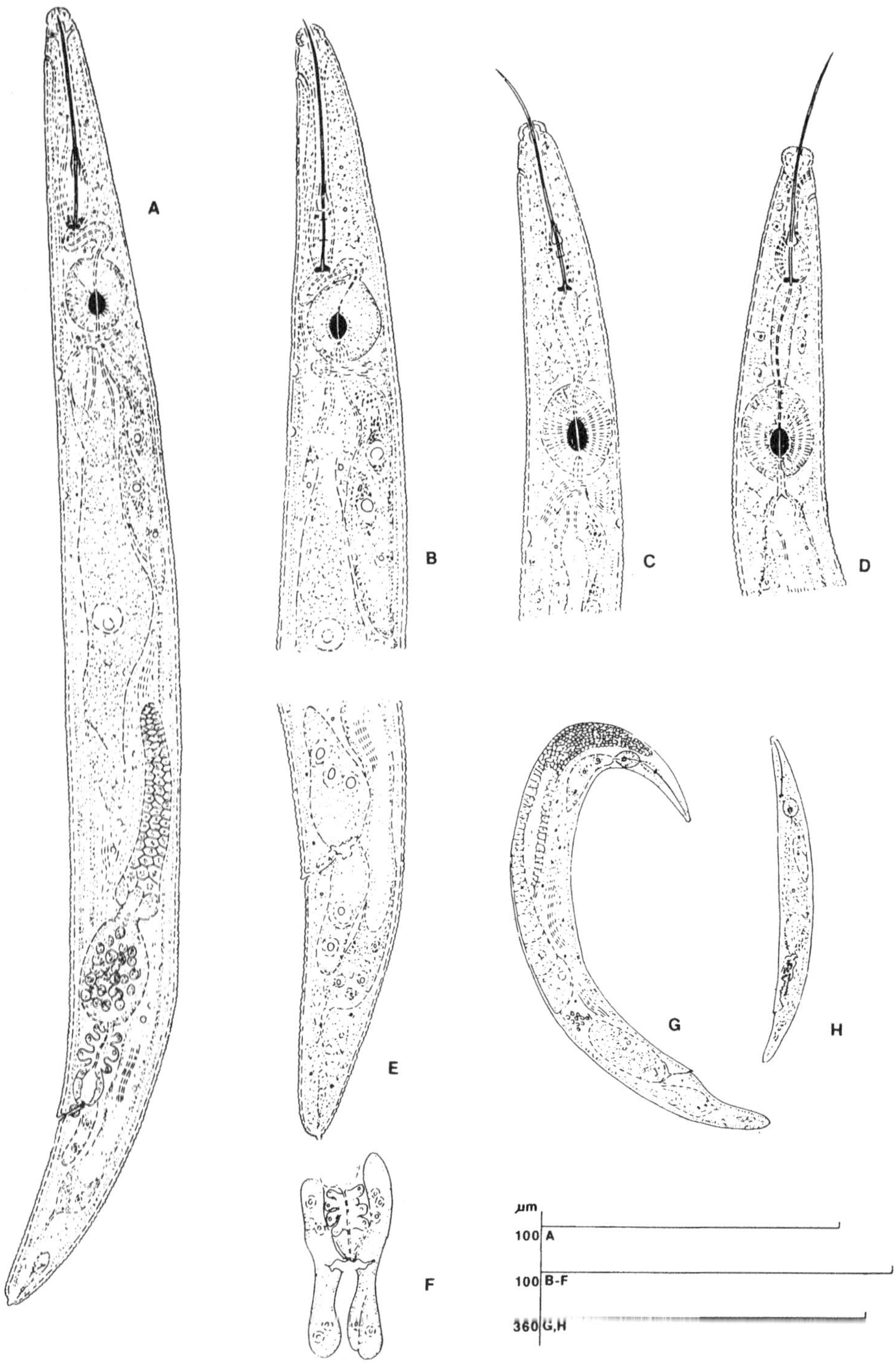
A
B
C
D
E
F
G
H
µm
100 A
100 B-F
360 G,H

Subfamily Parasitaphelenchinae Rühm, 1956

DIAGNOSIS: Parasitaphelenchidae. Long, slender nematodes about 1.4–2.0 mm in size. Vulva very posterior, situated at about 85–92 % of the body length. Female tail short, conoid. Spicules partially fused. Small terminal bursa enveloping the tail spike. Obligate endoparasites of insects.

TYPE GENUS: *Parasitaphelenchus* Fuchs, 1930

OTHER GENERA: Monotypic.

GENUS *PARASITAPHELENCHUS* FUCHS, 1930 (Fig. 4.36)

DIAGNOSIS: Parasitaphelenchinae.
Adults: Found in frass of beetle galleries. **Long and slender nematodes**, typically about 1.4–2.0 mm in size, but exceptionally up to 3.7 mm. **Cephalic region offset** with six rounded lips about as high as wide. **Stylet fairly long** (15–23 μm) **and slender** with small basal knobs or thickenings. Procorpus cylindrical and leading to a well developed, rounded-rectangular median bulb with central valve plates. **Oesophageal gland lobe long and slender. Vulva** with protuberant lips and **very posterior in position (V = 85–92%)**. Genital tract monoprodelphic, outstretched, developing oocytes mostly arranged in several rows. **Post-uterine sac usually long and well developed**. Anus and rectum visible. Tail short conoid. **Male** similar to female in general respects. **Tail short, conoid, often with a digitate terminus**. Spicules small, paired and **fused for most of their length along the mid-line. Rostrum prominent and pointed**; apex well developed and usually rounded, continuing the line of the shaft. Three pairs of papillae, one pair approximately adanal and the other two pairs either postanal or one pair preanal and the other postanal. **A terminal flap of cuticle or 'bursa' surrounds the offset tip of the tail**.

Parasitic juvenile: Found in haemocoel of host. Heat relaxed form ventrally arcuate. Cephalic region more or less continuous with, or slightly expanded from, body contour. **A cuticular hook-like structure may be present on the head together with a pronounced cuticular mucron on the tail. Stylet apparently absent**, although the oesophagus is of normal development. **Body contents obscured by numerous fat globules**. Anus indistinct, probably non-functional. Tail region dorsally convex-conoid or bluntly rounded.

Fig. 4.34. *Vampyronema daptria*. Female. A. Young adult; B–D. Oesophageal regions show the stylet in progressive stages of protraction; E,F. tail region showing the uterine gland complex in lateral aspect and dorsal (F) view; G,H. Comparison of size differences between gravid (G) and young adult (H) specimens. (After Anderson and Laumond, 1992. Courtesy *Fundamental and Applied Nematology*.)

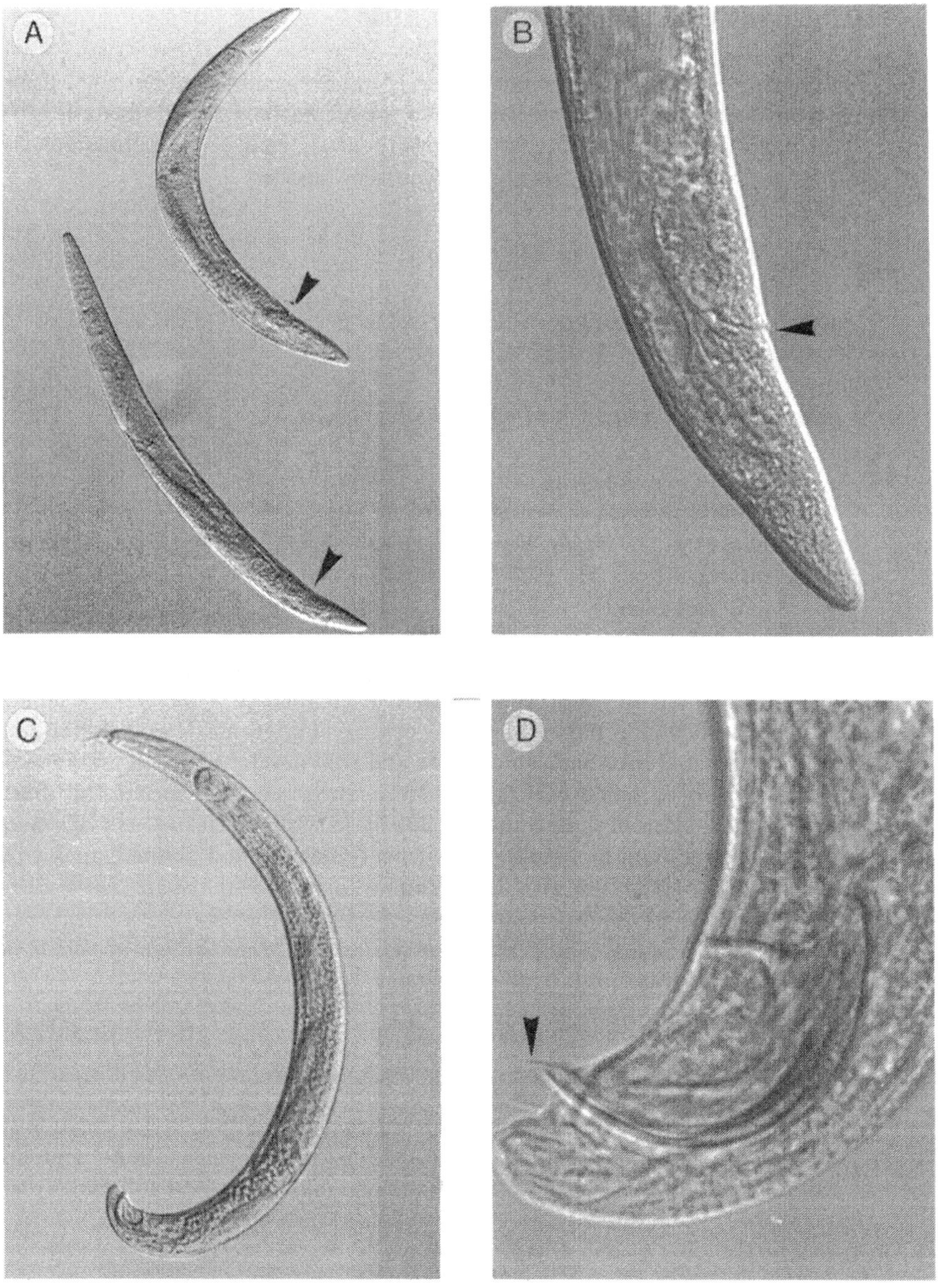

Fig. 4.35. *Vampyronema daptria.* A. Females (vulva is arrowed); B. Female posterior region (vulva is arrowed); C. Male; D. Male posterior region showing spicule and cloacal cylinder (arrowed). (Original, type material.)

Type species:

Parasitaphelenchus uncinatus (Fuchs, 1929) Fuchs, 1930
syn. *Tylenchus uncinatus* Fuchs, 1929
Aphelenchoides uncinatus (Fuchs, 1929) Filipjev, 1934
A. uncinatus ateri (Fuchs, 1937) T. Goodey, 1951
P. (P) uncinatus ateri (Fuchs, 1929) Fuchs, 1937
P. ateri (Fuchs, 1937) Rühm, 1956

Other species:

P. acroposthion (Steiner, 1932) Rühm, 1956
syn. *Aphelenchoides acroposthion* Steiner, 1932
P. acuminati Rühm, 1956
P. beccus Massey, 1974
P. dendroctoni Massey, 1966
P. gallagheri (Massey, 1960) J.B. Goodey, 1960
syn. *Aphelenchoides gallagheri* Massey, 1960
P. heterophallus (Steiner, 1934) Rühm, 1956
syn. *Aphelenchoides heterophallus* Steiner, 1934
P. oldhami Rühm, 1956
P. oregonensis (Steiner, 1942) Rühm, 1956
syn. *Aphelenchoides oregonensis* Steiner, 1942
P. papillatus Fuchs, 1937
syn. *Aphelenchoides papillatus* (Fuchs, 1937) T. Goodey, 1951
P. poligraphi Fuchs, 1937 (Rühm, 1956)
syn. *P. uncinatus poligraphi* Fuchs, 1937
Aphelenchoides uncinatus poligraphi (Fuchs, 1937) T. Goodey, 1951
P. procerus Massey, 1974
P. sexdentati Fuchs, 1937 (Rühm, 1956)
syn. *P. uncinatus sexdentati* Fuchs, 1937
Aphelenchoides uncinatus sexdentati (Fuchs, 1937) T. Goodey, 1951
P. strongylocephalus Rühm, 1956

Species inquirendae:

P. autographi Fuchs, 1930
syn. *Aphelenchoides* (*Schistonchus*) *autographi* (Fuchs, 1930) Filipjev, 1934
Schistonchus autographi (Fuchs, 1930) Filipjev, 1934
P. curvidentis Fuchs, 1930
syn. *Aphelenchoides* (*Schistonchus*) *curvidentis* (Fuchs, 1930) Filipjev, 1934
Schistonchus curvidentis (Fuchs, 1930) Filipjev, 1934
P. ligniperdae Fuchs, 1930
syn. *Aphelenchoides* (*Schistonchus*) *ligniperdae* (Fuchs, 1930) Filipjev, 1934

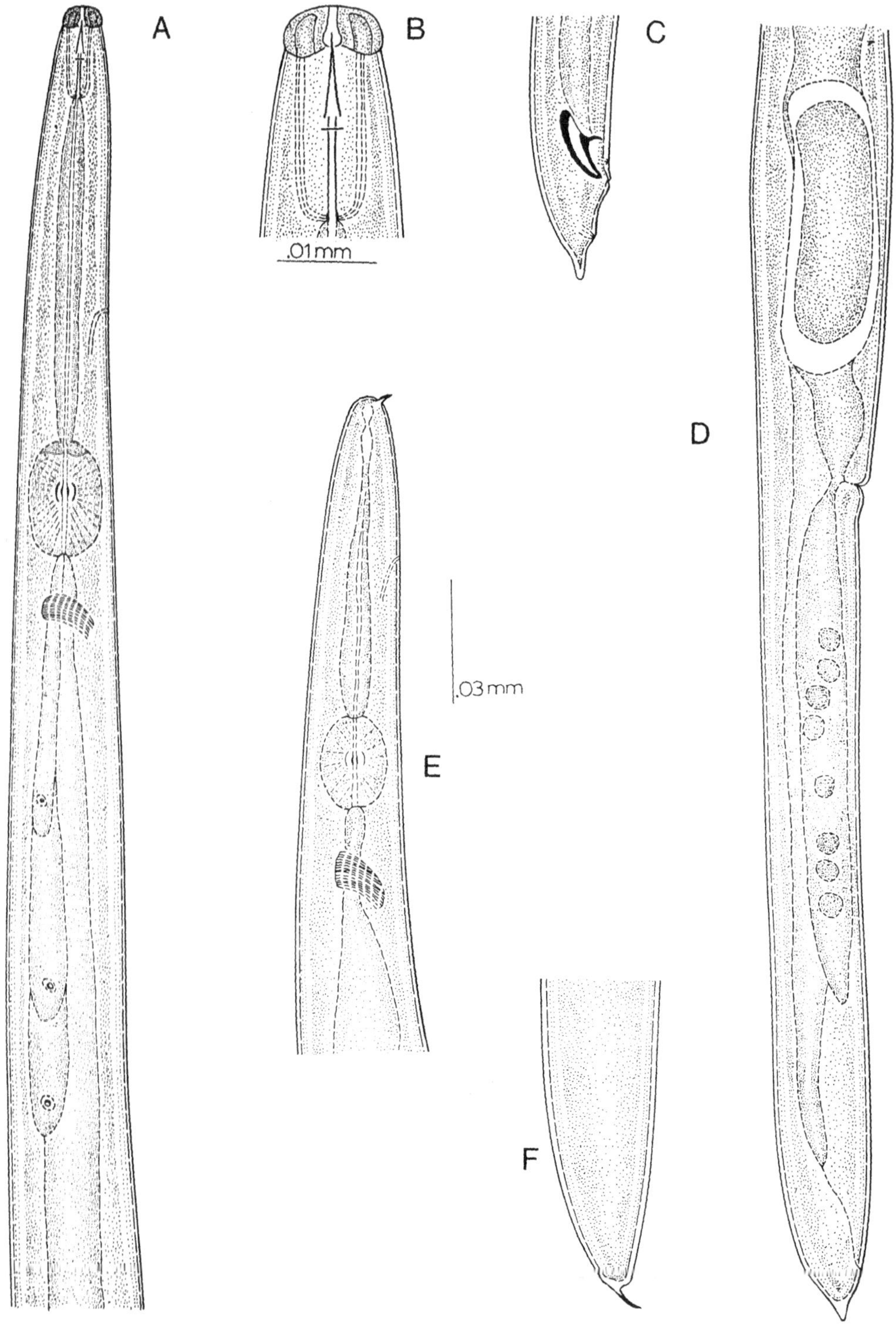

Fig. 4.36. *Parasitaphelenchus beccus.* A. Head and neck; B. Head; C. Male tail; D. Female tail; E. Juvenile head and neck; F. Juvenile tail. (After Massey, 1974. Courtesy USDA.)

Schistonchus ligniperdae (Fuchs, 1930) Filipjev, 1934
P. macrohami Lazarevskaya, 1961
P. pissodis notati Fuchs, 1930
syn. *Aphelenchoides (Schistonchus) pissodis notati* (Fuchs, 1930) Filipjev, 1934
Schistonchus pissodis notati (Fuchs, 1930) Filipjev, 1934
P. pissodis piceae Fuchs, 1930
syn. *Aphelenchoides (Schistonchus) pissodis piceae* (Fuchs, 1930) Filipjev, 1934
Schistonchus pissodis piceae (Fuchs, 1930) Filipjev, 1934
P. rhenanus Fuchs, 1930
syn. *Aphelenchoides (Schistonchus) rhenanus* (Fuchs, 1930) Filipjev, 1934
Schistonchus rhenanus (Fuchs, 1930) Filipjev, 1934

Nomen nudum: *P. bordzomica* Kakuliya, Devdariani and Maglakelidze, 1980

Bionomics: *Parasitaphelenchus* species are associated with various species of bark beetle (Coleoptera: Scolytidae). The adults are found in the frass of the gallery systems and are mycetophagous (Hunt and Hague, 1974). Eggs are laid in the frass and there are reported to be two post-hatch moults to reach the infective third stage (Rühm, 1956; Saunders and Norris, 1961). The third stage juveniles usually possess a sclerotized mouth hook and penetrate the haemocoel of the bark beetle grubs. Here they moult to the fourth stage which absorbs nourishment from the haemolymph and enlarges considerably (Rühm, 1956). Several hundred such juveniles can be found in the haemocoel of a single insect. The juveniles remain in the adult beetle until it has excavated a maternal gallery in a new site. On the death of the beetle they emerge from the mouth and anus before moulting to the adult (Hunt and Hague, 1974).

P. dendroctoni is unusual in that the parasitic fourth stage juveniles and adults are remarkably long. The free-living adults are over 3 mm long (females 2.5–3.7 mm; males 1.7–2.4 mm), and are present in the frass of *Dendroctonus adjunctus* in October and in May. Eggs are laid in the frass and development proceeds to the third stage which invades the host larvae. The parasitic fourth stage reaches a length of 1.5 mm, up to 25 nematodes being found in a single host. When mature, these enter the gut and subsequently moult to the adult in the frass (Massey, 1966).

Reports of the effect of *Parasitaphelenchus* on the host vary. Hunt and Hague (1974) noted no consistently observable host effects with *P. oldhami* infesting *Scolytus scolytus* and *S. multistriatus*, despite the haemocoel being almost solidly packed with juveniles on occasion. Tomalak *et al.* (1984) reported varying reductions in the size of the fat body and structural damage to the genitalia in *Tomicus piniperda* infested by *P. papillatus*. Lieutier (1984,

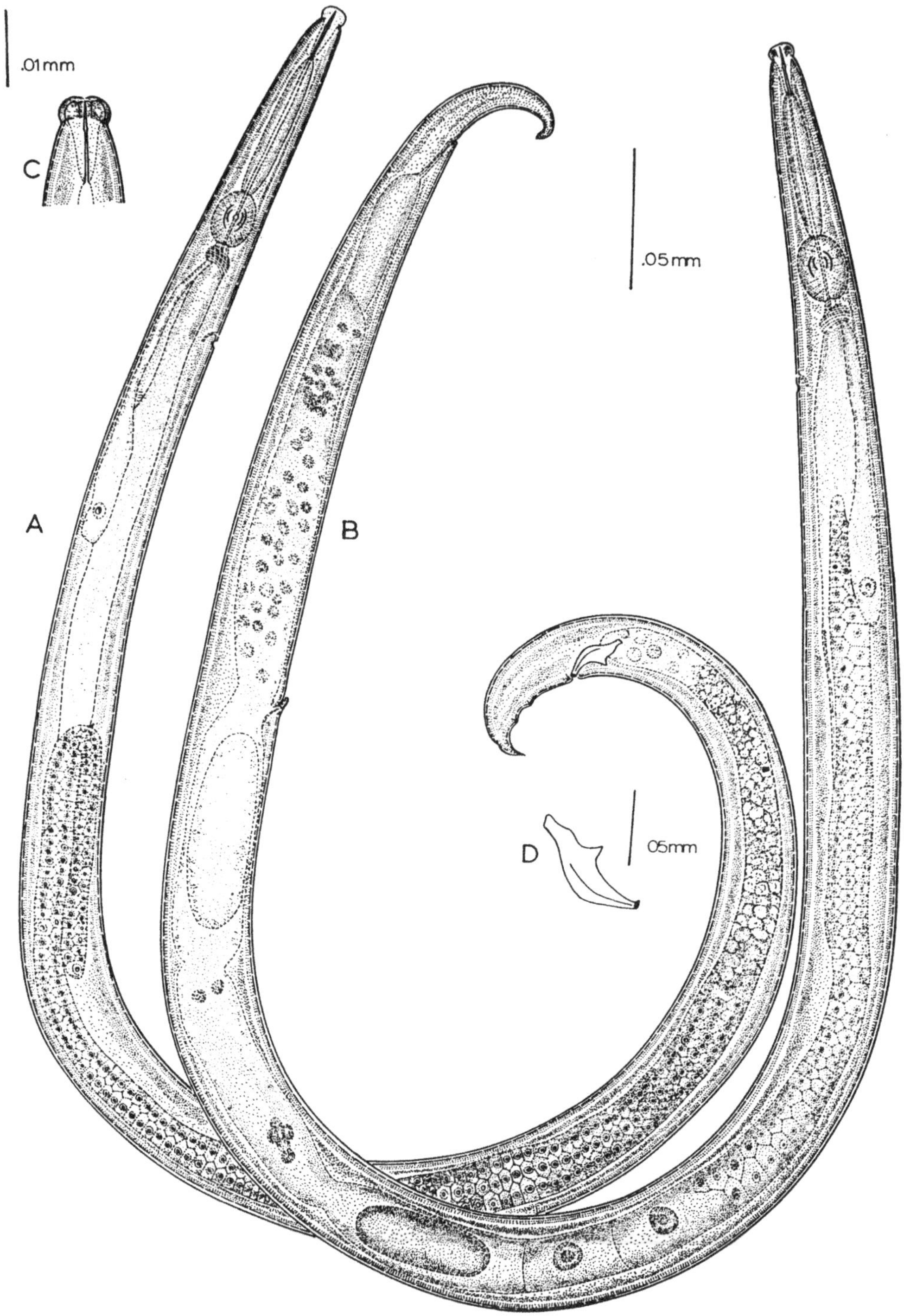

Fig. 4.37. *Bursaphelenchus corneolus.* A. Male; B. Female; C. Head; D. Spicule. (After Massey, 1974. Courtesy USDA.)

1985) and Lieutier *et al.* (1985) reported that *Parasitaphelenchus* infestations in *Ips sexdentatus* caused no difference in the protein spectra of the haemolymph or ovary tissues, nor in haemolymph volume, but that there was a decrease in the haemolymph amino acid concentration and in the fat body proteins. Severe effects on the host are unlikely as the nematodes depend on the beetle retaining the ability to migrate and complete its life cycle in order that they can in turn reproduce.

Distribution: Europe, CIS (formerly USSR), North America.

Etymology: Derived from the Greek *para* = beside or near, *sitos* = food (i.e. 'parasite') and *aphelenchus*.

Subfamily Bursaphelenchinae Paramonov, 1964

Diagnosis: Parasitaphelenchidae. Short to medium sized nematodes, usually fairly stout, but exceptionally very slender. Vulva usually at 70–80% of the body length. Female tail medium to long conoid. Spicules robust with a prominent rostrum. Terminal bursa present on tail tip. Usually ectophoretic, rarely endophoretic, associates of insects.

Type genus: *Bursaphelenchus* Fuchs, 1937

Other genus: *Rhadinaphelenchus* J.B. Goodey, 1960

Key to genera

1. Adult nematodes exceptionally slender (a ratio = >100). Endophoretic associates of palm weevils and phytoparasites of palm trees. . . . . *Rhadinaphelenchus*

– Adult nematodes not exceptionally slender. Usually ectophoretic, rarely endophoretic, associates of insects or phytoparasites of pine trees . . . . *Bursaphelenchus*

GENUS *BURSAPHELENCHUS* FUCHS, 1937 (Figs 4.37, 4.38)

syn. *Aphelenchoides* (*Bursaphelenchus*) Fuchs, 1937 (Rühm,1956)
Huntaphelenchoides Nickle, 1970
Omemeea Massey, 1971
Teragramia Massey, 1974
Ipsaphelenchus Lieutier and Laumond, 1978

Diagnosis: Bursaphelenchinae. Small to long nematodes (about 0.4 mm–1.5 mm), slender. Females tend to die ventrally arcuate on heat relaxation whilst the male tail curls more strongly. Cuticle finely annulated. The number

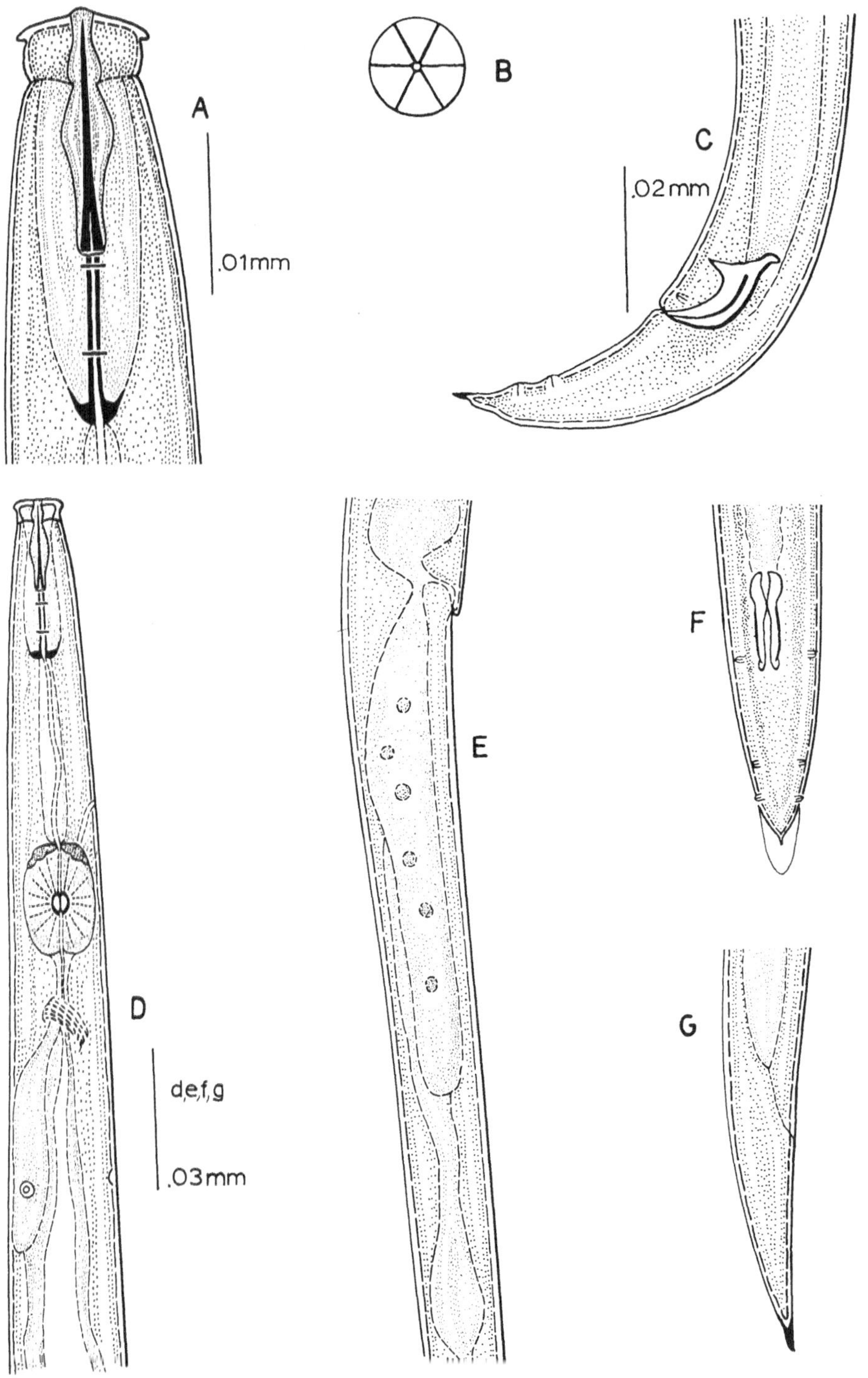

Fig. 4.38. *Bursaphelenchus* [= *Omemeea*] *maxbassiensis*. A. Head; B. Face view; C. Lateral view, male tail; D. Head and neck; E. Female midbody; F. Ventral view, male tail; G. Female tail. (After Massey, 1971. Courtesy *Journal of Nematology*.)

of incisures in the lateral field, when reported, varies from none to four. **Cephalic region relatively high, usually offset by a constriction**, sometimes weak, sometimes strong. **Stylet well developed**, usually 12–20 μm long but may be up to 26 μm. **Conus attenuating to a fine point, the shaft more robust and usually with basal swellings of** lesser or greater development. Procorpus cylindrical leading to a strong median bulb which is usually ovoid or elongate ovoid in shape. Oesophageal glands forming a dorsally overlapping lobe up to about nine body widths long. Nerve ring within one body width of the base of the median bulb. Excretory pore usually in the vicinity of the nerve ring but, exceptionally, may be anterior to the median bulb. Vulva posterior, usually at 70–80% of the body length, but may be more anterior or posterior in exceptional species. Vulval lips slightly protuberant but in a minority of species the anterior lip is directed posteriorly to form a vulval flap. Vagina at right angles to the body axis or directed anteriorly. Genital tract monoprodelphic, outstretched. Developing oocytes in two or even three rows, more rarely reported as a single row. **Post-uterine sac** rarely less than a body width long, more **usually three to six body widths in extent**, although longer still in some species. Some species have a pair of three pronged sclerotized structures at the junction of the uterus and postuterine sac (see Giblin and Kaya, 1984). Tail variable, usually medium conoid, but may be longer and more attenuate. Terminus rounded or pointed. **Male tail strongly curved ventrally, conoid; terminus variously pointed. A small, bursa-like, cuticular structure situated terminally**. Two or three pairs of caudal papillae variously arranged between the preanal, adanal and postanal areas. One or more pairs of small ventrosublateral papillae are reported within the bursal flap in some species. A number of species (e.g. *B. hylobianum* n.comb. and *B. kevini*) have nine papillae: a single ventromedian preanal; a preanal pair and three postanal pairs, the last of which is near the beginning of the bursal flap and may be indistinct. Spicules paired and usually separate, although reported to be partially fused proximally in some species. Recorded lengths, measured along the curved median line, vary from 10 to 30 μm, the majority of species being between 14–20 μm. The shape is distinctive yet quite variable, some being short and stout with the ventral limb well separated from the dorsal and extending further posteriorly, e.g. *B. hunti*, whilst others are much longer and so strongly curved that the transverse bar of the capitulum is almost parallel to the shaft axis, e.g. *B. fraudulentus*, *B. kolymensis*, *B. mucronatus* and *B. xylophilus*. The rostrum is usually prominent and pointed whilst the apex is well developed and often bluntly rounded, although in some species it is pointed, or even hooked as in *B. borealis*, *B. leoni* and *B. silvestris*. The distal end of each spicule in a few species (*B. mucronatus*, *B. xylophilus*) forms a swelling, or disc-like structure, which was named the cucullus by Tarjan and Aragón (1982).

TYPE SPECIES:

Bursaphelenchus piniperdae Fuchs, 1937

syn. *Aphelenchoides piniperdae* (Fuchs, 1937) T. Goodey, 1951

A. (*Bursaphelenchus*) *piniperdae* (Fuchs, 1937) T. Goodey, 1951

Other species:

B. bestiolus Massey, 1974
B. borealis Korenchenko, 1980
B. chitwoodi Rühm, 1956 (J.B. Goodey, 1960)
    syn. *Aphelenchoides (Bursaphelenchus) chitwoodi* Rühm, 1956
        B. pinasteri Baujard, 1980
B. corneolus Massey, 1966
B. crenati Rühm, 1956 (J.B. Goodey, 1960)
    syn. *A. (Bursaphelenchus) crenati* Rühm, 1956
B. digitulus Loof, 1964
B. eggersi Rühm, 1956 (J.B. Goodey, 1960)
    syn. *A. (Bursaphelenchus) eggersi* Rühm, 1956
B. eidmanni Rühm, 1956 (J.B. Goodey, 1960)
    syn. *A. (Bursaphelenchus) eidmanni* Rühm, 1956
B. elytrus Massey, 1971
B. eremus Rühm, 1956 (J.B. Goodey, 1960)
    syn. *A. (Bursaphelenchus) eremus* Rühm, 1956
B. eucarpus Rühm, 1956 (J.B. Goodey, 1960)
    syn. *A. (Bursaphelenchus) eucarpus* Rühm, 1956
B. fraudulentus Rühm, 1956 (J.B. Goodey, 1960)
    syn. *A. (Bursaphelenchus) fraudulentus* Rühm, 1956
B. fungivorus Franklin and Hooper, 1962
    syn. *Huntaphelenchoides fungivorus* (Franklin and Hooper, 1962) Nickle, 1970
B. gonzalezi Loof, 1964
    syn. *Huntaphelenchoides gonzalezi* (Loof, 1964) Nickle, 1970
B. hunanensis Yin, Fang and Tarjan, 1988
B. hunti (Steiner, 1935) Giblin and Kaya, 1983
    syn. *Aphelenchoides hunti* Steiner, 1935
        Huntaphelenchoides hunti (Steiner, 1935) Nickle, 1970
B. hylobianum (Korenchenko, 1980) n.comb.
    syn. *Parasitaphelenchus hylobianum* Korenchenko, 1980
B. idius Rühm, 1956 (J.B. Goodey, 1960)
    syn. *A. (Bursaphelenchus) idius* Rühm, 1956
B. incurvus Rühm, 1956 (J.B. Goodey, 1960)
    syn. *A. (Bursaphelenchus) incurvus* Rühm, 1956
B. kevini Giblin, Swan and Kaya, 1984
B. kolymensis Korenchenko, 1980
B. leoni Baujard, 1980
B. maxbassiensis (Massey, 1971) Baujard, 1989
    syn. *Omemeea maxbassiensis* Massey, 1971
B. mucronatus Mamiya and Enda, 1979
B. newmexicanus Massey, 1974
B. nuesslini Rühm, 1956 (J.B. Goodey, 1960)
    syn. *A. (Bursaphelenchus) nuesslini* Rühm, 1956
B. pityogeni Massey, 1974
B. poligraphi Fuchs, 1937

syn. *A. poligraphi* (Fuchs, 1937) T. Goodey, 1951
A. (*Bursaphelenchus*) *poligraphi* (Fuchs, 1937) T. Goodey, 1951
B. ratzeburgii Rühm, 1956 (J.B. Goodey, 1960)
syn. A. (*Bursaphelenchus*) *ratzeburgii* Rühm, 1956
B. sachsi Rühm, 1956 (J.B. Goodey, 1960)
syn. *A.* (*Bursaphelenchus*) *sachsi* Rühm, 1956
B. scolyti Massey, 1974
B. seani Giblin and Kaya, 1983
B. sexdentati Rühm, 1960
syn. *A.* (*Bursaphelenchus*) *sexdentati* Rühm, 1960
B. bakeri Rühm, 1964 (= junior objective synonym)
B. naujaci Baujard, 1980
B. silvestris (Lieutier and Laumond, 1978) Baujard, 1980
syn. *Ipsaphelenchus silvestris* Lieutier and Laumond, 1978
B. steineri Rühm, 1956 (J.B. Goodey, 1960)
syn. *A.* (*Bursaphelenchus*) *steineri* Rühm, 1956
B. sutoricus Devdariani, 1974
B. sychnus Rühm, 1956 (J.B. Goodey, 1960)
syn. *A.* (*Bursaphelenchus*) *sychnus* Rühm, 1956
B. talonus (Thorne, 1935) Massey, 1956
syn. *A. talonus* Thorne, 1935
A. (*Bursaphelenchus*) *talonus* (Thorne, 1935) Rühm, 1956
B. teratospicularis Kakuliya, 1966
B. tritrunculus Massey, 1974
B. typographi (Kakuliya, 1967) Ebsary, 1991
syn. *Devibursaphelenchus typographi* Kakuliya, 1967
B. varicauda Thong and Webster, 1983
B. wilfordi Massey, 1964
B. willi (Massey, 1974) Baujard, 1989
syn. *Teragramia willi* Massey, 1974
B. xerokarterus Rühm, 1956 (J.B. Goodey, 1960)
syn. *A.* (*Bursaphelenchus*) *xerokarterus* Rühm, 1956
B. xylophilus (Steiner and Buhrer, 1934) Nickle, 1970
syn. *A. xylophilus* Steiner and Buhrer, 1934
Paraphelenchoides xylophilus (Steiner and Buhrer, 1934) Haque, 1967
B. lignicolus Mamiya and Kiyohara, 1972

Species inquirendae:

B. conjunctus (Fuchs, 1930) Andrássy, 1958 nec *Aphelenchoides* (*B.*) *conjunctus* of Rühm, 1956 and *B. conjunctus* of J.B. Goodey, 1960 (= *B. ruehmi* Baker, 1962)
syn. *Parasitaphelenchus conjunctus* Fuchs, 1930
Aphelenchoides (*Schistonchus*) *conjunctus* (Fuchs, 1930) Filipjev,1934
Schistonchus conjunctus (Fuchs, 1930) Filipjev, 1934
B. conurus (Steiner, 1932) J.B. Goodey, 1960
syn. *Aphelenchoides conurus* Steiner, 1932
B. cryphali (Fuchs, 1930) Rühm, 1956

syn. *Parasitaphelenchus cryphali* Fuchs, 1930
Aphelenchoides cryphali (Fuchs, 1930) Fuchs, 1937
A. (Schistonchus) cryphali (Fuchs, 1930) Fuchs, 1937
A. (*Bursaphelenchus*) *cryphali* (Fuchs, 1930) Rühm, 1956
Schistonchus cryphali (Fuchs, 1930) Fuchs, 1937

Species incertae sedis:
B. lignophilus (Körner, 1954) Meyl, 1961
syn. *A. lignophilus* Körner, 1954
Laimaphelenchus lignophilus (Körner, 1954) J.B. Goodey, 1960

Species indeterminatae:
B. ruehmi Baker 1962
syn. *Aphelenchoides* (*B.*) *conjunctus* of Rühm, 1956 .
B. conjunctus of J.B. Goodey. 1960
B. ruehmi J.B. Goodey, 1963 (= junior objective homonym)

Nomina nuda:
B. georgicus Kakuliya, Devdariani and Maglakelidze, 1980
B. populneus Kakuliya, Devdariani and Maglakelidze, 1980
B. tbilisensis Kakuliya, Devdariani and Maglakelidze, 1980
B. wekuae Kurashvili, Kakuliya and Devdariani, 1980
B. welchi Devdariani, 1974

NOTES:

1. J.B. Goodey (1960) reestablished *Bursaphelenchus* at generic level and proposed numerous new combinations. This action was also taken by Meyl in a paper published the following year (Meyl, 1961). However, following the Principle of Coordination (Article 43 of the Code), Rühm (1956), in describing species in the subgenus *Aphelenchoides* (*Bursaphelenchus*), simultaneously made the generic combination with *Bursaphelenchus*. I have, however, placed the name of the author who raised the combination to generic rank in parentheses, thus following the scheme used for the alteration of rank in higher taxa.
2 The combination *B. lignophilus* (Körner, 1954) was proposed by Meyl (1961). The species was previously regarded as belonging to *Laimaphelenchus* by J.B. Goodey (1960) and is best regarded as *species incertae sedis*.
3. *B. hylobianum* (Korenchenko, 1980) n.comb. is proposed for *Parasitaphelenchus hylobianum* Korenchenko, 1980. The good illustration and description, particularly of the male tail, spicule and disposition of the nine caudal papillae, clearly identify the species as belonging to *Bursaphelenchus*, Korenchenko (1980) presumably placing it in *Parasitaphelenchus* because the juveniles were found in the haemocoel, rather than under the elytra, of *Hylobius albosparsus*.

4. Rühm (1960) proposed the combination *Aphelenchoides* (*Bursaphelenchus*) *sexdentati* which he subsequently renamed (1964) as *B. bakeri* because of homonymy with *Aphelenchoides uncinatus sexdentati* (Fuchs, 1937) T. Goodey, 1951, a species originally described in the genus *Parasitaphelenchus*. Under Article 57c of the Code, *A.* (*B.*) *sexdentati* is a junior secondary homonym of *A. uncinatus sexdentati*. However, under Article 59d, the junior secondary homonym should be reinstated if the two species-group taxa are not considered congeneric, as is the case here, *B. bakeri* thus becoming a junior objective synonym.

Remarks: *Bursaphelenchus* is home to a considerable assemblage of species, some of which have been placed in separate genera on such characters as head shape, stylet length, excretory pore position, tail form and spicule shape (keys to species were provided by Tarjan and Aragón (1982) and by Yin *et al.* (1988). In this account I have mostly followed Baujard (1980, 1989) and interpreted *Bursaphelenchus* as having a broad suite of characters, the various other genera being regarded as synonyms. However, I feel that a discussion of the other nominal genera is both appropriate and beneficial.

Huntaphelenchoides was erected by Nickle (1970b) for *B. hunti* and *B. fungivorus* on the basis of the characteristic spicule shape described as being wide with a prominent rostrum and with the tip of the ventral limb well-separate from and terminating posterior to, the dorsal limb. Subsequently *B. gonzalezi* was transferred to this genus and, on spicule shape, other species such as *B. kevini* and *B. seani*, could also belong to this group. However, Giblin and Kaya (1983c), on describing *B. seani* and noting its similarities with *H. hunti*, *H. fungivorus* and *H. gonzalezi*, regarded *Huntaphelenchoides* as a synonym of *Bursaphelenchus*. Nickle and Hooper (1991) stated in their generic diagnosis that the spicules of *Huntaphelenchoides* are fused, but did not discuss the implications further. If this is the case for all species in the group (both *B. kevini* and *B. seani* are reported to have separate spicules, although their shape is very similar to that of *Huntaphelenchoides*), this character could possibly justify separate generic status, although partially fused spicules have been reported for other *Bursaphelenchus* species (Rühm, 1956) and also in the closely related *Rhadinaphelenchus cocophilus* (Giblin-Davis *et al.*, 1989).

Omemeea was erected by Massey (1971) for the type and only species, *O. maxbassiensis*, which was characterized by a sclerotized, umbrella-like lip region, a long stylet (23–26 μm) with well developed basal knobs and the excretory pore anterior to the median bulb. Baujard (1989) pointed out that other *Bursaphelenchus* species had a stylet nearly as long (in fact *B. hunanensis* has a stylet of the same length – 20–26 μm, see Yin *et al.*, 1988), that an anteriorly located excretory pore was not unique to *O. maxbassiensis* and consequently regarded *Omemeea* as a junior synonym of *Bursaphelenchus*.

Massey (1974) proposed *Teragramia* to hold *T. willi* on the basis of the shape and size of the spicules, vulva and tail shape in both sexes. Baujard (1989) pointed out that the supposedly large spicule size (20 μm) was not unique in *Bursaphelenchus* and that tail shape in the genus was variable. He therefore justifiably regarded *Teragramia* as a synonym of *Bursaphelenchus*.

Ipsaphelenchus was proposed by Lieutier and Laumond (1978) for *I. silvestris* which was characterized as having a vulval flap; possessing a hook-shaped spicule apex and lacking a terminal bursa. Baujard (1980), on reexamining the type material, pointed out that a terminal bursa was, in fact, present and synonymized *Ipsaphelenchus* with *Bursaphelenchus*. Incidentally, the characteristic hamatiform spicule apex is also found in *B. leoni* and *B. borealis*, both of which also have a vulval flap, and *Ipsaphelenchus* could, perhaps, be justified on these two characters via an emended diagnosis.

Bionomics: Virtually all the nominal species are associated with bark beetles (Coleoptera: Scolytidae), although associations with other insects, mostly beetles, have been reported. Many species have specialized third stage juveniles, or dauerlarvae, which are usually ectophoretic (rarely endophoretic in the reproductive tract) and use the insects as a transport device. They are commonly found under the elytra and also intersegmentally on the abdomen. The adults and juveniles are mycetophagous.

Two species have been recorded from bees (Giblin, 1982; Giblin and Kaya, 1981, 1983a,b, 1984; Giblin *et al.*, 1981, 1983, 1984b). *B. seani* is a phoretic associate of *Anthophora bomboides* (Hymenoptera: Anthophoridae) found in the brood cells and the reproductive tract of adult bees. It reproduces readily on fungal cultures with a life cycle time of four to five days at 25°C (Giblin and Kaya, 1983a). *B. kevini* is associated with sweat bees (Giblin and Kaya, 1984; Giblin *et al.*, 1984b). The dauerlarvae of *B. kevini* were found in the bursa copulatrix and oviduct of overwintering adult *Halictus farinosus* Smith (Hymenoptera: Halictidae) and may sequester compounds from the host (Giblin *et al.*, 1984b). Adults and juveniles fed on fungal hyphae isolated from the brood cells and had a life cycle duration of eight days at 25°C.

Giblin (1985) reported a *Bursaphelenchus* species associated with nitidulid beetles (Coleoptera: Nitidulidae). The third stage dauerlarvae occurred in adult beetles in the oviduct and ovipositor sac of female *Urophorus humeralis* and in the internal sac of the male aedeagus. Venereal transmission of juveniles from male to female was demonstrated, but not the reverse.

Several *Bursaphelenchus* species are associated with longhorn beetles (Coleoptera: Cerambycidae). One of the latter, *B. xylophilus*, is involved in a serious wilt disease of pines (see below), but also feeds on fungi and has a life cycle duration of five days at 25°C (Mamiya, 1975).

Distribution: North America, Palaearctic region, Hong Kong (introduced *B. xylophilus*).

Etymology: Derived from the Latin *bursa* = a small, skin purse, and *aphelenchus*.

Species description

***Bursaphelenchus xylophilus* (Steiner and Buhrer, 1934) Nickle, 1970** (Fig. 4.39)

syn. *Aphelenchoides xylophilus* Steiner and Buhrer, 1934
Bursaphelenchus lignicolus Mamiya and Kiyohara, 1972

Measurements: (after Nickle *et al.*, 1981)
Female lectotypes (n = 5): L = 0.52 (0.45–0.61) mm; a = 42.6 (37–48); b = 9.6 (8.3–10.5); c = 27.2 (23–31); V = 74.7 (73–78)%; stylet = 12.8 (12.6–13.0) μm.
Male lectotypes (n = 5): L = 0.56 (0.52–0.60) mm; a = 40.8 (35–45); b = 9.4 (8.4–10.5); c = 24.4 (21–29); stylet = 13.3 (12.6–13.8) μm; spicule = 21.2 (18.8–23.0) μm.
(After Mamiya and Kiyohara, 1972) Female (n = 40): L = 0.81 (0.71–1.01) mm; a = 40.0 (33–46); b = 10.3 (9.4–12.8); c = 26.0 (23–32); V = 72.7 (67–78)%; stylet = 15.9 (14–18) μm.
Male (n = 30): L = 0.73 (0.59–0.82) mm; a = 42.3 (36–47); b = 9.4 (7.6–11.3); c = 26.4 (21–31); stylet = 14.9 (14–17) μm; spicules = 27.0 (25–30) μm.

Description: (after Nickle *et al.*, 1981)
With the general characters of the genus.
Female: Cephalic region high, offset, with six lips. Stylet with small basal swellings. Oesophageal gland lobe slender and about three to four body widths long. Excretory pore located at about the level of the oesophago-intestinal junction, occasionally at the same level as the nerve ring. Hemizonid about two-thirds of a body width behind the median bulb. **Vulva posterior, the anterior lip overhanging to form a flap**. Genital tract monoprodelphic, outstretched. Developing oocytes mostly in single file. Post-uterine sac well developed, extending for three quarters or more of the distance to the anus. **Tail subcylindroid with a broadly rounded terminus. Mucron usually absent**, but some populations, especially from Japan, have a short, 1–2 μm, mucron.

Male: Similar to female in general respects. **Spicules large, strongly arcuate so that the prominent transverse bar is almost parallel to the body axis when the spicules are retracted**. The apex is bluntly rounded and the rostrum prominent and **pointed. The distal tip of each spicule is expanded into a disc-like structure** which was named the cucullus by Yin *et al.* (1988). Tail arcuate with a pointed, talon-like terminus bearing a small bursa. **Seven caudal papillae present; one pair adanal, a single preanal ventromedian papilla and two postanal pairs near the tail spike and just anterior to the start of the bursa**.

Note: *Aphelenchoides xylophilus* was first described by Steiner and Buhrer (1934) and then largely forgotten until Nickle (1970b) transferred it to *Bursaphelenchus*. Mamiya and Kiyohara (1972) isolated the causal nematode of pine wilt disease in Japan and named it *B. lignicolus*. Later, Mamiya and Enda (1979) described *B. mucronatus*, a closely related species with a

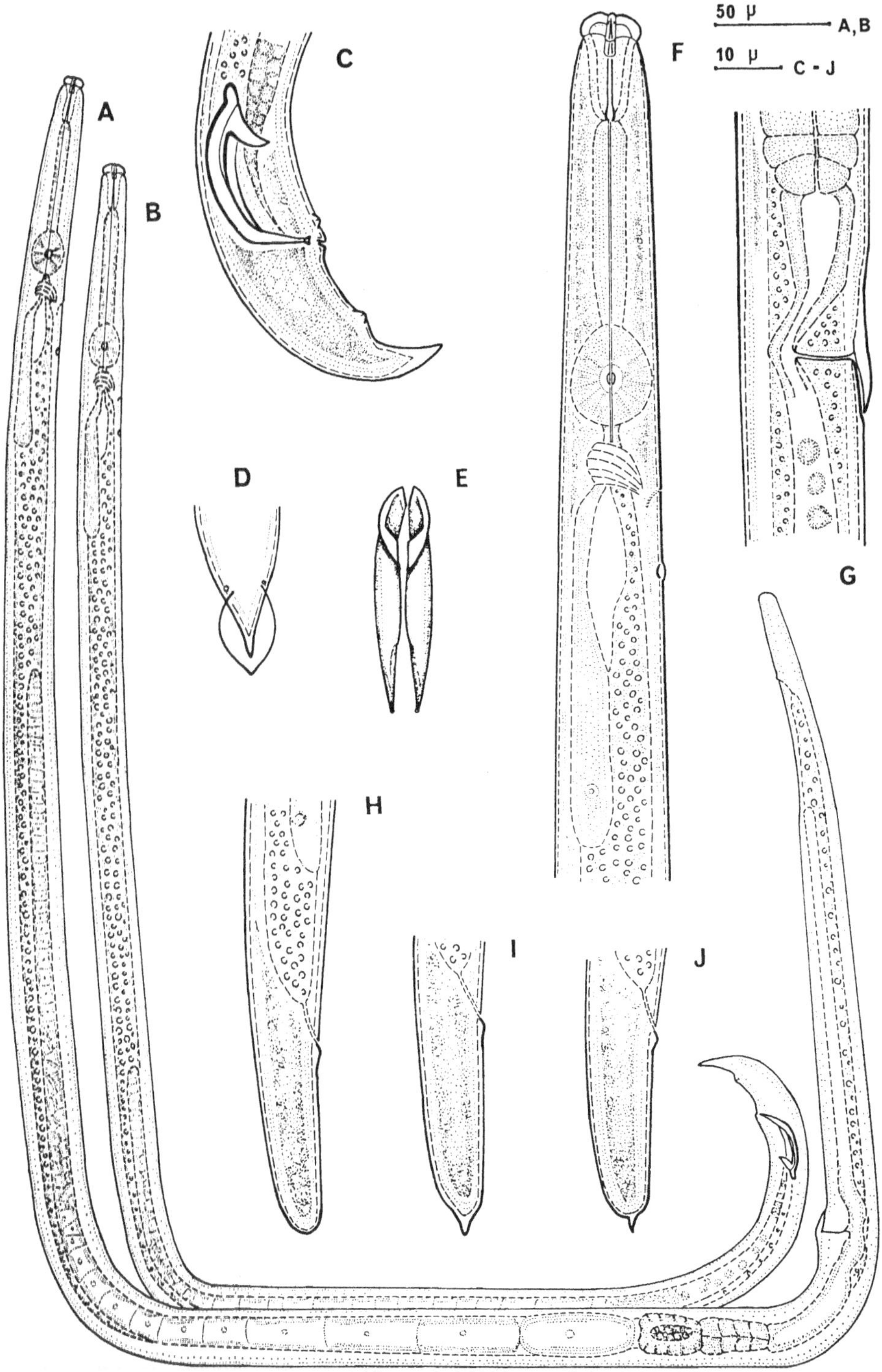

Fig. 4.39. *Bursaphelenchus xylophilus.* A. Female; B. Male; C. Male tail; D. Ventral view of male tail, tip with caudal alae; E. Ventral view of spicules; F. Female, anterior portion; G. Female vulva; H-J. female tail. (After Mamiya and Kiyohara, 1972. Courtesy *Nematologica*.)

strongly mucronate tail. Subsequently, Nickle *et al.* (1981), after examining type material, decided that *B. lignicolus* was synonymous with *B. xylophilus*. Some confusion in differentiating *B. xylophilus* from *B. mucronatus* on morphological grounds alone continued to cause problems and the two nomina were even considered conspecific. However, biochemical and cross-hybridization studies have now confirmed their separate identity (Webster *et al.*, 1990; Abad *et al.*, 1991; Riga *et al.*, 1992).

Bionomics: *Bursaphelenchus xylophilus* is known as the 'pine wilt', or 'pine wood', nematode because of the symptoms induced in susceptible species of the genus *Pinus*. Initially, the tree responds to infection by a reduced oleoresin flow in reaction to wounding. This is presumably due to the production of tyloses within the resin canals in response to nematode invasion. Bolla *et al.* (1989) reported changes in resin acid profiles in infected pines. Unique oleoresins and dehydroabietic acid were produced in reponse to nematode invasion, the latter compound showing nematicidal activity. In healthy pines abietic acid, a compound lacking such activity, was predominant. Subsequent to invasion, transpiration from the needles falls and then ceases altogether and the needles wilt and turn yellow. Initially, such symptoms may be localized near the site of infection, but eventually they spread throughout the crown of the tree (Mamiya, 1983; Malek and Appleby, 1984). Tree death can occur in 30–40 days after infection in warm climates, but takes longer in cooler areas. In the countries of the Far East where *B. xylophilus* occurs, the damage to pine plantations is devastating as the tree species are highly susceptible to this introduced nematode and the high ambient temperatures combined with low rainfall facilitate rapid progress of the disease (Yokobori, 1986). However, in the USA, although the nematode is widespread in native plantations, it usually only causes damage to amenity trees or in plantations of non-indigenous species. Native European species, including *Pinus sylvestris* and *P. pinaster*, have proved very susceptible to *B. xylophilus* and would be greatly at risk should the pine wilt nematode become established in Europe.

B. xylophilus, in common with most of the 'bud and leaf' nematodes of the genus *Aphelenchoides*, can be either mycetophagous or phytoparasitic (Wingfield, 1983). In the mycetophagous life cycle, *B. xylophilus* is vectored to new sites by a number of beetle species. The major vectors are two longhorn beetles (Coleoptera: Cerambycidae); *Monochamus alternatus* in Japan and *M. carolinensis* in the USA. Many other cerambycids and also some buprestids (Coleoptera: Buprestidae) and curculionids (Coleoptera: Curculionidae) are known to carry the nematode (Linit, 1988), but appear to be inefficient vectors. Sato *et al.* (1987) recorded 12.5–95% of *Monochamus saltuarius* carrying *B. xylophilus* with from one to nearly 10 000 nematodes per beetle. The fourth stage juveniles of *B. xylophilus* are carried under the elytra and in the tracheae of the beetles (Mamiya and Enda, 1972) and migrate onto the bark when the beetle oviposits. They then gain access to the tree via the gallery systems of the developing beetle larvae. Here they moult to the adult and feed on fungi, particularly *Ceratocystis* species (known colloquially as blue-stain fungi), which are also transmitted by the beetles during oviposition. Eventually, possibly in response to food shortage, a specialized third stage juvenile dispersal stage is produced which congregates in the vicinity of the pupation chambers formed by the mature beetle larvae. The inner surfaces of the pupal chambers are covered with the fruiting bodies, or perithecia, of the *Ceratocystis* fungi and the nematodes accumulate at their tips. When the beetle emerges, it brushes against the perithecia and collects both fungal spores and nematodes, the latter congregating in the tracheae and beneath the elytra, before flying off in search of a fresh breeding site. This mycetophagous life-cycle is the norm in North America and can continue indefinitely in the absence of a susceptible host tree to initiate the phytoparasitic cycle.

On emerging from a tree, the vector beetles require a maturation feeding period in order to allow their reproductive tracts to fully develop. This normally takes place on young shoots of pine trees. The juveniles of *B. xylophilus* can enter the feeding wounds and gain access to the resin canals (Fig. 4.40) of the shoots where they feed on the epithelial cells. This process only seems to occur in susceptible tree species, although the mechanism for resisting attack in non-susceptible trees is not understood. Linit (1990) found that transmission frequency during the maturation feed increased with the age of the beetle, but that the actual number of nematodes invading the wound was low. Once inside the resin canals, the nematodes cause, either directly or indirectly, a dramatic decrease in oleoresin production, thus allowing the nematodes to migrate throughout the dying tree and reach enormous populations. Beetles are attracted to dying trees and lay eggs, the resulting offspring eventually carrying the *B. xylophilus* to fresh sites. This type of life cycle predominates where the host trees are susceptible, as in Japan and other countries of the Far East, but also occurs in North America, largely on introduced susceptible tree species. Under appropriate conditions, i.e. plantations of susceptible trees, high ambient temperatures and abundant vectors, pine wilt disease can spread extremely rapidly and cause great economic loss. Highly susceptible host species include: *Pinus densiflora, P. echinata, P. lambertiana, P. nigra, P. pinaster, P. radiata, P. sylvestris,* and *P. thunbergii.*

B. xylophilus is amphimictic and can be readily cultured *in vitro* on fungal cultures, particularly *Ceratocystis* spp. Life cycle times are 12 days at 15°C, 6 days at 20°C and 3 days at 30°C, thus facilitating extremely rapid build-up of populations.

Distribution: *B. xylophilus* appears to be a native of North America and is presumed to have been transported in infected timber to Kyushu, Japan around the turn of the century (Nickle *et al.*, 1981; Mamiya, 1983; Malek and Appleby, 1984), where the pine wilt disease was first recorded in 1913. *B. xylophilus* is widely distributed in Japan and is now also recorded from southern China, Hong Kong, Korea and Taiwan. In North America it is found in Canada and the USA, where it is widespread in native coniferous stands. Several authors, e.g. Rutherford and Webster, 1987; Rutherford *et al.*, 1990, have proposed that a mean summer temperature below 20°C is inimical to disease development and that the 20°C isotherm could be a northern limiter

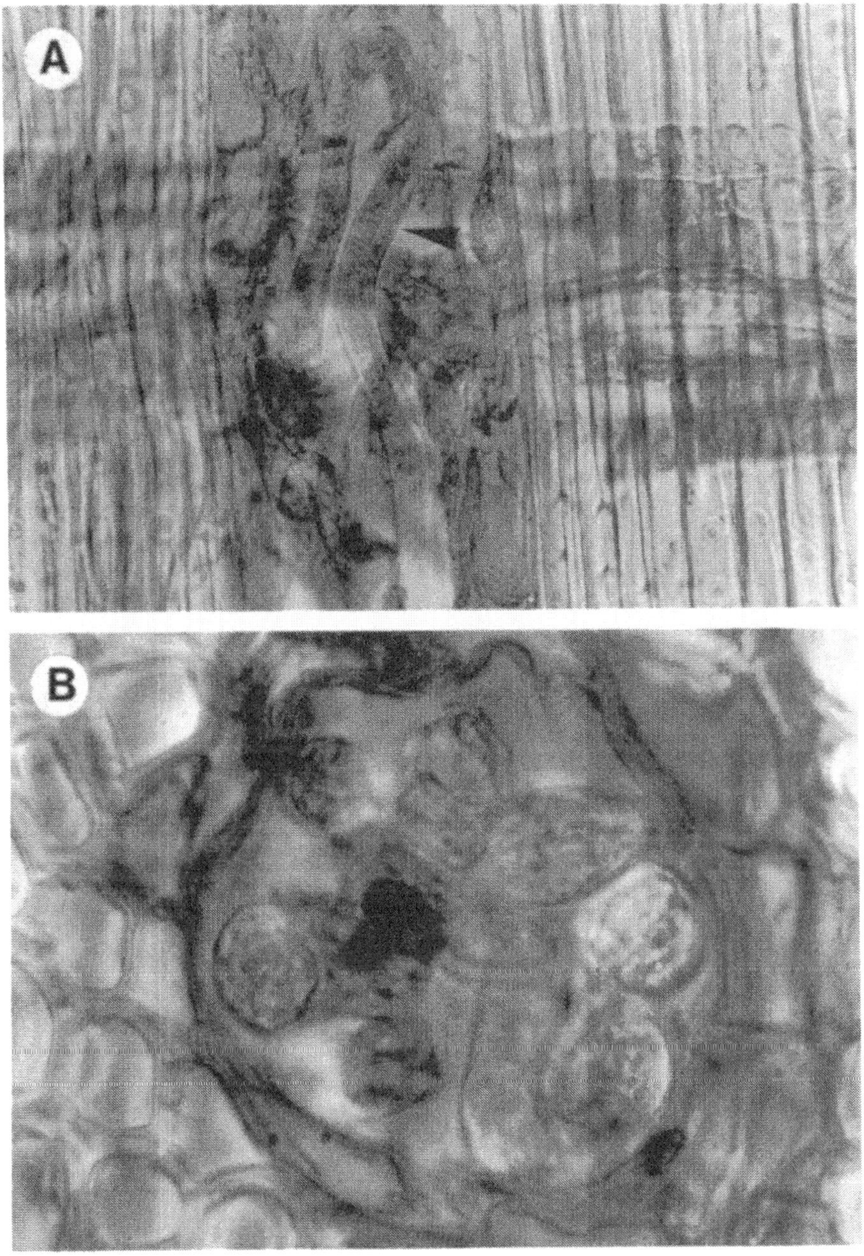

Fig. 4.40. *Bursaphelenchus xylophilus* in resin canals of pine. A. Radial section of stem from a nematode inoculated seedling. Nematode (arrowed) in an axial resin canal; B. Cross section of stem from a naturally infested Japanese red pine tree. Nematodes fill an axial resin canal. (After Mamiya and Kiyohara, 1972. Courtesy *Nematologica*.)

in Europe and North America. However, Magnusson (1986) considered that the disease could still be a threat in Scandinavia despite low summer temperatures.

The nematode is reportedly absent from Europe, the record of Baujard *et al.* (1979) being erroneous, but closely related species of *Bursaphelenchus* do occur. *B. fraudulentus* is recorded from deciduous trees in Germany (Rühm, 1956) and *B. mucronatus* from conifers in Austria, Finland, France, Norway and Sweden. *B. mucronatus* also occurs in Japan (the type locality), North America and CIS (formerly USSR). In addition, a poorly known species, *B. kolymensis*, is recorded from the genus *Larix* in the eastern CIS (Korenchenko, 1980). As previously mentioned, *B. xylophilus*, *B. mucronatus* and *B. fraudulentus* may form a species complex (often referred to as the 'pine wood nematode species complex' or PWNSC) with differing geographical distributions and host preferences. De Guiran and Brugieur (1989) proposed a supraspecies concept to embrace the various populations of the pine wood nematode.

Control: Control is restricted to cultural practices involving the removal and destruction of dead and dying trees from the plantation in association with chemical control of the beetle vectors by aerial application of insecticide (Ikeda, 1984). Individual trees of high amenity value or, as in the case of Japan, of religious significance, can be given a prophylactic treatment. Efficient quarantine procedures are needed to eliminate the risk of the nematode being introduced on infected timber. This includes heat treatment of timber and/or wood chips to 60°C or fumigation with suitable compounds (Kinn, 1986; OEPP/EPPO, 1990). Such precautions are particularly necessary for Europe where suitable host vectors are indigenous and could pick up *B. xylophilus* from infected round wood.

GENUS *RHADINAPHELENCHUS* J.B. GOODEY, 1960 (Fig. 4.41)

Diagnosis: Bursaphelenchinae. **Medium sized to long nematodes (0.8–1.4 mm).** Heat relaxed form more or less straight or slightly ventrally arcuate, the male tail region curling more sharply. **Body very slender (a = 60–96 in the female and 65–179 in the male).** Cuticle finely annulated, the annules being 0.6–1 μm apart. Lateral field marked by four incisures. **Cephalic region high, slightly narrower than the adjoining body, smooth with flattish sides and top. Cephalic framework well developed.** Stylet about 11–13 μm long, slender and with basal knobs. Procorpus elongate, cylindrical, leading to the rounded-rectangular median bulb, the valve plates of which are slightly posterior to the centre. Oesophageal gland lobes obscure, elongate. Nerve ring one to two body widths behind the bulb and just anterior to the excretory pore. Vulva post-median, at about 65–70% of the body length. **The dorsal lip is strongly developed and forms a flap overhanging the strongly sclerotized posterior lip.** Vagina thick walled. Genital tract monoprodelphic, outstretched. Developing oocytes in a single row. **Post-uterine sac very long, reaching for about 75% of the vulva/anus distance and often containing spermatozoa. Tail elongate, narrow, cylindroid to a rounded terminus and about 10–17 anal body widths**

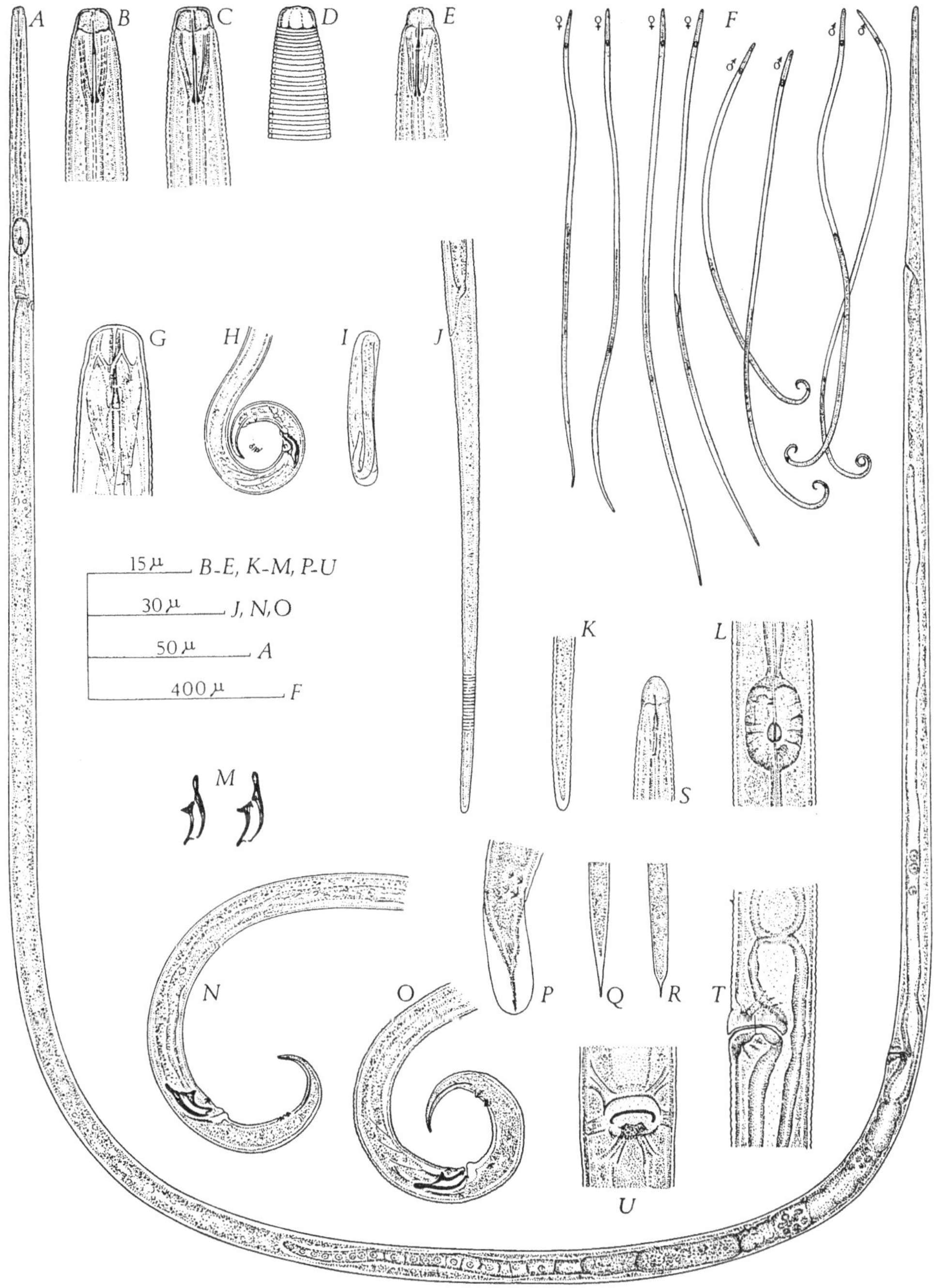

Fig. 4.41. *Rhadinaphelenchus cocophilus.* A,F. Adults; B–D, G. Female heads; E. Male head end; H,N,O. Male tail ends; I. Egg; J. Female tail; K. Female tail tip; L. Female median oesophageal bulb; M. Spicules; P. Bursa in dorsal view; Q,R. Juvenile tail tips; S. Juvenile head end; T,U. Vulva in lateral and ventral view, respectively. (G, H, I after Cobb, 1919; rest after Brathwaite and Siddiqi, 1975.)

long. Male spicules rosethorn-shaped. Distal tip appears notched. Rostrum small, pointed. **Apex well developed and continues the line of the dorsal limb.** Gubernaculum absent, but the dorsal wall of the spicular pouch is thickened. **Bursa terminal and extending for 40–50% of the tail length.** Seven caudal papillae present according to a recent SEM study (Giblin-Davis *et al.*, 1989). **Tail strongly curled ventrally, initially cylindrical and then tapering to become conoid with a pointed tip. Juveniles have a high, domed head.** Tails of second and third stage juveniles are conoid with a mucronate tip. **Fourth stage juveniles show dimorphic tails**: male juveniles having a sharply pointed tip whereas female juveniles have a rounded tip.

Type species:
Rhadinaphelenchus cocophilus (Cobb, 1919) J.B. Goodey, 1960
syn. *Aphelenchus cocophilus* Cobb, 1919
A. (*Chitinoaphelenchus*) *cocophilus* (Cobb, 1919) Micoletzky, 1922
Aphelenchoides cocophilus (Cobb, 1919) T. Goodey, 1933
Bursaphelenchus cocophilus (Cobb, 1919) Baujard, 1989
Chitinoaphelenchus cocophilus (Cobb, 1919) Micoletzky, 1922

Other species: Monotypic.

Remarks: The genus *Rhadinaphelenchus* shares many characteristics with *Bursaphelenchus* and has long invited synonymy. Baujard (1989) pointed out these morphological and biological similarities and proposed the combination *B. cocophilus*. His arguments are persuasive, e.g. *B. wilfordi* Massey, 1974 is slender with an 'a' ratio of 60; some species have a very long postuterine sac, e.g. *B. naujaci* Baujard, 1980 and *B. xylophilus* (Steiner and Buhrer, 1934) Nickle, 1970; a long cylindroid tail is found in *B. steineri*; a number of species possess a vulval flap. Giblin-Davis *et al.* (1989), in a paper studying the morphology of *R. cocophilus* using the light microscope and the SEM, also comment on the numerous similarities between *Rhadinaphelenchus* and *Bursaphelenchus* and conclude that the primary morphological distinguishing feature between the genera is the exceptionally slender form of *Rhadinaphelenchus*. However, this could be an adaptation to a parasitic habit in coconut tissue. Giblin-Davis *et al.* (1989) also suggested that the spicules could be proximally fused and that this, if confirmed by TEM studies, would be unique in the aphelenchoids. Although the bulk of the evidence points to *Rhadinaphelenchus* and *Bursaphelenchus* being congeneric, I have retained the former as valid because of a combination of the economic importance of such a well-known genus and the extensive, associated, literature; the exceptionally slender form of both sexes; the host range restricted to palms and the fact that *Bursaphelenchus* itself is a rather uneasy amalgam of diverse species which could yet fragment into a number of subgenera or genera, thus leading to the possible reinstatement of a prematurely synonymized *Rhadinaphelenchus*.

Bionomics: An obligate phytoparasite of palms, particularly the coconut palm, *Cocos nucifera* and vectored by the palm weevil, *Rhynchophorus palmarum*.

Distribution: Regions of Central America and the West Indies.

Etymology: Derived from the Greek *rhadinos* = slender or slim, and *aphelenchus*.

Species description

Rhadinaphelenchus cocophilus **(Cobb, 1919) J.B. Goodey, 1960** (Fig. 4.1)

syn. *Aphelenchus cocophilus* Cobb, 1919
A. (*Chitinoaphelenchus*) *cocophilus* (Cobb, 1919) Micoletzky, 1922
Aphelenchoides cocophilus (Cobb, 1919) T. Goodey, 1933
Bursaphelenchus cocophilus (Cobb, 1919) Baujard, 1989
Chitinoaphelenchus cocophilus (Cobb, 1919) Micoletzky, 1922.

Measurements:
(After Goodey, 1960) Female: L = 0.97–1.18 (1.05) mm ; a = 78–96 (87); b = 8.7; c = 9.5–13.2 (11.6); V = $^{25-41}64–68^{18-21}$ ($^{36}66^{20}$).
Male: L = 0.84–1.16 (1.02) mm; a = 100–179 (120); b = 6.5; c = 24–35 (28); T = 50–68 (59).
(After Cobb, 1919 in Goodey, 1960) Female: L = 1 mm; a = 71; b = 7.7; c = 12.5; V = $^{28}68^{13}$.
Male: L = 1 mm; a = 100; b = 7.7; c = 20; T = 45.
(After Lordello and Zamith, 1954) Female: L = 0.83–1.11 mm; a = 60–96; b = 11–14; c = 9.4–10.7; V = 66–69; stylet = 15.3 μm.
Male: L = 0.82–1.42 mm; a = 92–143; b = 11–19; c = 22–47; stylet = 10.7–13.8 μm.
(After Thorne, 1961) Female: L = 1.2 mm; a = 69; b = 8.4; c = 12; V = $^{35}65^{20}$.
Male: L = 1.0 mm; a = 65; b = 7.8; c = 19; T = 42.

Note: Morphometric comparisons of geographic and host isolates of *R. cocophilus* were given by Gerber *et al.* (1989). They concluded that the differences could be attibuted to intraspecific variability.

Description: (modified from Brathwaite and Siddiqi, 1975)
Female: **Body about 1 mm long and very slender** (a = 60–96), arcuate to nearly straight when heat relaxed. Cuticle thin, marked with transverse striae 0.6–1 μm apart; lateral fields with four incisures and a faint median line suggesting a fifth incisure, occupying 0.25 of body-width; deirids and phasmids absent. **Cephalic region smooth, high, anteriorly flattened with rather straight sides and slightly narrower than the body. Cephalic framework prominent, sclero-**

tized. Stylet 11–13 μm long, attenuated, knobbed at base although the knobs may be obscure, especially on immature specimens; conus less than half the stylet length and sharply pointed; protractor muscles of stylet prominent, attached to basal plate of labial framework. Procorpus elongate-cylindrical; median bulb oval, usually about twice as long as wide, with prominent **valve-plates just posterior to centre**; dorsal oesophageal gland orifice midway between anterior margin of bulb and valve plates. Oesophageal glands overlapping intestine dorsally, usually obscure. Nerve ring a wide band surrounding isthmus within a bulb length of the posterior margin of the bulb. Excretory pore a little behind nerve ring and anterior to hemizonid which is about three annules long. Intestine with small granules and indistinct lumen. **Vulva slit-like, appearing as an open C in ventral view, slightly overhung by a wide, thick dorsal lip; posterior lip** is also thick and **heavily sclerotized**. Vagina thick-walled, slightly curved as it leads inwards to a distance of about half the body-width. Genital tract monoprodelphic, outstretched; oocytes in a row. **Postvulval uterine sac elongate**, extending to about 75% of vulva-anus distance, often with a few large, spheroid sperms. Rectum about 1.5 anal body widths long; anus distinct. **Tail elongate-subcylindrical with a rounded, unstriated terminus, 10–17 anal body-widths long**.

Male: Body ventrally arcuate, more strongly curved in tail region. Cephalic region, stylet and oesophagus as in female. Testis single, anteriorly outstretched, with spermatogonia in a row. **Spicules** paired, **small**; dorsal limb 9–11 μm long with an elongated rounded apex and ending distally before the ventral limb whose distal end appears to recurve to join the dorsal limb so that the entire **spicule appears notched distally**; the ventral element has a distinct rostrum proximally and appears to be connected to the dorsal limb through a transverse bar with a central hole for passage of nerves. According to Giblin-Davis *et al.* (1989) the spicules are fused proximally, but separate distally. There is no true gubernaculum, but the dorsal wall of the spicule pouch is thickened to form an apophysis. **Tail strongly curved ventrally (may form a complete circle, or more), subcylindrical in anterior half, then conoid to a pointed terminus. Bursa terminal**, prominent in dorsal or ventral view (not easily detectable in lateral view as it does not project beyond tail contour) with finely striated margins, **enveloping approximately the distal half of the tail**. There are two pairs of distinct ventrosubmedian papillae near base of bursa and a preanal pair, about 0.5 of the spicule length anterior to cloaca; a fourth pair of ventrolateral papillae behind the cloaca has also been reported (Thorne, 1961; Nickle, 1970b), but these are very obscure and often not detectable. Giblin-Davis *et al.* (1989), using the SEM, reported the presence of seven papillae: a single preanal ventromedian; one pair subventral preanal or adanal, and two pairs postanal.

Juveniles: The juveniles have high, dome-shaped cephalic regions which are not offset from the body. Tails of second and third stage juveniles have conoid or sharply mucronate tips, and those of fourth stage juveniles have dimorphic tips: in female juveniles they are rounded as in the female, and in males sharply drawn out.

Bionomics: *Rhadinaphelenchus cocophilus*, the causal agent of red ring disease of coconut (Fig. 4.42), was first described as *Aphelenchus cocophilus* by Cobb (1919) from root and trunk material sent to him by Mr William Nowell from Grenada, West Indies. However, according to Nowell (1923), red ring disease was first reported by Hart from the Cedros Peninsula of Trinidad in 1905. The disease is now known to be widespread in the Caribbean and Latin American regions and can cause extensive losses, an average 10–15% reduction in yield due to palm death being reported by Esser and Meredith (1987).

R. cocophilus is spread naturally by an insect vector, namely the palm weevil, *Rhynchophorus palmarum* L. (Coleoptera: Curculionidae). The association of the weevil and the spread of red ring was first suggested by Ashby (1921). Hagley (1962) showed that 95% of red ring infected palms were also colonized by palm weevils and that nearly 40% of adult weevils carried living *R. cocophilus* externally. Subsequently, Hagley (1963) correlated the incidence of red ring with the seasonal abundance of *R. palmarum*. Gerber and Giblin-Davis (1990) found more than 90% of freshly emerged weevils carrying *R. cocophilus* internally with over 47% being infested by in excess of 1000 nematodes each. Transmission of the nematode via the root system has been shown experimentally both from soil inoculum and, recently, via root contact (Warwick and Bezerra, 1992), but this is unlikely to be of much practical significance compared to the insect vector.

R. cocophilus is found as the third stage juvenile in large numbers in the haemocoel, gut and ovipositor of *Rhynchophorus palmarum*. When an infested weevil oviposits in the tissue of a leaf axil in the palm crown, the nematodes are also deposited, a localized inoculum of at least 5000 being necessary to initiate an infection according to Griffith (1968). The nematodes migrate intercellularly through all the parenchymatous tissue of the palm, but in the trunk they appear to be restricted to a narrow band of discoloured tissue, the 'red ring', which is about 3–5 cm beneath the trunk surface. This ring is reddish or orange-red in hue and approximately 2–4 cm wide. Nematodes are concentrated within the ring, but are also numerous in tissue 1 cm outside, and 2.5 cm inside, the ring periphery. Most of the nematodes within the ring tissue are juvenile, the adults being most abundant at the margins, particularly the internal periphery, of the ring. Dean and Velis (1976) and Dean (1979) report that in palms of the Panama Tall type the ring is often more diffuse and not as red as in Jamaica Tall and that in extreme cases the whole central area may discolour. Although initially intercellular, lysigenous cavities are formed as the disease progresses and these contain numerous nematodes. The life cycle is rapid, typically lasting 9–10 days and Esser (1969) reported that 10 g of tissue could contain 50 000 nematodes.

The third stage juvenile nematodes enter the palm weevil grubs which are burrowing within the palm tissue via the spiracles and mouth and then migrate to the haemocoel. They remain within the insect as it completes its metamorphosis and, in the adult female weevil, they concentrate in the ovipositor. Some researchers have claimed that those beetles carrying a sufficient load of nematodes to initiate red ring are considerably smaller in size than their brethren,

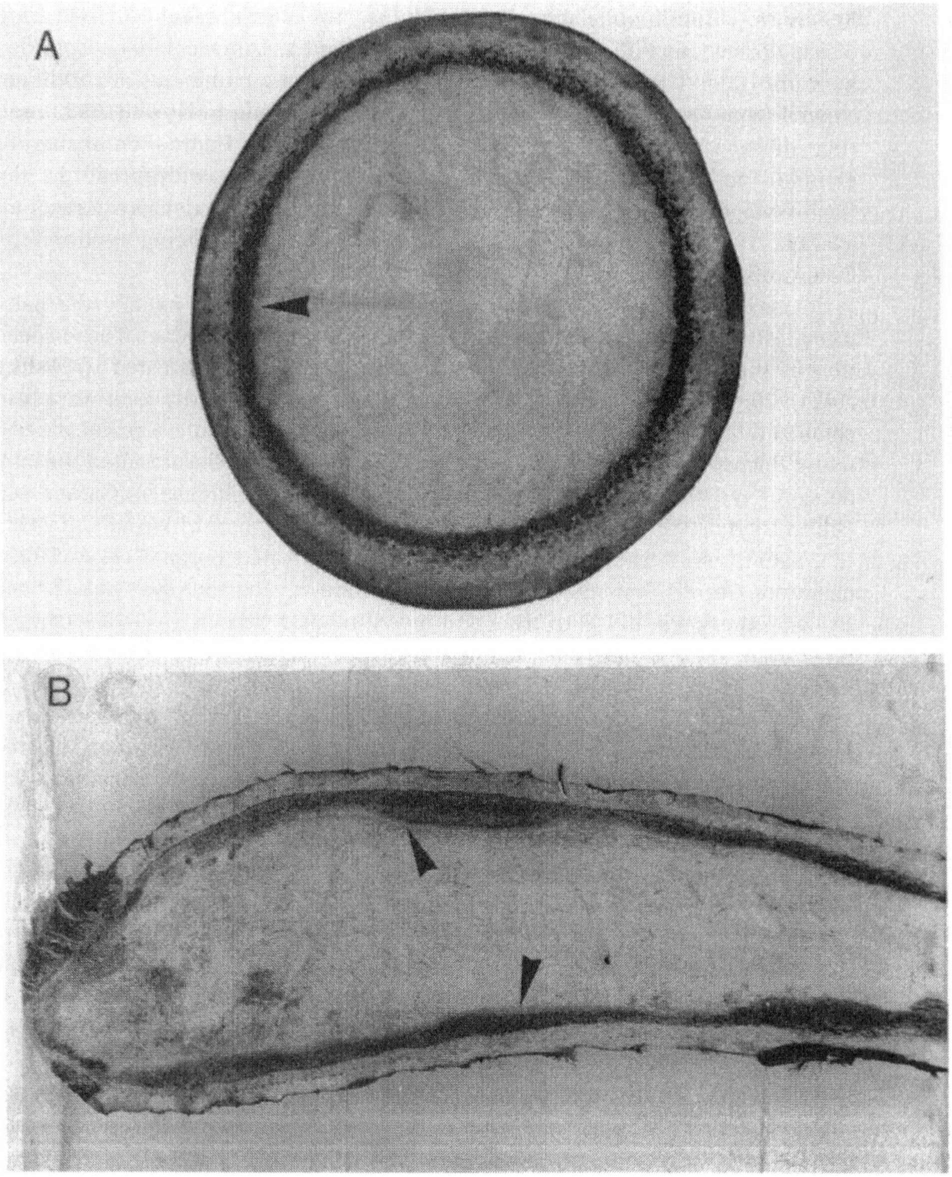

Fig. 4.42. Red ring disease. A. Transverse section through coconut palm stem; B. Longitudinal section through coconut palm stem. Both show the discoloured tissue (arrowed). (After Brathwaite and Siddiqi, 1975. Photographs courtesy R. Griffith.)

but other reports disagree. It has also been reported that the weevils show a degree of resistance to the nematode, being able to destroy small infestations of juveniles.

In addition to the internal ring of reddish orange or brownish tissue in the stem, the disease is characterized by a number of external symptoms. Firstly, the leaves begin to yellow and die back, often breaking near the petiole base and hanging limply down the trunk. Subsequently, the crown often snaps off, although this is due mainly to the burrowing activities of the weevil larvae. Shedding of flowers and premature dropping of nuts has also been reported. Infested trees show a marked reduction in water uptake compared to healthy palms due to numerous tyloses occluding the xylem vessels and die 3–4 months after infection. As the nematodes do not enter the xylem vessels, such a reaction is presumably a defence response by the host, perhaps initiated by toxins, and is similar in some repects to the effect of *B. xylophilus* on pines.

In addition to the main host, *Cocos nucifera*, red ring disease is known to infest over a dozen other species of Arecaceae including the oil palm, *Elaeis guineensis* Jacq., on which it can be devastating (Oostenbrink, 1963) and the date palm, *Phoenix dactylifera* L. Recently, red ring disease has increased in economic importance as a result of lethal yellows disease in Central America. In an attempt to control lethal yellows, resistant dwarf palms have been planted in large numbers and these, being in the susceptible age range of 3–10 years, often succumb to red ring.

For more information on red ring, the following reviews should be consulted: Brathwaite and Siddiqi, 1975; Dean, 1979; Griffith and Koshy, 1990.

Distribution: The disease, restricted to the Caribbean and Latin America, is currently reported from Belize, Brazil, Colombia, Costa Rica, Dominican Republic, Ecuador, El Salvador, French Guyana, Grenada, Guatemala, Guyana, Honduras, Mexico, Nicaragua, Panama, Peru, St Vincent, Surinam, Trinidad and Tobago and Venezuela. A report of *Rhadinaphelenchus* from China (Yin and Feng, 1981) is probably spurious.

Control:. Control of red ring is two pronged and involves destroying the weevil vector, either by spraying insecticide into the palm crowns or by employing baiting techniques with insecticide treated coconut tissue, supplemented by efficient hygiene techniques whereby infected palms are destroyed before the weevil vectors can complete their life cycle, migrate and thus disseminate the nematodes. Palms are most susceptible to red ring between 3 10 years old and so particular care during this period is beneficial. Details on control techniques are given by: Brathwaite and Siddiqi, 1975; Dean, 1979; Morin *et al.*, 1986; Griffith, 1987; Bala, 1988 and Griffith and Koshy, 1990.

Family Entaphelenchidae Nickle, 1970

Diagnosis: Aphelenchoidoidea. Endoparasites in the haemocoel of insects (Coleoptera). There are three adult forms: male; immature female; mature parasitic female. Mature parasitic female oviparous or ovoviviparous. Vulva posterior, usually about 80–90%, but more anterior in *Roveaphelenchus*. Male spicules robust, often with a well developed, pointed rostrum. Bursa absent. Stylet 8–22 μm long, with or without knobs.

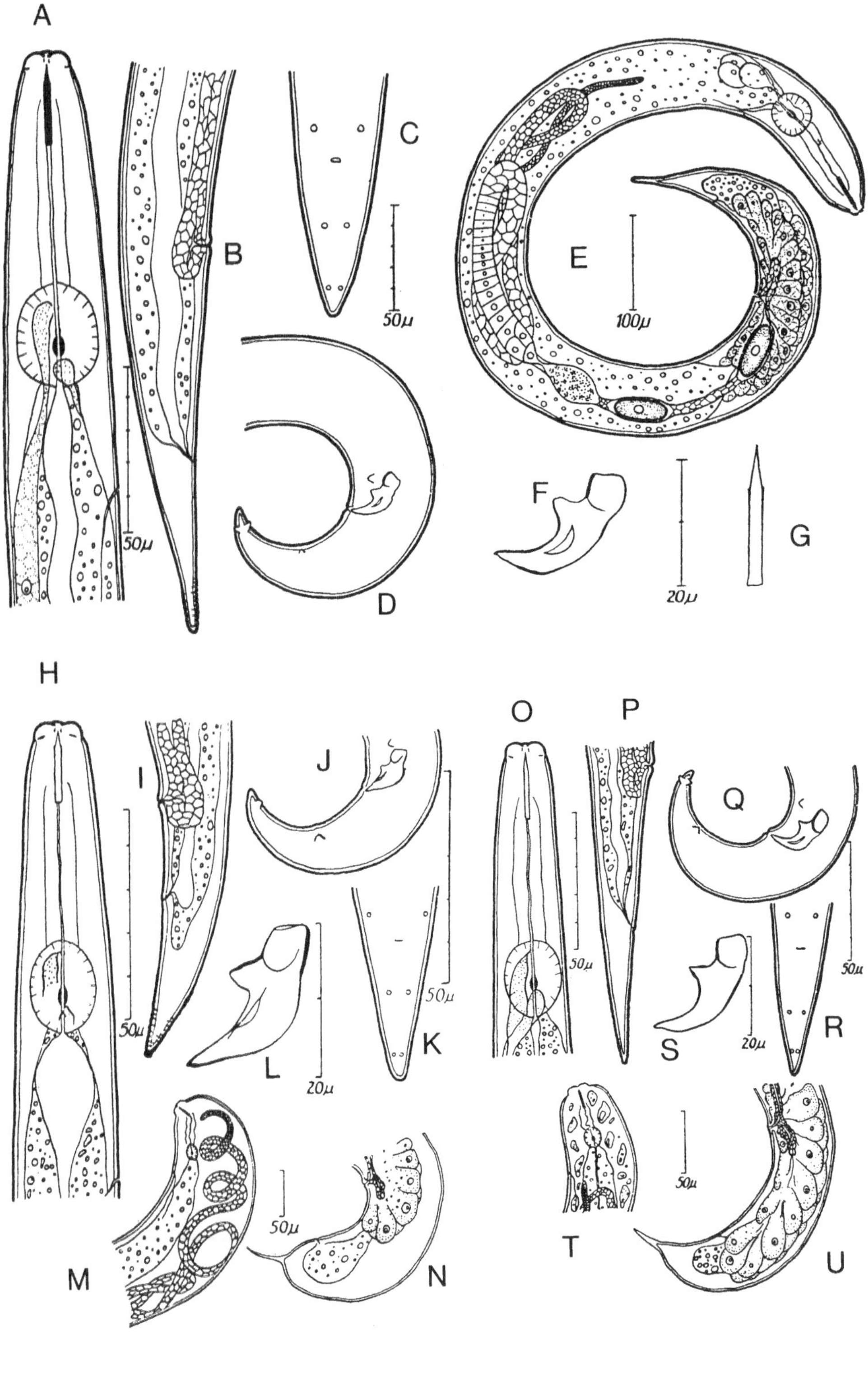
A
B
C
50µ
D
E
100µ
F
G
20µ
50µ
H
I
J
K
L
50µ
20µ
M
N
50µ
O
P
Q
R
S
50µ
20µ
50µ
T
U
50µ

Type genus: *Entaphelenchus* Wachek, 1955

Other genera:
Peraphelenchus Wachek, 1955
Praecocilenchus Poinar, 1969
Roveaphelenchus Nickle, 1970

Type subfamily: Entaphelenchinae Nickle, 1970 (Siddiqi, 1980)

Key to genera

1. Only mature parasitic female found in insect haemocoel. Parasites of staphylinid beetles . . . . *Entaphelenchus*

– Three adult forms found within insect haemocoel. Parasites of silphid, curculionid and staphylinid beetles 2

2. Mature parasitic female oviparous, male corkscrew-shaped on death, postuterine sac long . . . . *Peraphelenchus*

– Mature parasitic female ovoviviparous, postuterine sac absent in mature parasitic female 3

3. Vulva in mature parasitic female very posterior, about 96–97% . . . . *Roveaphelenchus*

– Vulva more anterior about 80–84% . . . . *Praecocilenchus*

GENUS *ENTAPHELENCHUS* WACHEK, 1955 (Fig. 4.43)

Diagnosis: Entaphelenchinae. **Three adult forms.**
Immature free-living female: Body length less than 1 mm. Cuticle smooth but with indications of annules at either end. Lips well developed, slightly offset. **Stylet 17–22 μm long**, fairly robust and lacking basal knobs. Oesophagus consisting of a short, cylindrical procorpus leading to an almost spherical median bulb with the valve plates post-median in position. Oesophageal gland lobe well developed. Excretory pore about a body width behind the bulb. **Vulva posterior at about 85%** or more of the body length. **Genital tract** monoprodelphic, out-

Fig. 4.43. *Entaphelenchus* spp. *E. oxyteli.* A. Female, oesophagus; B. Female, posterior end; C. Male, posterior end, ventral view; D. Male, posterior end; E. Parasitic female, entire; F. Spicule; G. Stylet. *E. philonthi.* H. Female, oesophagus; I. Female, posterior end; J. Male, posterior end; K. Male, posterior end, ventral view; L. Spicule; M. Parasitic female, anterior end; N. Parasitic female, posterior end. *E. bledii.* O. Female, oesophagus; P. Female, posterior end; Q. Male, posterior end; R. Male, posterior end, ventral view; S. Spicule; T. Parasitic female, anterior end; U. Parasitic female, posterior end. (After Wachek, 1955. Courtesy *Parasitologische Schriftenreihe.*)

stretched, **immature. Post-uterine sac very short**, less than a body width long. Tail conoid, dorsally convex, leading to a finely rounded terminus.

Free-living male: Similar to the immature female in general respects. **Spicules paired with a large, pointed rostrum and a rounded, well developed apex** continuing the line of the shaft. Usually three pairs of caudal papillae, one pair pre-anal and the others postanal. No bursa. Tail conoid, strongly curved ventrad on heat relaxation.

Mature parasitic female: **Body 1.3–2.5 mm long, sausage-shaped. Oesophageal glands short and swollen. Genital tract fully developed, convoluted**, developing oocytes mostly in multiple rows. **Perivaginal glands greatly developed** and almost filling the body cavity in the vaginal region and for a distance of one to two body widths anteriorly and posteriorly. Oviparous. As the body swells the tail tends to become more hemispheroid with a spicate tip.

Type species: *Entaphelenchus oxyteli* Wachek, 1955

Other species:
E. aliantae Wachek, 1955
E. bledii Wachek, 1955
E. philonthi Wachek, 1955
E. platysethi Wachek, 1955
E. xantholini Wachek, 1955

Bionomics: All nominal species are obligate parasites of various staphylinid beetles (Coleoptera: Staphylinidae) from the genera *Alianta*, *Bledius*, *Oxytelus*, *Philonthus*, *Platystethus* and *Xantholinus*, in which the mature female and juveniles are found in the haemocoel. The juveniles exit the host at the third stage and moult to the free-living males and females. After mating, the inseminated female penetrates the pupal stage of the host, increases greatly in size and becomes sexually mature. Most infested insects contained one or two parasitic females (although as many as four were recorded) and up to 6 000 juveniles. Of 229 insects examined, 13% were infested by *Entaphelenchus*, the fat body of such individuals being reduced (Wachek, 1955).

Giblin and Kaya (1983b) reported an undescribed species of Entaphelenchidae from the haemocoel of *Platysoma punctigerum* (Coleoptera: Histeridae). Phelps and Mokry (1976) reported the presence of immature entaphelenchid nematodes in *Culicoides babrius* and *C. ravus* (Diptera: Ceratopogonidae) from Zimbabwe. However, this report needs confirmation and may be spurious.

Distribution: Germany.

Etymology: Derived from the Greek *entoma* = insect and *aphelenchus*.

GENUS *PERAPHELENCHUS* WACHEK, 1955 (Fig. 4.44)

DIAGNOSIS: Entaphelenchinae. **Three adult forms.**
Immature female: Medium sized nematodes about 0.6–0.7 mm long. **Dying C-shaped** on heat relaxation. Cuticle smooth. Cephalic region slightly offset, low and rounded. **Stylet** 17–18 μm long, **with conus bent ventrad anteriorly and with wide lumen**. No basal knobs. Procorpus cylindrical, leading to a well developed, **rounded trapezoid-shaped median bulb** with central valve plates. Oesophageal gland lobe well developed and extending for about five body widths. Nerve ring behind median bulb; excretory pore about one body width behind the bulb. **Vulva posterior** at about 80% of the body length, lips not protuberant. **Genital tract** monoprodelphic, outstretched; **immature with few developing oocytes, but containing numerous sperm in the uterus**. Post-uterine sac about a body width or more long. **Tail short, conoid, dorsally concave to a spicate terminus**. Anus and rectum visible with the intestine extending into the tail.

Male: similar to the immature female in general respects, 0.8–0.9 mm long. **Corkscrew-shaped** on heat relaxation. The spicules are variously reported as single (Wachek, 1955), paired but fused, or paired, but possibly fused at their tips (Nickle, 1970b). **The rostrum is low and rounded and the apex prominent.** Two pairs of postanal caudal papillae. Bursa absent. Tail conoid and sharply curved ventrad.

Mature female: Body about 0.75–0.8 mm long and almost twice the diameter of the immature female. Ovary short. Oviparous.

TYPE SPECIES: *Peraphelenchus necrophori* Wachek, 1955

OTHER SPECIES: Monotypic.

BIONOMICS: The type species is an obligate parasite of *Necrophorus investigator* Zettst., *N. vespillo* L., and *N. vespilloides* Hrbst. (Coleoptera: Silphidae). The mature females lay eggs in the hosts gonad which then develop to adult males and immature females. After mating, the corkscrew-shaped males die and the inseminated immature females leave the host via the rectum and enter a new host larva. An average of two or three mature parasitic females were found in each host and these produced 40–60 descendants. Of 423 insects examined, 17% were infested by *Peraphelenchus necrophori*, but no damage to the host was noticed (Wachek, 1955).

DISTRIBUTION: Germany.

ETYMOLOGY: Derived from the Greek *peiro* = to bore through, and *aphelenchus*.

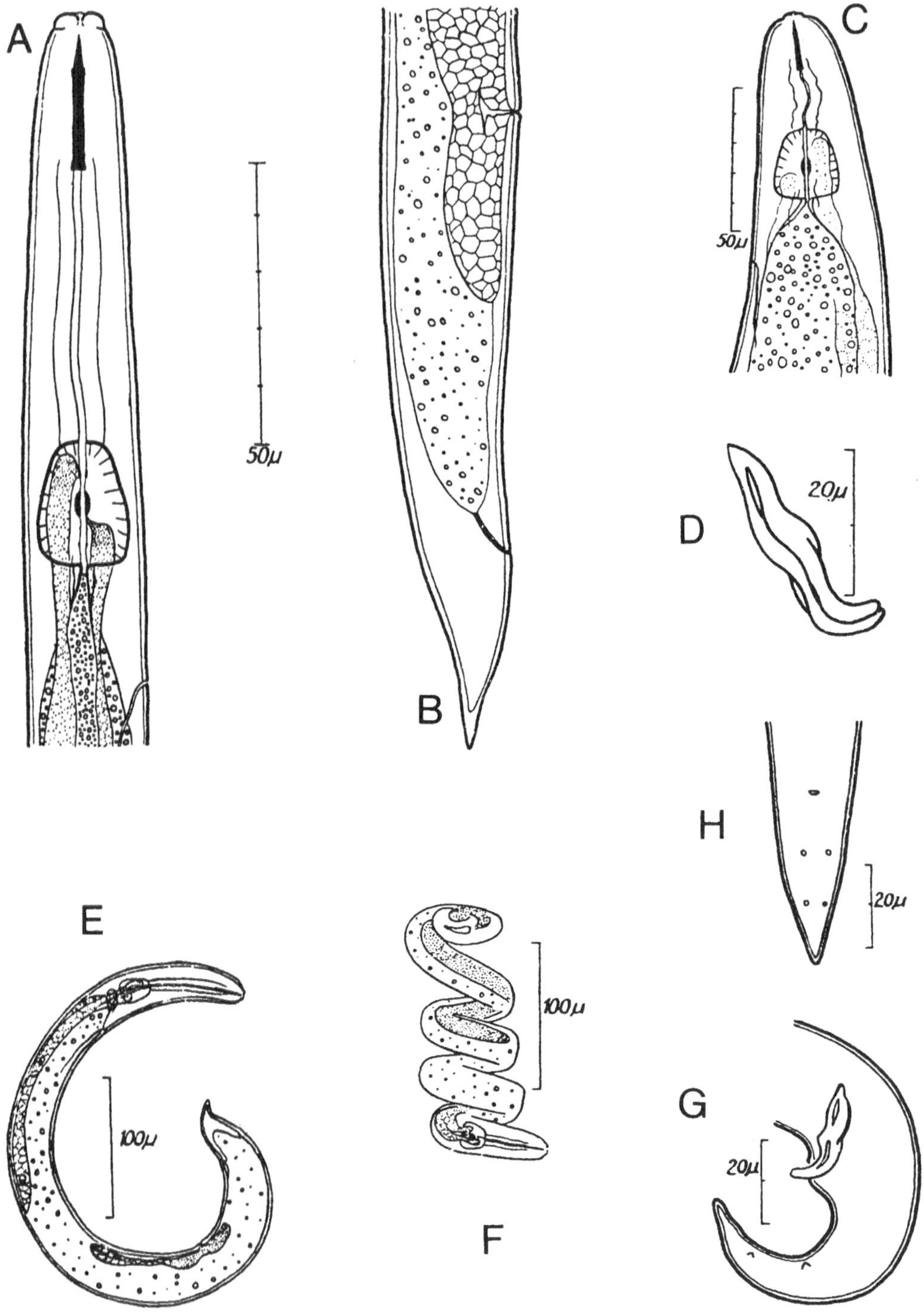

Fig. 4.44. *Peraphelenchus necrophori*. A. Female, oesophagus; B. Female, posterior end; C. Parasitic female, anterior end; D. Spicule; E. Female, entire; F. Male, entire; G. Male, posterior end; H. Male, posterior end, ventral view. (After Wachek, 1955. Courtesy *Parasitologische Schriftenreihe*.)

GENUS *PRAECOCILENCHUS* POINAR, 1969 (Fig. 4.45)

DIAGNOSIS: Entaphelenchinae. **Three adult forms.**
Mature parasitic female: Fairly long to long nematodes (0.95–2.26 mm), ventrally curved on heat relaxation. Cuticle faintly striated. **Cephalic region continuous with body contour, rounded and with the lips fused**. Stylet faint, about 9 μm long in the type species and lacking basal knobs. Procorpus short, cylindrical, leading to a well developed median bulb with central valve plates. Oesophageal glands well developed and apparently separate. Excretory pore not seen. Intestine containing needle-like crystals in the type species. Anus not visible. **Vulva well posterior (V = 79–84%) with strongly protuberant lips** in more mature specimens. Genital tract monoprodelphic, outstretched. **Post-uterine sac absent. Females ovoviviparous with the juveniles developing to sexual maturity within the uterus**. Tail bluntly rounded.

Intrauterine/free-living female: **Very small nematodes** (0.26–0.33 mm). Cephalic region rounded and continuous with the body contour. Stylet weakly developed, about 9 μm long, and lacking basal knobs. Procorpus cylindrical, leading to a well developed, almost spherical median bulb. **Oesophageal gland lobe well developed and extending posteriorly almost to the tip of the ovary** (i.e. about five body widths). Excretory pore just over a body width behind the bulb. Vulva well posterior with non-protuberant lips. Genital tract monoprodelphic, outstretched, with the developing oocytes in two rows. Tail short, conoid, terminating in a point.

Intrauterine male: Small nematodes (0.4–0.56 mm). Cuticle faintly striated. **Cephalic region rounded, slightly expanded**. Stylet less developed than in the intrauterine female but of about the same length. Oesophageal structure similar to that of the intrauterine female, but weaker. Spicules paired, rosethorn-shaped with a pronounced, rounded, apex and a pointed rostrum. **Tail short, conoid, with a pointed tip**.

TYPE SPECIES: *Praecocilenchus rhaphidophorus* Poinar, 1969

OTHER SPECIES: *P. ferruginophorus* Rao and Reddy, 1980

BIONOMICS: The type species was reported from the haemocoel of *Rhynchophorus bilineatus* (Montrouzier), a palm weevil attacking coconut in Indonesia and Papua New Guinea (Poinar, 1969). The nematode was found in both sexes of the adult beetle, but not in the larvae. An individual beetle could contain hundreds of female worms of both the intrauterine and mature parasitic form. Males were less common in the haemocoel, but occurred in the uterus of the mature parasitic female. Intrauterine males and females appeared to mate *in utero* and also, probably, in the haemocoel. Nematodes were found packed into the ovipositor of female weevils, although free-living females were also found in the faecal material. Males were rarely found outside the beetles. Poinar (1969) reported that the ovaries of parasitized

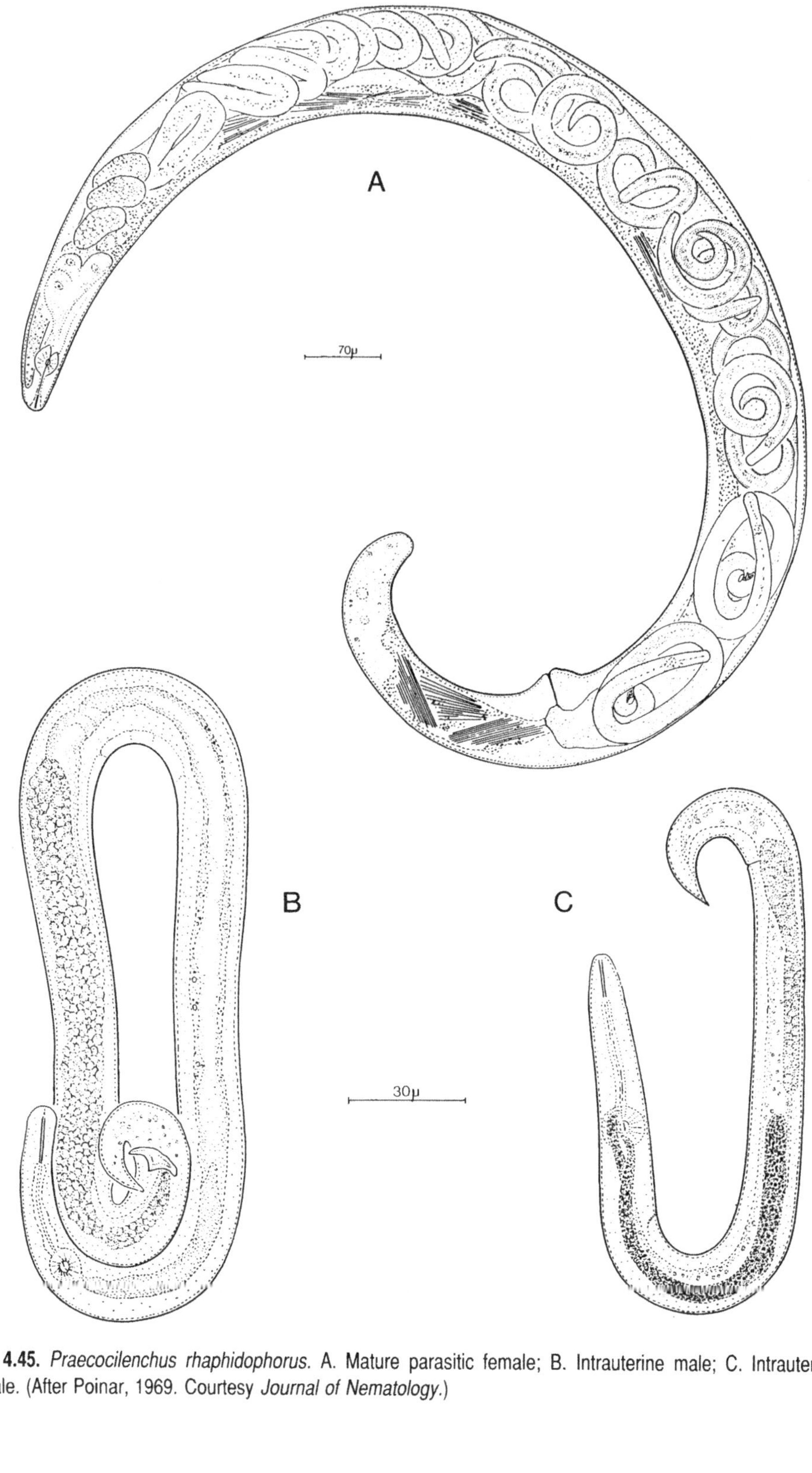

Fig. 4.45. *Praecocilenchus rhaphidophorus.* A. Mature parasitic female; B. Intrauterine male; C. Intrauterine female. (After Poinar, 1969. Courtesy *Journal of Nematology.*)

weevils were smaller, but accurate determination of the effect on the host remains to be done. Bedford (1974) reported that 15% of *R. bilineatus* surveyed in New Britain were parasitized by *P. rhaphidophorus* and that parasitized females produced fewer mature eggs than unparasitized weevils. Poinar (1974) reported that weevils could display immunity to *P. rhaphidophorus* by means of melanotic encapsulation. *P. ferruginophorus* was described from *Rhynchophorus ferrugineus* by Rao and Reddy (1980) and may be synonymous with *P. rhaphidophorus* as it only differs in lacking needle-like crystals in the mature females and, more importantly, by the in tandem arrangement of the oesophageal glands.

DISTRIBUTION: Keravat, New Britain, the type locality, and Kerala, India.

ETYMOLOGY: Derived from the Latin *praecocis* = premature and *aphelenchus*.

GENUS *ROVEAPHELENCHUS* NICKLE, 1970 (Fig. 4.46)

DIAGNOSIS: Entaphelenchinae. **Three adult forms found in the body cavity of the host.**

Female: Medium sized, vermiform nematodes (0.5–0.58 mm long). Heat relaxed form arcuate or S-shaped. Cuticle finely annulated. Cephalic region low, flattened and more or less continuous with the body contour. Cephalic framework weakly sclerotized. Stylet 11.7–13.4 μm long, delicate. No knobs or basal swellings. Procorpus slender, cylindrical; leading to a weakly developed, ovoid median bulb. Oesophageal glands forming a dorsal lobe about three body widths long. **Excretory pore about two to three body widths behind the median bulb with the excretory pore about a body width anterior. Excretory canal prominent. Vulva posterior (68–73%) with non-protuberant lips. Genital tract** monoprodelphic **with a degenerate ovary. Uterus contains a single juvenile or embryo.** No postuterine sac. Anus not visible. **Tail** (i.e. post-vulval region) **long cylindroid, terminating in a truncate tip with four prominent mucrons.**

Obese parasitic female: **Much larger than the vermiform female** (1.54–2.2 mm long), **swollen** and milky-white in colour. Stylet short (11.3–14.7 μm long). Median bulb well developed but oesophageal glands degenerate. **Vulva very posterior (V = 96–97%). Genital tract** monoprodelphic, **filling most of body cavity and with the ovary reflexed.** Developing oocytes in multiple rows. **Uterus contains eggs and pre-adult stage juveniles. Tail** short, conoid, with the **tip often dorsally hooked.**

Male: Small nematodes, about 0.34–0.36 mm long, vermiform. **Heat relaxed form with the tail coiled around on itself.** Cephalic region continuous, cephalic framework very weakly sclerotized. Stylet 10–11.8 μm long, weakly developed and lacking knobs or basal swellings. Procorpus slender, cylindrical. Median

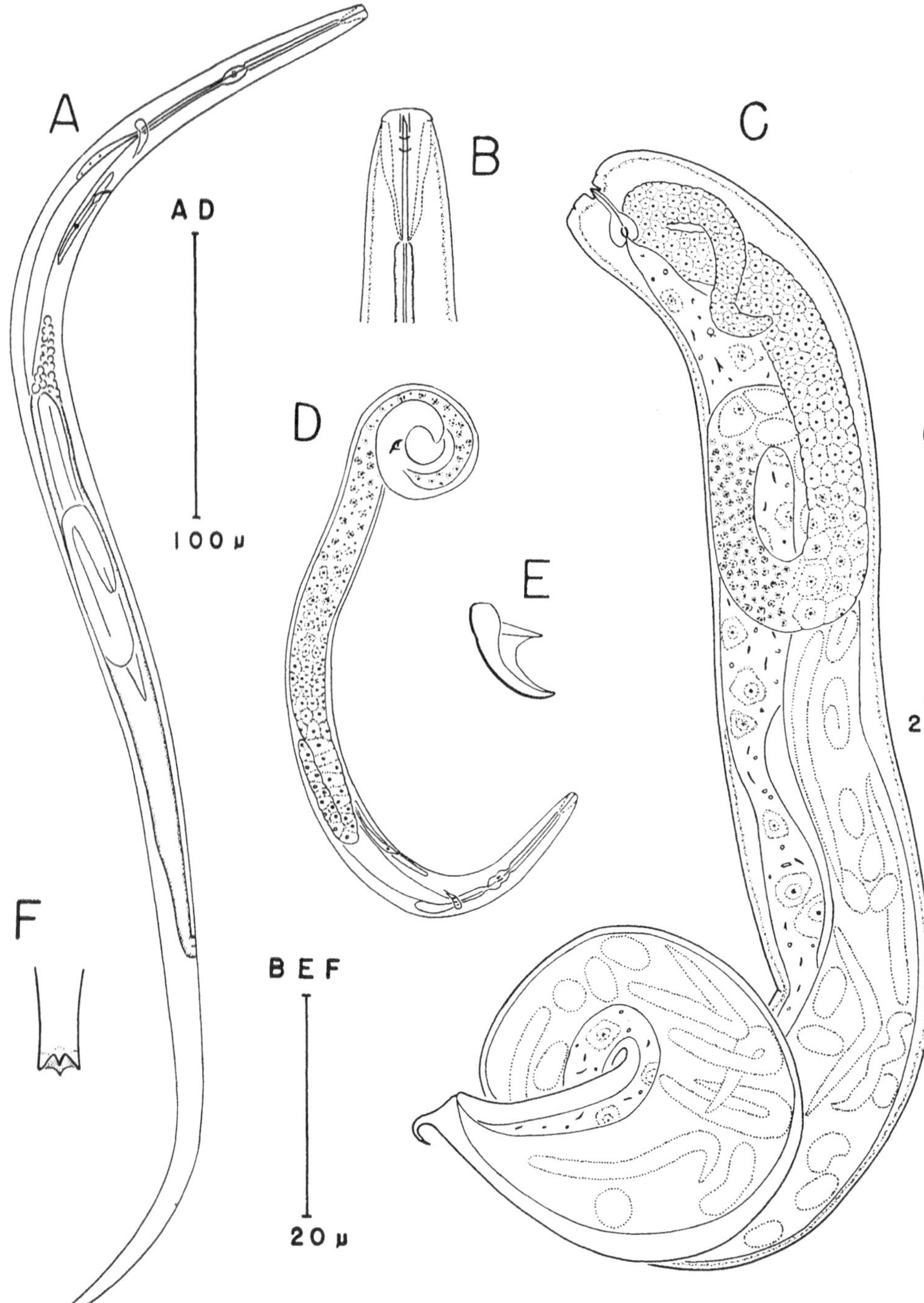

Fig. 4.46. *Roveaphelenchus jonesi.* A. Female, entire; B. Female, anterior end; C. Mature parasitic female; D. Male, entire; E. Spicule; F. Female tail tip. (After Nickle, 1970. Courtesy *Proceedings of the Helminthological Society of Washington.*)

bulb ovoid, often appearing fiddle-shaped in fixed specimens. Oesophageal gland lobe less than two body widths long. Excretory pore and nerve ring situated in similar relative positions to those of the female. **Spicules rosethorn-shaped with a broadly rounded apex and a strongly developed, pointed rostrum**. Testis filling most of body cavity. Tail short, conoid, with a rounded tip.

Type species: *Roveaphelenchus jonesi* Nickle, 1970

Other species: Monotypic.

Bionomics: All three adult forms were found in the haemocoel of the rove beetle, *Aleochara tristis* Gravenhorst (Coleoptera: Staphylinidae) by Nickle (1970a). No details of the biology have been published, but presumably the vermiform female is the infective stage which then produces the male and obese, parasitic female stage.

Distribution: The type locality is given as Lincoln, Nebraska, USA.

Etymology: A combination of *rove* = a colloquial name for the host beetle, and *aphelenchus*.

Genus Incertae Sedis

GENUS *OLIGOAPHELENCHOIDES* POINAR, 1977

Diagnosis: ? Bursaphelenchinae. **Fossil nematodes**. Body 1–1.25 mm long. **Females straight or slightly ventrally arcuate; males with the tail region sharply recurved**. Cuticle smooth. Cephalic region slightly offset. No internal structures visible. Female tail medium conoid to a rounded tip; **male tail conoid with a partly offset terminal region** resembling that of *Bursaphelenchus*.

Type species: *Oligoaphelenchoides atrebora* Poinar, 1977

Other species: Monotypic.

Remarks: Identified from amber from the Palo-Blanco mine, Tapilula, Chiapas, Mexico and dated to La Quinta Formation of the Late Oligocene (Poinar, 1977b). The adult nematodes were associated with eggs, juveniles and fungal hyphae. The amber is estimated to be about 25 million years old and has also yielded numerous fossil arthropods (Hurd *et al.* 1962). Jansson (1986) found that fossils of *O. atrebora* were parasitized by fungi strongly reminiscent of present day nematophagous species.

Distribution: Chiapas, Mexico.

Etymology: Derived from *oligocene*, the geological period of the fossil, and *aphelenchoides*.

II Family Longidoridae

5 Historical Review

The first longidorid nematode was probably described by Grube (1849) when he named *Anguillula linea*, not realizing that the nomen was a junior objective homonym of *A. linea* Oken, 1815. Thorne (1937) made the combination *Xiphinema lineum*, but it is now regarded as a *species inquirenda*. The honour of describing the first longidorid to be currently regarded as valid belongs to Bütschli (1874) when he named *Dorylaimus maximus*, the longest member of the group (Fig. 5.1). This species was subsequently placed under *Longidorus* before being transferred first to *Paralongidorus* (Siddiqi *et al.*, 1963) and then to *Siddiqia* (Khan *et al.*, 1978). De Man (1876) described *Dorylaimus elongatus* (Fig. 5.2), now the type species of *Longidorus* and von Linstow (1879) proposed *Dorylaimus tenuis*, currently regarded as a synonym of *L. elongatus*. Cobb (1893b, Fig. 5.3) described *Tylencholaimus ensiculiferus* from Fiji (= *X. ensiculiferum*, a species which was unknown and largely ignored until redescribed from topotype specimens by Southcy and Luc (1974)). In 1913, Cobb erected the genus *Xiphinema* with the new species *X. americanum* as type and sole species, yet inexplicably overlooked the previously described *ensiculiferum* from his 1893 paper. He did, however, make the prophetic comment: 'Genus represented in many parts of the world. . . .There are beyond doubt dozens of species, possibly hundreds.'

The habitat and distribution of *X. americanum* was given as about the roots of various plants on the Atlantic and Pacific slopes of the USA. Micoletzky (1922) proposed *Longidorus* as a new subgenus of *Dorylaimus*, but it was not until the elevation by Filipjev (1934b) that *Longidorus* achieved full generic rank.

The number of nominal species slowly increased with papers such as those by: T. Goodey (1936); Schuurmans Stekhoven and Teunissen (1938); Loos (1949); Thorne and Allen (1950); Lordello (1951a,b); Schuurmans Stekhoven (1951) and Meyl (1953). The year 1958 saw the first paper on longidorids by Michel Luc and marked the beginning of a nonpareil sequence of high quality

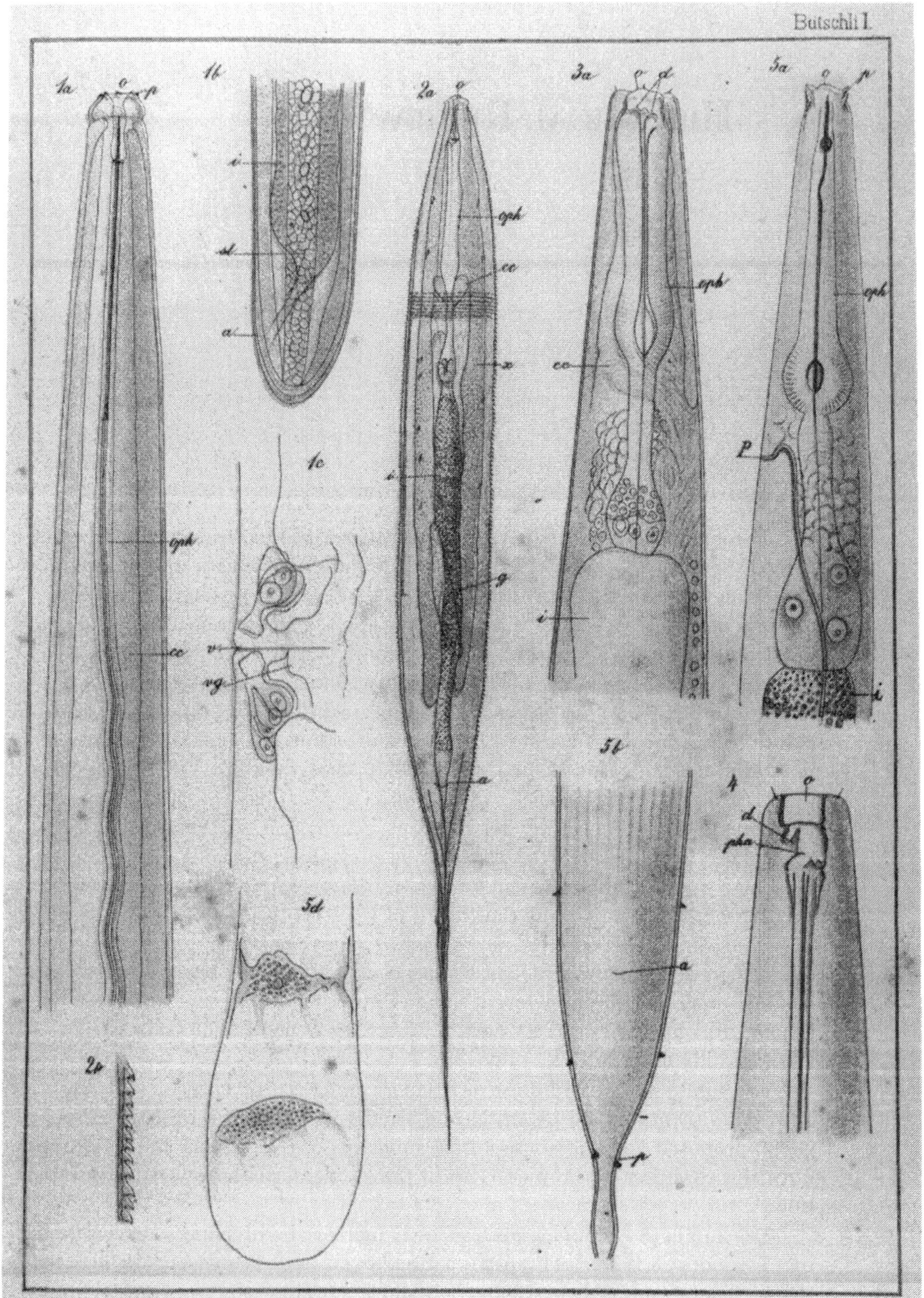

Fig. 5.1. Plate I from O. Bütschli's 1874 paper showing *Dorylaimus* [now *Paralongidorus (Siddiqia)*] *maximus* (1a, b, c).

contributions, largely on the taxonomy of *Xiphinema*, which still continues some 35 years after its inception (Luc, 1958, 1961, 1973, 1975, 1981a,b; Luc and Dalmasso, 1964, 1975a,b; Luc and Doucet, 1984; Luc and Hunt, 1978; Luc and Southey, 1980; Luc and Williams, 1978; etc.).

The early sixties also saw an excellent paper from Hooper (1961) who described and illustrated five new species of *Longidorus* from England, the beginning of another long-running sequence of quality papers by Heyns and co-workers from South Africa (Heyns, 1962, 1965a,b, 1966a,b,c, etc.) and valuable contributions by Coomans, Loof and Siddiqi, amongst others.

With the pace quickening, the numbers of new species proposed began to cause problems in the construction of conventional dichotomous identificaton keys. Cohn and Sher (1972) reviewed *Xiphinema* and proposed dividing the burgeoning number of species into eight subgeneric groups (*Xiphinema*, *Basiphinema*, *Diversiphinema*, *Elongiphinema*, *Halliphinema*, *Krugiphinema*, *Radiphinema*, *Rotundiphinema*) to facilitate identification. Although Roy and Gupta (1974) accepted these subgenera and, in addition, proposed the subgenus *Filiphinema*, this system was not favourably received by other workers, partly, perhaps, because the only genuinely definable subgroup is the X. *americanum* complex. Luc and Dalmasso (1975a) advocated a polytomous approach using a lattice of characters. This scheme worked very competently with adequately characterized species and was readily updatable, but a profusion of poorly described species, largely from India, caused considerable problems until Luc *et al.* (1985) synonymized many of them. The other area of difficulty stemmed from the contentious paper of Lamberti and Bleve-Zacheo (1979) where the X. *americanum* complex was subdivided into a large number of taxonomically adjacent nomina. Loof and Luc (1990) produced an updated polytomous key to *Xiphinema*, but pointedly left out the intractable X. *americanum* complex. However, Lamberti and Carone (1991) have subsequently attempted to provide a usable dichotomous key to the complex.

The taxonomy of the Longidorinae has also proved controversial. Siddiqi *et al.* (1963) proposed the genus *Paralongidorus* to embrace those species of *Longidorus* with an amphidial opening in the form of a transverse slit, as opposed to the minute pore of *Longidorus*, *sensu stricto*. Their type species was *P. sali*, which has a fairly short amphidial slit and a much narrower and more tapering amphidial fovea compared to some of the other species included in the genus which have a much wider slit and a broad, funnel-shaped amphid. For some time there was debate as to the importance of whether the base of the amphid pouch was bilobed or not and how this feature correlated with the amphidial aperture character. In some species, the bilobed appearance of the fovea is incontrovertible whereas in others it is more open to interpretation.

In an effort to overcome these problems, Khan *et al.* (1978), in an important paper, proposed the genus *Longidoroides* to hold those species with short slit-like amphidial openings and pouch-like amphids, leaving *Longidorus*, *sensu stricto* to retain those with a pore-like opening and pouched amphid. *Longidoroides* has been widely accepted, yet currently contains some nominal species having an amphidial slit and amphid comparable in size and shape respectively to the type of *Paralongidorus*. Khan et *al.* (1978) also proposed *Siddiqia*

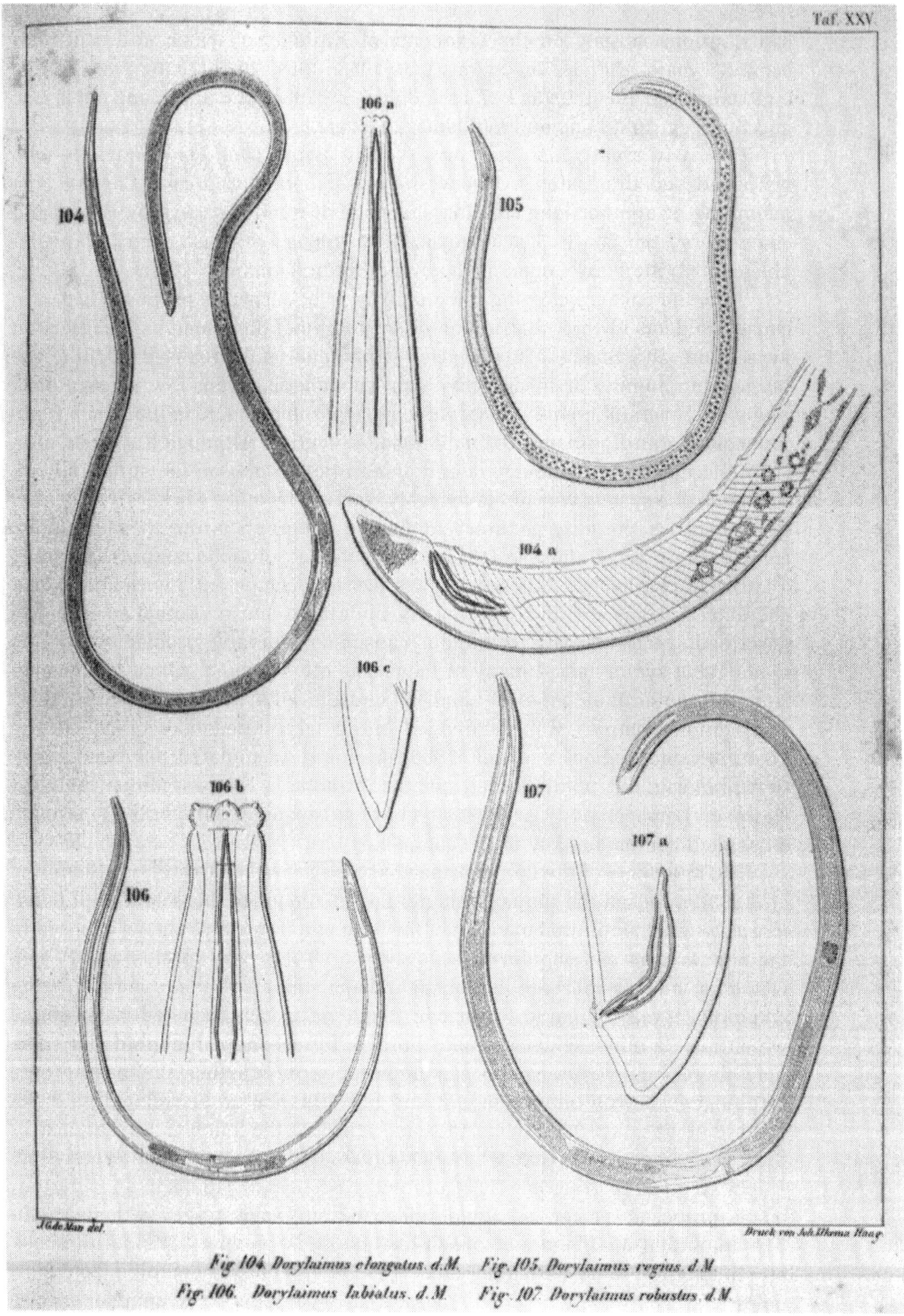

Fig. 5.2. Plate XXV from J.G. de Man's 1876 paper showing *Dorylaimus* (now *Longidorus) elongatus* (104, 104a), the type species of the Longidoridae.

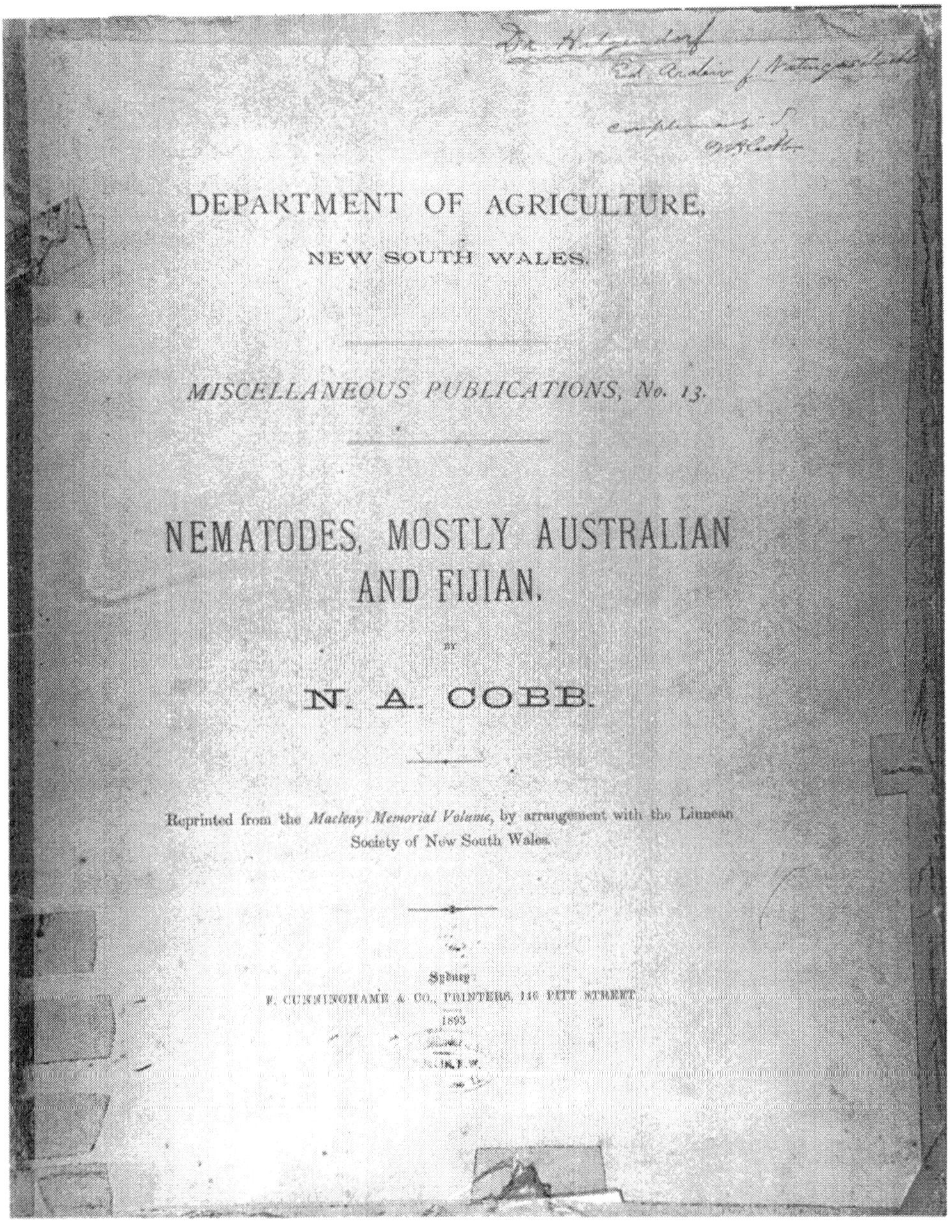

Fig. 5.3. Title page of N.A. Cobb's 1893 publication (signed by the author) in which he described *Tylencholaimus* (now *Xiphinema*) *ensiculiferus.*

for those species of *Paralongidorus* having a characteristically offset and expanded cephalic region. Such nematodes are often rather long compared to *Paralongidorus*, *sensu stricto*. The genus *Siddiqia* has not been universally accepted. Jacobs and Heyns (1982a) originally accepted it as valid whilst Luc and Doucet (1984) regarded it as a synonym of *Paralongidorus*, a point of view subsequently accepted by Jacobs and Heyns (1987a) and Jairajpuri and Ahmad (1992). The species placed in *Siddiqia* certainly form a convenient, easily recognizable and consistent group and in my opinion warrant subgeneric rank at the very least.

Khan (1982) proposed the genus *Inagreius* (sometimes spelled *Inagrius* in later papers), to contain two species with intermediate characters between *Siddiqia* and *Longidoroides* (the offset head of the former, and the bilobed amphidial fovea of the latter genus) and later (Khan, 1987a) added two more species. *Inagreius* was synonymized under *Longidoroides* by Luc and Doucet (1984) on the basis of the bilobed fovea, the only remaining diagnostic feature of the offset head character having been invalidated by their synonymy of *Siddiqia* with *Paralongidorus*. Stegarescu (1980) proposed *Brevinema*, characterized in part by a posterior guide ring, but this is inadequately defined and is, perhaps, best regarded as a *genus inquirendum* as guide ring position is somewhat variable, although Coomans (1985) regarded it as a synonym of *Longidorus*. *Neolongidorus* was erected by Khan (1987b), but only differs from *Longidorus* in having a more posterior guide ring and should therefore be regarded as a junior synonym as proposed by Xu and Hooper (1990).

Monteiro (1976) proposed *Xiphidorus*, an interesting new genus from Brazil with *Xiphidorus yepesara* as the type and only species. Subsequently, several authors have added a handful of species, all coming from either Argentina or Brazil (Chaves and Coomans, 1984; Luc and Doucet, 1984; Uesugi *et al*., 1985; Rashid *et al*., 1986a). The genus embraces features of both *Xiphinema* and *Longidorus*, although the characteristics of the former genus are usually regarded as being ancestral.

As of 1992, there are over 350 valid species of Longidoridae, about 200 belonging to the Xiphinematinae and its sole genus, *Xiphinema* with the next largest genus being *Longidorus* with over 90 species. *Paralongidorus* (*sensu lato*) comprises almost 50 species, the smallest genera being *Longidoroides* with about 15 and *Xiphidorus* with eight species. Although the rate of naming new species has slowed somewhat in the last few years, there are undoubtedly many more yet to describe. As for the genera, the only certainty is that there are either too many ... or too few!

6 General Morphology

Longidorids are long to very long nematodes ranging in length from as little as 1.5 mm in some *Xiphinema* species to over 12 mm in some *Paralongidorus*. They are languid in movement and, due to their large size, are particularly susceptible to mechanical damage. As a consequence they are, like trichodorids, best extracted from soil by passive techniques such as sieving, particularly immersion sieving (Whitehead and Hooper, 1970), which minimize mechanical injury. On heat relaxation they die in the form of a spiral, open spiral, C-shaped, ventrally arcuate or virtually straight. The tail in Longidorinae is typically short, dorsally convex-conoid to a rounded terminus, or cylindroid with a more or less hemispheroid contour. However, in *Xiphinema* its form is extremely variable, e.g. short hemispheroid, with or without a terminal peg; short conoid, with or without a ventral peg or digitiform process; medium to long conoid or initially conoid and then long attenuate to filiform. The tail form in each sex is similar.

Cuticle and associated structures: The cuticle is smooth and comprises two main layers (see Heyns and Jacobs (1989) for details of the ultrastructure). At the tail terminus the cuticle may be very thick, the inner layer being striated and responsible for the additional development. Laterally, the cuticle is pierced by one or two (sometimes three) rows of body pores which rise from the hypodermal glands in the lateral chords and run for the entire body length. In addition there is a dorsomedian series which is restricted to the neck region and a ventromedian series which extends, more or less regularly, from the head to the oesophageal base or even as far as the tail region.

The cephalic region is formed from six fused lips and may be continuous with the body contour or offset, either weakly or strongly (Fig. 6.1). Swart and Heyns (1987a) reported on the comparative cephalic morphology of various longidorids from South Africa using the SEM. The oral aperture is small and the head displays the usual 6 + 10 arrangement of papillae. The amphids

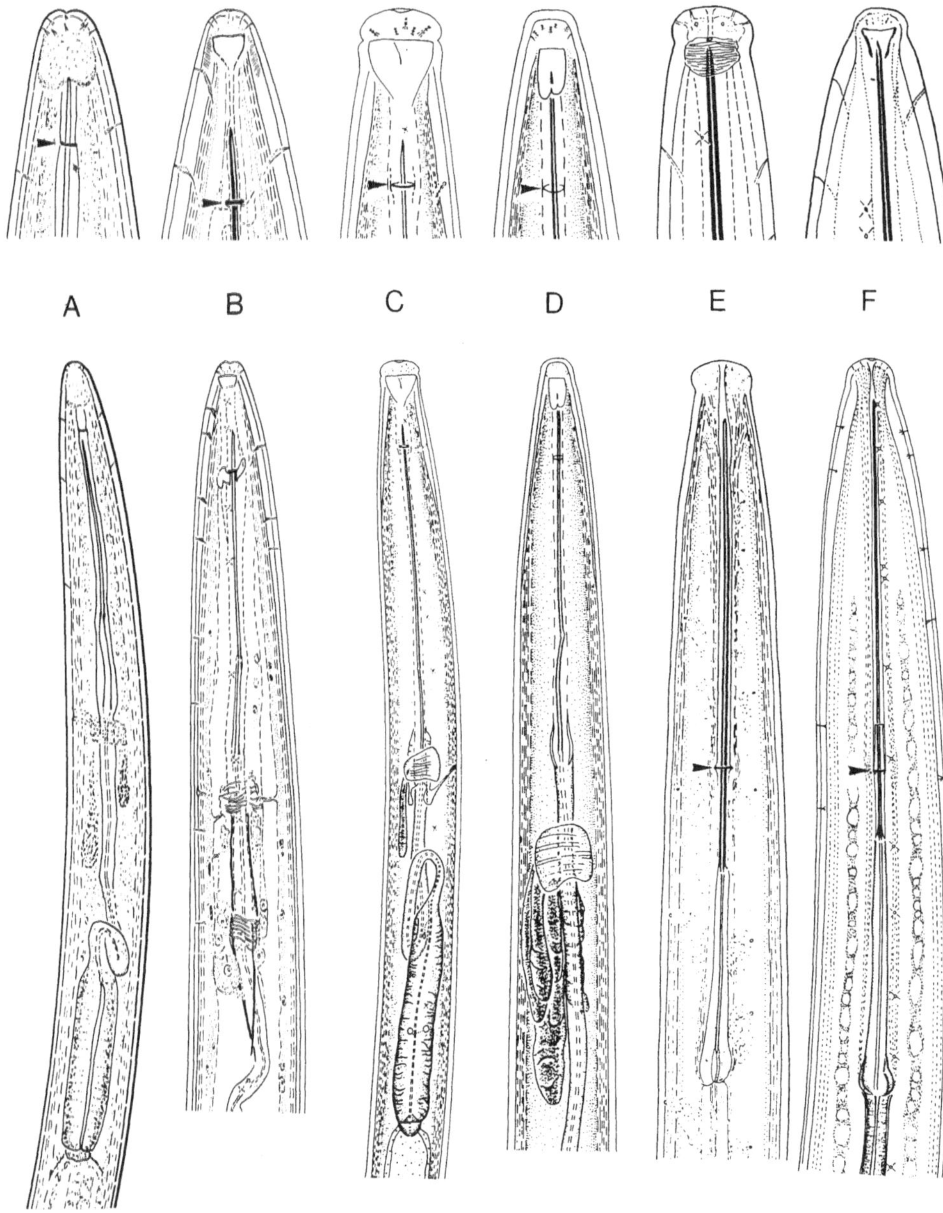

Fig. 6.1. Generic comparison of head and odontostylet region of Longidoridae. A. *Longidorus*; B. *Paralongidorus (Paralongidorus)*; C. *Paralongidorus (Siddiqia)*; D. *Longidoroides*; E. *Xiphidorus*; F. *Xiphinema*. Guide ring is shown arrowed for each genus. (After various authors. Not to scale.)

are well developed and post-labial (Fig. 6.1). The apertures may take the form of minute pores as in *Longidorus*, short transverse slits, as in *Longidoroides*, or very broad transverse slits almost encircling the head as in some *Paralongidorus* and *Xiphinema*. The amphid fovea itself (Fig. 6.1) may be pouch-like as in *Longidorus* for example, or stirrup- or funnel-shaped as in *Xiphinema*. Sometimes the amphid may appear to be shallowly or deeply bilobed proximally (Fig. 6.1, A, D). This has been used as a taxonomic character, but in some species the former condition appears to stem from the pushing forward of the amphidial duct during fixation and in such cases is probably not reliable enough to be diagnostic.

Buccal cavity and armature: Anteriorly the invagination of the cuticle into the oral opening forms the cheilostome. In longidorids this invagination can be very long as, for example, in *Xiphinema* and at its junction with the stomodaeum it forms a heavily sclerotized guiding ring. This ring (Fig. 6.1) may be near the lip region as in *Longidorus* and related genera, or well posterior near the odontostyle/odontophore junction as in *Xiphidorus* and *Xiphinema*. In *Longidorus* the cuticle of the proximal stomodaeum forms a guiding sheath which runs parallel to the odontostyle and attaches to the guide ring. However, in *Xiphinema* this sheath runs anterior to the guide ring and then folds back before making the attachment. This anterior extension into the cheilostome gives rise to the expression 'tubular guiding sheath' and the anterior fold to the terms 'second' or 'moveable guide ring'; terms much used in the descriptive taxonomy of *Xiphinema*.

The buccal armature in longidorids is formed differently to the stylet of aphelenchs or tylenchs and is more properly referred to as the odontostyle (Fig. 6.2, A, D). It is a greatly elongated, needle-like structure which in cross-section can be seen to be formed from a tube of cuticle rolled, so as to leave a fine groove, connected to a central lumen, and running dorsally for the entire length except for the extreme distal tip which is solid. The proximal end of the odontostyle may be simple or apparently forked (Fig. 6.2, E) at its junction with the spear extension, the odontophore. The odontostyle is secreted by a specialized cell located in the ventrosublateral region of the anterior oesophagus (see Grootaert and Coomans, 1981, for a detailed account of its formation in *Labronema*). Juvenile longidorids always have a second, or replacement odontostyle, ready to move forward when the functional odontostyle is shed at the next moult. At the last moult the odontostyle forming cell often generates just the distal tip of yet another odontostyle and this tip, or mucro, as it is often called, is then readily visible in the subventral oesophageal tissue of the adult, a little posterior to the odontophore base. The odontophore is probably formed by fusion of the three sectors of the posterior stomodaeum and may range from being simple, as in *Longidorus*, through to being heavily flanged posteriorly, as in *Xiphinema* (Figs 6.1; 6.2,D). The three sectors of the odontophore have sinuses containing cytoplasm and nerve fibres and may be gustatory in function (Robertson, 1975; Robertson and Taylor, 1975a). The odontostyle and odontophore, considered as one structure, are often referred to as the odontostylet, or, abbreviated further, as the stylet. The ultrastructure of the odontostyle, odontophore and associated structures has received much attention since the work of Taylor and Robertson (1971) and Robertson and Taylor (1975a). Jacobs *et al.* (1987) provided detailed studies of the pharyngeal region, illustrated by transmission electron micrographs, of a range of longidor-

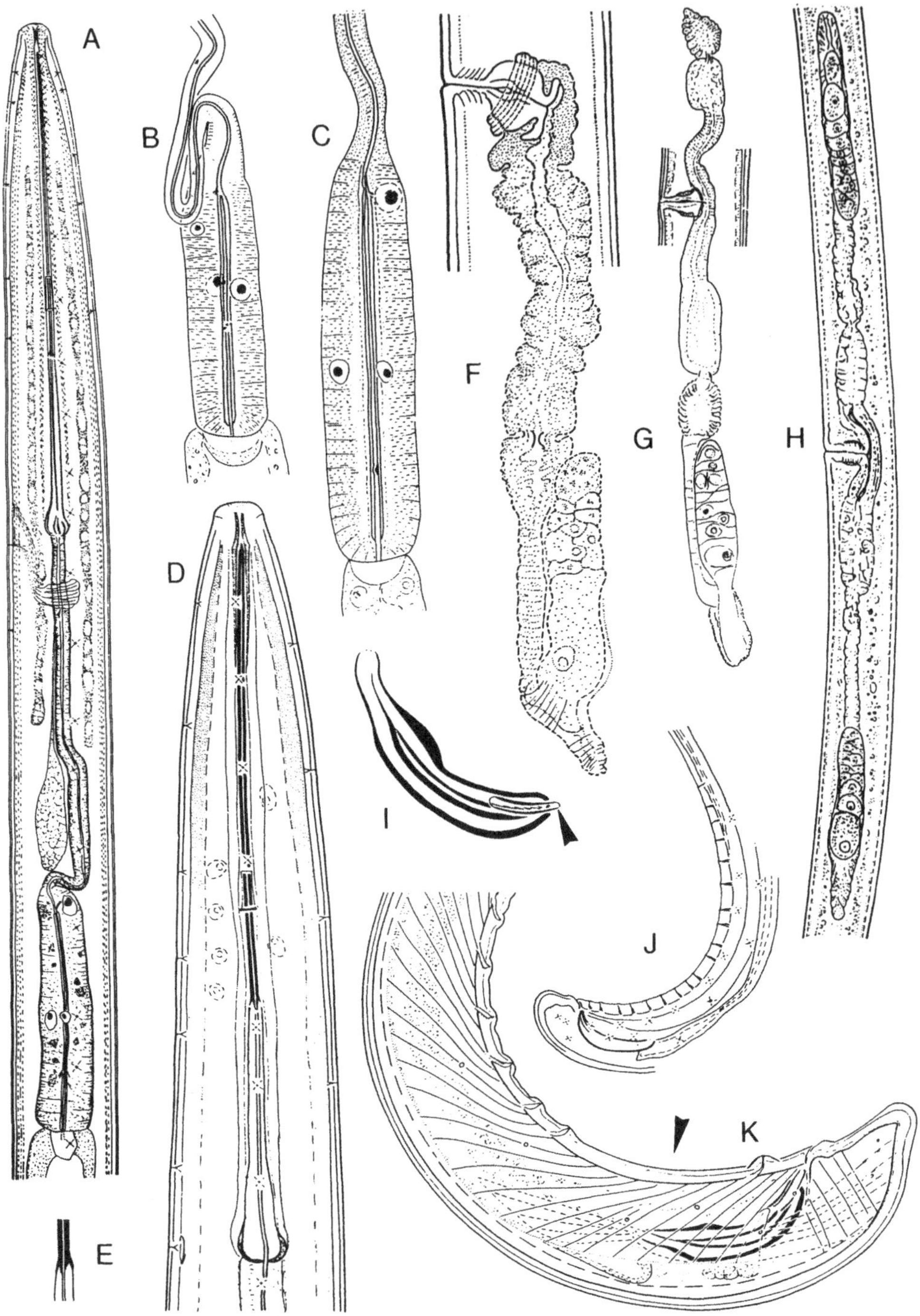

Fig. 6.2. General morphology. A. Oesophageal region of *Xiphinema belmontense*; B. Oesophageal bulb and gland nuclei of *Xiphidorus minor*; C. Oesophageal bulb and gland nuclei of *Xiphinema krugi*; D. Odontostylet of *Xiphinema ifacolum*; E. Forked odontostyle base of *Xiphidorus achalae*; F. Mono-opisthodelphic genital tract of *Xiphinema brasiliense*; G. Pseudo-mono-opisthodelphic genital tract of *Xiphinema krugi*; H. Amphididelphic genital tract of *Paralongidorus (Siddiqia) bullatus;* I. Spicule and

ids. Other studies have involved the use of the scanning electron microscope on excised odontostylets (see Swart and Heyns, 1987b; Heyns *et al.*, 1987; Cho and Robbins, 1990).

The odontostylet is protracted by eight muscles arranged in four pairs which are attached in the region of the odontophore base and then run in parallel to attachment points in the cephalic region. The muscles may spiral between the guide ring and the odontophore base (Wright *et al.*, 1983). In *Xiphinema* the exceptionally long cheilostome is prevented from buckling during odontostylet protraction by the dilatores buccae, a series of oblique muscles running posteriorly from the cheilostome to the body wall. The odontostylet is retracted by special muscles, the arrangement of which varies depending on genus. In *Xiphinema*, which has the most primitive arrangement, the retractor muscles extend posteriorly from the odontophore base before dividing to form four muscles which diverge and attach to the body wall in the region of the anterior section of the oesophageal bulb. In *Longidorus* there are three different arrangements of the retractors: (i) the muscles sheath the narrow, anterior, part of the oesophagus and then posteriorly diverge as four muscles to the body wall; (ii) one set of retractors are arranged as above, but a second set of four extend from the odontophore base, diverge and pass outside of the nerve ring to the body wall; (iii) the muscles sheath the anterior oesophagus, passing between the oesophagus and the nerve ring and then diverge as four muscles to the body wall as in *Xiphinema* (see Robertson and Taylor, 1975a).

Alimentary tract: The oesophagus comprises two parts: a narrow, tubular anterior section which has a circular, cuticle-lined lumen and often loops back on itself when the odontostylet is in the retracted position and a posterior, expanded, cylindrical bulb with a triradiate lumen and containing powerful muscles and the oesophageal glands (Fig. 6.2, A). Typically, dorylaims have five oesophageal glands, but the longidorids have lost one, possibly the anterior, pair of ventrosublateral glands, although their orifi can still be seen. The sclerotized orifice of the dorsal gland opens just anteriorly to the three pairs of longitudinal valve platelets which run for most of the bulb length and delineate the pumping chamber. The single pair of ventrosublateral glands lies about halfway down the bulb, their orifi debouching to the bulb lumen considerably further posteriorly in the posterior third. In Xiphinematinae the dorsal gland nucleus is larger than the ventrosublateral nuclei, almost spherical in shape and situated almost adjacent to its outlet (Fig. 6.2, C). In contrast, the genera of the Longidorinae and Xiphidorinae have the dorsal gland nucleus typically smaller than those of the ventrosublaterals, somewhat elongate in shape and located at some distance posterior to the orifice (Fig. 6.2, B). The dorsal gland is larger than the ventrosublaterals and is drained by six ducts; two are dorsal and extend

accessory guiding piece (arrowed) of *Xiphidorus tucumanensis*; J. Male tail region with supplement series lacking a pre-cloacal hiatus in *Paralongidorus (Paralongidorus) cebensis;* K. Male tail region with supplement series showing a pre-cloacal hiatus (arrowed) in *Xiphidorus tucumanensis.* (After various authors. Not to scale.)

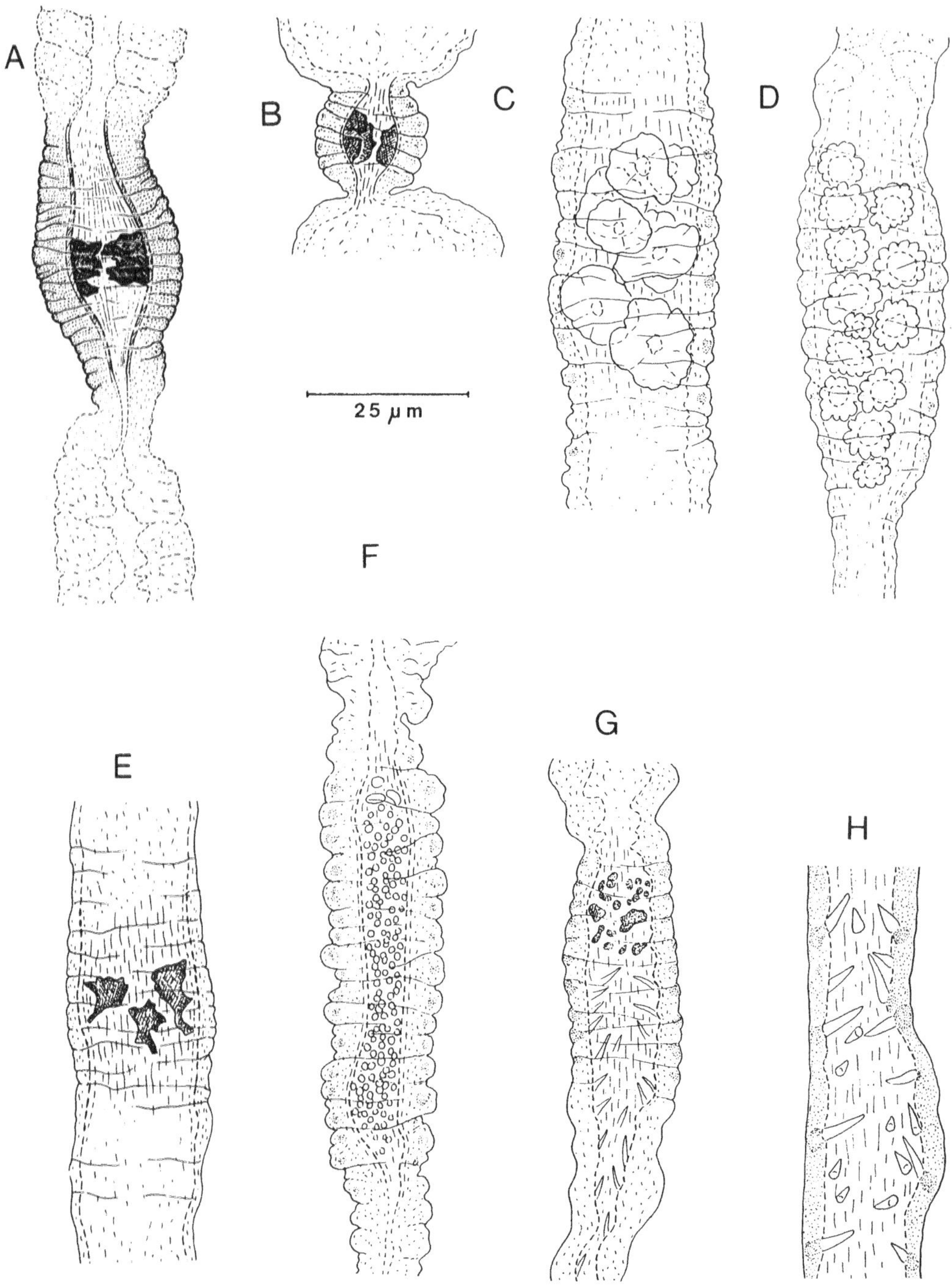

Fig. 6.3. General morphology of the Z organ. True Z Organ. A. *Xiphinema ifacolum*; B. *X. ebriense*. Pseudo Z Organ. C. *X. ingens*; D. *X. diversicaudatum*; E. X. *pini*; F. *X. marsupilami*; G. *X. malagasi*; H. *X. spinuterus*. (After Luc, 1975. Courtesy ORSTOM.)

for virtually the entire bulb length whilst the remaining four are ventral and run about halfway down the bulb (Robertson and Wyss, 1979; 1983). A triradiate oesophago-intestinal valve-like structure, called the cardia, extends as an inverted cone of tissue into the intestinal lumen.

The intestine is not lined with cuticle, but by epithelial cells, thus reflecting its endodermal origin. It is a simple tube, the microvillous epithelial cells often being packed with food reserves. Posteriorly there is a distinct junction with the prerectum which extends for several body widths before another sphincter introduces the cuticle-lined rectum. The anus is a transverse slit which is opened by the dilator ani, an H-shaped muscle.

Female genital system: The vulva is in the form of a broad transverse slit, usually median in position, but exceptionally postmedian as in some *Paralongidorus* for example, or premedian as in all monodelphic and some didelphic *Xiphinema* species. The vagina is short and barrel-shaped with pronounced circular constrictor muscles and oblique dilatory muscles extending to the body wall. The genital tract is typically amphididelphic reflexed, but can be considerably modified from the basic form in *Xiphinema* where the anterior branch progressively regresses leading initially to a pseudomonodelphic state, where the anterior branch is degenerate yet still visible, and ultimately to the true mono-opisthodelphic condition (Fig. 6.2, F–H).

Each genital tract consists of a uterus, oviduct and telogonic ovary. The uterus may be of relatively simple structure as found in Longidorinae or considerably more complex as in some Xiphinematinae. At its simplest, it forms a long tube with fairly thick walls, but in some species, e.g. *Xiphinema americanum* and its relatives, the uterus is very short. Where the uteri join the vagina, the tissues are usually organized to form a symmetrical, thick-walled, bean-shaped structure called the ovejector. In the true monodelphic species the ovejector is asymmetric as it lacks the anterior development. In *Xiphinema* and some *Xiphidorus* species, the uterine tube may contain various sclerotized structures ranging from tooth-like structures or spines attached to the lumen wall and pointing away from the vulva, to granules of divers form. Associated with these structures may be a thickening of the uterine wall and weak circular muscles. This type of uterine differentiation is often referred to as a pseudo-Z organ to distinguish it from the Z organ proper as first described by Luc (1958), which is a more advanced ovoid to spheroid structure with thick walls and circular and longitudinal muscles, separated from the undifferentiated uterus by sphincters and containing a small number of strongly sclerotized tooth-like structures. The true Z organ (Fig. 6.3, A, B) is fairly rare in *Xiphinema*, but the pseudo-Z organ is more prevalent (Fig. 6.3, C–H). The function and significance of these structures is open to speculation. Bleve-Zacheo *et al.* (1984, 1985) reported on the ultrastructure of the pseudo-Z organ and Z organ in *X. diversicaudatum* and *X. ifacolum* respectively whilst Kruger (1988) reviewed in some detail the uterine differentiation in fourteen species of *Xiphinema*.

The uterus may further differentiate to form a spermatheca-like structure and also a thick-walled area with glandular cells which may be involved in eggshell secretion. There is a sphincter at the junction of the uterus and oviduct, the tissue on either side of which may be differentiated to form the pars dilatata uterus and the pars dilatata oviductus, respectively. The oviduct is a thin walled, elastic tube which may be modified to act as a spermatheca adjacent to the sphincter. It leads to the telogonic ovary which is reflexed with the developing oocytes arranged in a single row.

Male genital system: The male genital system is diorchic and opposed, the posterior testis being reflexed towards the cloaca. Both testes join a common vas deferens considerably anterior to the cloaca. Posteriorly the vas deferens has a series of ejaculatory glands. The spicules are paired, massive and ventrally curved. Distally, on the outer face of each spicule, is a small sclerotized structure called the accessory guiding piece, or crura (Fig. 6.2, I). The spicules are protruded by the contraction of a large protractor muscle running from the proximal end of each spicule to the terminal or subdorsal body wall. A retractor muscle attaches to the proximal end of each spicule and runs anteriorly to the subdorsal body wall. Oblique copulatory muscles, attached to the dorsolateral and subventral regions of the body wall, extend anteriorly from the cloacal region for a considerable distance and are responsible for the curvature of the male tail during copulation and when heat relaxed (Fig. 6.2, K). A number of innervated copulatory supplements extend anteriorly from the cloaca. There is a prominent adanal pair and then a ventromedian sequence of from one to twenty supplements which may take the form of a staggered, partially double row, posteriorly. The ventromedian series may commence almost adjacent to the adanal pair (Fig. 6.2, J) or be separated by a pronounced hiatus (Fig. 6.2, K). The cuticle of the tail is perforated by several caudal pores and there is usually a higher density of lateral body pores within the range of the oblique copulatory muscles. A bursa is invariably absent.

7 Bionomics

Longidorids are usually regarded as having four juvenile stages, but there is increasing evidence for only three in some species (Yagita, 1975; Shishida, 1983; Alkemada and Loof, 1989; Halbrendt and Brown, 1992). Each juvenile stage has a replacement odontostyle, located in an oesophageal pouch, which moves forward to replace the old, sloughed odontostyle, at each moult. Reproduction can be amphimictic or parthenogenetic. Life cycle times are generally long and vary, for example, from nine weeks at 28°C and 30°C respectively for *L. africanus* (Kolodge *et al.*, 1987) and *L. elongatus* (Wyss, 1970a), to seven to nine months at 20–23°C for *X. index* (Cohn and Mordechai, 1969). Life cycle times appear to be shorter for *Longidorus* than for *Xiphinema*, but can be over two years in *Paralongidorus* (Stirling and Shannon, 1986).

In general, longidorids are more abundant in sandy or coarse-structured soils, although the smaller species can be more abundant in heavier soils. Populations are generally highest at depths below 20 cm and depths of over one metre can be achieved. The nematodes appear to dislike fluctuations in soil moisture content and this, together with the hazards of mechanical cultivation in agricultural soils, probably accounts for their depth preference. In Europe, there is evidence that the July soil isotherm has a close association with longidorid distribution (Boag *et al.*, 1991). Longidorids can tolerate both very dry soils and waterlogged mud. *Xiphinema bergeri*, a parasite of rice roots in West Africa, lives for a third of the year in flooded soil, yet survives the remainder in sun-baked conditions (Luc, 1973). A similar phenomenon is found in three species of *Paralongidorus*, one on rice in Australia (Stirling and McCulloch, 1984) and two on deepwater rice in Bangladesh (Hunt and Rahman, 1991). Adult longidorids can live for a number of years and the nematodes can survive without host plants for periods of up to six or seven years (Dalmasso, 1969). A good general account of the ecology of longidorids was given by Weischer (1975).

Longidorids are polyphagous ectoparasites of plant roots. In general, *Long-*

idorus and related genera are more often found in association with herbaceous plants whereas *Xiphinema* is better adapted to feed on the roots of woody plants. The first reports of damage to plant roots were made by Horner and Jensen (1954) and Schindler (1957) for *Longidorus* and *Xiphinema* respectively. The first report of virus transmission by longidorids was that of Hewitt *et al.* (1958).

Agar plate studies on the feeding of *Xiphinema* species were carried out by a number of workers in the sixties and seventies, e.g. Fisher and Raski (1967); Sutherland (1969); Cohn (1970); Cotten (1974) and Weischer and Wyss (1976). These studies initially concentrated on nematodes known to be virus vectors such as *X. diversicaudatum* and *X. index*. Some early observations on the feeding of *Longidorus* (Cohn, 1970; Wyss, 1970a) lacked detail, a situation later remedied by Towle and Doncaster (1978) working with *L. caespiticola* feeding on rye grass roots growing in agar. They used cinemicrography to facilitate detailed observations of the feeding cycle.

The feeding cycle comprises five phases: exploration; penetration; salivation; ingestion and withdrawal. The exploration phase involves the nematode lips being rubbed over the root surface and lasts for a shorter time on roots which have already been fed upon (Towle and Doncaster, 1978). Penetration differs between *Longidorus* and *Xiphinema* – in the former the odontostyle penetrates towards the centre of the root until it is fully protracted (Towle and Doncaster, 1978) whereas in *Xiphinema* the odontostyle usually penetrates only a short distance and is followed by a feeding phase, before moving deeper into the tissues, the process then being repeated (Fisher and Raski, 1967; Sutherland, 1969; Trudgill, 1976; Wyss, 1978a,b; Hunt and Towle, 1979). Salivation is typified by intermittent bouts of slow contractions of the anterior region of the basal bulb and ingestion then follows for periods of several hours. The length of the ingestion period suggests that nutrients are also being withdrawn from the cells surrounding the one containing the odontostyle aperture, a phenomenon reported by McElroy and van Gundy (1968) in *Hemicycliophora arenaria*. Wyss (1978a) and Hunt and Towle (1979), working with *X. index* and *X. vulgare* repectively, recorded extended ingestion periods of several hours and speculated that this was a consequence of the odontostyle penetrating phloem tissue. Wyss *et al.* (1988), working with *X. index* feeding on *Ficus carica* roots, used video-enhanced contrast light microscopy to study details of the oesophageal bulb function. Withdrawal occurs with the odontostyle still protracted and may be followed by oesophageal pumping, possibly to clear the feeding apparatus. *Longidorus* appears to show a distinct preference for feeding on the root tip (Fig. 7.1). Attacked root tips may become hook-shaped before ceasing growth or swell to form terminal galls. Lateral root initials then form and are in turn attacked. *Xiphinema* does feed near the root tip, but also attacks further back along the root, often in the root hair zone. Both species aggregate at suitable feeding sites.

Root pathology is similar for both *Longidorus* and *Xiphinema* with darkening of tissues, cortical hyperplasia, lateral root proliferation and tip galling being the norm (Cohn, 1975). Cells with thickened walls and granular cytoplasm are associated with the feeding sites and phellogen formation may also be initiated. A necrotic response may occur in the vicinity of the feeding site (Wyss, 1978b; Hunt and Towle, 1979). As the meristematic activity is retarded, the vascular differentiation extends to the vicinity of the root tip (Davis and Jenkins, 1960; Cohn and Orion, 1970; Cohn, 1975). Wyss *et al.* (1988), using video-enhanced contrast light microscopy, provided detailed observations of the root cell responses to feeding by *Xiphinema index*. Andrés *et al.* (1989) reported on details of the ultastructure of cereal and legume roots parasitized by *Longidorus belloi*.

The discovery in the late 1950s that longidorids were implicated in the transmission of plant viruses (Hewitt *et al.*, 1958; Harrison and Cadman, 1959; Jha and Posnette, 1959) stimulated a prolonged period of research, notably by B.D. Harrison and co-workers, van Hoof, and C.E. Taylor and co-workers. Longidorids vector nepoviruses, in contrast to trichodorids which transmit tobraviruses. The major longidorid vectored nepoviruses are grapevine fanleaf, arabis mosaic, tobacco ring spot, tomato ring spot and raspberry ring spot. *Longidorus* species known to transmit viruses include *L. apulus, L. attenuatus, L. breviannulatus, L. elongatus, L. macrosoma and L. martini; Xiphinema* species acting as vectors include X. *americanum*, X. *coxi*, X. *diversicaudatum*, X. *index*, X. *italiae* and X. *vuittenezi*.

The specific virus retention sites differ between the two genera. In *Longidorus*, retention occurs on the lining of the guide sheath surrounding the odontostyle whereas in *Xiphinema* the oesophageal lumen and lining of the odontophore act as retention sites (Taylor and Robertson, 1969, 1970a; McGuire *et al.*, 1970; Raski *et al.*, 1973). In both genera the virus particles are lost during moulting and the nematodes must therefore feed on infected cells before transmission is again possible (Harrison and Winslow, 1961; Taylor and Raski, 1964; Taylor and Robertson, 1970a).

Retention time has been reported for nearly a year in starved X. *americanum* (Bergeson *et al.*, 1964) and for up to 8 months in X. *diversicaudatum* allowed to feed on virus-immune hosts (Harrison and Winslow, 1961). *Longidorus* species appear to have shorter retention times and this may be related to their more restricted retention site. Virus dissociation is possibly triggered by the passage of saliva (Taylor and Robertson, 1975) and transmissibility is dependent on virus serotype.

Viruses transmitted by longidorids are responsible for greater economic loss than the direct feeding damage to the root system caused by the nematodes. Chemically controlling the vectors is both costly and difficult as elimination of all virus retaining nematodes is required. The development of virus immune crops is therefore a logical and potentially more effective solution.

Good general reviews of longidorids as virus vectors include those of Taylor and Robertson (1975), Martelli (1975), Taylor (1978), Lima (1986) and Martelli and Taylor (1990).

The biogeography of the Longidoridae, although there is still a paucity of data from some regions, is reasonably clear and was discussed in some detail by Coomans (1985). The genus *Xiphinema* is cosmopolitan and is found, with the exception of Antarctica, in all the major regions that once constituted the supercontinent Pangaea. Numerically the majority of species have been

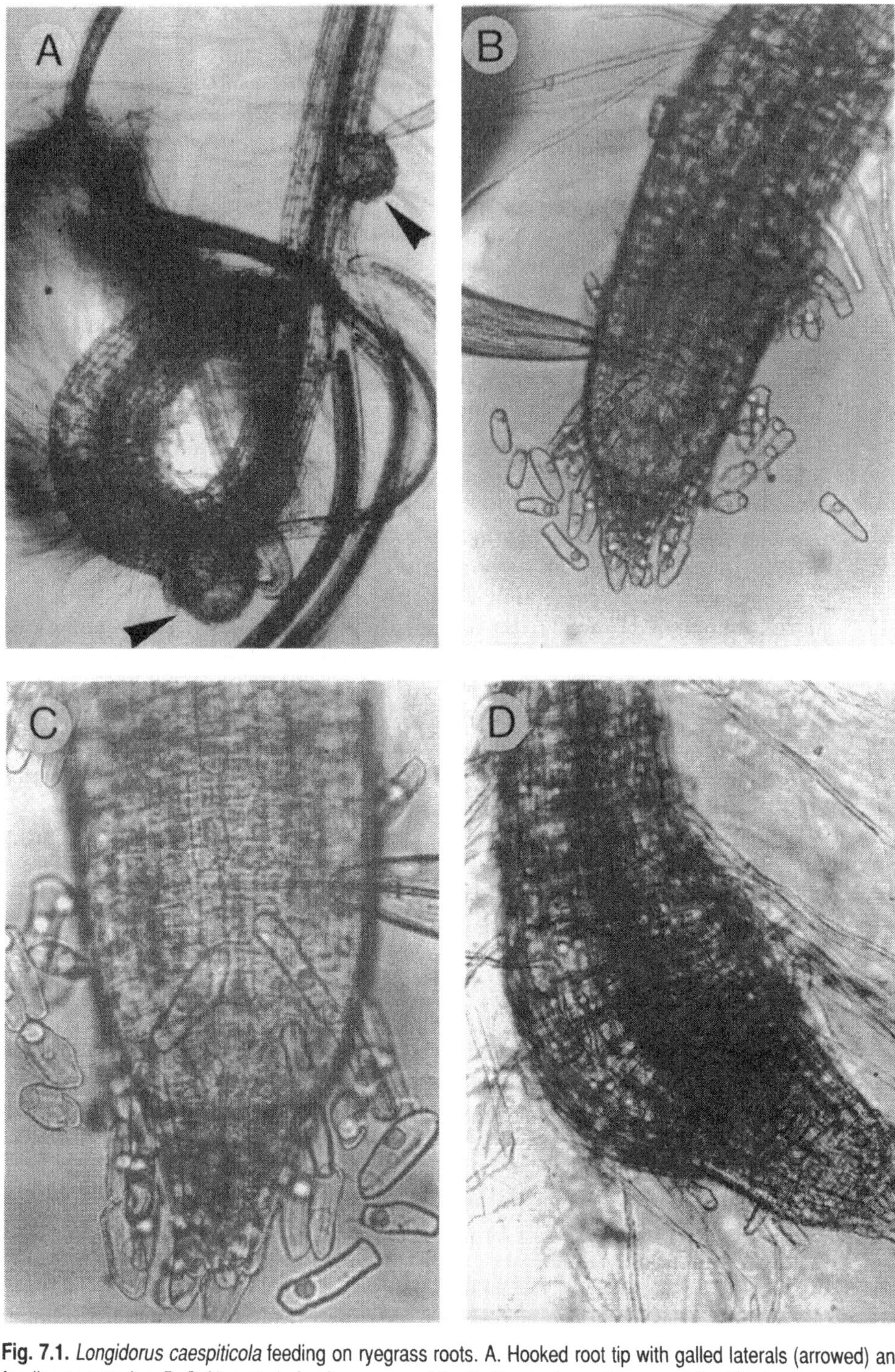

Fig. 7.1. *Longidorus caespiticola* feeding on ryegrass roots. A. Hooked root tip with galled laterals (arrowed) and feeding nematodes; B, C. Nematode feeding near root tip; D. Developing gall near root tip. (Photographs courtesy A. Towle.)

described from those areas of which Gondwanaland was formerly comprised, i.e. Africa, India and South America, with fewer species from Laurasia (Asia, Europe, North America). Coomans (1985) speculated that *Xiphinema* originated in Gondwanaland, probably in the Permian period and spread to Laurasia before the break-up of Pangaea into the northern and southern continental masses. The main speciation, based on current data, appears to be in Africa where numerous parthenogenetic species occur, although this speciation could partly be a result of relatively recent climatic changes when southern Africa experienced a drier phase subsequent to the ending of the last ice age, thus leading to the isolation and separate development of once continuous populations.

The Longidorinae, containing *Paralongidorus*, the most primitive genus, together with *Longidorus* and *Longidoroides* (plus *Xiphidorus* if the Xiphidorinae is not regarded as valid) seems to have evolved after the separation of Africa and South America in the Cretaceous period. *Paralongidorus* (*sensu lato*) is known mainly from Africa, Asia, Australasia and Europe, but not South America, *Longidoroides* from India and southeast Africa (an undescribed species has been reported from Brazil) whilst *Longidorus* has a similar geographic spread to *Paralongidorus*, but with greatest speciation in Europe. The preponderance of *Paralongidorus* and *Longidoroides* species from India and southeast Africa suggests that these genera evolved before the separation of Africa and India.

Xiphidorus is restricted to South America, Coomans (1985) speculating on its development from ancestral stock after the separation of Africa and South America in the Cretaceous period. The genus may well represent a relict transitional stage in the evolution of the Longidorinae from the Xiphinematinae and its systematic position, discussed in more detail in the following section, has yet to be agreed upon.

8 Systematics and Taxonomy

The systematics of the longidorids has seen a certain degree of flux in the last twenty or so years. Khan and Ahmad (1975) started the process when they proposed the superfamily Longidoroidea, separating it from Dorylaimoidea on the basis of the long, attenuate, odontostylet. They also proposed raising the Xiphinematinae of Dalmasso (1969) to family rank alongside the redefined Longidoridae.

Khan *et al.* (1978), in a thorough review of the group, went further and proposed that *Xiphidorus*, a genus with characters somewhat intermediate between the Longidoridae and Xiphinematidae, as redefined by Khan and Ahmad (1975), should be given its own family, the Xiphidoridae. They also split the Longidoridae *sensu* Khan and Ahmad into the Longidorinae, comprising *Longidorus* and *Longidoroides* and the Paralongidorinae for *Paralongidorus* and *Siddiqia*, the subfamilies being based on the shape of the amphid fovea, i.e. whether pouch-like, as in Longidorinae or stirrup-shaped as in Paralongidorinae. Their paper marked the apogee of the trend to inflationary systematics, although their scheme may in fact be more reasonable than is currently accepted.

Robbins and Weiner (1978), on describing the unusual new genus *Californidorus*, proposed the subfamily Californidorinae within the Longidoridae. However, Jairajpuri (1982a), on examining type material, placed the genus in Nordiidae on the basis of five oesophageal glands being present as opposed to three in the longidorids. This move was accepted by Siddiqi (1983b), who also showed that the system of retractor muscles was not longidorid in pattern; by Luc and Doucet (1984), also on examination of type material, and, with some reservations, by Robbins (1985) when he reviewed the genus and described three more species. Coomans (1985) commented in greater detail on these differences.

Luc and Doucet (1984) reviewed the prevailing systematic state and proposed a drastic simplification. In their scheme, only the family Longidoridae was accepted and was placed under the superfamily Dorylaimoidea. No subfamilies were recognized due to the perceived variation in character sets amongst the genera. Coomans (1985), on the other hand, accepted the Longidoridae with the subfamilies Longidorinae and Xiphinematinae, but divided the former into the tribes Longidorini and Xiphidorini, a view adopted by Heyns *et al.* (1987). Jairajpuri and Ahmad (1992), recognizing the Longidoroidea, followed a slightly different scheme and placed the Xiphidorinae under the Xiphinematidae, thus emphasizing the morphological similarities of *Xiphidorus* and *Xiphinema*. I agree with Luc and Doucet (1984) and Coomans (1985) in rejecting the superfamily concept and accept the Longidoridae *sensu* Luc and Doucet. However, I feel that the subfamily concept has some value in the group and recognize Longidorinae and Xiphinematinae, based largely on the location, size and shape of the dorsal oesophageal gland nucleus, but also on the absence or presence of a hiatus in the male ventromedian series of copulatory supplements. I also consider that the idea of a separate subfamily for *Xiphidorus*, the Xiphidorinae, has some merit and reflects the restricted geographical distribution and separate evolutionary line of this morphologically distinct genus. The systematic position of *Xiphidorus* is problematic, combining as it does, features of both Longidorinae and Xiphinematinae. The former group characters include the amphidial pouch, the disposition and relative size of the oesophageal gland nuclei and the odontostylet retractor muscles; those relating to Xiphinematinae being the morphology of the odontostylet, the guide ring near the odontostyle/odontophore junction, the development of uterine differentiation in the form of spines in some species and the small number of male copulatory supplements together with a hiatus between the adanal pair and the ventromedian series. Although the view of Chaves and Coomans (1984) and Coomans (1985) that the *Xiphinema*-like characters of the odontostylet, particularly the forked odontostyle base, are plesiomorphic, is a persuasive one, particularly with the discovery of some *Paralongidorus* species also exhibiting a strong furcation of the odontostyle base, there is also the character of the presence or absence of a pronounced hiatus between the adanal pair of male copulatory supplements and the ventromedian series to be considered. Both *Xiphidorus* and *Xiphinema* posess such a hiatus whereas *Longidorus*, *Paralongidorus* and *Longidoroides* do not. I therefore intend placing *Xiphidorus* in the Xiphidorinae, thus reflecting its separate distribution from Longidorinae (*sensu stricto*) and unique combination of morphological characters from both Longidorinae (*sensu stricto*) and Xiphinematinae, characters which are otherwise difficult to reconcile.

Whether we should use morphological characters on their own or in concert with an assumed phylogeny is problematic. The prime objective of any systematic scheme is surely to facilitate identification; that the scheme should also reflect evolutionary trends is desirable, yet secondary. Achieving the correct balance between these two, often conflicting, demands, is a compromise between the requirements of a natural, phylogenetic, Darwinian system and the utilitarianism of the phenetic approach. Coomans (1985), for example, is an ardent exponent of the former whilst Vinciguerra (1988), applying Expert System Technology, demonstrates the elegant simplicity of the latter, whilst simultaneously acknowledging its shortcomings.

With regard to the delineation of genera, the main diagnostic characters involve the type of amphidial aperture and fovea, the morphology of the odontostylet and guiding apparatus and the relative size and position of the oesophageal gland nuclei. Increasing knowledge of character variability has resulted in a decline in value of some of these characters, e.g. the forked odontostyle base used to be confined to *Xiphinema*, but is now known to occur in *Xiphidorus* and, to some extent, in certain *Paralongidorus* species. However, the tubular guiding sheath extending well anterior to the guide ring when the stylet is in the fully retracted position is still a valid character for differentiating *Xiphinema*. Whether the amphidial slit is a pore or a transverse slit is also valuable, although when a pore becomes a minute slit is a moot point. The shape of the amphidial pouch or fovea is possibly less reliable as the amphid duct may push forward and produce a 'bilobed amphid' in certain species possessing pouch or cup-like fovea. The subjective value given to characters such as these largely influences which genera are regarded as valid. Such values are not absolute and vary, partly with taxonomic fashion, partly with increased knowledge.

The genus *Inagreius* was erected by Khan (1982) on the shape of the amphid fovea. It was regarded as a synonym of *Longidoroides* by Luc and Doucet (1984), a proposal which has received general acceptance (Coomans, 1985; Jairajpuri and Ahmad, 1992). *Longidoroides* was proposed by Khan *et al.* (1978) based on pouch-like amphids with small slit-like openings. However, since then the genus has accumulated a number of species with a broader slit and also some with narrow, tapering fovea. The type of *Paralongidorus*, *P. sali* also has a small slit-like amphidial aperture and narrow, tapering fovea and on this basis some *Longidoroides* species are indistinguishable from the type of *Paralongidorus* and should therefore be regarded as belonging to the latter genus.

Siddiqia, proposed by Khan *et al.* (1978) has some claim to validity, despite being usually regarded as a synonym of *Paralongidorus* (Luc and Doucet, 1984; Coomans, 1985; Jacobs and Heyns, 1987a; Jairajpuri and Ahmad, 1992). The very distinctive expanded head offset by a groove, coupled with a broad, slit-like amphidial opening and stirrup-shaped fovea are consistent characters. Recently, scanning electron microscope studies of the cephalic region of some species revealed four small flap-like structures located in the sublateral sectors (Swart and Heyns, 1987a). I have seen such structures under the light microscope on a number of *Siddiqia* species, but unfortunately they were not incontrovertibly visible on the limited material available for study of *Siddiqia citri*, the type of the genus. Luc and Doucet (1984) regarded *Siddiqia* as a synonym of *Paralongidorus*, dismissing the character of offset head on the basis that *Xiphinema* contained species with continuous or offset heads. Such an argument is not really pertinent - *Xiphinema* embraces the entire range of head shapes from continuous to expanded offset whereas that of *Paralongidorus*, sensu stricto, is always continuous and *Siddiqia* invariably offset and expanded.

Neolongidorus was erected for certain species of *Longidorus* with a posterior guide ring (Khan, 1987b). The guide ring position is fairly variable in the Longidorinae and I concur with Xu and Hooper (1990) in regarding *Neolongi-*

dorus as a synonym of *Longidorus*. *Brevinema*, erected by Stegarescu (1980), also has a posteriorly located guide ring and is more problematic. The genus is poorly defined and is best regarded as *genus inquirendum*.

Two outline classifications are given below: one indicating the inflationary concept as advocated by Khan *et al.* (1978) and the other, which I intend following in this book, is a modification of the conservative schemes of Luc and Doucet (1984) and Coomans (1985).

OUTLINE CLASSIFICATION OF KHAN, CHAWLA AND SAHA, 1978
(updated to include subsequent genera)

DORYLAIMIDA Pearse, 1942

Dorylaimina Pearse, 1936

Longidoroidea Thorne, 1935 (Khan and Ahmad, 1975)

Longidoridae Thorne, 1935 (Meyl, 1961)

Longidorinae Thorne, 1935
Longidorus Micoletzky, 1922 (Filipjev, 1934)
Longidoroides Khan, Chawla and Saha, 1978
Inagreius Khan, 1982
Neolongidorus Khan, 1987

Paralongidorinae Khan, Chawla and Saha, 1978
Paralongidorus Siddiqi, Hooper and Khan, 1963
Siddiqia Khan, Chawla and Saha, 1978

Xiphidoridae Khan, Chawla and Saha, 1978
Xiphidorus Monteiro, 1976

Xiphinematidae Dalmasso, 1969 (Khan and Ahmad, 1975)
Xiphinema Cobb, 1913

NOTE: The subfamily Xiphineminae was proposed by Dalmasso (1969) and has often been used in this form by later authors. However, this derivation contravenes the requirements of the Code (Article 29b), which states that the stem for the formation of a family-group name is obtained by deleting the case ending from the genitive singular, in this case *nematos*. The correct derivation, Xiphinematinae or Xiphinematidae, is used herein and does not affect the original authorship and date.

PROPOSED OUTLINE CLASSIFICATION

DORYLAIMIDA Pearse, 1942

Dorylaimina Pearse, 1936

Dorylaimoidea de Man, 1876 (Thorne, 1934)

Longidoridae Thorne, 1935 (Meyl, 1961)

Longidorinae Thorne, 1935
Longidorus Micoletzky, 1922 (Filipjev, 1934)
Longidoroides Khan, Chawla and Saha, 1978
Paralongidorus Siddiqi, Hooper and Khan, 1963
P. (Paralongidorus) Siddiqi, Hooper and Khan, 1963
P. (Siddiqia) Khan, Chawla and Saha, 1978 n.rank

Xiphidorinae Khan, Chawla and Saha, 1978
(Jairajpuri and Ahmad, 1992)
Xiphidorus Monteiro, 1976
Xiphinematinae Dalmasso, 1969
Xiphinema Cobb, 1913

Family Longidoridae Thorne, 1935 (Meyl, 1961)

DIAGNOSIS: Dorylaimoidea. **Long to very long, slender nematodes ranging from 1.5 to 12 mm in length.** Cuticle smooth. Cephalic region rounded, continuous with body contour or offset. **Lips amalgamated** with the usual 6 + 10 circlets of papillae. Amphidial apertures ranging from small pores to broad transverse slits. Amphids large, pouch-like or stirrup-shaped. Lateral chords broad with one to three rows of lateral body pores. Dorsal and ventral series of body pores usually present. **Odontostyle greatly elongate, attenuate, 50–220 μm long. Odontophore elongate,** sometimes with three strong, basal flanges. **Junction of cheilostome and stomodaeum marked by a strongly sclerotized guide ring** varying in position from near the lip region to near the odontostyle base. Oesophagus in two parts: an anterior, narrow, tubular section and posterior, short cylindrical, bulb containing longitudinal valve plates. **Three oesophageal glands – one dorsal and two ventrosublateral.** Female vulva located anteriorly to post-median in position, usually median. Genital tracts typically amphididelphic reflexed, but may be pseudomonodelphic or mono-opisthodelphic. Uterus of some species of *Xiphinema* and *Xiphidorus* with various sclerotized structures in the lumen and/or attached to the walls. Male genital tracts diorchic, opposed. Spicules large, dorylaimoid with lateral accessory guiding pieces. Copulatory supplements in the form of an adanal pair and then a ventromedian series of one to 20 papillae. Tail shape variable, but generally similar in each sex.

TYPE GENUS: *Longidorus* Micoletzky, 1922 (Filipjev, 1934)

TYPE SUBFAMILY: Longidorinae Thorne, 1935

OTHER SUBFAMILIES:
Xiphidorinae Khan, Chawla and Saha, 1978 (Jairajpuri and Ahmad, 1992)
Xiphinematinae Dalmasso, 1969

Key to subfamilies

1. Dorsal gland nucleus elongate, smaller than those of the ventrosublateral glands and located at some distance posterior to its orifice. 2

– Dorsal gland nucleus round, larger than those of the ventrosublateral glands and located adjacent to its orifice. . . . . Xiphinematinae

2. Amphidial aperture a minute slit (pore-like), odontostyle with furcate base, guide ring located near to odontostyle/odontophore junction, male copulatory supplements few in number (less than eight) and with an hiatus between the adanal pair and the ventromedian series . . . . Xiphidorinae

– Not with the above combination of characters . . . . Longidorinae

Subfamily Longidorinae Thorne, 1935

Diagnosis: Longidoridae. Body length from 2 to 12 mm. **Amphidial apertures either pore-like or transverse slits.** Amphidial fovea pouch-like or stirrup shaped. **Odontophore flanges usually not developed,** but exceptionally (e.g. in a few *Paralongidorus*) they are moderately developed. **Guide ring** single, **usually in anterior third of odontostyle**, but rarely located at up to 60% of its length. **Guiding sheath never extending well anterior to guide ring when the odontostylet is fully retracted. Dialatores buccae absent. Dorsal oesophageal gland nucleus smaller than the nuclei of the ventrosublateral glands and located some distance posterior to the orifice. Female genital tracts always amphididelphic,** reflexed; **never reduced. No uterine differentiation in the form of spines,** etc. **Male ventromedian series of copulatory supplements** up to 20 in number and **commencing without a hiatus from the adanal pair.**

Type genus: *Longidorus* Micoletzky, 1922 (Filipjev, 1934)

Other genera:
Longidoroides Khan, Chawla and Saha, 1978
Paralongidorus Siddiqi, Hooper and Khan, 1963
Paralongidorus (*Paralongidorus*) Siddiqi, Hooper and Khan, 1963
Paralongidorus (*Siddiqia*) Khan, Chawla and Saha, 1978 n.rank

Genus inquirendum: *Brevinema* Stegarescu, 1980

Key to genera

1. Amphidial aperture pore-like . . . . *Longidorus*

– Amphidial aperture slit-like 2

2. Amphidial fovea pouch-like, initially as broad as the aperture for some distance posteriorly and with a broadly rounded base .... *Longidoroides*

– Amphidial fovea narrowing almost immediately posterior to the aperture, base not broadly rounded..3

3. Cephalic region markedly offset and expanded from body ....*Paralongidorus* (*Siddiqia*)

– Cephalic region continuous with body contour and not expanded........ .... *Paralongidorus* (*Paralongidorus*)

GENUS *LONGIDORUS* MICOLETZKY, 1922 (FILIPJEV, 1934) (Fig. 8.1)

syn. *Neolongidorus* Khan, 1987
? *Brevinema* Stegarescu, 1980

DIAGNOSIS: Longidorinae. Body long to very long (3 to >10 mm) and slender. Heat relaxed form varying from more or less straight to C-shaped. Lateral chords broad and with one or two rows of lateral body pores. Cephalic region rounded; continuous or offset. Lips fused and with the usual 6 + 10 arrangement of papillae. **Amphidial apertures in the form of small, inconspicuous pores which lead back to well developed pouch-like amphid fovea. Odontostyle elongate, needle-like; not heavily sclerotized. Guiding apparatus with a simple ring usually situated within a couple of head-widths of the anterior end,** but exceptionally further posterior, perhaps at up to 40% of the odontostyle length. **Junction of odontostyle and odontophore simple. Odontophore** about two thirds of the odontostyle in length, **moderately sclerotized,** thickening slightly in the posterior region, but **lacking basal flanges.** Odontostylet protractor muscles attached to base of odontophore and running parallel to the cephalic region. Oesophagus comprising a narrow, cylindrical anterior section, which is looped back on itself when the odontostylet is in the retracted position, and a posterior bulboid expansion which is muscular and glandular with valve plates running for almost the full length. There are three glands: dorsal and two ventrosublateral. **The nucleus of the dorsal gland is situated some distance posteriorly to the orifice** and is smaller than the ventrosublateral nuclei. Nerve ring located around the narrow anterior section of the oesophagus; a second nerve ring, located more posteriorly, occurs in some species. Hemizonid prominent. Intestine simple, prerectum well developed and several anal body widths long. Anus in the form of a transverse slit. **Vulva** a transverse slit, **median in position.** Vagina well developed, muscular, at right angles to the body axis and leading to a substantial ovejector. **Genital tract amphididelphic, reflexed.** Tail short, dorsally convex-conoid to a finely rounded terminus, or broadly rounded. Several pairs of caudal pores present. Male genital tract diorchic, opposed, the posterior testis being reflexed. Both testes join a common vas deferens well anterior to the cloaca. Spicules dorylaimoid, paired, massive,

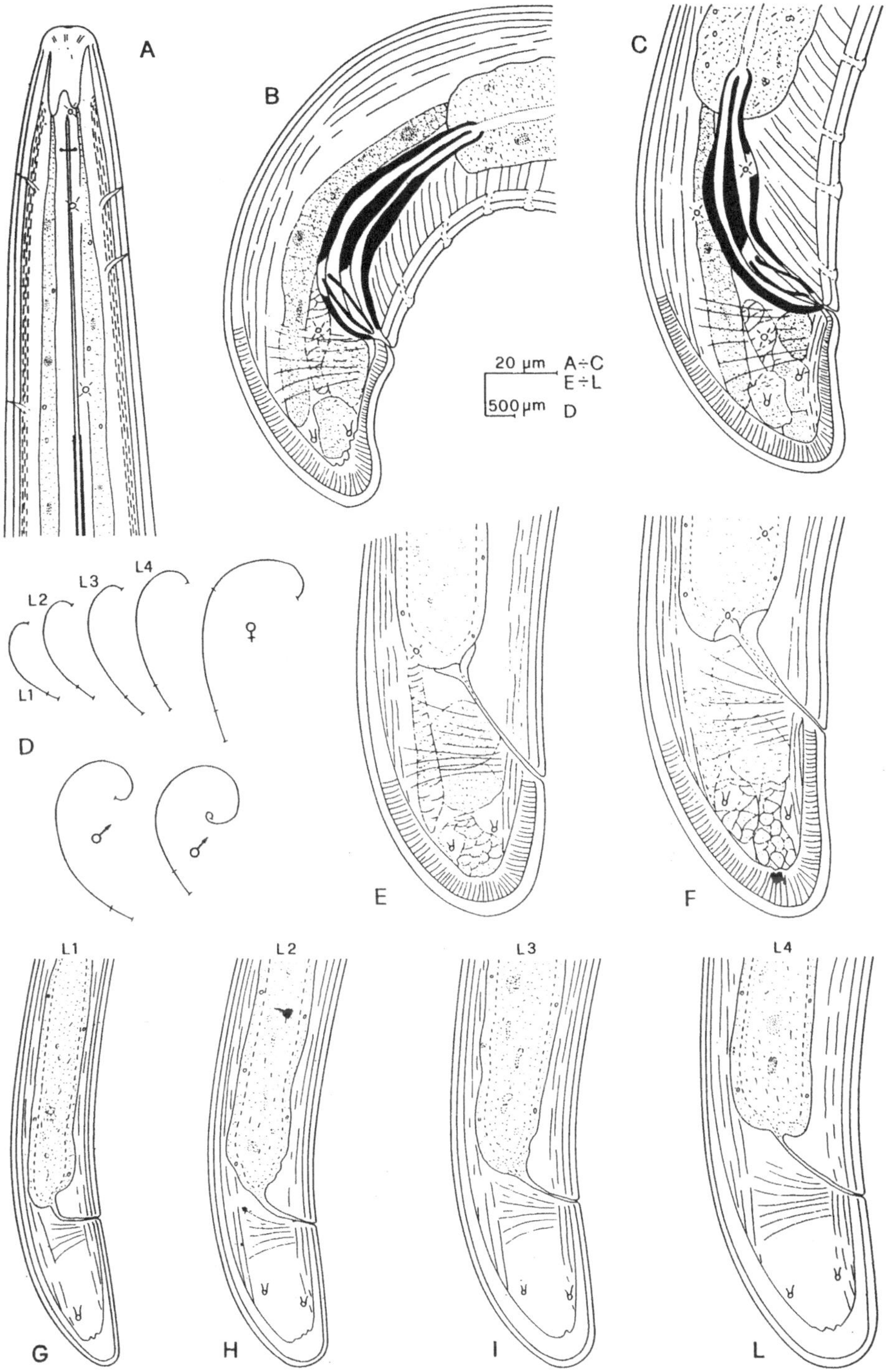

Fig. 8.1. *Longidorus athesinus*. A. Female lip region in lateral view; B, C. Male posterior regions; D. Posture of juvenile and adult stages; E, F. Female posterior regions; G,H,I,L. Tails of I, II, III, and IV juvenile stages, respectively. (After Lamberti *et al*., 1992. Courtesy *Nematologia Mediterranea*.)

ventrally arcuate and with short accessory guiding pieces located distally. Oblique copulatory muscles prominent and extending for several body widths anterior to the cloaca. **Copulatory supplements consisting of an adanal pair** (some species have two or three pairs) **and then a ventromedian series of up to 20 extending anteriorly without a hiatus between the adanal pair and the series.** In some species the ventromedian series may in part form a double, staggered row. Tail similar in shape to that of the female.

Type species:

Longidorus elongatus (de Man, 1876) Micoletzky, 1922
  syn. *Dorylaimus elongatus* de Man, 1876
    D. (Longidorus) elongatus de Man, 1876 (Micoletzky, 1922)
    Trichodorus elongatus (de Man, 1876) Filipjev, 1921
    D. tenuis von Linstow, 1879
    L. menthasolanus Konicek and Jensen, 1961
    L. monohystera Altherr, 1953

Other species:

L. aetnaeus Roca, Lamberti, Agostinelli and Vinciguerra, 1986
L. africanus Merny, 1966
L. alvegus Roca, Pereira and Lamberti, 1989
L. ampullatus Jacobs and Heyns, 1987
L. apulus Lamberti and Bleve-Zacheo, 1977
L. arenosus Kankina and Ivanova, 1986
L. athesinus Lamberti, Coiro and Agostinelli, 1992
L. attenuatus Hooper, 1961
L. auratus Jacobs and Heyns, 1987
L. belloi Andrés and Arias, 1988
L. belondiroides Heyns, 1966
L. breviannulatus Norton and Hoffmann, 1975
L. caespiticola Hooper, 1961
L. carpetanensis Arias, Andrés and Navas, 1986
L. closelongatus Stoyanov, 1964
L. cohni Heyns, 1969
L. congoensis Aboul-Eid, 1970
L. conicaudatus Khan, 1987
L. conicaudoides Jacobs and Heyns, 1989
  syn. *L. conicaudatus* Jacobs and Heyns, 1987 (= junior homonym)
    nec *L. conicaudatus* Khan, 1987
L. crassus Thorne, 1974
L. curvatus Khan, 1987
L. cylindricaudatus Kozlowska and Seinhorst, 1979
L. diadecturus Eveleigh and Allen, 1982
L. distinctus Lamberti, Choleva and Agostinelli, 1983
L. dunensis Brinkman, Loof and Barbez, 1987
L. edmundsi Hunt and Siddiqi, 1977

L. *eridanicus* Roca, Lamberti and Agostinelli, 1984
L. *euonymus* Mali and Hooper, 1974
L. *fangi* Xu and Cheng, 1990
L. *fasciatus* Roca and Lamberti, 1981
L. *fragilis* Thorne, 1974
L. *fursti* Heyns, Coomans, Hutsebaut and Swart, 1987
L. *globulicauda* Dalmasso, 1969
L. *goodeyi* Hooper, 1961
L. *henanus* Xu and Cheng, 1992
L. *heynsi* Andrássy, 1970
L. *himalayensis* (Khan, 1987) Xu and Hooper, 1990
    syn. *Neolongidorus himalayensis* Khan, 1987
L. *indicus* Prabha, 1973
L. *intermedius* Kozlowska and Seinhorst, 1979
L. *iranicus* Sturhan and Barooti, 1983
L. *ishrati* Javed, 1983
L. *iuglandis* Roca, Lamberti and Agostinelli, 1984
L. *jiangsuensis* Xu and Hooper, 1990
L. *jonesi* Siddiqi, 1962
    syn. *Neolongidorus jonesi* (Siddiqi, 1962) Khan, 1987
L. *juvenilis* Dalmasso, 1969
L. *juveniloides* Jacobs and Heyns, 1987
L. *kakamus* Jacobs and Heyns, 1987
L. *kuiperi* Brinkman, Loof and Barbez, 1987
L. *laevicapitatus* Williams, 1959
L. *leptocephalus* Hooper, 1961
L. *litchii* Xu and Cheng, 1992
L. *longicaudatus* Siddiqi, 1962
L. *lusitanicus* Macara, 1986
L. *macromucronatus* Siddiqi, 1962
L. *macrosoma* Hooper, 1961
L. *macroteromucronatus* Altherr, 1974
L. *magnus* Lamberti, Bleve-Zacheo and Arias, 1982
L. *major* Roca and D'Errico, 1987
L. *makatinus* Jacobs and Heyns, 1987
L. *martini* Merny, 1966
L. *mirus* Khan, Chawla and Seshadri, 1972
L. *mobae* Jacobs and Heyns, 1987
L. *moesicus* Lamberti, Choleva and Agostinelli, 1983
L. *monile* Heyns, 1966
L. *moniloides* Heyns, 1966
L. *nevesi* Macara, 1986
L. *nirulai* Siddiqi, 1965
L. *olegi* Kankina and Metlitskaya, 1983
L. *orientalis* Loof, 1983
L. *orongorongensis* Yeates, Van Etteger and Hooper, 1992
L. *paraelongatus* Altherr, 1974

L. paramirus Darekar and Khan, 1983
L. paramonile Jacobs and Heyns, 1982
L. pawneensis Luc and Coomans, 1988
L. picenus Roca, Lamberti and Agostinelli, 1984
L. pini Andrés and Arias, 1988
L. pisi Edward, Misra and Singh, 1964
syn. *Brevinema pisi* (Edward, Misra and Singh, 1964) Stegarescu, 1980
L. siddiqii Aboul-Eid, 1970
Xiphinema brevicaudatum of Siddiqi, 1959 nec Schuurmans Stekhoven, 1951
L. latocephalus Lamberti, Choleva and Agostinelli, 1983
L. poessneckensis Altherr, 1974
L. profundorum Hooper, 1966
L. protae Lamberti and Bleve-Zacheo, 1977
L. proximus Sturhan and Argo, 1983
L. pseudoelongatus Altherr, 1976
L. psidii Khan and Khan, 1972
L. reneyii Raina, 1966
L. rotundicaudatus Jacobs and Heyns, 1987
L. saginus Khan, Seshadri, Weischer and Mathen, 1971
L. sylphus Thorne, 1939
syn. *L. striola* Merzheevskaya, 1951
L. taniwha Clark, 1963
L. tardicauda Merzheevskaya, 1951
L. tarjani Siddiqi, 1962
L. unedoi Arias, Andrés and Navas, 1986
L. vineacola Sturhan and Weischer, 1964

Species inquirendae:
L. brevicaudatus (Schuurmans Stekhoven, 1951) Siddiqi, 1962
syn. *Xiphinema brevicaudatum* Schuurmans Stekhoven, 1951
Neolongidorus brevicaudatus (Schuurmans Stekhoven, 1951) Khan, 1987
L. meyli Sturhan, 1963 (probably a synonym of *Paralongidorus (Siddiqia) maximus)*

Bionomics: Migratory ectoparasites, mainly of roots of herbaceous plants, their feeding resulting in galling and stunting of the root system. Most species appear to be polyphagous and a number of species have been implicated in the transmission of nepoviruses.

Distribution: The majority of species have been described from regions with a temperate or mediterranean type climate. *Longidorus* is found worldwide, with the probable exception of South and Central America where *L. laevicapitatus*, for example, is a recent introduction. Hunt and Siddiqi (1977) described *L. edmundsi* from littoral sand in St. Lucia and Dominica in the West Indies, but this could also be an introduction as very similar forms have since been described from comparable edaphic situations in Italy and Holland.

Etymology: Derived from the Latin *longus* = long and the Greek *dorys* = spear.

Species descriptions

Longidorus attenuatus Hooper, 1961 (Fig. 8.2)

Measurements: (After Hooper, 1961) Holotype female: L = 6.0 mm; a = 201; b = 18.9; c = 107; V = $^{8}49^{7}$; tail length = 55 μm; odontostyle (ods) = 75 μm; odontophore (odp) = 43 μm; oral aperture to guide ring (ogr) = 29 μm.
Male (n = 2): L = 5.5, 6.3 mm; a = 119, 171; b = 12.9, 17.1; c = 99, 114; ods = 80, 84 μm; odp = 38, 40 μm; ogr = 30, 32 μm.
Specimens from Germany and England. Female (n = 32): L = 5.2–7.5 (6.3) mm; a = 109–210 (151); b = 12.7–22.5 (15.5); c = 82–147 (112.6); V = $^{6-12}44\text{–}55^{6-10}$ ($^{8}49^{8}$); ods = 73–93 (81) μm; odp = 36–45 (42) μm; ogr = 27–32 (31) μm.
Juveniles: (After Hooper, 1961): First stage (7): L = 1.0–1.3 (1.1) mm; a = 56–78 (64); b = 4.8–6.7 (6.0); c = 28–39 (32); ods = 44–49 (47) μm; odp = 18–30 (24) μm; replacement ods (rep. ods) = 46–55 (51) μm. Second stage (4): L = 1.5–1.8 (1.7) mm; a = 54–72 (67); b = 6.2–6.5 (6.3); c = 31–35 (34); ods = 43–52 (48) μm; odp = 27–32 (30) μm; rep.ods = 57–67 (62) μm. Third stage (6): L = 2.3–2.8 (2.7) mm; a = 68–95 (80); b = 5.1–9.5 (8.3); c = 35–60 (47); odp = 32–36 (35) μm; rep.ods = 63–71 (68) μm. Fourth stage (8): L = 3.8–4.5 (4.2) mm; a = 83–120 (101); b = 9.9–14.4 (12.1); c = 47–76 (67); ods = 67–71 (69) μm; odp = 35–49 (40) μm; rep.ods = 75–86 (79) μm.

Description: (modified after Brown and Boag, 1977)
Female: Body long and narrow; when relaxed by heat it forms an open C-shape with slightly greater curvature in the posterior half. **Cephalic region flattened, characteristically expanded and definitely offset from the neck contour;** about 75% of the width of the neck at the guide ring. Amphids large, **symmetrically bilobed** at their base, extending half way to the guide ring and almost encircling the head. Amphid pores very small. Odontostyle long, attenuate, about twice the length of the odontophore with which it has a simple junction. Oesophagus typical of the genus. Oesophago-intestinal valve roundly conoid. Lateral chords about half mid-body width, tapering to a thin line in the oesophageal region where lateral pores are serially arranged. The oesophageal region has five to eight ventral body pores. Vulva a transverse slit one quarter of the mid-body width wide. **The tail is dorsally convex, conoid, 1.5–1.75 times the anal body width long.** A pair of caudal pores is present on each side of tail.

Male: Rare. **Spicules 45 μm long;** lateral guiding pieces with bifid distal ends. An adanal pair of supplements plus a ventral series of ten present. Tail dorsally convex, ventrally concave, narrowly conoid.

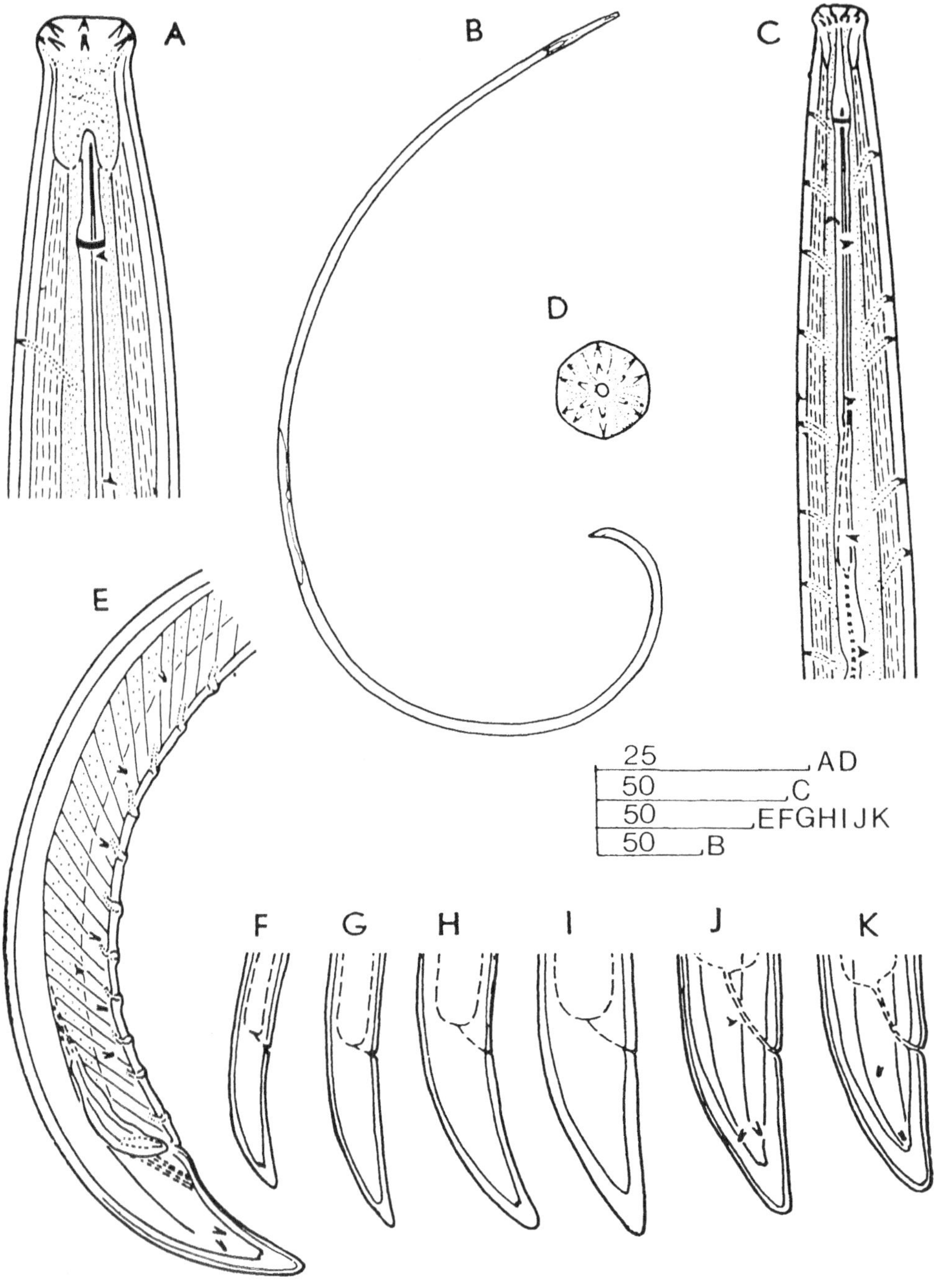

Fig. 8.2. *Longidorus attenuatus.* A. Head, lateral; B. Adult female; C. Anterior oesophageal region, ventral; D. Lip region, *en face*; E. Male tail, lateral; F–K. Tails; F. J1; G. J2; H. J3; I. J4; J, K. Female tails, lateral. (B, after Siddiqi, the rest after Hooper, 1961.)

Juveniles: Lip region and amphids similar to adults. Tail of first stage is conical, about four anal body widths long.

Bionomics: This species is found most frequently in open, well-drained, sandy soils (Whitehead and Hooper, 1970; Taylor and Brown, 1976). Depth distribution depends on crop and time of year with few nematodes being found in the top 5 cm in the summer months (Whitehead *et al.*, 1971). Reproduction is parthenogenetic, the life cycle lasting in excess of one year (Flegg, 1968). *L. attenuatus* feeds mainly on root tips and is a contributory factor in Docking disorder of sugarbeet (Whitehead and Hooper, 1970). It is polyphagous, but mainly attacks herbaceous plants, the host range including barley, brassicas, carrot, clover, grapevine, lucerne, potato, strawberry, sugarbeet and wheat. *L. attenuatus* is a vector of tomato black ring virus (TomBRV) to sugarbeet.

Distribution: Widespread in Europe and the CIS (formerly USSR). Recent distribution reviews include those of: Brown and Taylor, 1987; Brown *et al.*, 1990 and Martelli and Taylor, 1990.

Longidorus elongatus (de Man, 1876) Micoletzky, 1922 (Fig. 8.3)

syn. *Dorylaimus elongatus* de Man, 1876
Trichodorus elongatus (de Man, 1876) Filipjev, 1921
Dorylaimus (Longidorus) elongatus de Man, 1876 (Micoletzky, 1922)
Dorylaimus tenuis Linstow, 1879
Longidorus monohystera Altherr, 1953
Longidorus menthasolanus Konicek and Jensen, 1961

Measurements: (After Hooper, 1961) Neotype female: L = 4.87 mm; a = 122; b = 18.7; c = 128; V = 51; tail length = 38 μm; odontostyle (ods) = 95 μm; oral aperture to guide ring (ogr) = 29 μm.
Female (n = 84, from 10 different British populations): L = 4.5–6.4 (5.5) mm; a = 76–123 (92); b = 9.7–17.5 (13.3); c = 73–141 (99); V = $^{5-11}45-53^{5-12}(^{9}49^{8})$; odontostyle (ods) = 81–102 (94) μm; odontophore (odp) = 34–71 (50) μm; ogr = 30–36 (33) μm.
Juveniles:
Second stage (16): L = 1.5–2.1 (1.8) mm; a = 47–63 (56); b = 5.3–8.4 (6.2); c = 27–43 (37); ods = 59–67 (63) μm; odp = 34–49 (43) μm; rep ods = 70–77 (73) μm.
Third stage (12): L = 2.5–3.2 (2.7) mm; a = 51–85 (67); b = 6.5–11.1 (8.1); c = 38–56 (51); ods = 69–76 (73) μm; odp = 45–56 (50) μm; rep ods = 79–86 (83) μm.
Fourth stage (5): L = 3.7–4.1 (3.9) mm; a = 62–81 (69); b = 10.6–12.7 (11.6);

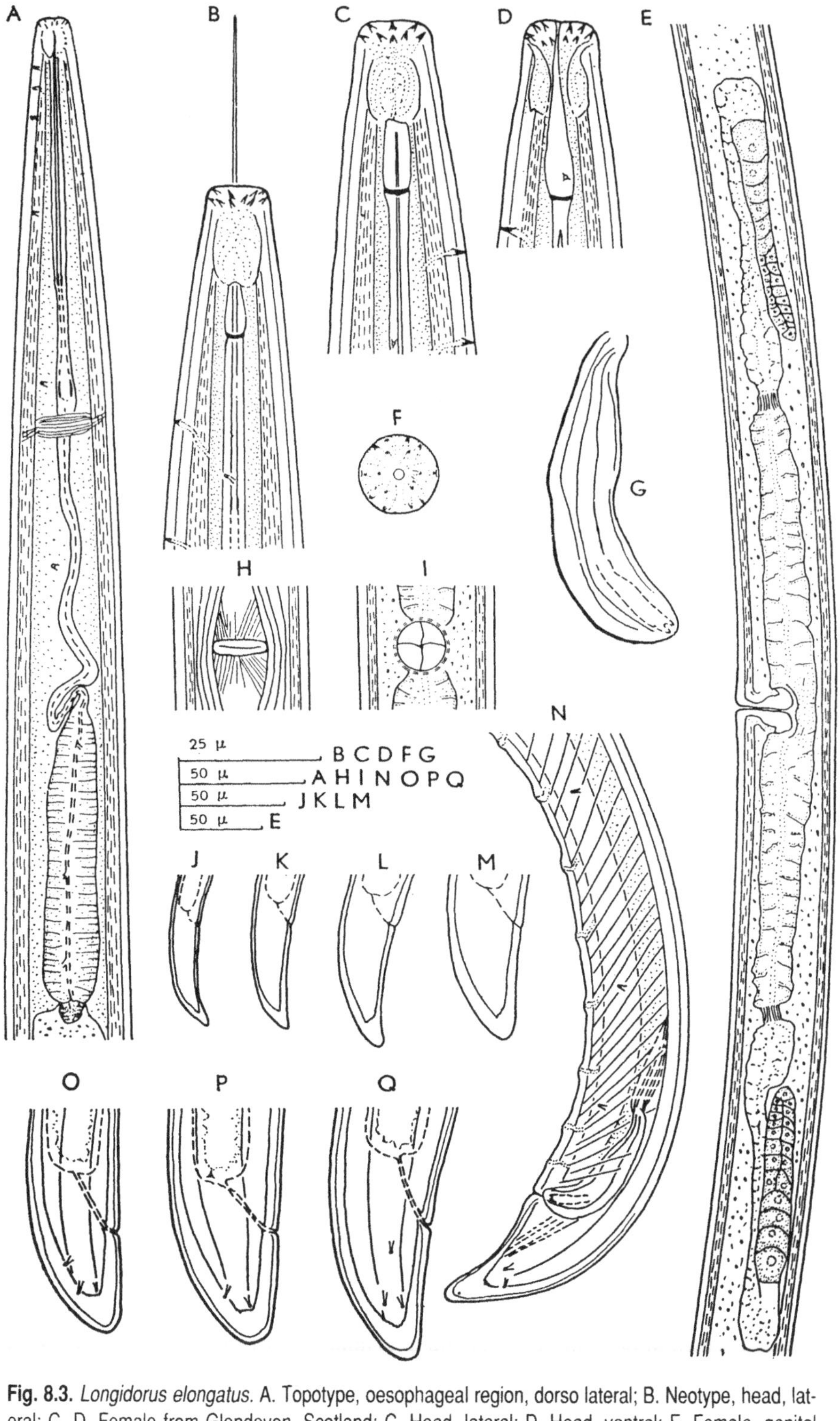

Fig. 8.3. *Longidorus elongatus.* A. Topotype, oesophageal region, dorso lateral; B. Neotype, head, lateral; C, D. Female from Glendevon, Scotland; C. Head, lateral; D. Head, ventral; E. Female, genital tracts, lateral; F. Lip region *en face*; G. Spicules and lateral guiding piece, lateral; H. Vulva, ventral;

c = 52–81 (66); ods = 80–86 (83) μm; odp = 51–60 (55) μm; rep ods = 90–95 (92) μm.

Description: (modified after Hooper, 1973)

Female: Body long and narrow, open C-shaped to slightly coiled when killed by heat; narrows anteriorly to a **cephalic region that is continuous with or very slightly offset from the neck contour. Cephalic region flattened anteriorly** about half to two-thirds the body width at the guide ring with 16 papillae directed away from the oral aperture. Amphids large, pouch-like, **extending about half the distance from the oral aperture to the conspicuous guide ring,** the base of the pouch is slightly bilobed in lateral view; amphid apertures pore-like, very indistinct, at base of lip region. Odontostyle long and narrow with a simple junction with the odontophore; odontophore about half the odontostyle length and widening and thickening posteriorly, but not very conspicuous. Anterior oesophagus a narrow tube often bent or looped as it joins the wider, muscular, posterior oesophageal bulb that is about five times as long as wide. Oesophago-intestinal valve bluntly convex-conoid. Dorsal oesophageal and subventral gland openings and their associated glands at about 15 and 59% of the distance from anterior end of oesophageal bulb; suggestion of second subventral opening at 85% of its length. A nerve ring surrounds the anterior oesophagus just behind the base of the odontophore and a second nerve ring about a body width behind the first is sometimes visible. Prerectum about ten anal body widths long, rectum just less than one anal body width. Tail dorsally convex, ventrally flattened or very slightly concave, about 1–1.3 anal body widths long with a roundly conoid terminus and two or three lateral pores present on each side. Cuticle generally smooth, inner layer thickens in the neck region and on the tail and has very faint transverse striations that have a radial appearance on the tail. Glandular hypodermal lateral chords present about a quarter of the mid-body width wide but narrowing to a thin line in the oesophageal region. Lateral pores arise from the chords throughout the length of the body and are mainly in a staggered double row but merge to a single row in the oesophageal region. **Usually four to six ventral pores present in the oesophageal region** and occasionally a single dorsal pore somewhat posterior to the guide ring. Vulva a transverse slit about a third of the mid-body width wide. Genital tracts paired, opposed, reflexed; reflexed part consists mainly of the ovary in which the larger oocytes are in a single row, the rest multiple.

Male: Anterior region similar to female, posterior third curved ventrally when killed by heat. Paired spicules prominent, about 58 μm long, separate, arcuate with bluntly rounded distal ends. Lateral guiding pieces present with furcate distal ends; tail conoid to bluntly conoid, dorsally convex, ventrally concave with two or three lateral pores on each side. An adanal pair plus a **ventral**

I. Vagina, ventral; J–M. Juvenile tails; J. J1; K. J2; L. J3; M. J4; N. Male, Elgin, Scotland, tail, lateral; O. Neotype, female, tail, lateral; P. Topotype, female, tail, lateral; Q. Female, tail, lateral. (After Hooper, 1961. Courtesy *Nematologica*.)

series of seven to 10 supplementary papillae present. Testes paired, opposed, anterior testis almost reaching to middle of body. **Males are usually very rare or absent in most populations,** but in some populations they may be almost as common as females.

Juveniles: Anterior region similar to adults but a replacement odontostyle is always present in the anterior oesophagus and in the first stage its anterior end lies inside the odontophore of the functional odontostyle. The juvenile tails become blunter and shorter relative to anal body width as they progress from the first to fourth stage, but the actual tail length stays about the same.

BIONOMICS: *L. elongatus* occurs in various soil types ranging from sandy to sandy loams to fen peat. Depth distribution (see Boag *et al.*, 1987) is related to the penetration of the host root system, soil moisture and time of year. Boag *et al.* (1991) showed that the distribution of *L. elongatus* in Europe was related to the July soil isotherm. Reproduction is usually parthenogenetic although some populations have abundant males. At 30°C the life cycle can be completed in 9 weeks (Wyss, 1970a), but under field conditions there is probably one generation per year (Thomas, 1969). *L. elongatus* is a polyphagous ectoparasite attacking plants such as carrot, peppermint, raspberry, redcurrant, strawberry and sugarbeet. Direct feeding activities can severely damage carrot, peppermint, strawberry and sugarbeet, the affected root systems being stunted with galled root tips. *L. elongatus* has been implicated in Docking disorder of sugarbeet (Whitehead and Hooper, 1970) and transmits raspberry ring spot virus (RRV), tomato black ring virus (TomBRV) and spoon leaf virus. RRV is readily transmitted to raspberry even though the plant is a poor host to the nematode.

DISTRIBUTION: Widespread in temperate regions (CIS, Europe, India, New Zealand, North America, South Africa). Recent distribution reviews include those of: Brown and Taylor, 1987; Brown *et al.*, 1990, Martelli and Taylor, 1990 and Robbins and Brown, 1991.

LONGIDORUS LAEVICAPITATUS WILLIAMS, 1959 (Fig. 8.4)

MEASUREMENTS: (Ex sugarcane soil, Mauritius; after Williams, 1959)
Female (n = 10): L = 2.0–3.2 (2.5) mm; a = 53–63 (57); b = 7.0–11.5 (9.4); c = 61–76 (68); V = 46–49 ($^{8}47^{9}$).
(Additional measurements from type specimens.)
Male (n = 6): lip region width (lw) = 9 μm; oral aperture to guide ring (ogr) = 21.5–22.5 (22.3) μm; body width at guide ring (bwgr) = 16–19 (17.3) μm; length of amphid pouch = 12–15 (14.2) μm; odontostyle (ods) (n = 3) = 60–67 (63); c' = 1.1–1.3 (1.2).
(Ex sugarcane soil, Natal; after Jacobs and Heyns, 1982)
Female (n = 78): L = 2.04–3.28 mm; a = 55–83; b = 7.3–11.6; c = 68–88; c' = 1.0–1.6; V = 45–50; ods = 47–66; ogr = 22–27 μm; lw = 9–12 μm.
(Ex yam soil, Nigeria.)

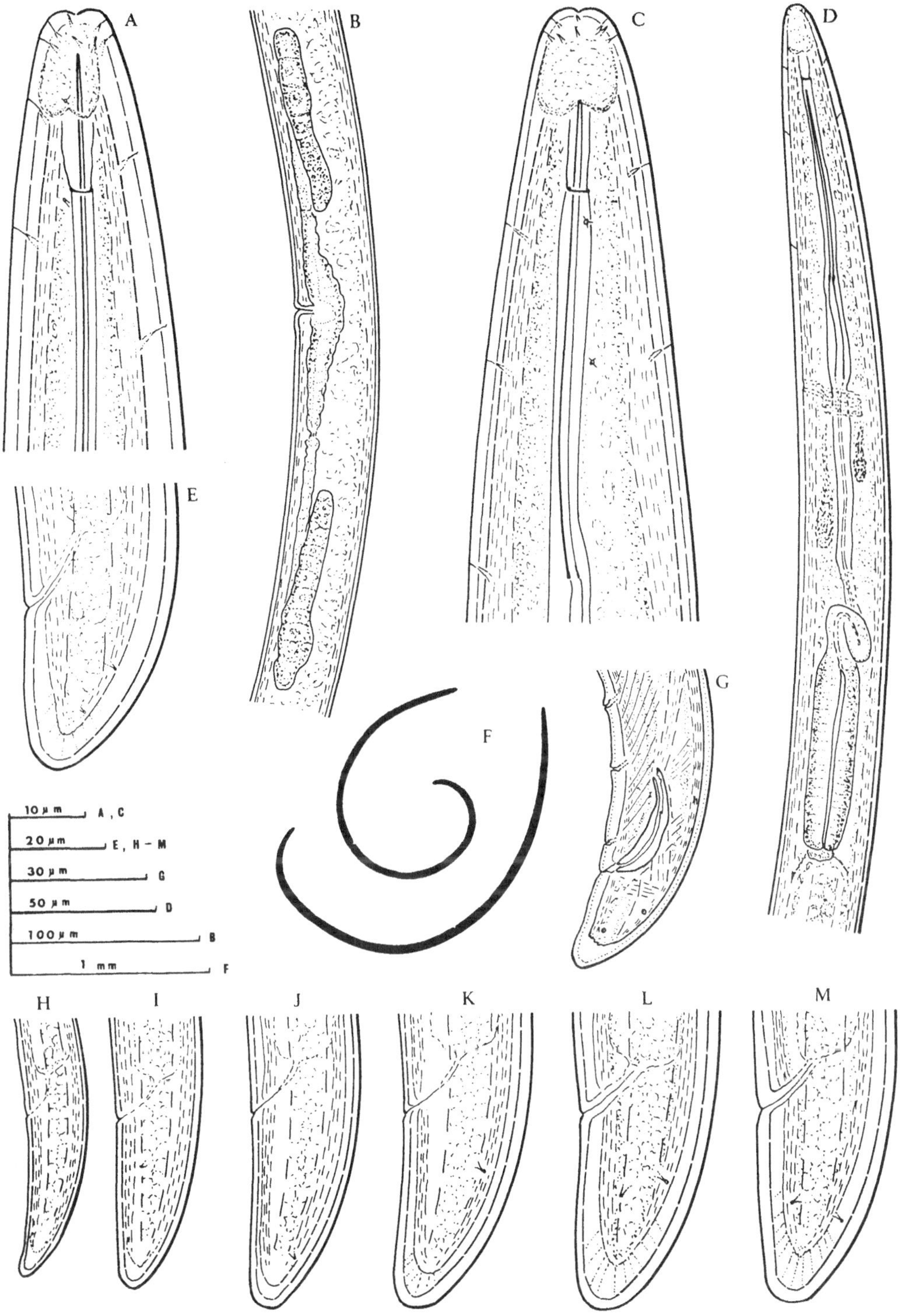

Fig. 8.4. *Longidorus laevicapitatus.* A. Anterior end, sublateral; B. Female genital tracts; C. Anterior end, lateral; D. Oesophageal region, lateral; E. Female tail (slightly flattened); F. Death shape of heat killed females; G. Male tail, lateral; H–M. Tails, lateral, of: H, first stage juvenile; I, second stage; J, third stage; K, fourth stage; L, M, females. (After Hooper, 1985.)

Female (n = 13): L = 2.16–2.90 (2.53 ± 0.62) mm; a = 58–70 (61 ± 1.0); b = 7.6–13.4 (9.6 ± 0.4); c = 58–73 (65 ± 1.4); c′ = 1.2–1.6 (1.4 ± 0.1); V = 46–50 (47 ± 0.3); ods = 55–64 (60 ± 0.7); ogr = 21–25 (22.5 ± 0.4) μm; bwgr = 15–19 (17.3 ± 0.3) μm.
Juveniles: First stage (n = 13). L = 0.75–0.81 (0.78 ± 0.68) mm; a = 41–67 (43 ± 0.5); b = 4.4–6.0 (5.1 ± 0.1); c = 24–32 (27 ± 0.6); c′ = 2.0–2.7 (2.4 ± 0.1); ods = 38–44 (40 ± 0.7) μm; replacement odontostyle (rep. ods) = 39–46 (43 ± 0.5) μm; ogr = 15–17 (16 ± 0.2) μm; bwgr = 10–12 (11.4 ± 0.24) μm.
Second stage (n = 9). L = 1.13–1.26 (1.22 ± 0.25) mm; a = 44–55 (50 ± 1.0); b = 5.9–7.3 (6.4 ± 0.2); c = 29–45 (36 ± 1.8); c′ = 1.5–2.4 (1.8 ± 0.1); ods = 38–44 (42 ± 0.6); rep. ods = 46–55 (51 ± 1.1) μm; ogr = 16–19 (17.4 ± 0.2) μm; bwgr = 13–16 (14 ± 0.3) μm.
Third stage (n = 9). L = 1.72–1.93 (1.84 ± 0.22) mm; a = 51–61 (57 ± 1.2); b = 6.2–9.9 (7.9 ± 0.4); c = 41–54 (46 ± 1.4); c′ = 1.4–1.8 (1.5 ± 0.1); ods = 47–55 (51 ± 0.8); rep. ods = 55–64 (59 ± 1.0) μm; ogr = 18–24 (20 ± 0.7) μm; bwgr = 14–17 (16 ± 0.3) μm.
Fourth stage (n = 8). L = 1.98–2.39 (2.15 ± 0.42) mm; a = 48–62 (56 ± 1.6); b = 7.5–11.1 (8.7 ± 0.5); c = 47–55 (52 ± 0.8); c′ = 1.3–1.8 (1.6 ± 0.1); ods = 52–60 (56 ± 0.8); rep. ods = 59–66 (62 ± 1.0) μm; ogr = 19–24 (22 ± 0.5) μm; bwgr = 14–17 (16 ± 0.4) μm.
(Ex sugarcane soil, Guadeloupe; after Dalmasso, 1967)
Male (n = 3): 2.09–2.14 (2.12) mm; a = 44–61 (50); b = 7.4–8.8 (8.0); c = 58–62 (59); c′ = 1.0–1.3 (1.1); ods = 57–61 (59) μm; ogr = 24 μm; T = 61–65 (63); spicule length (calculated from fig.) = 46 μm.

Description: (modified after Hooper, 1985)
One of the smaller species in the genus.
Female: Body C-shaped to spiral in form when killed by heat and virtually cylindrical from base of oesophagus to anus; tapering gradually anteriorly to the **cephalic region** which is **smoothly rounded and continuous with the body contour.** Cuticle and hypodermis with faint transverse striations which are more easily seen in the neck and tail regions; cuticle in two layers, the outer layer being thicker throughout most of the body but the inner layer becomes thicker in the neck region and particularly on the tail where the terminal region has radial striations. Lateral chord about one third the body width throughout most of the body with two irregular rows of pores; narrows anteriorly to a single line in the anterior oesophageal region where there are also some ventral and dorsal body pores. **Cephalic region some 9 μm wide, about half the body width at the guide ring.** amphids pouch-like, extending about two thirds the distance to the guide ring; base of pouch indented to slightly bilobed, amphidial apertures not visible. Guide ring not very distinct, about 23 μm from the oral opening. Odontostyle, odontophore and oesophagus typical for the genus. **Tail one to one and a half anal body widths long, dorsally convex, conoid, with a bluntly rounded terminus;** usually one pair of lateral pores on each side of the tail. Vulva about median in position with genital tracts amphididelphic, reflexed.

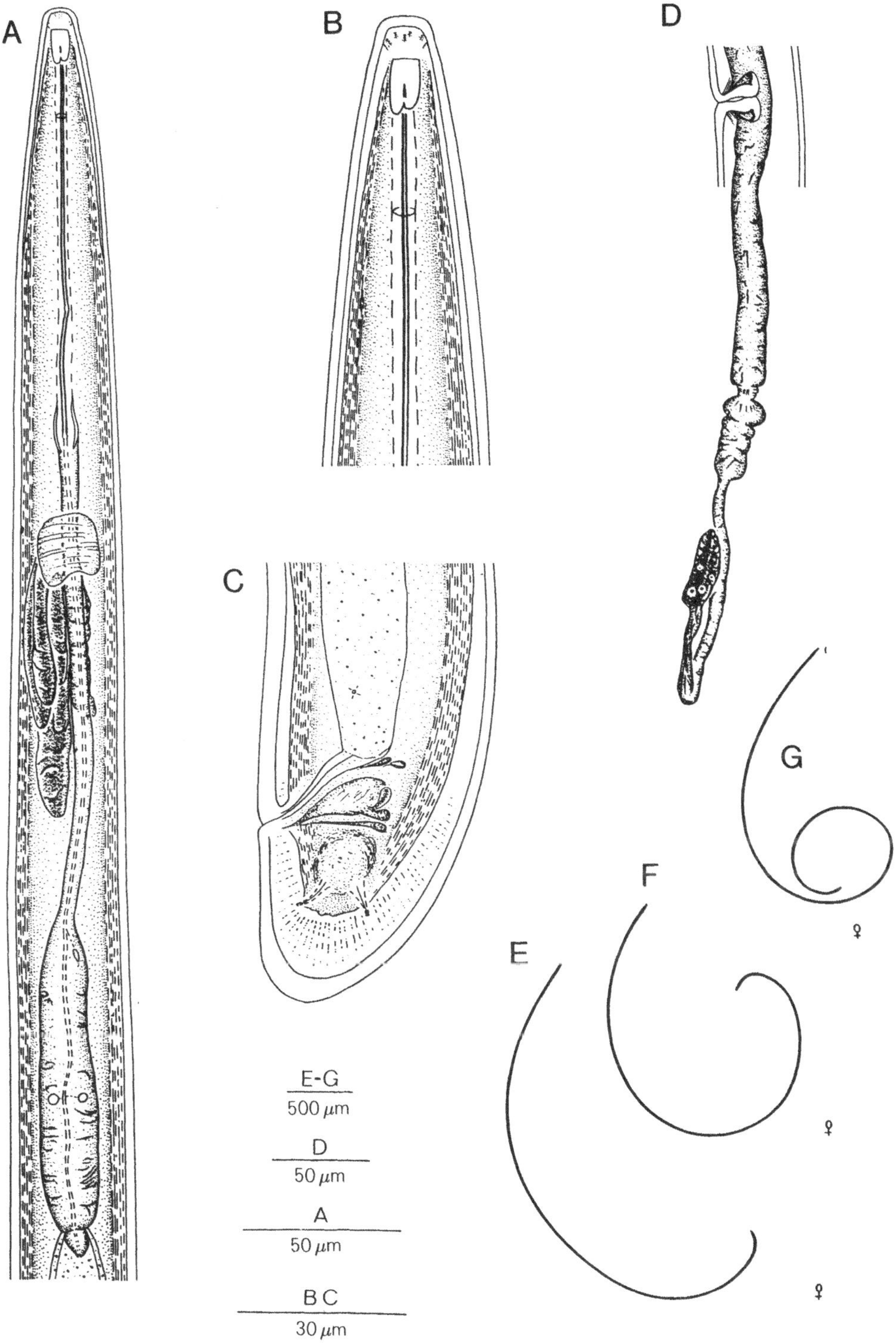

Fig. 8.5. *Longidoroides pulcher.* A. Anterior part of body; B. Head, lateral view; C. Tail of female; D. Posterior genital tract of female; E–G. Relaxed body posture. (After Jacobs and Heyns, 1982. Courtesy *Phytophylactica*.)

Male: (After Dalmasso, 1967). Very rare. Three males recorded from amongst many females in a sample from Guadeloupe, French West Indies. Tail dorsally convex, roundly conoid with a pair of lateral papillae. **The copulatory supplements consist of an adanal pair plus four to five in a ventromedian row.** Spicules about 45 μm long and of the usual longidorid shape, but with a small, internal, dorsal diverticulum which looks thorn-like in some views; lateral accessory guiding pieces not seen.

Juveniles: All stages assume an open C-shape when killed by heat. The first stage has the anterior part of the replacement odontostyle inserted into the base of the odontophore. Tail at first stage, relative to anal body width, narrower and longer than in female ($c' = 2.0$–2.7) and more conoid; the cuticle at the terminus much thickened and sometimes with a slight indentation on the dorsal side giving the terminus a pegged appearance. The tail becomes relatively blunter and shorter in later stages.

BIONOMICS: *L. laevicapitatus* is often found in fairly heavy, fine textured, soils (Spaull and Heyns, 1991), usually in association with sugarcane. It is parthenogenetic and males are rare. In addition to sugarcane, it has been recorded associated with many crops, e.g. avocado, banana, citrus, cocoa, coffee, pineapple, potato and yam. Kaiser *et al.* (1978) recorded it in association with tomato black ring virus (TBRV) infected potato, but did not demonstrate transmission.

DISTRIBUTION: Widely distributed, probably by association with sugarcane, its major host, in the tropical and subtropical regions (Africa, Caribbean, Central and South America, Israel, Mauritius, South Pacific, etc.).

GENUS *LONGIDOROIDES* KHAN, CHAWLA AND SAHA, 1978 (Fig. 8.5)

DIAGNOSIS: Longidorinae. Body length ranging from 2.5 to 9 mm, but usually in the range of 4–5 mm. Heat relaxed form usually C-shaped or an open spiral. Cephalic region continuous with the body contour, exceptionally expanded, but not offset by a distinct groove. **Amphidial apertures in the form of short to medium length transverse slits. Amphidial fovea pouch-like, as broad or broader than the slit for at least a slit width (usually more) posteriorly and hardly tapering to a broad, rounded, usually bilobed, base. Odontostyle** needle-like, attenuate, 55 μm to over 100 μm long and **with a simple base** at its junction with the odontophore. **Odontophore lacking basal flanges.** Guide ring anteriorly located near the lip region and within the anterior third of the odontostyle. **Dialatores buccae absent. Dorsal oesophageal gland nucleus smaller than the nuclei of the ventrosublateral glands and located posterior to its orifice.** Vulva approximately median in position (44–58%). Genital tracts amphididelphic, reflexed. No uterine sclerotizations reported. Tail hemispheroid to short or medium conoid in shape. Male genital tract diorchic, opposed. Spicules dorylaimoid in form with lateral accessory guiding pieces. Copulatory supplements comprising an adanal pair, sometimes followed by two or three staggered, slightly ventrosublateral pairs and then a **ventromedian series of up to 11 supplements commencing without a pronounced hiatus from the adanal pair.** Tail similar to that of female in form.

TYPE SPECIES:
Longidoroides afzali (Khan, 1964) Khan, Chawla and Saha, 1978
syn. *Longidorus afzali* Khan, 1964
Paralongidorus afzali (Khan, 1964) Siddiqi and Husain, 1965

OTHER SPECIES:
L. bikanerensis Lal and Mathur, 1986
L. boshi (Khan, Saha and Seshadri, 1972) Khan, Chawla and Saha, 1978
syn. *P. boshi* Khan, Saha and Seshadri, 1972
L. cedari Khan, Chawla and Saha, 1978
L. costatus Jacobs and Heyns, 1987
L. droseri (Sukul, 1972) Khan, Chawla and Saha, 1978
syn. *P. droseri* Sukul, 1972
L. latilabiatus Jacobs and Heyns, 1982
L. longiurus Chawla and Samathanam, 1980
L. lutosus (Heyns, 1965) Khan, Chawla and Saha, 1978
syn. *Longidorus lutosus* Heyns, 1965
Paralongidorus lutosus (Heyns, 1965) Aboul-Eid, 1970
L. pini Jacobs and Heyns, 1987
L. pulcher Jacobs and Heyns, 1982
L. pulcheroides Jacobs and Heyns, 1987
L. spiralis (Khan, Saha and Seshadri, 1972) Khan, Chawla and Saha, 1978
syn. *Paralongidorus spiralis* Khan, Saha and Seshadri, 1972
L. strelitziae (Heyns, 1966) Khan, Chawla and Saha, 1978
syn. *Longidorus strelitziae* Heyns, 1966
P. strelitziae (Heyns, 1966) Aboul-Eid, 1970
L. utriculoides (Corbett, 1964) Khan, Chawla and Saha, 1978
syn. *Longidorus utriculoides* Corbett, 1964
P. utriculoides (Corbett, 1964) Siddiqi and Husain, 1965

NOTES: *L. droseri* (Sukul, 1972) is spelled both as *dorseri* and *droseri* in the original description. As the plant with which it was associated is *Drosera*, the latter spelling is the author's intentional nomen, *dorseri* therefore being an incorrect original spelling (Article 32c, ii) and a *lapsus calami* which has unfortunately been repeated by some authors.

REMARKS: This emended diagnosis of *Longidoroides* necessitates the relocation of several species. *Longidoroides clavicaudatus* Jacobs and Heyns, 1982 is transferred to *P. (Paralongidorus)* as it has narrow, tapering, amphid fovea rather than broad pouches. *Longidoroides hooperi* (Heyns, 1966) is, as noted by Heyns (1966b), more problematic in its placement with its combination of expanded head but short amphidial slit. However, the subgenus *P. (Siddiqia)* does contain some species with the latter character, e.g. *P. (S.) dasturi*, and is the most appropriate location for the species at present. *Siddiqia seclipsi* Khan, Singh and Singh, 1981 was placed in *Longidoroides* by Luc and Doucet (1984) on the basis of the reportedly bilobed amphid fovea. However, the fovea are tapering and funnel-shaped rather than pouch-like, the fovea base is similar in form both to *P. (P.) clavicaudatus* and *P. (S.) capensis* (both of which, incidentally, have a median cuticular rib in the fovea wall which gives a bilobed appearance posteriorly near the funnel base), and also to those species formerly placed in *Inagreius*, and the species is otherwise typical of *Siddiqia* to which it is returned. [The problems posed by the shape of the fovea base can be additionally exemplified by *Xiphinema neodimorphicaudatum* which was described and illustrated as having a cup-shaped fovea with a bilobed base (Khan, 1982)!]. The validity of *Longidoroides* is still open to doubt. One possible 'solution' might well be to ignore altogether the shape of the amphidial fovea and lump all the species of *Longidoroides* under *Paralongidorus*, either with or without the subgenus *Siddiqia* based solely on head shape. However, I feel that such a move is premature pending more detailed study, particularly of the Indian species, and that a certain degree of tolerance to heterogenous or 'fuzzy' genera in this group is the more constructive approach.

Bionomics: Presumably migratory ectoparasites of higher plants and found in the rhizosphere of sugarcane, vines, pines, groundnut, etc., but no further details reported.

Distribution: The nominal species are mostly described from South Africa, Malawi and India. An undescribed species has been reported from Brazil (Rashid *et al.*, 1986a) and represents an unusual departure from the existing known distribution. A report of *L. afzali* from the USA (Norton *et al.*, 1984) requires confirmation (see Robbins and Brown, 1991). The current distribution pattern suggests that the group evolved in what is now south east Africa and India, possibly subsequent to the break-up of Gondwanaland.

Etymology: Derived from *longidorus* and the Greek *oides* = likeness of form.

GENUS *PARALONGIDORUS* SIDDIQI, HOOPER AND KHAN, 1963

NOTE: On the basis of the characteristically offset and expanded cephalic region, I propose considering *Siddiqia* as a subgenus under *Paralongidorus* in this classification and the species are arranged accordingly.

Subgenus *Paralongidorus (Paralongidorus)* Siddiqi, Hooper and Khan, 1963 (Fig. 8.6)

Diagnosis: Longidorinae. Long, slender nematodes. Heat relaxed form more or less straight to C-shaped. **Cephalic region continuous with body contour,** not

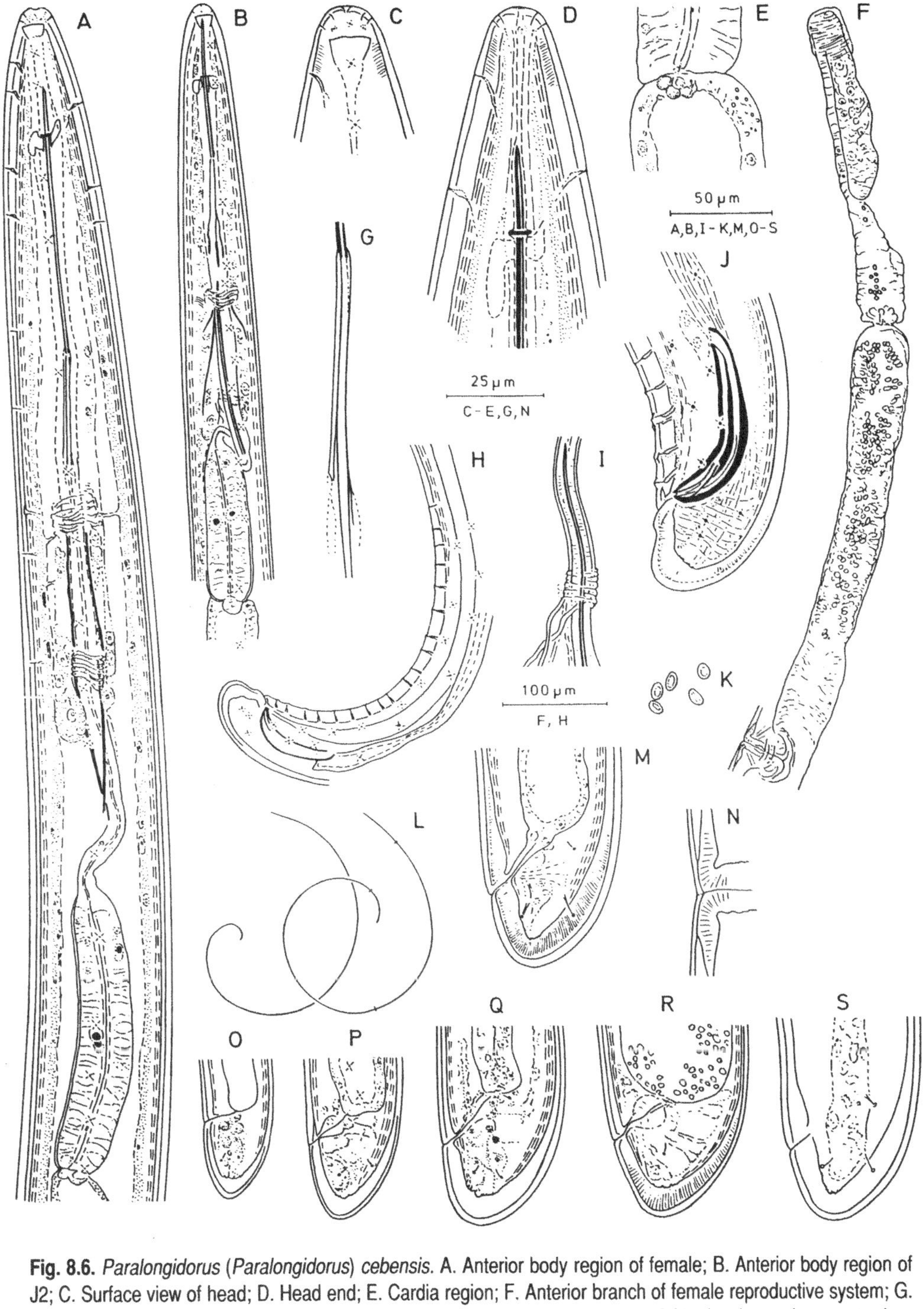

Fig. 8.6. *Paralongidorus* (*Paralongidorus*) *cebensis*. A. Anterior body region of female; B. Anterior body region of J2; C. Surface view of head; D. Head end; E. Cardia region; F. Anterior branch of female reproductive system; G. Odontophore; H. Posterior part of male; I. Part of anterior slender oesophagus of J4, showing replacement odontostyle and oesophageal retractors; J. Male tail; K. Sperm cells in vesicula seminalis of male; L. Body posture of male and female; M. Female tail; N. vulva showing radial striations; O. Tail of J2; P. Tail of J2/3?; Q. Tail of J3; R. Tail of J4; S. Tail of female in surface view; F,M,S. Holotype female; other drawings from paratypes. (After Heyns and Coomans, 1989. Courtesy *Phytophylactica.)*

expanded and offset by a distinct groove. **Amphidial apertures in the form of transverse slits. Amphidial fovea elongate, funnel-shaped.** Odontostyle long, attenuate and may be strongly sclerotized. Junction with odontophore ranges from plain to markedly furcate. Odontophore usually lacking well developed basal flanges, but showing distinct development in exceptional species. **Guiding ring markedly posterior to the lip region,** but not usually at more than a third of the odontostyle length. **Guiding sheath not extending anterior to the guide ring. Dilatores buccae absent.** Odontostylet retractor muscles passing between the nerve ring and the oesophagus before diverging to the body wall. Oesophagus comprising a narrow tubular section anteriorly which expands abruptly into a posterior bulboid section. **Nucleus of dorsal oesophageal gland small, rounded; situated some distance behind its orifice.** Nuclei of anterior ventrosublateral glands more developed than that of the dorsal gland. Female vulva median to post-median in position. Genital tracts amphididelphic, reflexed. Uterine spines or other sclerotizations not reported. Tail short, rounded; may be conoid or hemispheroid. Male genital tract diorchic, opposed. Spicules paired, massive; dorylaimoid in form with accessory guiding pieces. **Copulatory supplements consisting of an adanal pair and a ventromedian series which commences without a hiatus from the adanal pair.** Tail similar to that of the female.

TYPE SPECIES: *Paralongidorus (P.) sali* Siddiqi, Hooper and Khan, 1963

OTHER SPECIES:

P. (P.) agni Sharma and Edward, 1985
P. (P.) australis Stirling and McCulloch, 1984
P. (P.) buckeri Sharma and Edward, 1985
P. (P.) cebensis Heyns and Coomans, 1989
P. (P.) clavicaudatus (Jacobs and Heyns, 1982) n. comb.
  syn. *Longidoroides clavicaudatus*. Jacobs and Heyns, 1982
P. (P.) esci Khan, Chawla and Saha, 1978
P. (P.) fici Edward, Misra and Singh, 1964
P. (P.) flexus Khan, Seshadri, Weischer and Mathen, 1971
P. (P.) lutescens Hunt and Rahman, 1991
P. (P.) lutosus (Heyns, 1965) Aboul-Eid, 1970
  syn. *Longidorus lutosus* Heyns, 1965
P. (P.) microlaimus Siddiqi, 1964
P. (P.) nudus (Kirjanova, 1951) Lamberti, 1975
  syn. *Longidorus nudus* Kirjanova, 1951
P. (P.) oryzae Verma, 1973
P. (P.) rotundatus Khan, 1987
P. (P.) sacchari Siddiqi, Hooper and Khan, 1963
P. (P.) sandellum (Heyns, 1966) Chaves and Coomans, 1984
  syn. *Xiphinema sandellum* Heyns, 1966
    Longidorus sandellum (Heyns, 1966) Khan, Chawla and Saha, 1978
    Brevinema sandellum (Heyns, 1966) Chaves and Coomans, 1984
P. (P.) similis Khan, Chawla and Prasad, 1972
P. (P.) zenobiae Hunt and Rahman, 1991

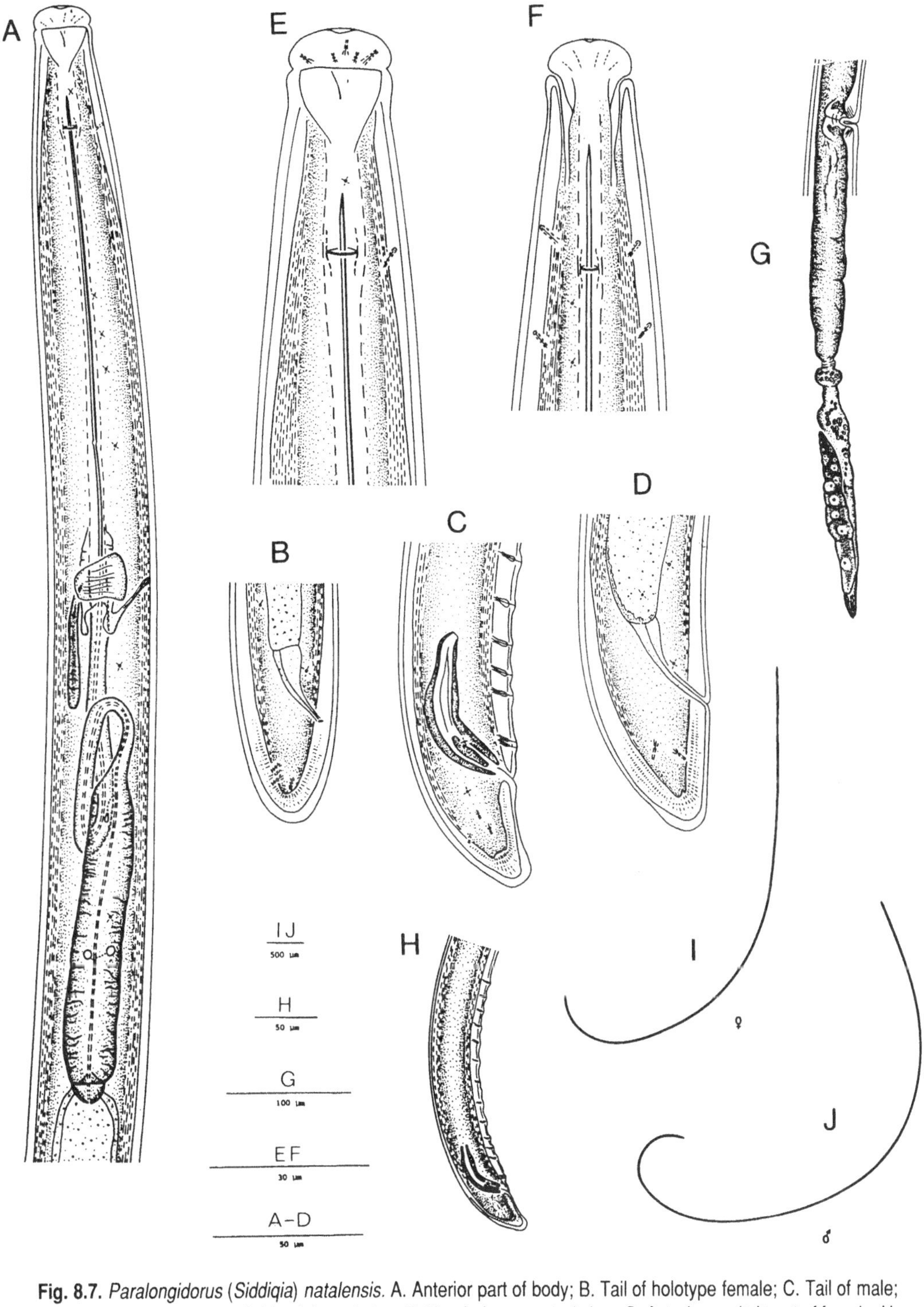

Fig. 8.7. *Paralongidorus* (*Siddiqia*) *natalensis.* A. Anterior part of body; B. Tail of holotype female; C. Tail of male; D. Tail of paratype female; E. Head, lateral view; F. Head, dorso-ventral view; G. Anterior genital tract of female; H. Posterior part of male; I, J. Relaxed body posture. (After Jacobs and Heyns, 1982. Courtesy *Phytophylactica.*)

NOTE: *P. (P.) rotundatus* Khan, 1987 is also spelled as *rosundatus* in the same paper. The latter nomen should be regarded as a *lapsus calami* as the name clearly refers to the rounded tail of the nematode. Khan (1987) illustrates the cephalic region as being rounded truncate and offset by a groove. However, he makes no mention of the head being offset in the description and compares the species to *P. (P.) esci* which has a continuous cephalic region.

BIONOMICS: Ectoparasites found in the rhizosphere of a variety of plants. This genus has not been implicated in vectoring plant viruses.

DISTRIBUTION: Widespread in Africa and Asia. *P. (P.) microlaimus* has been recorded from the USA (Robbins, 1978) and represents an unusual departure from the typical Old World distribution.

ETYMOLOGY: Derived from the Greek *para* = near to, and *longidorus*.

SUBGENUS *PARALONGIDORUS (SIDDIQIA)* KHAN, CHAWLA AND SAHA, 1978 N.RANK. (Fig. 8.7)

syn. *Siddiqia* Khan, Chawla and Saha, 1978
Inagreius Khan, 1982

DIAGNOSIS: Longidorinae. Similar to *Paralongidorus (Paralongidorus)*, but differing in the following respects. Body usually longer, up to 12 mm. **Cephalic region expanded, markedly offset by a constriction and usually with a second depression just behind the groove. Amphidial aperture broad, slit-like, almost extending across the entire lip width. Amphidial fovea stirrup-shaped, tapering towards the base** which may appear to be bilobed in a few species. In the latter group, however, the distance from the slit to the bilobed region is less than the width of the slit (cf. *Longidoroides* where the distance is usually considerably longer, occasionally equal).

TYPE SPECIES:
Paralongidorus (Siddiqia) citri (Siddiqi, 1959) n.comb.
syn. *Xiphinema citri* Siddiqi, 1959
Longidorus citri (Siddiqi, 1959) Thorne, 1961
Paralongidorus citri (Siddiqi, 1959) Siddiqi, Hooper and Khan, 1963
Siddiqia citri (Siddiqi, 1959) Khan, Chawla and Saha, 1978

OTHER SPECIES:
P. (S.) beryllus Siddiqi and Husain, 1965 n.comb.
syn. *P. beryllus* Siddiqi and Husain, 1965
Siddiqia beryllus (Siddiqia and Husain, 1965) Khan, Chawla and Saha, 1978
Inagreius beryllus (Siddiqi and Husain, 1965) Khan, 1982
Longidoroides beryllus (Siddiqi and Husain, 1965) Luc and Doucet, 1984

P. (S.) buchae Lamberti, Roca and Chinappen, 1985 n.comb.
syn. *P. buchae* Lamberti, Roca and Chinappen, 1985
P. (S.) bullatus Sharma and Siddiqi, 1990 n.comb.
syn. *P. bullatus* Sharma and Siddiqi, 1990
P. (S.) capensis (Heyns, 1966) n.comb.
syn. *P. capensis* Heyns, 1966
S. capensis (Heyns, 1966) Khan, Chawla and Saha, 1978
P. (S.) dasturi (Ganguly, Patil and Khan, 1981) n.comb.
syn. *S. dasturi* Ganguly, Patil and Khan, 1981
P. dasturi (Ganguly, Patil and Khan, 1981) Luc and Doucet, 1984
P. (S.) deborae (Jacobs and Heyns, 1982) n.comb.
syn. *S. deborae* Jacobs and Heyns, 1982
P. deborae (Jacobs and Heyns, 1982) Luc and Doucet, 1984
P. (S.) epimikis (Dalmasso, 1969) n.comb.
syn. *P. epimikis* Dalmasso, 1969
S. epimikis (Dalmasso, 1969) Khan, Chawla and Saha, 1978
P. (S.) erriae (Heyns, 1966) n.comb.
syn. *P. erriae* Heyns, 1966
S. erriae (Heyns, 1966) Khan, Chawla and Saha, 1978
P. (S.) eucalypti (Fisher, 1964) n.comb.
syn. *P. eucalypti* Fisher, 1964
S. eucalypti (Fisher, 1964) Khan, Chawla and Saha, 1978
P. (S.) eugeni (Khan, 1987) n.comb.
syn. *Inagreius eugeni* Khan, 1987
Longidoroides eugeni (Khan, 1987) Jairajpuri and Ahmad, 1992
P. (S.) fischeri Heyns, 1972 n.comb.
syn. *P. fischeri* Heyns, 1972
S. fischeri (Heyns, 1972) Khan, Chawla and Saha, 1978
P. (S.) georgiensis (Tulaganov, 1937) n.comb.
syn. *Longidorus georgiensis* Tulaganov, 1937
P. georgiensis (Tulaganov, 1937) Siddiqi, 1965
S. georgiensis (Tulaganov, 1937) Khan, Singh and Singh, 1981
P. (S.) gloriosus (Khan, 1982) n.comb.
syn. *Inagreius gloriosus* Khan, 1982
Longidoroides gloriosus (Khan, 1982) Luc and Doucet, 1984
P. (S.) hooperi (Heyns, 1966) n.comb.
syn. *P. hooperi* Heyns, 1966
Longidoroides hooperi (Heyns, 1966) Jacobs and Heyns, 1982
S. hooperi (Heyns, 1966) Khan, Chawla and Saha, 1978
P. (S.) inagreina (Chawla and Samathanam, 1981) n.comb.
syn. *S. inagreina* Chawla and Samathanam, 1981
P. inagreina (Chawla and Samathanam, 1981) Luc and Doucet, 1984
P. (S.) indicus (Phukan and Sanwal, 1983) n.comb.
syn. *S. indicus* Phukan and Sanwal, 1983
P. indicus (Phukan and Sanwal, 1983) Luc and Doucet, 1984
P. (S.) major (Verma, 1973) n.comb.
syn. *P. major* Verma, 1973

S. major (Verma, 1973) Khan, Singh and Singh, 1981

P. (S.) maximus (Bütschli, 1874) n.comb.

syn. *Dorylaimus maximus* Bütschli, 1874

D. (Longidorus) maximus (Bütschli, 1874) Micoletzky, 1922

Longidorus maximus (Bütschli, 1874) Thorne and Swanger, 1936

P. maximus (Bütschli, 1874) Siddiqi, 1964

S. maxima (Bütschli, 1874) Khan, Chawla and Saha, 1978

P. (S.) mediensis (Ganguly, Patil and Khan, 1981) n.comb.

syn. *S. mediensis* Ganguly, Patil and Khan, 1981

P. mediensis (Ganguly, Patil and Khan, 1981) Luc and Doucet, 1984

P. (S.) namibiensis (Jacobs and Heyns, 1987) n.comb.

syn. *P. namibiensis* Jacobs and Heyns, 1987

P. (S.) natalensis (Jacobs and Heyns, 1982) n.comb.

syn. *S. natalensis* Jacobs and Heyns, 1982

P. natalensis (Jacobs and Heyns, 1982) Luc and Doucet, 1984

P. (S.) paramaximus (Heyns, 1965) n.comb.

syn. *P. paramaximus* Heyns, 1965

S. paramaximus (Heyns, 1965) Khan, Chawla and Saha, 1978

P. (S.) remyi (Altherr, 1963) n.comb.

syn. *Longidorus remyi* Altherr, 1963

P. remyi (Altherr, 1963) Siddiqi and Husain, 1965

S. remyi (Altherr, 1963) Khan, Chawla and Saha, 1978

P. (S.) rex Andrássy, 1986 n.comb.

syn. *P. rex* Andrássy, 1986

P. (S.) seclipsi (Khan, Singh and Singh, 1981) n.comb.

syn. *S. seclipsi* Khan, Singh and Singh, 1981

Longidoroides seclipsi (Khan, Singh and Singh, 1981) Luc and Doucet, 1984

P. (S.) spasskii (Heyns, 1972) n.comb.

syn. *P. spasskii* Heyns, 1972

S. spasskii (Heyns, 1972) Khan, Chawla and Saha, 1978

P. (S.) spaulli (Jacobs and Heyns, 1982) n.comb.

syn. *S. spaulli* Jacobs and Heyns, 1982

P. spaulli (Jacobs and Heyns, 1982) Luc and Doucet, 1984

P. (S.) teres (Khan, 1987) n.comb.

syn. *Inagreius teres* Khan, 1987

Longidoroides teres (Khan, 1987) Jairajpuri and Ahmad, 1992

P. (S.) xiphinemoides Heyns, 1965 n.comb.

syn. *P. xiphinemoides* Heyns, 1965

S. xiphinemoides (Heyns, 1965) Khan, Chawla and Saha, 1978

NOTES:

1. Khan, Singh and Singh (1981), in a key to the species of *Siddiqia*, quoted Khan, Chawla and Saha (1978) as the combining authority for *S. georgiensis* and *S. major*. However, these combinations were not made in the latter paper and so the correct authority for each combination is Khan *et al.* (1981).

2. *P. (S.) remyi* (Altherr, 1963) is usually spelled *remeyi* in the literature. In the original publication, the spelling is consistently *remyi*.
3. Khan (1987), in his description of two new species of *Inagreius*, consistently mis-spells the genus as *Inagrius*. This should be regarded as an incorrect subsequent spelling (Article 33c) and neither combination enters into homonymy.

Remarks: The genus *Inagreius* was originally proposed by Khan (1982) to hold species with characters of the cephalic region and amphid somewhat intermediate between *Siddiqia* and *Longidoroides*. The species formerly placed in *Inagreius* display characters such as the form of the cephalic region, which is typical of *Paralongidorus* (*Siddiqia*), and the form of the amphid which is similar to *P. (Siddiqia)*, yet shallower and with an apparently bilobed base as found in *Longidoroides*. Indeed, the species *beryllus, eugeni, gloriosus* and *teres* fit more comfortably in the latter subgenus than in *Longidoroides,* as proposed by Luc and Doucet (1984) and Jairajpuri and Ahmad (1992), who regarded the bilobed base to the amphidial fovea as being of greatest importance. I feel that the overall amphid shape, and particularly the form of the cephalic region, indicate a closer relationship with *P. (Siddiqia)* to which they are here transferred.

Bionomics: Ectoparasites recorded from the rhizosphere of a wide variety of plants. They are not known to be virus vectors.

Distribution: Recorded widely from Africa (particularly southern Africa), Asia (India, Pakistan, Australia) and Europe.

Etymology: Derived from M.R. *Siddiqi*, a nematologist's name.

Species description

Paralongidorus (Siddiqia) maximus **(Bütschli, 1874) n.comb.** (Fig. 8.8)

syn *Dorylaimus maximus* Bütschli, 1874
Dorylaimus (Longidorus) maximus Bütschli, 1874 in Micoletzky, 1922
Longidorus maximus (Bütschli, 1874) Thorne and Swanger, 1936 nec
Longidorus maximus of Meyl (1954) (= *Longidorus meyli* Sturhan, 1963)
Paralongidorus maximus (Bütschli, 1874) Siddiqi, 1964
Siddiqia maxima (Bütschli, 1874) Khan, Chawla and Saha, 1978

Measurements: After Sturhan (1963) from Southern Germany: Female (n = 46–83): L = 7.60–12.35 (9.84) mm; a = 77–133 (107); b = 12.8–23.7 (15.6); c = 178–320 (250); V = $^{5.0-13.3}38.3\ (36.1-40.4)^{5.3-15.1}$; odontostyle (ods) = 152–187 (171.5) μm; odontophore (odp) = 42–70 (53.5) μm; oral aperture to guide ring (ogr) = 37–47 (41) μm.
Male (n = 2): L = 8.55–10.70 mm; a = 97–110; b = 12.9–14.1; c = 190–237; ods = 171–173 μm; odp = 52.5–56 μm; ogr = 39.5–43 μm.

First-stage juveniles (n = 15–78): L = 1.3–2.3 (1.7) mm; a = 39–49 (44); b = 3.5–8.8 (5.9); c = 35–52 (41); ods = 73–89 (80) μm; odp = 24–31 (28) μm; replacement odontostyle (rep. ods) = 84–106 (94) μm; ogr = 15–25 (19) μm.
Second-stage juveniles (n = 15–78): L = 2.1–4.0 (2.7) mm; a = 49–65 (52); b = 5.0–8.5 (6.5); c = 56–88 (69); ods = 79–103 (93) μm; odp = 32–43 (37) μm; rep. ods = 112–150 (123) μm; ogr = 20–26 (23) μm.
Third-stage juveniles (n = 25–81): L = 3.1–6.1 (4.3) mm; a = 52–80 (65); b = 6.9–11.3 (9.1); c = 83–138 (109); ods = 109–129 (122) μm; odp = 33–51 (41) μm; rep. ods = 132–163 (143) μm; ogr = 26–31 (29) μm.
Fourth-stage juveniles (n = 20–65); L = 5.3–8.4 (7.0) mm; a = 78–97 (87); b = 10.2–15.3 (12.0); c = 135–207 (173); ods = 134–169 (149) μm; odp = 41–62 (48) μm; rep. ods = 153–190 (170) μm; ogr = 32–42 (35) μm.
Female (n = 2, from oak tree soil, Perthshire, Scotland): L = 8.42–9.54 mm; a = 87–91; b = 17.4; c = 225–290; V = $^{8.0-8.8}36.6-39.7^{10.3-10.7}$; ods = 168–182 μm; odp = 44–52 μm; odontostylet = 220–226 μm; ogr = 39–43 μm.

Description: (modified from Heyns, 1975)
Female: **Large, 7–12 mm long.** Heat-relaxed specimens ventrally curved in an open spiral, with most of the curvature in the posterior half. Cuticle 7–10 μm thick on front part of neck adjacent to lip region, 4–7 μm around midbody, and 10–18 μm towards tail terminus. Surface of cuticle with minute criss-cross lines, best visible towards extremities; subcuticle with fine irregular transverse striae, 0.7–1.0 μm apart. **Cephalic region rounded**, 35–39 μm wide, less than half this in height. When seen **in lateral view the cephalic region is expanded and set off by a deep constriction close behind the level of the amphidial apertures**, but when seen dorsoventrally there is a slight constriction at the amphidial aperture and a shallow depression further back. Cephalic region with the usual complement of six inner and ten outer labial papillae. **Amphids stirrup-shaped, their apertures about two-thirds as wide as the lip region.** Lateral chord one-fifth to one-quarter as wide as the body, narrowing from about the first third of the oesophagus to the front end where it is visible only as a narrow line. Lateral pores in a single irregular line over the entire length of the body, about 370–520 in number, of which about 15–36 occur in the oesophageal area. Ventral pores present in the oesophageal area, about 15–20 in number. In the first third of this area there are also 6–10 dorsal pores. **Odontostyle 152–187 μm long** and about 3 μm wide towards its base; **odontophore 42–71 μm long, and total stylet length 220–263 μm.** Dorsal oesophageal gland nucleus at anterior end of bulb, first ventrosublateral pair of nuclei near middle of bulb, and second vestigial pair near base of bulb, rarely distinct. Cardia convex-conoid, up to 36 μm long, but sometimes barely visible. Intestinal walls with darkish granules which become smaller and less distinct in starved specimens. **Prerectum** variable, **from four to 17 anal body widths in length**, usually extending posteriorly over the rectum in specimens with shorter prerectums, indicating that these may have been pushed back by the intestine. **Length of rectum one half to almost equal to anal body width. Tail short, bluntly rounded**, only 26–38 μm long, which is about half the anal

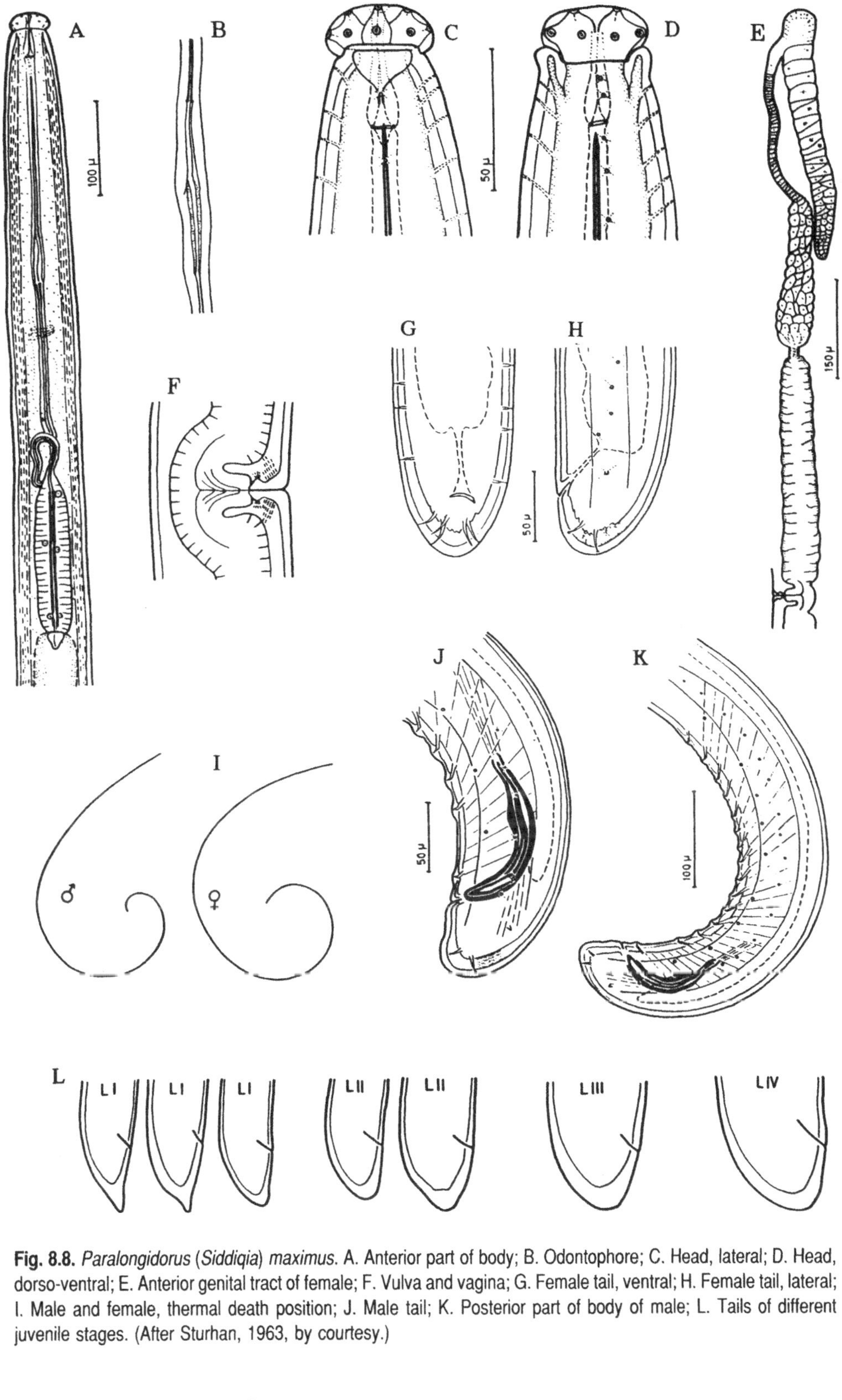

Fig. 8.8. *Paralongidorus* (*Siddiqia*) *maximus*. A. Anterior part of body; B. Odontophore; C. Head, lateral; D. Head, dorso-ventral; E. Anterior genital tract of female; F. Vulva and vagina; G. Female tail, ventral; H. Female tail, lateral; I. Male and female, thermal death position; J. Male tail; K. Posterior part of body of male; L. Tails of different juvenile stages. (After Sturhan, 1963, by courtesy.)

body width ($c' = 0.44–0.60$); with two or three pairs of subterminal caudal papillae, one pair of which is usually subventral, the other subdorsal or lateral. Last pore of lateral series at level of anus. Vulva a transverse slit, about one quarter the corresponding body width in length. Vagina muscular, reaching about two-thirds across the body. Genital tracts amphididelphic, reflexed. Oviduct with a slender distal and a wide basal part, the latter separated from the uterus by a distinct sphincter. Never more than two eggs in each uterus. Average size of eggs 260 x 65 μm (n = 6).

Male: **Extremely rare**, not playing a role in reproduction. Anterior part of body similar to female. Posterior part more strongly ventrally curved. **Spicules strong, 100–106 μm long.** Lateral accessory guiding pieces 32–35 μm long, with pointed distal ends. Copulatory supplements consist of an adanal pair and 14–15 in a ventromedian series. The ventromedian series occupies a distance of about seven to eight anal body widths anterior to the anus. Over this same distance there occurs a series of nine ventrosubmedian pores which are slightly bigger than the lateral pores. Testes apparently absent (Sturhan, 1963).

Juveniles: Anterior part of body similar to, but smaller than, that of adult, and with somewhat different proportions. Tail of first stage juvenile relatively long and somewhat digitate, becoming relatively shorter and more rounded after each moult, while the absolute length of the tail remains more or less constant from the first-stage juvenile to adult (Sturhan, 1963).

BIONOMICS: *P. (S.) maximus* mostly occurs in open textured sandy or sandy loam soils with highest densities at 20–40 cm. Males are extremely rare and reproduction is parthenogenetic. The life cycle is long, the optimal conditions for reproduction being long, warm summers (Sturhan, 1963). *P. (S.) maximus* is extremely polyphagous, attacking a broad range of woody and herbaceous plants, although Cruciferae and grasses appear to be poor hosts. Direct feeding on the roots causes galling, distortion and stunting with necrosis and secondary invasion by microbial pathogens. Virus transmission has not been demonstrated.

DISTRIBUTION: Largely restricted to western Europe (Brown and Taylor, 1987).

Subfamily Xiphidorinae Khan, Chawla and Saha, 1978 (Jairajpuri and Ahmad, 1992)

DIAGNOSIS: Longidoridae. Body length 2–6.3 mm. Amphidial apertures minute slits, sometimes described as pore-like. Amphidial fovea pouch-like. **Odontostyle well developed with a furcate base, odontophore with the basal flanges moderately developed. Guide ring single, prominent, located near the odontostyle/odontophore junction. Guide sheath not extending anterior to the guide ring when the odontostylet is retracted.** Dilatores buccae absent. Dorsal oesophageal gland nucleus smaller than the ventrosublateral gland nuclei and located some distance posterior to its orifice. Female genital tracts amphididelphic, reflexed. **Uterine differentiation in the form of spines reportedly present in some species. Male copulatory supplements comprising an adanal pair and then a short ventromedian series of four to eight separated from the pair by a pronounced hiatus.**

TYPE GENUS: *Xiphidorus* Monteiro, 1976

OTHER GENERA: Monotypic.

GENUS *XIPHIDORUS* MONTEIRO, 1976 (Fig. 8.9)

DIAGNOSIS: Xiphidorinae. Body long and slender, 2–6.3 mm in length. Heat relaxed form C-shaped to an open spiral. **Cephalic region expanded**, offset from the body contour by a depression. **Amphidial apertures** usually described as either **minute slits or pore-like. Amphidial fovea** well developed, **pouch-like**; bilobed in some species. **Odontostyle** 70–120 μm long, attenuate, needle-like, **with a forked base** at its junction with the odontophore. **Odontophore with moderately developed basal flanges.** Odontostylet about 100–180 μm long. **Guiding ring single, thick and posteriorly located near the odontostyle base. Guiding sheath not extending anterior to guide ring when the stylet is in the retracted position**, but extending slightly anteriorly when the odontostyle is protracted. **Dialatores buccae absent.** Odontostylet retractor muscles passing between the nerve ring and the oesophagus before diverging to the body wall. Oesophagus comprising a narrow, tubular section anteriorly and a posterior, bulboid expansion. **Nucleus of dorsal gland small, rounded; situated some distance posterior to its orifice.** Nuclei of ventrosublateral glands more developed than that of the dorsal gland. Female vulva pre-median to slightly post-median in position (43–55%). Genital tract amphididelphic, reflexed. Uterine spines have been reported in X. *achalae* (Luc and Doucet, 1984) and two other species (Kruger and Heyns, 1990). Male genital tract diorchic, opposed. Spicules paired, massive; dorylaimoid in form with accessory guiding pieces. **Copulatory supplements consisting of an adanal pair and then an anterior series of four to eight ventromedian supplements. There is a hiatus in the series between the adanal pair and the first of the ventromedian series.** Tail short, conoid, with a rounded or digitate terminus.

TYPE SPECIES: *Xiphidorus yepesara* Monteiro, 1976

OTHER SPECIES:
X. *achalae* Luc and Doucet, 1984
X. *amazonensis* Uesugi, Huang and Cares, 1985
X. *balcarceanus* Chaves and Coomans, 1984
X. *minor* Rashid, Coomans and Sharma, 1986
X. *parthenus* Monteiro, Lordello and Nakasono, 1981
X. *saladillensis* Chaves and Coomans, 1984
X. *tucumanensis* Chaves and Coomans, 1984

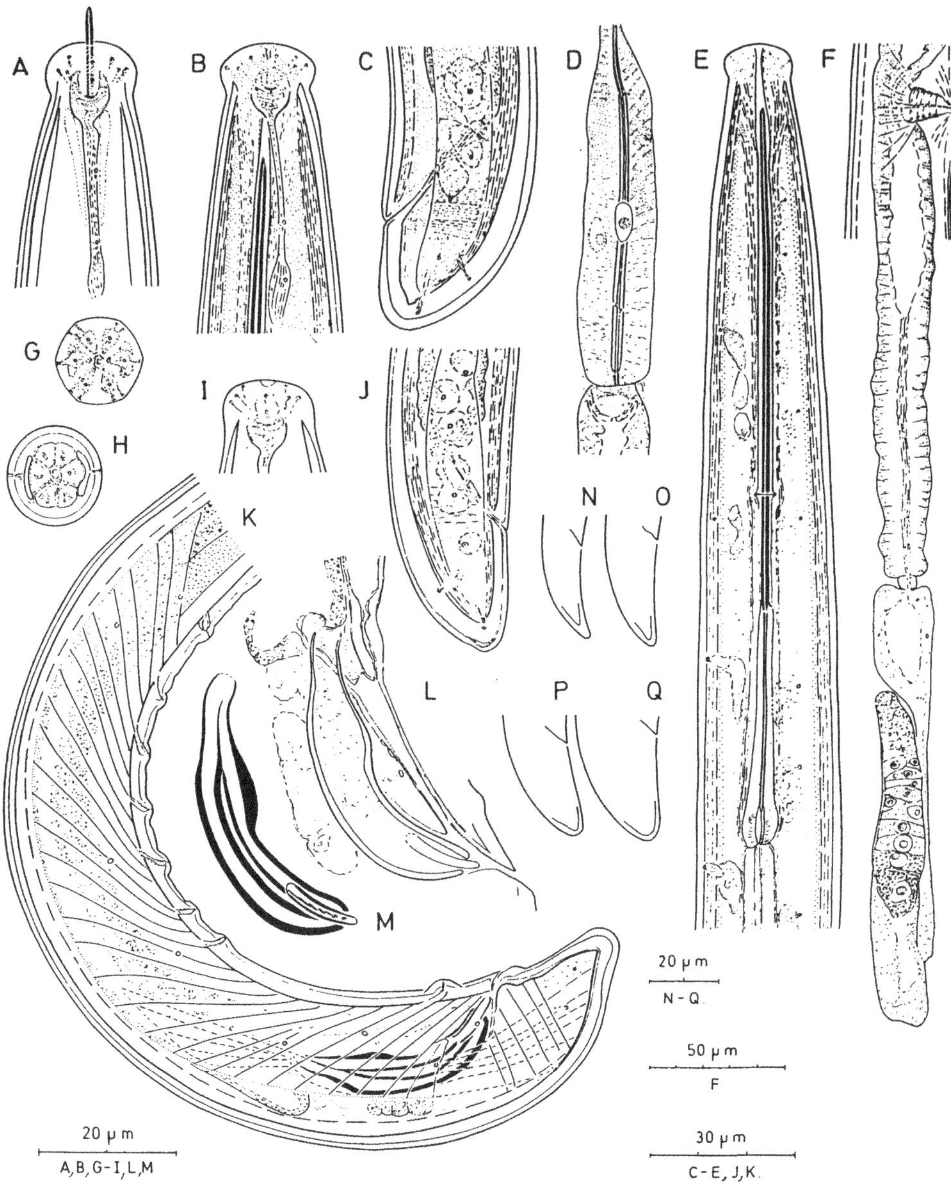

Fig. 8.9. *Xiphidorus* spp. *X. balcarceanus*. A,B. Head region; C. Female tail; D. Oesophageal bulb; E. Odontostylet region; F. Female reproductive system (posterior branch); G. *En face* view; H. Optical section at the level of the amphidial pouches; I. Lip region. *X. tucumanensis*. J. Female tail; K. Male posterior end; L. Cloacal region; M. Spicule and lateral guiding piece. *X. balcarceanus*. N. Tail J1; O. Tail J2; P. Tail J3; Q. Tail J4. (After Chaves and Coomans, 1984. Courtesy *Revue de Nématologie*.)

Remarks: The papers by Chaves and Coomans (1984), Coomans (1985) and Luc and Doucet (1984) provide exemplary accounts of this unusual genus and those of Heyns *et al.* (1989), Hutsebaut *et al.* (1990) and Kruger and Heyns (1990) provide additional morphological and ultrastructural data.

Bionomics: The type species was found around roots of *Passiflora edulis* in Brazil. Other species are recorded from the rhizosphere of various plants (grass, banana, cassava, maize, potato, sugarcane, etc.), often at depths exceeding 40 cm. *Xiphidorus amazonensis* was found in the rhizosphere of a variety of plants in an area subject to flooding for six months of the year in Brazil. *Xiphidorus* species are ectoparasitic and presumably feed in a similar manner to *Longidorus* and *Xiphinema*. No data are available on their potential as virus vectors.

Distribution: Restricted to South America, the nominal species being reported from Argentina (four species) and Brazil (four species). In addition, I have seen an undescribed species from Bolivia. The phylogeny of this group has been speculated upon by Coomans (1985) and Jacobs and Heyns (1987b). If, as seems likely, the genus is an ancient one, there remains the possibility of finding relict species in areas such as southern Africa and possibly Australia, areas which were once contiguous before the breakup of Gondwanaland.

Étymology: Derived from Xiph*inema* and Longi*dorus*, thus reflecting its intermediate suite of characters.

Subfamily Xiphinematinae Dalmasso, 1969

Diagnosis: Longidoridae. Body length from 1.5 to 6 mm. **Amphidial apertures broad, transverse slits. Amphidial fovea stirrup-shaped. Odontostyle heavily sclerotized with a furcate base. Odontophore with heavily sclerotized, well developed basal flanges. Guide ring located near odontostyle/odontophore junction**, usually appearing 'double' when the odontostylet is retracted. **Guiding sheath extending anterior to guide ring when odontostylet is retracted. Dilatores buccae present. Dorsal oesophageal gland nucleus** slightly oval in shape and **larger than the nuclei of the ventrosublateral glands and located adjacent to its orifice. Female genital tract may be** amphididelphic or **monodelphic** with a concomitant shift in the vulva position. **Male ventromedian copulatory supplements numbering up to seven; hiatus present between adanal pair and commencement of series.**

Type genus: *Xiphinema* Cobb, 1913

Other genera: Monotypic.

GENUS *XIPHINEMA* COBB, 1913 (Figs 8.10, 8.11, 8.12)

DIAGNOSIS: Xiphinematinae. Body long to very long, 1.5–6 mm, and fairly stout. Heat relaxed form straight, ventrally arcuate, C-shaped or an open spiral. Cuticle smooth. Lateral chords broad with one or two rows of lateral body pores. Dorsal and ventral series of body pores may be present, particularly in the oesophageal region. Cephalic region rounded, continuous or offset. Lips fused with the usual 6 + 10 circlets of papillae. **Amphidial apertures broad slits extending for almost the entire lip width. Amphid fovea stirrup- or funnel- shaped. Odontostyle elongate, needle-like; heavily sclerotized. Guiding apparatus tubular with a strongly sclerotized posterior ring and, apparently, a lightly sclerotized anterior ring** (really just a fold in the guiding sheath). **The guide ring proper is posteriorly located near the odontostyle/odontophore junction. Proximal end of odontostyle appearing forked** at its junction with the **odontophore** which is **strongly developed with three massive posterior flanges** to which the protractor muscles attach. Oesophagus comprising a narrow, cylindrical anterior section, which is normally looped back on itself, leading to an expanded cylindroid expansion containing the glands. **Dorsal gland nucleus located at the same level as the orifice, more developed than ventrosublateral nuclei.** Nerve ring around the anterior section of the oesophagus. Hemizonid prominent. Intestine simple, pre-rectum well developed and several anal body widths long. Anus in the form of a transverse slit. Vulva located anteriorly to post-median, in the form of a transverse slit. Vagina well developed, muscular; at right angles to the body axis or posteriorly directed in some forms with an anterior vulva. Ovejector prominent. **Genital tract variable**; often amphididelphic reflexed, but as the vulva migrates anteriorly the anterior branch progressively regresses, first becoming non-functional, then a remnant and finally completely absent (= mono-opisthodelphic). **Some species display sclerotized structures in the uterus.** Rarely these sclerotizations are found in a Z organ, a specialized structure with thick walls and circular muscles which is constricted at both ends by a sphincter. More commonly the sclerotizations take the form of spines or variously shaped structures in the uterus. **Tail form very variable**, e.g. short hemispheroid, with or without a digitate process; medium to long conoid; initially conoid and then attenuating to a filiform terminal section. Male genital tract diorchic, opposed. Spicules paired, massive, dorylaimoid with distal accessory guiding pieces. Oblique copulatory muscles prominent and extending anteriorly from the cloaca. **Copulatory supplements consisting of an adanal pair followed by a hiatus and then a ventromedian series of up to seven papillae.** Tail of similar form to that of the female.

TYPE SPECIES:

Xiphinema americanum Cobb, 1913
syn. *Tylencholaimus americanus* (Cobb, 1913) Micoletzky, 1922

OTHER SPECIES:

X. abrantium Roca and Pereira, 1991
X. aceri Chizhov, Tiev and Turkina, 1986

X. *adenohysterum* Lamberti, Castillo, Gomez-Barcina and Agostinelli, 1992
X. *aequum* Roca and Lamberti, 1988
X. *algeriense* Luc and Kostadinov, 1982
X. *arcum* Khan, 1964

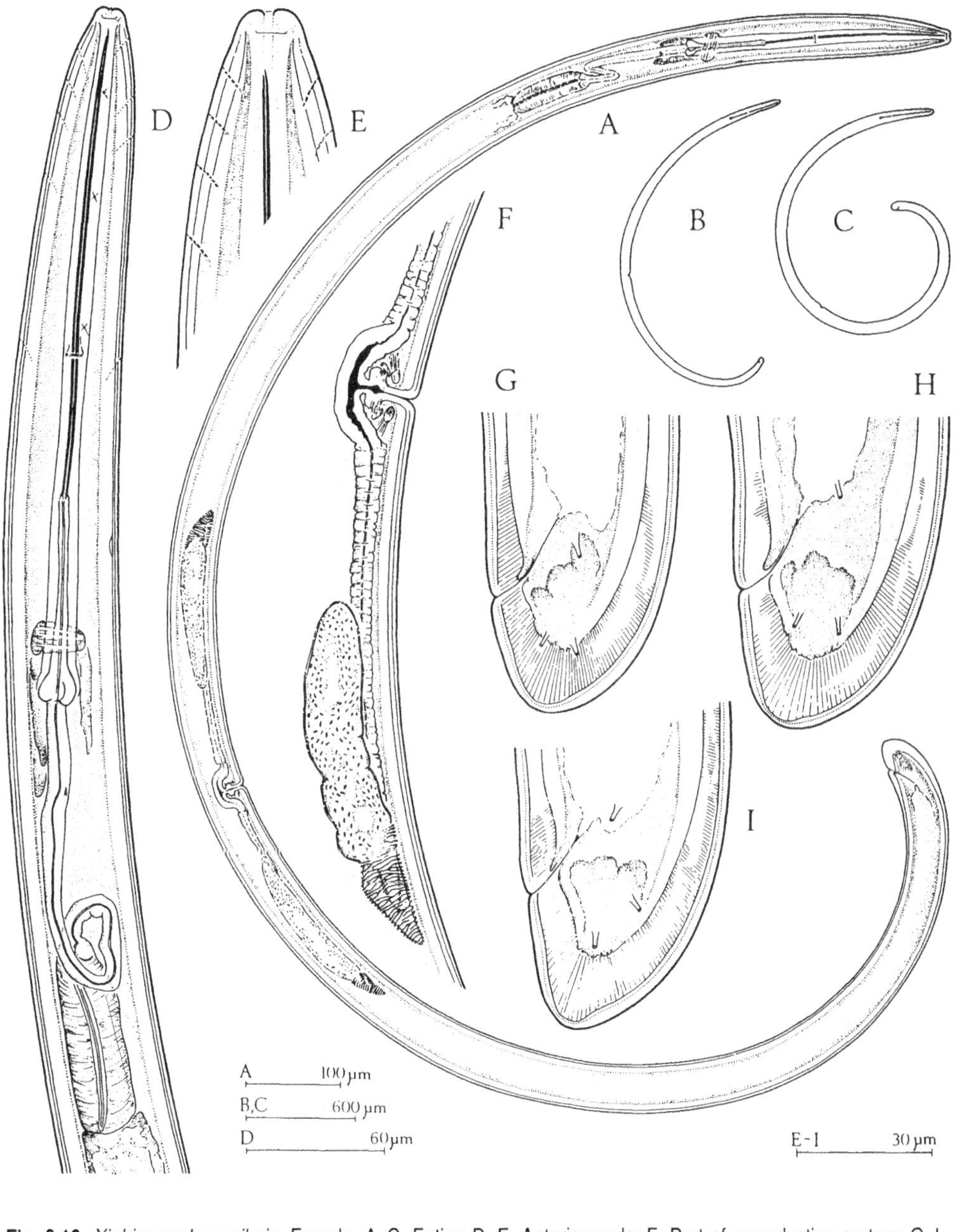

Fig. 8.10. *Xiphinema bacaniboia.* Female. A–C. Entire; D, E. Anterior ends; F. Part of reproductive system; G–I. Tails. (After Orton Williams, 1984. Courtesy *Systematic Parasitology*.)

X. *artemisiae* Chizhov, Tiev and Turkina, 1986
X. *bacaniboia* Orton Williams, 1984
X. *bajaji* Jairajpuri and Lamberti, 1981
    syn. X. *luci* Bajaj and Jairajpuri, 1979, (homonym of X. *luci* Lamberti and Bleve-Zacheo, 1979)
X. *bakeri* Williams, 1961
X. *barbercheckae* Coomans and Heyns, 1985
X. *barense* Lamberti, Roca, Agostinelli and Bleve-Zacheo, 1986
X. *basilgoodeyi* Coomans, 1965
X. *basiri* Siddiqi, 1959
    syn. X. *cobbi* Sharma and Saxena, 1981
        X. *hayati* Javed, 1983
X. *belmontense* Roca and Pereira, 1992
X. *bergeri* Luc, 1973
X. *bolandium* Coomans and Heyns, 1985
X. *bourkei* Stocker and Kruger, 1988
X. *brasiliense* Lordello, 1951
    syn. X. *itanhaense* Carvalho, 1962
        X. *mammillocaudatum* Khan, 1982
X. *brevicolle* Lordello and Da Costa, 1961
    syn. X. *americanum* apud Carvalho (1955, 1962)
        X. *saopaoloense* Khan and Ahmad, 1975
X. *brevistylus* Jairajpuri, 1982
X. *bricolense* Ebsary, Vrain and Graham, 1989
X. *californicum* Lamberti and Bleve-Zacheo, 1979
X. *capense* Coomans and Heyns, 1985
X. *caprivense* Hutsebaut, Heyns and Coomans, 1989
X. *cavenessi* Luc, 1973
X. *chambersi* Thorne, 1939
X. *citricolum* Lamberti and Bleve-Zacheo, 1979
X. *clavatum* Heyns, 1965
X. *clavicaudatum* Huang, Uesugi and Raski, 1987
X. *codiaei* Singh, 1976
X. *cohni* Lamberti, Castillo, Gomez-Barcina and Agostinelli, 1992
    syn. X. *turcicum* sensu Cohn, 1969 (nec Luc and Dalmasso, 1964)
X. *colombiense* Hunt, 1982
X. *conurum* Siddiqi, 1964
X. *coomansi* Kruger and Heyns, 1986
X. *coronatum* Roca, 1991
X. *costaricense* Lamberti and Tarjan, 1974
X. *coxi coxi* Tarjan, 1964
X. *coxi europaeum* Sturhan, 1985
X. *dentatum* Sturhan, 1978
X. *diannae* Kruger and Heyns, 1987
X. *diffusum* Lamberti and Bleve-Zacheo, 1979
X. *dimidiatum* Loof and Sharma, 1979
X. *dissimile* Roca, Pereira and Lamberti, 1988

X. *diversicaudatum* (Micoletzky, 1927) Thorne, 1939

syn. X. *paraelongatum* Altherr, 1958

Dorylaimus elongatus apud Micoletzky, 1923

Dorylaimus (*Longidorus*) *diversicaudatus* Micoletzky, 1927

Longidorus diversicaudatus (Micoletzky, 1927) Thorne and Swanger, 1936)nec X. *diversicaudatum* apud Luc (1958) (= X. *seredouense*)

X. *diversicaudatum* apud Cohn (1969) (= X. *israeliae*)

X. *diversum* Roca, Lamberti, Santos and Abrantes, 1989

X. *dolosum* Bos and Loof, 1985

X. *douceti* Luc, 1973

X. *dracomontanum* Hutsebaut, Heyns and Coomans, 1989

X. *ebriense*, Luc, 1958

X. *elitum* Khan, Chawla and Saha, 1978

X. *elongatum* Schuurmans Stekhoven and Teunissen, 1938

syn. X. *campinense* Lordello, 1951

X. *hydrabadense* Quraishi and Das, 1984

X. *nagarjunense* Khan, 1982

X. *pratense* Loos, 1949

X. *uasi* Edward and Sharma, 1982

X. *ensiculiferum* (Cobb, 1893) Thorne, 1937

syn. *Tylencholaimus ensiculiferus* Cobb, 1893

X. *ensiculiferoides* Cohn and Sher, 1972 nec X. *ensiculiferoides* apud Loos (1949) and Williams (1959) (= X. *krugi*)

X. *ensiculiferum* apud Luc (1961) (= X. *hygrophilum*)

X. *ensiculiferum* apud Carvalho (1955) (= X. *surinamense*)

X. *erriae* Hutsebaut, Heyns and Coomans, 1988

X. *exile* Roca, Lamberti, Santos and Abrantes, 1989

X. *fagesi* Germani, 1990

X. *fatikae eburnense* Coomans, Rashid and Luc, 1990

X. *fatikae fatikae* Bos and Loof, 1985

X. *filicaudatum* Loof and Maas, 1972

X. *flagellicaudatum* Luc, 1961

X. *floridae* Lamberti and Bleve Zacheo, 1979

X. *fluminense* Huang, Uesugi and Raski, 1987

X. *fortuitum* Roca, Lamberti and Agostinelli, 1988

X. *georgianum* Lamberti and Bleve-Zacheo, 1979

X. *globosum* Sturhan, 1978

X. *guirani* Luc and Williams, 1978

nec X. *guirani* apud Lamberti and Bleve-Zacheo (1979) (= X. *pseudoguirani*)

X. *guillaumeti* Germani, 1989

X. *hallei* Luc, 1958

nec X. *hallei* apud Heyns (1962, 1971) (= X. *mluci*)

X. *hardingi* Joubert, Kruger and Heyns, 1988

X. *heynsi* Siddiqi, 1979

X. *hispanum* Lamberti, Castillo, Gomez-Barcina and Agostinelli, 1992

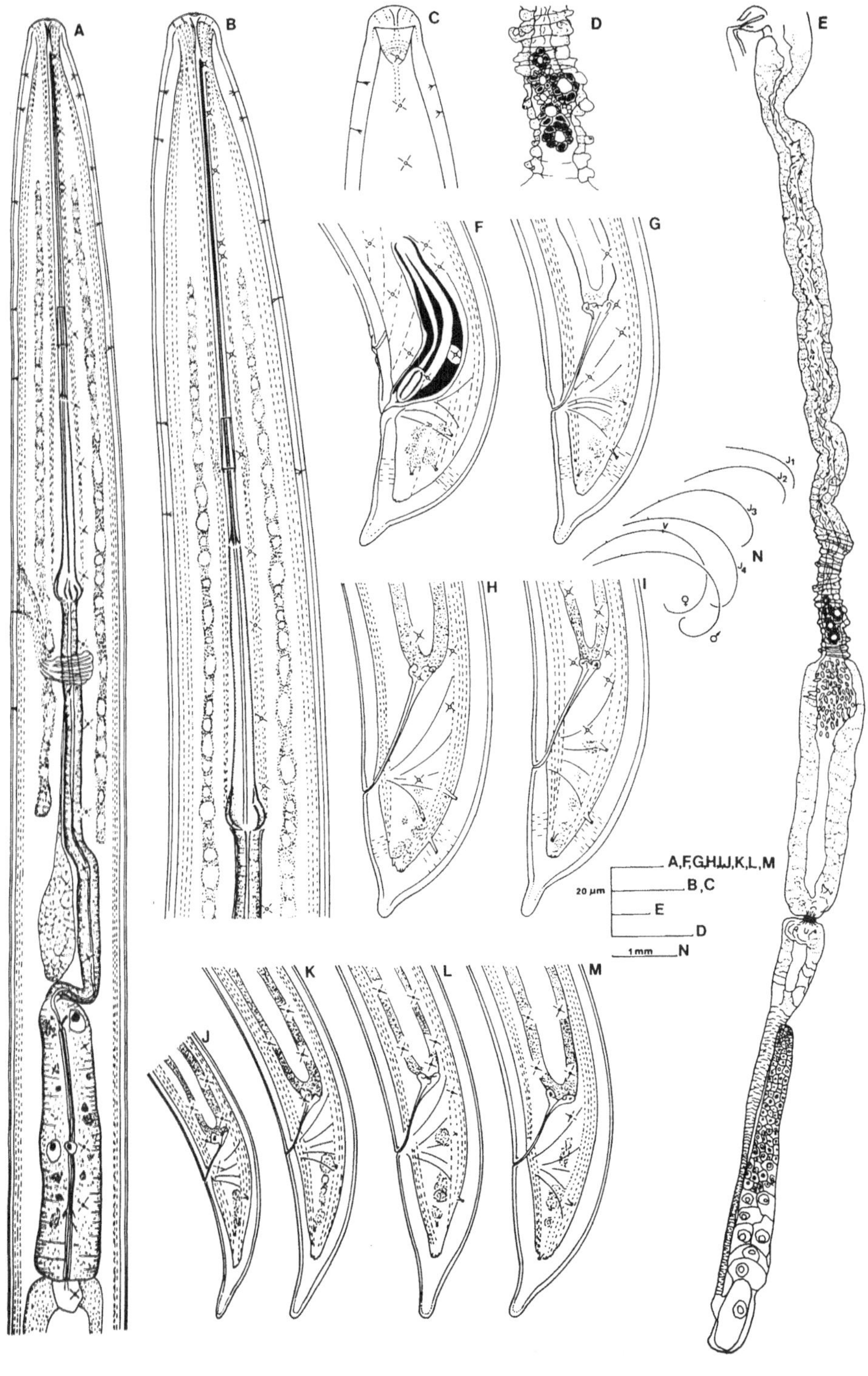

Fig. 8.11. *Xiphinema belmontense.* A,B. Female, anterior region; C. Head end (surface view); D.

X. *hygrophilum* Southey and Luc, 1974
    syn. X. *ensiculiferum* apud Luc (1961)
X. *ifacolum* Luc, 1961
X. *imambaksi* Loof and Maas, 1972
X. *imitator* Heyns, 1965
X. *inaequale* Khan and Ahmad, 1977
    syn. X. *neoamericanum* Khan and Ahmad, 1975 (= homonym of X. *neoamericanum* Saxena, Chabbra and Joshi, 1973)
X. *incertum* Lamberti, Choleva and Agostinelli, 1983
X. *incognitum* Lamberti and Bleve-Zacheo, 1979
X. *index* Thorne and Allen, 1950
X. *ingens* Luc and Dalmasso, 1964
X. *insigne* Loos, 1949
    syn. X. *indicum* Siddiqi, 1959
        X. *neodimorphicaudatum* Khan, 1982
        X. *tugewai* Darekar and Khan, 1983
X. *intermedium* Lamberti and Bleve-Zacheo, 1979
X. *israeliae* Luc, Brown and Cohn, 1982
    syn. X. *diversicaudatum* apud Cohn, (1969))
X. *italiae* Meyl, 1953
    syn. X. *arenarium* Luc and Dalmasso, 1964
        X. *bulgariense* Stoyanov, 1964
            nec X. *italiae* apud Chaves and Geraert, 1977 (= X. *savanicola*)
X. *jomercium* Joubert, Kruger and Heyns, 1988
X. *judex* Hutsebaut, Heyns and Coomans, 1989
X. *karachiense* Nasira, Firoza and Maqbool, 1992
X. *kosaigudense* Quraishi and Das, 1984
X. *krugi* Lordello, 1955
    syn. X. *denoudeni* Loof and Maas, 1972
        X. *loosi* Southey and Luc, 1974
        X. *ensiculiferum* apud Loos (1949) and Williams (1959)
X. *lacrimaspinae* Hutsebaut, Heyns and Coomans, 1988
X. *laevistriatum* Lamberti and Bleve-Zacheo, 1979
X. *lafoense* Roca, Pereira and Lamberti, 1988
X. *lambertii* Bajaj and Jairajpuri, 1977
X. *limbeense* Brown, Luc and Saka, 1983
X. *limpopoense* Heyns, 1977
X. *llanosum* Siddiqi and Lenne, 1990
X. *longicaudatum* Luc, 1961
X. *longidoroides* Luc, 1961
X. *loteni* Heyns, 1986
X. *louisi* Heyns, 1979
X. *luci* Lamberti and Bleve-Zacheo, 1979

Pseudo-Z organ; E. Posterior branch of the genital tract; F. Male posterior region (allotype); G–I. Female posterior region (G, holotype); J–M. Posterior region of juveniles, J1-J4 respectively; N. Posture of juveniles and adult stages. (After Roca and Pereira, 1992. Courtesy *Fundamental and Applied Nematology*.)

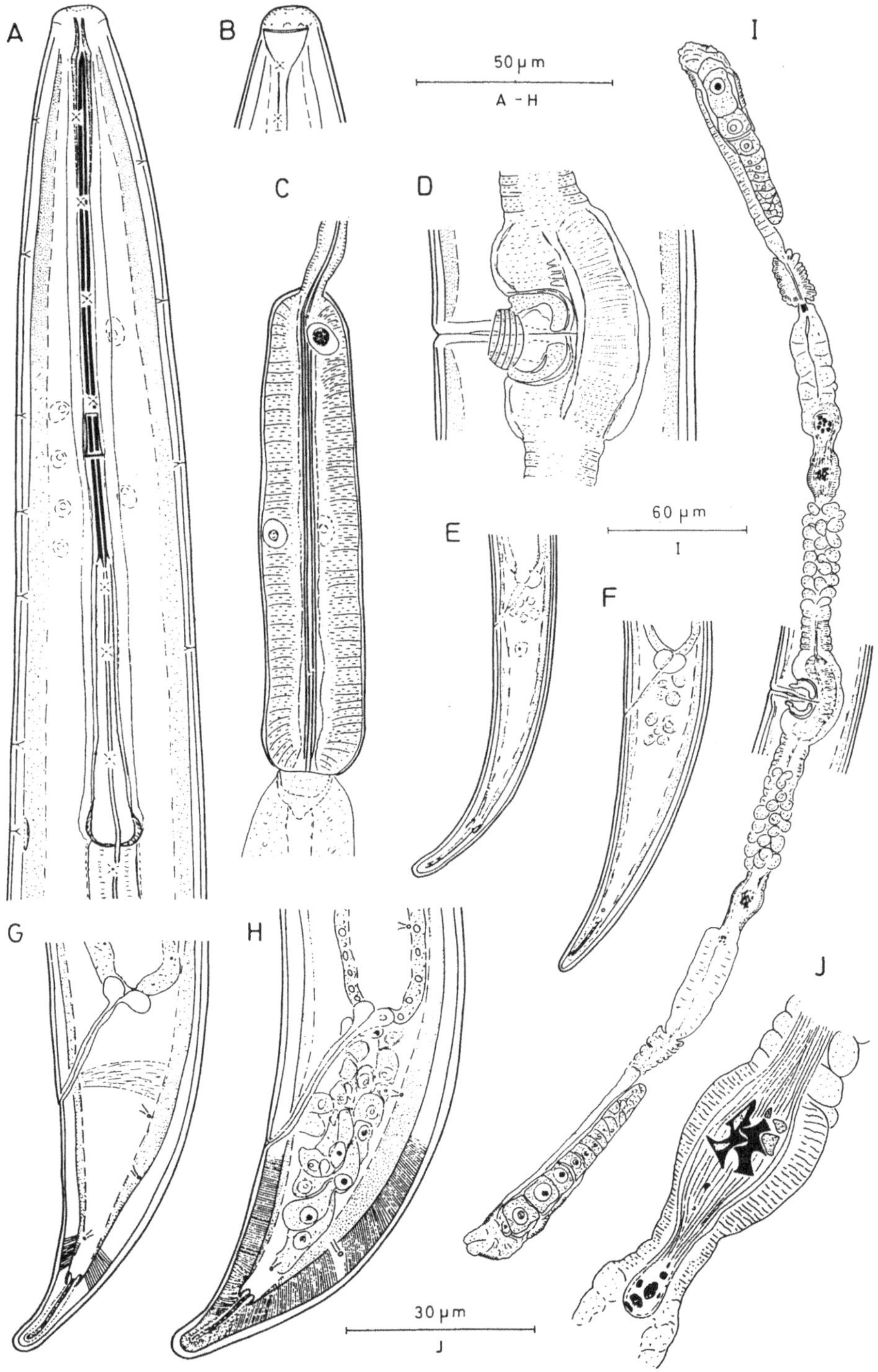

Fig. 8.12. *Xiphinema ifacolum.* A. Female odontostylet region; B. Head end (surface view); C. Oesophageal bulb; D. Vagina and vulva region; E. Tail J1; F. Tail J2; G. Tail J4; H. Female tail; I. Female reproductive system; J. Z organ. (After Rashid *et al.*, 1986. Courtesy *Nematologia Mediterranea.)*

X. *lusitanicum* Sturhan, 1983
X. *machoni* Hunt, 1980
X. *macroacanthum* Lamberti, Roca and Agostinelli, 1989
X. *macrogastrum* Lamberti, Castillo, Gomez-Barcina and Agostinelli, 1992
X. *macrostylum* Esser, 1966
X. *madeirense* Brown, Faria, Lamberti, Halbrendt, Agostinelli and Jones, 1993
X. *magaliesmontanum* Kruger and Heyns, 1986
X. *majus* Bos and Loof, 1985
X. *malagasi* Luc, 1973
X. *malawiense* Brown, Luc and Saka, 1983
X. *malutiense* Heyns, 1976
    syn. X. *vanderlindei* Heyns, 1962 *partim*
X. *mammatum* Siddiqi, 1979
X. *mammillatum* Schuurmans Stekhoven and Teunissen, 1938
X. *mampara* Heyns, 1979
X. *mampara* f. *bisexuale* Hutsebaut, Heyns and Coomans, 1989
X. *mampara* f. *major* Hutsebaut, Heyns and Coomans, 1989
X. *mampara* f. *minor* Hutsebaut, Heyns and Coomans, 1989
X. *manubriatum* Luc, 1975
X. *marsupilami* Luc, 1973
X. *melitense* Lamberti, Bleve-Zacheo and Arias, 1982
X. *meridianum* Heyns, 1971
X. *michelluci* Siddiqi, 1979
X. *miekeae* Heyns, 1991
X. *mluci* Heyns, 1976
    syn. X. *hallei* apud Heyns (1962) and (1971)
X. *monohysterum* Brown, 1968
X. *natalense* Heyns and Vermeulen, 1982
X. *neobasiri* Siddiqi, 1979
X. *neovuittenezi* Dalmasso, 1969
X. *nigeriense* Luc, 1961
X. *nuragicum* Lamberti, Castillo, Gomez-Barcina and Agostinelli, 1992
    syn. X. *turcicum* sensu Prota *et al.*, 1971 (nec Luc and Dalmasso, 1964)
X. *occiduum* Ebsary, Potter and Allen, 1984
X. *opisthohysterum* Siddiqi, 1961
X. *orbum* Siddiqi, 1964
X. *ornativulvatum* Kruger and Heyns, 1987
X. *ornatizulu* Hutsebaut, Heyns and Coomans, 1989
X. *orthotenum* Cohn and Sher, 1972
X. *oryzae* Bos and Loof, 1985
X. *oxycaudatum* Lamberti and Bleve-Zacheo, 1979
X. *pachtaicum* (Tulaganov, 1938) Kirjanova, 1951
    syn. *Longidorus pachtaicus* Tulaganov, 1938
        X. *mediterraneum* Martelli and Lamberti, 1967;
        X. *neoelongatum* Bajaj and Jairajpuri, 1977
X. *pachydermum* Sturhan, 1983
X. *pacificum* Ebsary, Vrain and Graham, 1989

X. *papuanum* Heyns and Coomans, 1983
X. *paramonovi* Romanenko, 1981
  syn. X. *paramericanum* Romanenko, 1973 (*nomen nudum*)
X. *parasetariae* Luc, 1958
  syn. X. *attorodorum* Luc, 1961
    X. *campinense* apud Luc (1958)
X. *parasimplex* Kruger, Kilian and Heyns, 1992
X. *paritaliae* Loof and Sharma, 1979
X. *parvistilus* Heyns, 1971
X. *parvum* Lamberti, Cicancio, Agostinelli and Coiro, 1992
X. *paulistanum* Carvalho, 1965
X. *peruvianum* Lamberti and Bleve-Zacheo, 1979
X. *phoenicis* Loof, 1983
X. *pini* Heyns, 1965
X. *pinoides* Joubert, Kruger and Heyns, 1988
X. *pongolense* Hutsebaut, 1989
X. *porosum* Roca and Agostinelli, 1986
X. *pseudocoxi* Sturhan, 1985
X. *pseudoguirani* Lamberti, Cicancio, Agostinelli and Coiro, 1992
  syn. X. *guirani* apud Lamberti and Bleve-Zacheo, 1979
X. *pyrenaicum* Dalmasso, 1969
X. *radicicola* T. Goodey, 1936
  syn. X. *australiae* McLeod and Khair, 1971
    X. *pararadicicola* Phukan and Sanwal, 1982
    X. *chothecolla* Renubala, Gambhir and Dhanachand, 1991 n.syn.
    X. *clavatus* Renubala, Gambhir and Dhanachand, 1991 n.syn.
X. *rarum* Heyns, 1979
X. *riocaquetae* Hunt, 1982
X. *ripogranum* Hutsebaut, Heyns and Coomans, 1988
X. *rivesi* Dalmasso, 1969
X. *rotundatum* Schuurmans Stekhoven and Teunissen, 1938
X. *sahelense* Dalmasso, 1969
  syn. X. *amarantum* Macara, 1970
X. *savanicola* Luc and Southey, 1980
  syn. X. *italiae* apud Chaves and Geraert (1977)
X. *seredouense* Luc, 1975
  syn. X. *diversicaudatum* apud Luc (1958)
X. *setariae* Luc, 1958
  syn. X. *vulgare* Tarjan, 1964
X. *sheri* Lamberti and Bleve-Zacheo, 1979
X. *silvaticum* Luc and Williams, 1978
X. *simile* Lamberti, Choleva and Agostinelli, 1983
X. *simillimum* Loof and Yassin, 1971
X. *simplex* Hutsebaut, Heyns and Coomans, 1989
X. *smoliki* Luc and Coomans, 1988
X. *spaulli* Heyns and Vermeulen, 1982
X. *sphaerocephalum* Lamberti, Castillo, Gomez-Barcina and Agostinelli, 1992

X. *spinuterus* Luc, 1973
X. *stenocephalum* Luc and Baujard, 1983
X. *stockeri* Kruger and Heyns, 1985
X. *surinamense* Loof and Maas, 1972
    syn. X. *ensiculiferum* apud Carvalho (1955)
X. *swarti* Stocker and Kruger, 1988
X. *tarjanense* Lamberti and Bleve-Zacheo, 1979
X. *tarjani* Luc, 1975
X. *taylori* Lamberti, Cicancio, Agostinelli and Coiro, 1992
X. *tenue* Joubert, Kruger and Heyns, 1988
X. *tenuicutis* Lamberti and Bleve-Zacheo, 1979
X. *theresiae* Stocker and Kruger, 1988
X. *thorneanum* Luc, Loof and Coomans, 1986
    syn. X. *vuittenezi* apud Thorne (1974)
X. *thornei* Lamberti and Golden, 1986
X. *transkeiense* Joubert, Kruger and Heyns, 1988
X. *tropicale* Zullini, 1973
X. *turcicum* Luc and Dalmasso, 1964
X. *umobae* Heyns and Spaull, 1979
X. *utahense* Lamberti and Bleve-Zacheo, 1979
X. *vanderlindei* Heyns, 1962
X. *variabile* Heyns, 1966
X. *vitis* Heyns, 1974
X. *vuittenezi* Luc, Lima, Weischer and Flegg, 1964
    nec X. *vuittenezi* apud Thorne (1974) (= X. *thorneanum*)
X. *xenovariabile* Kruger and Heyns, 1985
X. *yapoense* Luc, 1958
X. *zulu* Heyns, 1965

Species inquirendae vel dubiae:

X. *chappuisi* (W. Schneider, 1935) Loof and Luc, 1990
    syn. *Dorylaimus* (*Longidorus*) *chappuisi* W. Schneider, 1935;
        Longidorella chappuisi (W. Schneider, 1935) Thorne, 1939
X. *cubense* Razjivin in Razjivin, O'Relly and Millan, 1973
X. *cylindricaudatum* Schuurmans Stekhoven and Teunissen, 1938
X. *digiticaudatum* Schuurmans Stekhoven, 1951
X. *dolichodorum* (de Man, 1907) Thorne and Swanger, 1936
    syn. *Dorylaimus dolichodorus* de Man, 1907
        Dorylaimus makrodorus Vanha, 1893, (= junior homonym of
        Dorylaimus macrodorus de Man, 1880
        Xiphinema makrodorum (Vanha, 1893) Thorne, 1939
X. *effilatum* Schuurmans Stekhoven, 1951
X. *grande* (Steiner, 1914) Steiner, 1914
    syn. *Tylencholaimus grandis* Steiner, 1914
X. *lineum* (Grube, 1849) Thorne, 1937
    syn. *Anguillula linea* Grube, 1849 nec Oken, 1815
        Dorylaimus lineus (Grube, 1849) Oerley, 1880

X. *neoamericanum* Saxena, Chabbra and Joshi, 1973
X. *obtusum* Thorne, 1939
X. *sharmai* Luc, Loof and Brown, 1985
    syn. X. *indicum* Sharma and Saxena, 1981 (junior homonym of
        X. *indicum* Siddiqi, 1959)
X. *truncatum* Thorne, 1939

NOTE: The above species list is modified and updated from Loof and Luc (1990), a paper which also contains a useful polytomous key. The subgeneric groupings of Cohn and Sher (1972) and Roy and Gupta (1974) have been omitted from the synonymy.

BIONOMICS: Migratory ectoparasites of plant roots. Most species appear to be polyphagous and a number of species are known to be economically important vectors of certain nepoviruses.

DISTRIBUTION: Worldwide, and including all the major geographic areas from the temperate regions to the tropics, although probably more abundant and diverse in warmer soils. The cosmopolitan distribution lends weight to the supposition that *Xiphinema* is an ancient genus that arose before the break-up of the supercontinent Pangaea into Gondwanaland and Laurasia.

ETYMOLOGY: Derived from the Greek *xiphidion* = sword or *xiphias* = sword-shaped and *nema* = thread.

SPECIES DESCRIPTIONS

XIPHINEMA AMERICANUM COBB, 1913 (Fig. 8.13)
syn. *Tylencholaimus americanus* (Cobb, 1913) Micoletzky, 1922

MEASUREMENTS: Female topotypes (n = 4, from bent grass, Rhode Island, USA): L = 1.66–1.71 (1.68) mm; a = 47–52 (50); b = 6.3–7.5 (6.6); c = 48–53 (51); c' = 1.6–1.8 (1.7); V = 51–54 (52.7); odontostyle = 70–74 (71.5) μm; odontophore = 45–47 (46) μm.
Neotype male (designated by Tarjan, 1958): L = 1.44 mm; a = 57.6; b = 6.2; c = 36; c' = 1.95; T = 46.5; odontostyle = 68 μm; odontophore = 38 μm.
(After Tarjan, 1956) Female (n = 18, from Rhode Island, USA): L = 1.4–1.9 (1.6) mm; a = 33.6–46.6 (42.3); b = 4.7–7.2 (6.3); c = 36.5–52.8 (44.7); V = 46–54 (51); odontostyle (average) = 72 μm; odontophore (average) = 47 μm.
Male (n = 3): L = 1.5–1.7 (1.6) mm; a = 39.7–51.6 (47.2); b = 6.1–6.3 (6.2); c = 37.8–50.1 (43.6); T = 40–54 (46); spicule = 27–30 μm.

DESCRIPTION: (modified from Siddiqi, 1973)
Female: (based on topotypes): Body forms an open spiral on heat relaxation;

cuticle finely striated transversely; lateral hypodermal chords about one fourth of body width; lateral body pores roughly in two lines. Cephalic region hemispherical, slightly expanded; labial papillae six in inner and ten in outer circlet, not elevated. Amphids postlabial, vase-shaped, with obscure crescentic apertures, about half as wide as lip region. **Odontostyle about 72 μm long** and furcate at base; **odontophore** distally flanged (flanges about 9 μm across), **45–47 μm long.** Enlarged basal portion of oesophagus about two body widths long. Oesophago-intestinal valve rounded. Hemizonid opposite nerve ring. **Vulva a transverse slit, about equatorial in position.** Vagina at right angles to body axis, about half a body width long, its inner half globular in shape. **Genital tracts amphididelphic, reflexed**; uteri elongate, narrow; **Z organ absent**; ovaries with a few oogonia with large deformed nuclei, infested with bacteria. Intestine not overlapping rectum. **Tail conoid with a greater curvature dorsally, a hemispherical terminus and two pairs of lateral pores just anterior to its middle, 1.6–1.8 anal body widths long.**

Male: Body as in female but with greater curvature in posterior third. Cephalic region hemispherical, slightly wider than adjacent body. Four ventral body pores anteriorly and a distinct convex lens-shaped hemizonid near base of odontophore. Guiding ring of odontostyle 51 μm from anterior end. Intestine with scattered granules and vacuoles. Testes two, outstretched, opposed. **Spicules rather straight, cephalated, 29 μm long.** A pair of ventrosubmedian supplements 7 μm in front of anus, five ventromedian supplementary papillae at 22 μm, 67 μm, 81 μm, 92 μm and 104 μm anterior to anus. **Tail tapering, ventrally arcuate, about two anal body widths long, with a hemispherical terminus and two pairs of pores just anterior to its middle. Males are extremely rare in this species.**

BIONOMICS: X. *americanum* is a member of the americanum-species complex, an assemblage comprising some 38 nominal species which has not yet been satisfactorily resolved taxonomically, although Lamberti and Carone (1991) have provided a dichotomous key to the group. As a result, although X. *americanum* is a valid species, the reports of its occurrence, etc., in the literature may refer to another, closely related member of the complex. Preferred soil types are open textured silty loams and loamy sands, although it does occur in heavier soils as well. It is susceptible to prolonged freezing. Population increase is best in drier soils with a constant water content and at a temperature of 20–24°C. Reproduction is parthenogenetic and the life cycle takes at least one year. According to Halbrendt and Brown (1992) there are only three juvenile stages. They also speculate that the New World species of the americanum-complex, several of which are known to transmit nepoviruses, have only three juvenile stages whereas the Old World species, which have not been shown to transmit viruses, have four stages. The host range is wide and includes many woody and herbaceous plants – cereals, cherry, citrus, grapevine, lucerne, maize, rose, strawberry, soyabean, tobacco, tomato, etc. Root injury through direct feeding occurs on many hosts and X. *americanum* also vectors tomato ring spot virus (TomRSV), tobacco ring spot

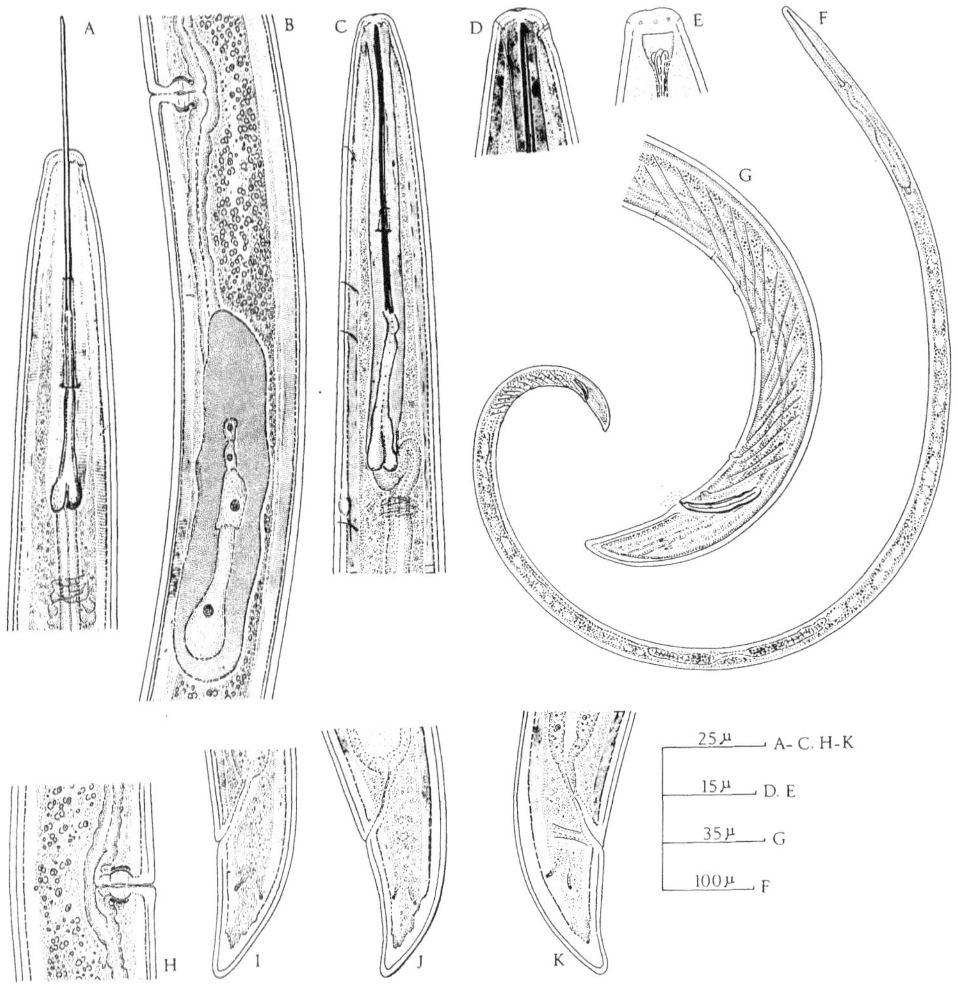

Fig. 8.13. *Xiphinema americanum.* C–G. Neotype male, remainder topotype females. A, C. Anterior ends; D, E. Head ends; G, I, J, K. Tail ends; B. Posterior female genital tract; F. Entire male; H. Vulval region. (After Siddiqi, 1973.)

virus (TRSV), peach rosette mosaic virus, cherry rasp leaf virus (CRLV) and peach yellow bud mosaic virus (PYBMV). Additional information is given in Anon. (1984).

Distribution: Many earlier records from Europe refer to other species such as *X. pachtaicum*, but *X. americanum* is widespread in the eastern USA and has also been reported from Australia, Canada, Europe, India, New Zealand and South Africa. Robbins and Brown (1991) review the distribution in North America and point out the need for a reassessment of the bionomics and distribution of *X. americanum*.

Xiphinema diversicaudatum **(Micoletzky, 1927) Thorne, 1939** (Fig. 8.14)
syn. *Dorylaimus* (*Longidorus*) *diversicaudatus* Micoletzky, 1927
Longidorus diversicaudatus (Micoletzky, 1927) Thorne and Swanger, 1936
Xiphinema paraelongatum Altherr, 1958
Xiphinema (*Diversiphinema*) *diversicaudatum* (Micoletzky, 1927) Cohn and Sher, 1972
Dorylaimus (*Longidorus*) *elongatus* of Micoletzky (1923), nec de Man, 1876

Measurements:
Lectotype male : L = 4.3 mm; a = 81; b = 8.6; c = 93 c′ = 1.1; T = 59; odontostyle = 134 μm; odontophore = 75 μm.
(After Micoletzky, 1923, 1927) Female (n = 1): L = 4.0 mm; a = 72; b = 10.1; c = 96; V = 47.7; odontostyle = 133 μm; odontophore = 78 μm. Male (n = 2): L = 4.07–4.30 mm; a = 74–81; b = 8.6–10.1; c = 90–103; T = 49; odontostyle = 128–132 μm.
(After Goodey *et al.*, 1960 British specimens: Female (n = 43): L = 4.0–5.5 (4.9) mm; a = 57–92 (74); b = 6.6–11.4 (9.1); c = 61–134 (78); V = 39–46 (43); odontostyle = 130–157 (143) μm; odontophore = 70–97 (85) μm.
Male (n = 33): L = 4.1–6.2 (4.9) mm; a = 57–96 (76); b = 7.4–11.3 (8.8); c = 55–100 (78); T = 47–67 (58); odontostyle = 131–153 (143) μm; odontophore = 72–90 (83) μm; spicules = 69–81 (76) μm; lateral guiding pieces of spicules = 16.5–20.7 (18) μm; anterior end to fixed ring of guiding sheath = 119–134 (126) μm.

Description: (modified from Pitcher, Siddiqi and Brown, 1974)
Male (lectotype): Body curved, slightly flattened due to pressure of the coverglass but overwise in excellent condition. **Body cuticle smooth, 3.3 μm thick** near middle: **external layer thinner than the inner.** Lateral body pores on right side numbering 22 in oesophageal region and 157 on rest of the body, in a line in oesophageal region but irregular posteriorly, forming a single or double row; three dorsal body pores in anteriormost region; ventrally, 13 pores in oesophageal region and about 50 between oesophagus and ventromedian supplements (this figure could be higher as the specimen is not exactly lateral for most of its length and some of the pores could have been overlooked). **Cephalic region low, smoothly rounded, continuous with body contour.** Amphids stirrup-shaped, just post-labial, with crescentic slits, **almost as wide as the lip region. Odontostyle 134 μm long**, distinctly furcate at base. **Odontophore 75 μm long,** with prominent tripartite basal flanges. Basal ring of guiding sheath 125 μm from anterior end. Oesophagus a narrow tube enlarging in its last quarter to form a muscular cylindrical bulb 100 μm long by 25 μm wide; a 3 μm long mucro resembling the odontostyle tip present in the wall of the oesophagus 32 μm behind odontophore. Retractor muscles of the odontostylet attached to the basal flanges anteriorly and extending to the body wall behind the middle of anterior slender part of oesophagus. Nerve ring a little behind base of odontophore. Oesophago-intestinal valve conoid-rounded. Testes two, on right of intestine, one anteriorly outstretched, other reflexed; vas deferens filled with spindle-shaped sperms. Paired submedian

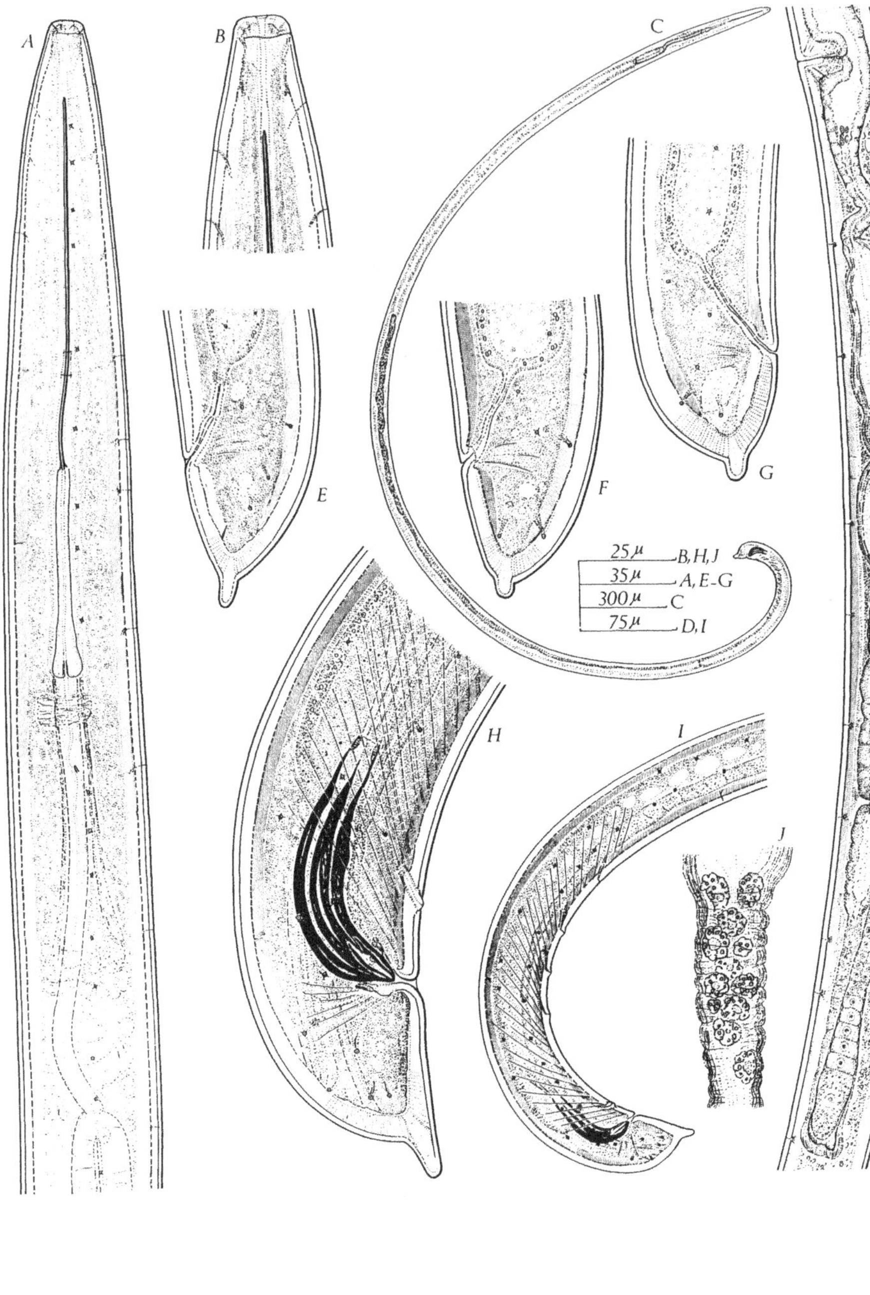

Fig. 8.14. *Xiphinema diversicaudatum*. A,B,C,H,I. Lectotype male; remainder females from hops, England. A, B. Anterior ends of body; C. Entire worm showing body posture; D. Posterior branch of reproductive system; E–I. Tail ends. J. Z organ with globular bodies. (After Siddiqi.)

supplementary papillae located 21 µm in front of anus; three well developed ventromedian supplementary papillae at 106 µm, 136 µm and 165 µm anterior to anus, a fourth rudimentary ventromedian supplement at 47 µm in front of the third. Strong diagonal copulatory muscles (34 bands on one side of body) present in region of supplements; they are responsible for the strong curvature of the tail end. **Spicules** robust, ventrally curved near middle, **75 µm along the chord**; lateral guiding piece 15 µm long. Prerectum 575 µm long. **Tail dorsally convex-conoid, ventrally somewhat flattened, with a terminal digitate, bluntly rounded 12 µm long peg; slightly over one anal body width long, with five pairs of body pores**; inner layer of cuticle bearing radial striations not extending into tail peg.

Female: (Based on populations from hops, England and from Belgium which have males identified as X. *diversicaudatum* after the lectotype described above, and partly after Goodey *et al.*, 1960): Body shape, cephalic region, odontostylet and oesophagus as described for male. Body cuticle 3.5–4.7 µm thick near middle: **external layer 0.30–0.45 as thick as the inner one.** There are four to six dorsal body pores on the front region and 50–64 well-spaced ventral body pores along the entire length. **Odontostyle and odontophore averaging 143 µm and 85 µm long, respectively.** Oesophageal bulb 3.3–5.0 times as long as wide, with dorsal oesophageal gland nucleus near its anterior end and nuclei of the anterior pair of subventral glands near its middle. Mucro in anterior part of oesophagus 2–6 µm long, 33–105 µm behind base of stylet. **Vulva a transverse slit, at 39–46 (43) % of body. Genital tracts amphidelphic, reflexed.** Uteri near vagina form a well developed ovejector. Each oviduct and uterus joined through a sphincter-Z. **A prominent pseudo-Z organ containing 10–20 (13) irregular globular bodies 3–10 µm in diameter present in each uterus.** Prerectum 7–12 times body width long; rectum shorter than anal body width. **Tail** shape as described for lectotype but ventral surface usually convex although less so than dorsal surface, **about one anal body width long; tail peg variable in length from 0–16 (9.6) µm; two to four (usually three) pairs of lateral body pores on tail.**

Juveniles: The four juvenile stages are recognized by the length of the body and of the funtional and replacement odontostyles; the length of the replacement odontostyle of one stage corresponding to the length of the functional odontostyle of the next stage. Only the preadult juveniles have tails similar to adults, earlier stages have more tapering tails lacking a distinct peg. **First stage juveniles have an elongate conoid, ventrally arcuate tail whose terminal fifth is hyaline.**

BIONOMICS: X. *diversicaudatum* is found in a wide variety of soils, particularly those under deciduous woodland, hedgerows and permanent pasture. Highest densities occur in the top 20 cm, but nematodes can be found down to at least one metre. The vertical distribution of this species is discussed by Boag *et al.* (1987) and the influence of the July soil isotherm on distribution in Europe by Boag *et al.* (1991). Reproduction is amphimictic, the life cycle taking up to three years to complete with individual females living for as long as five years (Flegg, 1968; Dalmasso, 1970). The host range mostly comprises woody plants such as apple, grapevine, hops, raspberry, rose, etc., but also includes asparagus, brassicas, carrot, clover and strawberry. Direct feeding results in galling and necrosis, but X. *diversicaudatum* also vectors arabis mosaic virus (AMV) and strawberry latent ring spot virus (SLRV).

Distribution: Recorded from the temperate regions of the world (Australia, CIS, Europe, New Zealand, North America). Recent distribution reviews include those of: Brown and Taylor, 1987; Brown *et al.*, 1990, Martelli and Taylor, 1990 and Robbins and Brown, 1991. The latter authors conclude that X. *diversicaudatum* is a comparatively rare, non-indigenous species in North America and was introduced from Europe, probably with imported rose plants.

Xiphinema index Thorne and Allen, 1950 (Fig. 8.15)

syn. *Xiphinema* (*Diversiphinema*) *index* (Thorne and Allen, 1950) Cohn and Sher, 1972

Measurements: Female syntypes (n = 6): L = 3.1 (2.91–3.28) mm; a = 62 (58–66); b = 6.8 (6.0–7.7); c = 84 (75–93); c′ = 1.12 (1.0–1.3); $V = {}^{14-10}39.4^{11-17}$ (38–40); odontostyle = 126 (119–129) μm; odontophore = 70 (63–78) μm; odontostylet = 196 (190–206) μm.
Lectotype female: L = 3.1 mm; a = 62; b = 7.3; c = 84; c′ = 1.1; $V = {}^{13}38.5^{19}$; odontostyle = 119 μm; odontophore = 72 μm; egg = 175 x 36 μm. After Thorne and Allen (1950): Female: L = 3.4 mm; a = 58; b = 7.6; c = 76; $V = {}^{15}38^{16}$.
Male: L = 3.6 mm; a = 63; b = 7.3; c = 88; T = 49.
(After Heyns, 1971) from South African vineyards: Female (n = 10): L = 3.25 (2.87–3.58) mm; a = 57 (54–61); b = 6.6 (6.2–8.0); c = 88 (72–98); c′ = 0.9 (0.7–1.1); V = 41 (40–42); odontostyle = 129 (123–134) μm; odontophore = 78 (74–81) μm; odontostylet = 206 (197–215) μm.

Description: (modified from Siddiqi, 1974)
Female: (Based on lectotype and syntypes): **Body** elongate-cylindrical, **about 3 mm long**, forming an open spiral with a greater curvature in posterior half; cuticle thick, finely striated superficially. Lateral body pores in a line in oesophageal region then in two irregular rows; dorsal body pores three or four in the anteriormost portion; ventral body pores spaced out along the entire body length, eight or nine in oesophageal region, 13 or 14 between oesophagus and vulva and 21 or 22 between vulva and anus. Cephalic region hemispherical, almost continuous with body contour except for a slight depression behind level of amphid apertures; labial papillae, six in inner and ten in outer circlet, not elevated. Amphids stirrup-shaped, just postlabial, with slit-like apertures almost as long as lip region width. **Odontostyle averaging 126 μm long**, furcate at base to give firm attachment to the **odontophore which averages 70 μm in length** and has three large basal flanges. Basal ring of guiding sheath 108

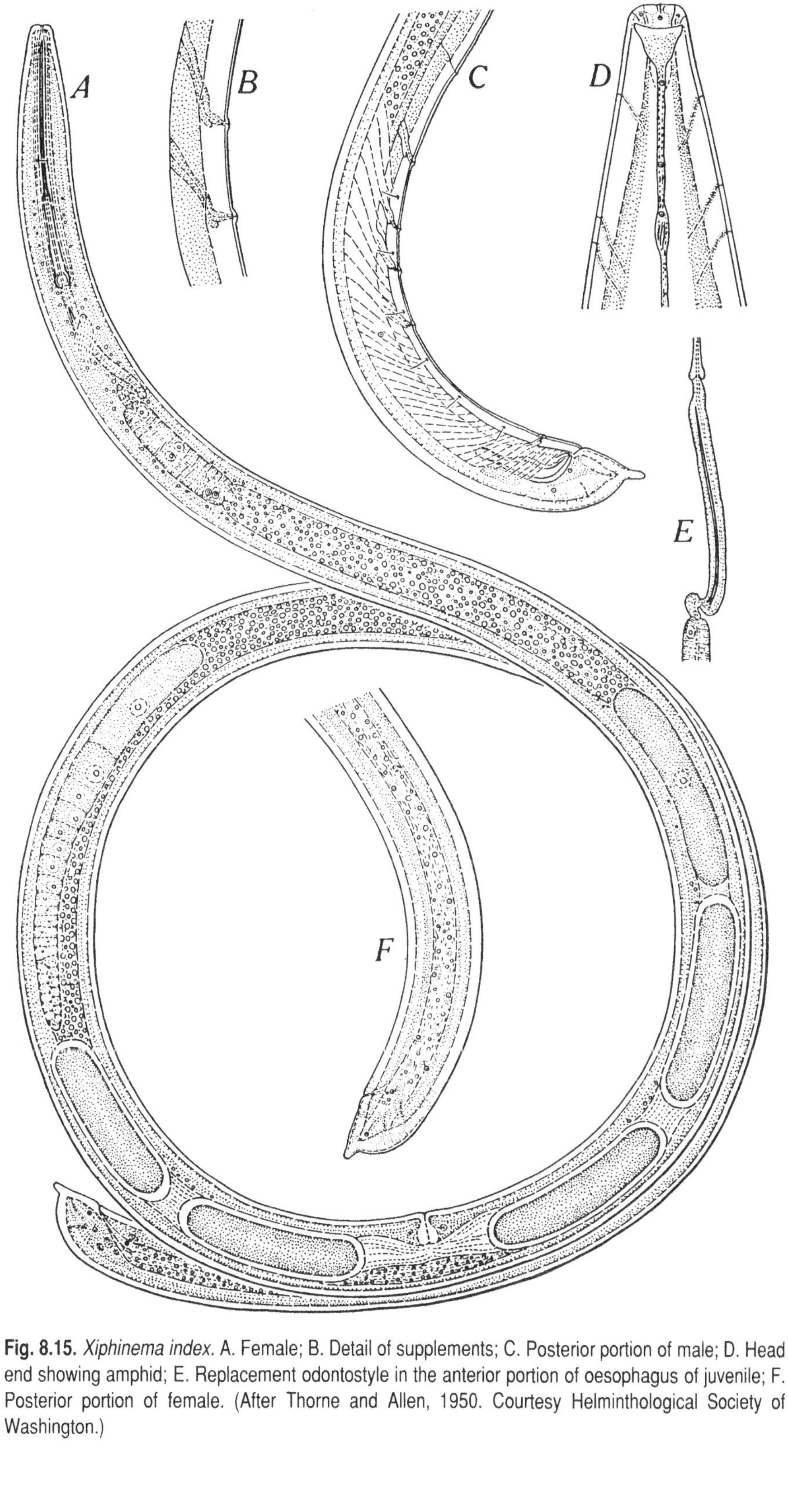

Fig. 8.15. *Xiphinema index.* A. Female; B. Detail of supplements; C. Posterior portion of male; D. Head end showing amphid; E. Replacement odontostyle in the anterior portion of oesophagus of juvenile; F. Posterior portion of female. (After Thorne and Allen, 1950. Courtesy Helminthological Society of Washington.)

(100–114) μm from anterior end. Retractor muscles of odontostylet attached anteriorly to its base and posteriorly to the body-wall near middle of anterior slender portion of oesophagus. Nerve ring a little behind base of odontophore with the hemizonid just anterior. A short mucro resembling the odontostyle tip in shape present in the oesophageal wall of female 30 μm behind base of odontophore. Posterior oesophageal part cylindroid, very muscular, about half as wide and 2.5 times as long as body width at its base; nucleus of dorsal gland and gland orifice near its anterior end; nuclei of ventrosublateral glands elongate, posteriorly tapering, at about its middle; those of posterior ventrosublateral glands not seen. Oesophago-intestinal valve small, conoid-rounded. **Vulva** a depressed transverse slit about half a body width long, **usually located at 38–40% of the body length.** Vagina at right angles to body axis, with well developed sphincter and dilator muscles, leading into a highly muscular uterine chamber or ovejector, about one body width long. **Genital tracts amphididelphic, reflexed at the oviduct. Z organ absent.** Ovaries elongate, symmetrical, with several oocytes arranged serially except in the region of multiplication near tip. Prerectum 320–390 μm long, its junction with intestine sometimes difficult to see. Intestine not extending over rectum; latter distinctly cuticularized in distal half; intestino-rectal junction with a sphincter muscle; anus distinct. **Tail mammiform, 1.0 to 1.3 times as long as anal body width, convex-conoid with a greater curvature dorsally and a digitate ventral or terminal peg 9 to 13 μm long**; inner protoplasmic core also convex-conoid; inner cuticle, thick, radially striated, not extending into tail peg; two or three pairs of caudal pores present behind anal level; females with pegless tail sometimes occur.

Male: **Extremely rare**, not essential for reproduction. Body shape, lip region, oesophagus, tail, etc. as in female. Testes two, one outstretched, the other reflexed. Spicules heavily built, not cephalated, 63 μm long; lateral guiding pieces of spicules cylindroid, 12 μm long, near distal ends of spicules. Diagonal copulatory muscles strongly developed. Besides the ventrosubmedian pair of preanal supplements, there is a series of five ventromedian papillae beginning about two spicule lengths in front of anus; anteriormost papilla poorly developed. A ventrosubmedian series of seven pairs of innervated papillae in tail region has also been reported (Thorne and Allen, 1950).

Bionomics: This species is most frequent and abundant in light to medium textured soils, but also occurs in heavier soils. Males are extremely rare and the life cycle is parthenogenetic. The life cycle has been recorded as short as 22–27 days at 24°C (Radewald and Raski, 1962), but Cohn and Mordechai (1969) reported longer times of 3–5 months at 28°C and 7–9 months at 20–23°C. Wyss *et al.* (1988) gave a detailed account of the feeding of this species on fig. The major economic host is grapevine, but it also attacks various woody plants including citrus, fig, mulberry, *Pistacia*, *Prunus* spp., *Pyrus* spp., rose, etc. In addition to the direct root damage caused by its feeding, such as necrosis, terminal galling and lack of lateral root development, X.

index vectors two viruses to grapevine, namely grapevine fanleaf virus (GFV) and grapevine yellow mosaic virus (GYMV).

Distribution: The recorded distribution closely shadows that of its major host, the grapevine, and includes almost every viticulture region in Africa, Australia, Europe, South America and the USA. Recent distribution reviews include those of: Brown and Taylor, 1987; Brown *et al.*, 1990, Martelli and Taylor, 1990 and Robbins and Brown, 1991.

Xiphinema italiae Meyl, 1953 (Fig. 8.16)

syn. *Xiphinema (Elongiphinema) italiae* (Meyl, 1953) Cohn and Sher, 1972
Xiphinema arenarium Luc and Dalmasso, 1964
Xiphinema bulgariense Stoyanov, 1964

Measurements:

(After Martelli *et al.*, 1966) Neotype female: L = 2.4 mm; a = 84; b = 6.4; c = 32; c′ = 4; V = 46; odontostyle = 89 μm; odontophore = 64 μm. Female topotypes (n = 5): L = 2.48 (2.3–2.8) mm; a = 80 (75–84); b = 6.8 (6.4–7.3); c = 34 (30–40); c′ = 3.8 (3.2–4.3); V = 46 (45–46); odontostyle = 91 (87–96) μm; odontophore = 59 (54–64) μm. Male (from Israel and France): L = 3.06 (2.4–3.35) mm; a = 86.5 (79–101); b = 7.9 (6.4–9.2); c = 48.8 (47–50); c′ = 2.25 (2–2.25); odontostyle = 101.3 (93–102) μm.

(After Martelli and Lamberti, 1967) combined measurements of populations from Italy, Israel and France: Female (n = 132): L = 3.0 (2.3–3.5) mm; a = 84 (75–114); b = 8.1 (6.4–12.7); c = 42 (30–56); c′ = 3.3 (2.1–4.3); V = 45 (41–50); odontostyle = 98 (87–112) μm; odontophore = 60 (54–68) μm.

(After Heyns, 1974) From South Africa. Female (n = 25) : L = 2.6 (2.3–2.8) mm; a = 68 (60–73); b = 7.4 (6.1–8.4); c = 40 (32–45); c′ = 2.6 (2.3–2.8); V = 46 (44–49); odontostyle = 85 (80–91) μm; odontophore = 62 (58–66) μm.

Male (n = 2): L = 2.3–2.4 mm; a = 62–63; b = 6.7–6.8; c = 46–49; c′ = 1.8–1.9; odontostyle = 88–90 μm; odontophore = 60–62 μm.

Description: (modified from Cohn, 1977)

Female: (based on topotypes and Israeli specimens) Body slender, **about 3 mm long**; almost straight in the first two-thirds of the body, then ventrally arcuate. Body cuticle smooth, 2–3 μm thick near middle, conspicuously thickened (5.0–6.5 μm) on dorsal side of tail; finely striated transversely, most conspicuously in caudal region. Lateral body pores in a single row, clearly discernible up to base of oesophagus, beyond which they are more difficult to see. **Cephalic region** gently rounded, 10–12 μm across, **set off by a slight constriction.** Labial papillae clearly visible. Amphids postlabial, stirrup-shaped, apertures slit-like, occupying between one-half and two-thirds of body diameter. **Odontostyle about 95 μm long** and furcate at base; **odontophore about 60 μm long** with prominent tripartite basal flanges, measuring 8–11 μm across. Basal ring of spear guiding sheath about 81 μm from anterior end. Nerve ring, 11–13 μm thick, located

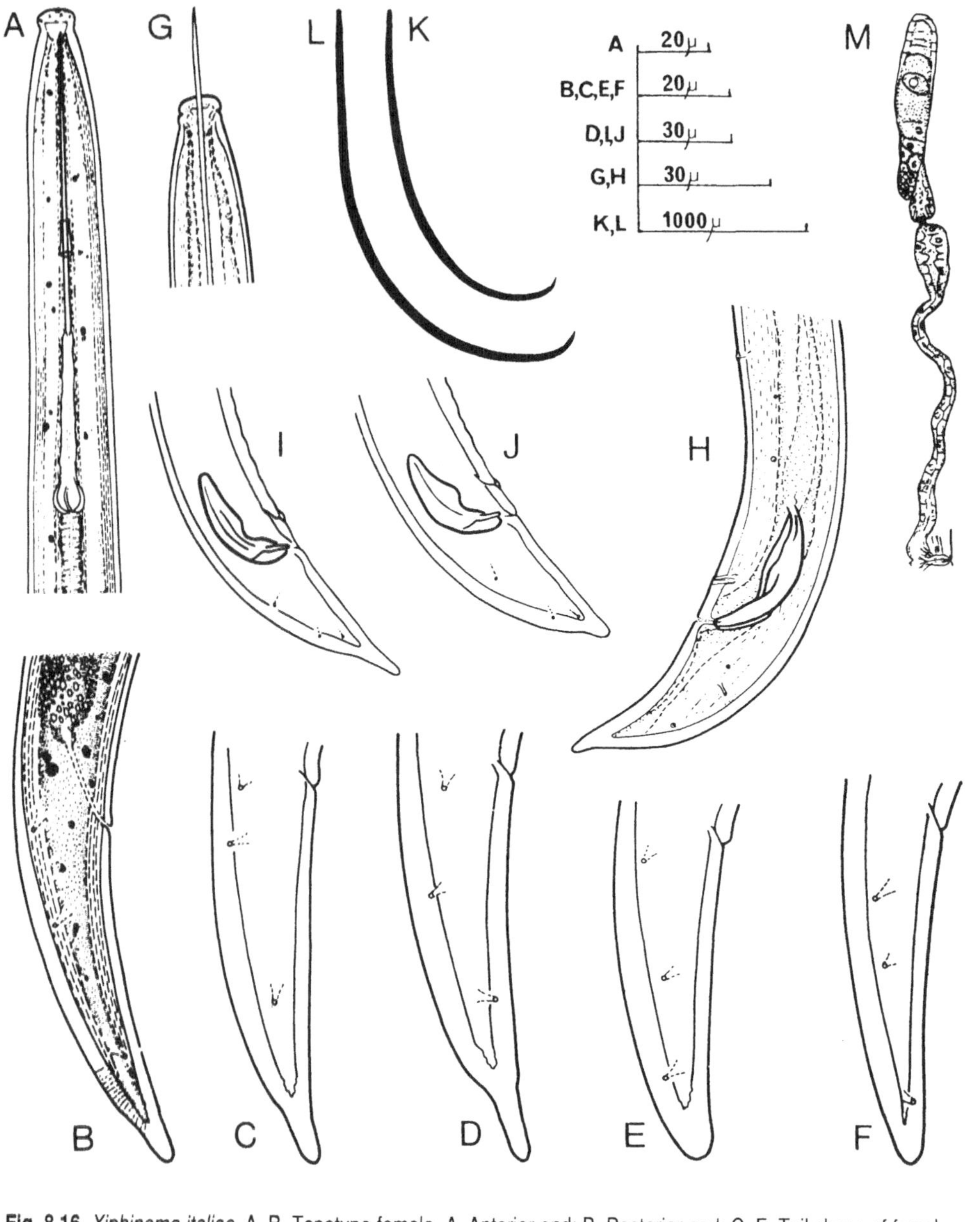

Fig. 8.16. *Xiphinema italiae.* A, B. Topotype female. A. Anterior end; B. Posterior end; C–F. Tail shape of females from different populations; G. Head end; H. Posterior end of male from French population; I,J. Male tail, South African populations; K,L. Female body postures; M. Female reproductive system, anterior branch. (A–H, K–L from Martelli *et al.*, 1966, courtesy *Nematologica*; I, J from Heyns, 1974, by courtesy; M from Dalmasso, 1969, by courtesy.)

behind base of odontophore; hemizonid about 170 μm from anterior end and hemizonion, often obscure, 213–235 μm from anterior end. Anterior part of oesophagus slender, coiled, with a triangular mucro, 2–3 μm long in oesophageal wall about 20 μm behind base of spear extension and pointing towards the head. Posterior part of oesophagus cylindrical, strongly muscular, measuring about 100 x 18 μm; nucleus of dorsal oesophageal gland near the anterior extremity of the bulb. Oesophago-intestinal valve often indistinct, conical, about 8 x 5 μm. **Vulva** a transverse slit, **slightly premedian (V = 43–49)**, leading to a vagina which extends one-third into the body. **Genital tracts amphidelphic, reflexed** at oviduct. **Z organ absent.** Uterus differentiated to form an elongated narrow section and a short dilated pouch, the two parts sometimes appear to be separated by a sphincter which is not always distinct. The oviduct is separated from the uterine pouch by a prominent sphincter and connects with the largest oocyte of the ovary. Uterine eggs, measuring about 112 x 25 μm, have been observed. Rectum oblique, 1–1.5 anal body widths long. **Tail shape** variable even within the same population, **most commonly elongate-conoid with slight dorsal and ventral constrictions towards the terminus**; sometimes bluntly conoid, and even almost subdigitate. **Tail length 2.8–4.3 times longer than anal body width** (2.3–2.8 times in South African specimens), with two or three pairs of caudal papillae.

Male: (based on specimens from Israel, France and South Africa) **Rare**, not essential for reproduction and, apparently, not functional. Body shape similar to that of female, but tail somewhat shorter, more conoid, with a smaller c' value which ranges from 2.0–2.3 in the Mediterranean specimens and from 1.8–1.9 in the South African ones. **Spicules arcuate**, measuring **44–48 μm along their axis**; lateral guiding pieces of South African specimens 13–14 μm long with dorsal projections. An adanal pair of supplements present and one or two additional ventromedian ones. Copulatory muscles strongly developed.

Juveniles: (based on a population near Bari, Italy) The four juvenile stages can be recognized by a progressive increase in length of body and of functional and replacement odontostyles. The c' value also decreases with age due to the greater development of the caudal region in diameter than in length.

Bionomics: X. *italiae* is restricted to coastal regions and areas with open, sandy soils, the highest populations occurring in the top 30 cm. Navas and Arias (1986) discussed the influence of ecological factors on the distribution of X. *italiae* in Spain. Males are rare and reproduction is parthenogenetic with a life cycle of 6 months on grapevine at 28°C (Cohn and Mordechai, 1969). The host range includes a wide variety of woody plants such as almond, apricot, citrus, conifers, mulberry, olive, peach and plum, but the major host is grapevine. X. *italiae* is ectoparasitic both at the root tip and along its length and causes darkening of the root tissue with some terminal galling. Attacked root systems are stunted. Grapevine fanleaf virus (GFV) is vectored by this nematode (see Cohn, 1977).

Distribution: Widespread around the Mediterranean Basin and also found in south-eastern Europe and South Africa. Recent distribution reviews include those of: Brown and Taylor, 1987; Brown *et al.*, 1990; Martelli and Taylor, 1990 and Spaull and Heyns, 1991.

Control of longidorids: The major reason for controlling longidorids is due to their importance as virus vectors. As they are polyphagous, crop rotation is not a promising approach and, as a result of their extended cycle and longevity, bare fallow is ineffective unless accompanied by repeated deep cultivation. Chemical control techniques involving fumigant nematicidal compounds such as DBCP, D-D or EDB have proved effective in reducing virus transmission, but many of these compounds have now been withdrawn because of concerns over their toxicity. Non-volatile nematicides can also be used and may be worthwhile for high value cash crops such as strawberry or other soft fruit, or where virus-free plants for certified stock are required. Virus resistant varieties, particularly for crops such as grapevine, remain the best solution.

III Family Trichodoridae

9 Historical Review

The first trichodorid was described by de Man (1880), when he named *Dorylaimus primitivus*, a species which subsequently became a member of *Trichodorus*, a new genus erected by Cobb (1913), when Micoletzky (1922) proposed the combination *T. primitivus*. On erecting *Trichodorus*, Cobb (1913) also described *T. obtusus*, but the identity of this species, originally found around the roots of various grasses at Arlington, Virginia, has unfortunately always been in doubt and it was regarded as a synonym of *T. primitivus* by Micoletzky (1922) and also by some later authors.

Despite this early start, it was not until the 1950s that *Trichodorus* started to receive significant attention. Christie and Perry (1951) reported on damage to plants caused by a trichodorid species which was later named *Trichodorus christiei* by Allen (1957a) and is now generally regarded as a synonym of *Paratrichodorus minor*. They coined the sobriquet 'stubby-root nematode', an apposite appellation reflecting the symptoms caused by its feeding. Seinhorst (1954) published the next trichodorid to be described when he named *Trichodorus pachydermus* (later to be transferred to *Paratrichodorus*), thus bringing the total number of nominal species to three. A further species, *Trichodorus minor* (now in *Paratrichodorus*) was described from Queensland, Australia by Colbran (1965). Allen (1957a) published ten new species, including the monodelphic *T. monohystera*, and gave an excellent review of the genus in a contribution which remained unsurpassed for nearly two decades. Allen considered erecting two subgenera under *Trichodorus* on the basis of the presence or absence of a bursa in the male, but refrained from formally proposing them as he felt that the intrageneric variation was not sufficiently well known to warrant such a step.

In the sixties a number of new species were described by Siddiqi (1960, 1962c, 1963) and Hooper (1962, 1963). Yokoo (1964a, 1966) described three species from Japan of which only one, *T. cedarus*, is now regarded as valid, and Mamiya (1967) added descriptive information for three species from Japanese forest nurseries. Seinhorst (1963) redescribed the male of *T. primitivus* and also proposed the new species *T. similis*. In the 1970s, Wyss (1970b) used photomicrographs to illustrate differences in spicule shape and vaginal form and also reiterated the observation of Allen (1957a) concerning the separation of the trichodorids into two groups, depending on the presence or absence of a bursa. Esser (1971) produced a useful species compendium of the genus, whilst Hooper (1972) named another two species of *Trichodorus*, *T. variopapillatus* and *T. velatus* from England and Loof (1973), studying the *T. aequalis* complex, proposed and described *T. hooperi*, naming it in recognition of Hooper's contribution to our knowledge of the group. Undoubtedly the major work of the decade was by Siddiqi (1974a) when he reviewed the systematics and morphology of the group, described three new species and proposed the new genus *Paratrichodorus* on the basis of the absence of a bursa in the male in concert with other characters of the male and female. He also divided *Paratrichodorus* into three subgenera, viz. *P.* (*Paratrichodorus*), *P.* (*Atlantadorus*) and *P.* (*Nanidorus*), based on characters of the oesophageal region, male ventromedian copulatory supplements and cervical lateral pores. This paper remains one of the benchmarks in trichodorid systematics. Loof (1975), in a comprehensive paper, also extensively reviewed and commented on the group whilst Hooper (1975) discussed trichodorid morphology and Robertson and Taylor (1975b) considered the ultrastructure of the feeding apparatus.

In 1976 Andrássy proposed *Monotrichodorus* for the monodelphic *Trichodorus monohystera*, a unique species originally described from South America by Allen (1957a). His diagnosis of this new genus redefined brevity, – 'very similar to *Trichodorus* but ovary one, straight' – but was expanded by Rodriguez-M. *et al.* (1978) when they also erected *Allotrichodorus*, another monodelphic genus originating from South America. Rodriguez-M. and Bell (1978b) reported that the subgenera of *Paratrichodorus* could be differentiated on the type of vulval aperture: *P.* (*Paratrichodorus*) with a longitudinal slit; *P.* (*Nanidorus*) with a transverse slit and *P.* (*Atlantadorus*) with an oval pore, thus supplementing Siddiqi's original diagnoses.

Partly on the basis of the vulval aperture character, Siddiqi (1980a) raised his subgenera of *Paratrichodorus* to generic rank, but Decraemer (1980) in a large, copiously illustrated paper dealing with the systematics and morphology of the trichodorids, together with detailed keys to the accepted species, rejected the concept of splitting *Paratrichodorus* (*sensu lato*) into either subgenera or genera. She only accepted four genera as valid, viz. *Trichodorus*, *Paratrichodorus*, *Monotrichodorus* and *Allotrichodorus* and her view of the systematics of the group has now been widely adopted. Decraemer and De Waele (1981), Sturhan (1985a) and Decraemer (1991) also rejected any subdivision of *Paratrichodorus*, but Jairajpuri and Ahmad (1992) went against this trend in accepting the three subgenera of Siddiqi (1974a).

The last decade has seen many more new species proposed (e.g. De Waele and Carbonell, 1983; Vermeulen and Heyns, 1984; Baujard and Germani, 1985; Rashid *et al.* 1986b; Siddiqi, 1992). Decraemer (1987, 1988, 1989a,b) has also continued to publish extensively on the variability within the group and the reliability of characters for specific and generic diagnoses and Decraemer

(1991), in an excellent overview, summarized the systematics and taxonomy of the group, together with details of the major economic species. Currently the group comprises about 84 valid species, the majority belonging to *Trichodorus* (43) and *Paratrichodorus* (28), but with 13 species attributed to either *Monotrichodorus* or *Allotrichodorus*, two monodelphic genera forming a distinct morphological and geographical grouping. The Trichodoridae, despite lacking the taxonomic charisma of some other groups, has nevertheless attracted a dedicated coterie of workers and has thus largely avoided some of the dilettante contributions that plague more popular groups.

10 General Morphology

Trichodorids are relatively distinctive under the stereomicroscope because of their plump, cigar-shaped bodies which are rounded at both ends, but the lack of an easily recognizable spear often leads the inexperienced to mistake them for non-parasitic nematodes. Live trichodorids are very sluggish in their movements and often appear to be moribund. When heat relaxed they either die straight or slightly ventrally arcuate (Fig. 10.1, H) with varying degrees of curvature in the male tail region. The body length varies from 0.5–1.5 mm, but is usually less than 1 mm. The tail is extremely short and rounded in both sexes, particularly so in the female where the anus is subterminal (Fig. 10.1, M).

Cuticle and associated structures: The cuticle is thick and smooth and is lacking in transverse striae when viewed under the light microscope. It is loose and wrinkly and may swell abnormally in some genera after fixation. The cephalic region is rounded and slightly offset, the outer ring of papillae sometimes being particularly prominent (Fig. 10.1, C). The amphidial apertures are located just posterior to the lip region and take the form of short, gaping, ellipsoid structures about one third to half the head width in size (Fig. 10.1, C). Internally the amphid is an elongate, funnel-shaped pocket or fovea which is separated from the sensilla sac, or fusus, by a constriction (Fig. 10.1, C) and not by a duct as in dorylaims. The excretory pore is small and fairly inconspicuous and is in the oesophageal region (Fig. 10.1, A) or, rarely, a few body widths posterior. The renette cell is small. In the female, lateral body pores ('advulval' pores) may be present or absent and their presence or absence within a body width of the vulva is regarded as a reliable generic character (Fig. 10.1, I, J). Very rarely, there may also be ventromedian body pores in the region of the vulva as found, for example, in *Paratrichodorus porosus* (Fig. 10.1, L). Lateral cervical body pores in the male can be present or absent and, when present, they may be situated near the onchiostyle base or more posterior near the nerve ring or excretory pore. In addition, males may also possess up to four ventromedian cervical papillae (Fig. 10.1, B, D). There is usually a pair of caudal pores (Fig. 10.1, U) which may be so close together as to give the impression of a single pore (e.g. as in *Monotrichodorus*) or, exceptionally, absent as in some *Paratrichodorus*.

Buccal cavity and armature: The stoma is sclerotized and forms a simple guiding tube around the anterior part of the onchiostyle, the proximal portion appearing as a ring or collar (Fig. 10.1, D). The onchiostyle differs in origin from both the odontostyle of the dorylaims and the stomatostyle of the tylenchs and aphelenchs. The anterior section consists of a dorsal tooth which is predominantly solid, but proximally hollow (Fig. 10.1, A, B, D). This is formed within the region of the stoma and is attached posteriorly to an extension which is fused to the dorsal wall of the pharynx and serves for attachment of the protractor musles. The onchiostyle is used as a pick to penetrate cell walls and no food passes through its structure in contrast to the hollow odontostyle and stomatostyle of the dorylaims and tylenchs/aphelenchs respectively. Onchiostyle length varies from about 25 μm to 150 μm.

Alimentary tract: The oesophagus comprises two parts: a narrow, tubular anterior section and a posterior, non-muscular bulboid expansion which accommodates the oesophageal glands. The shape of the oesophagus is noticeably different to that of dorylaims, the posterior section gradually expanding to form a spathulate bulb (Fig. 10.1, A, B) as opposed to the cylindrical form in the latter group. There are five glands. The dorsal gland is extensive and reportedly by far the largest in some species. Next in size are the two posterior ventrosublateral glands and then the much smaller anterior ventrosublateral pair, the nuclei of which are extremely difficult to differentiate from other, non-glandular, nuclei. The relative position of the gland nuclei varies somewhat depending on gland development, genus and species. In *Trichodorus* the dorsal gland nucleus is usually located posterior to the oesophageal enlargement and at the same level as the posterior ventrosublateral nuclei which in turn are situated anterior to the oesophago-intestinal junction. However, in *Paratrichodorus*, the dorsal gland nucleus is further anterior near the oesophageal expansion and the posterior ventrosublateral nuclei are more posterior near the oesophago-intestinal junction. In *Trichodorus*, the posterior oesophagus abuts the intestine whereas in *Paratrichodorus*, for example, it may extend as a ventral overlap, or lobe, over the intestine, the overlap being formed from the posterior ventrosublateral glands. According to Raski *et al.* (1969) the dorsal and posterior ventrosublateral glands have a radiating system of sinuses which drain the secretions to their respective orifi.

The intestine is a simple, relatively undifferentiated tube which may form an anterior extension overlapping the gland lobe both dorsally and subdorsally. The posterior intestine is not differentiated into a prerectum, as is the case in dorylaims. The rectum is lined with cuticle and oriented virtually parallel to

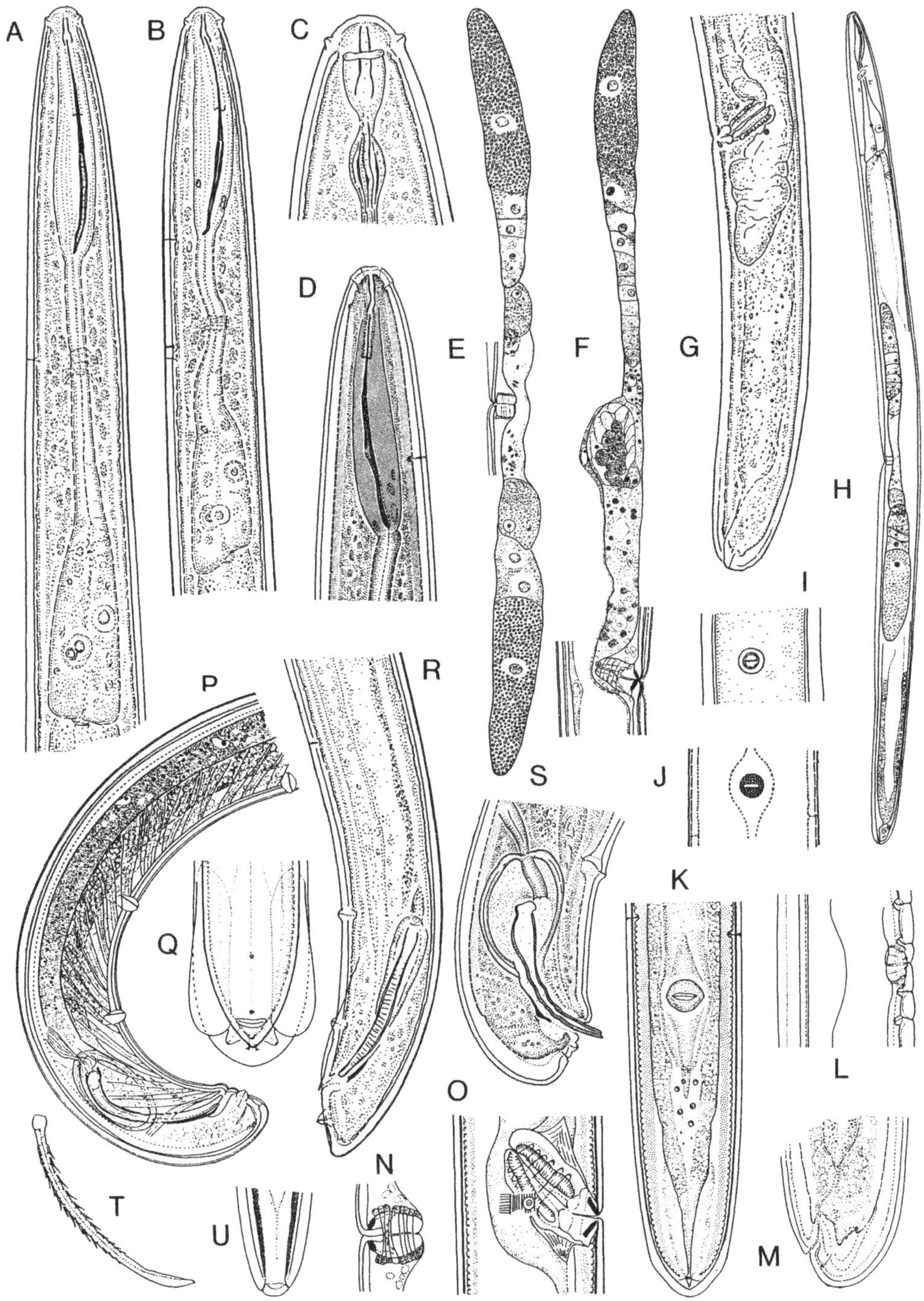

Fig. 10.1. General morphology of Trichodoridae. A,B. *Paratrichodorus paramirzai*, female and male oesophagus; C. Amphid; D. *Trichodorus* male, onchiostyle and cervical pore; E. *Paratrichodorus* minor, didelphic genital tract; F. *Monotrichodorus*, prodelphic genital tract; G. *Monotrichodorus*, female posterior end; H. *Paratrichodorus*, entire

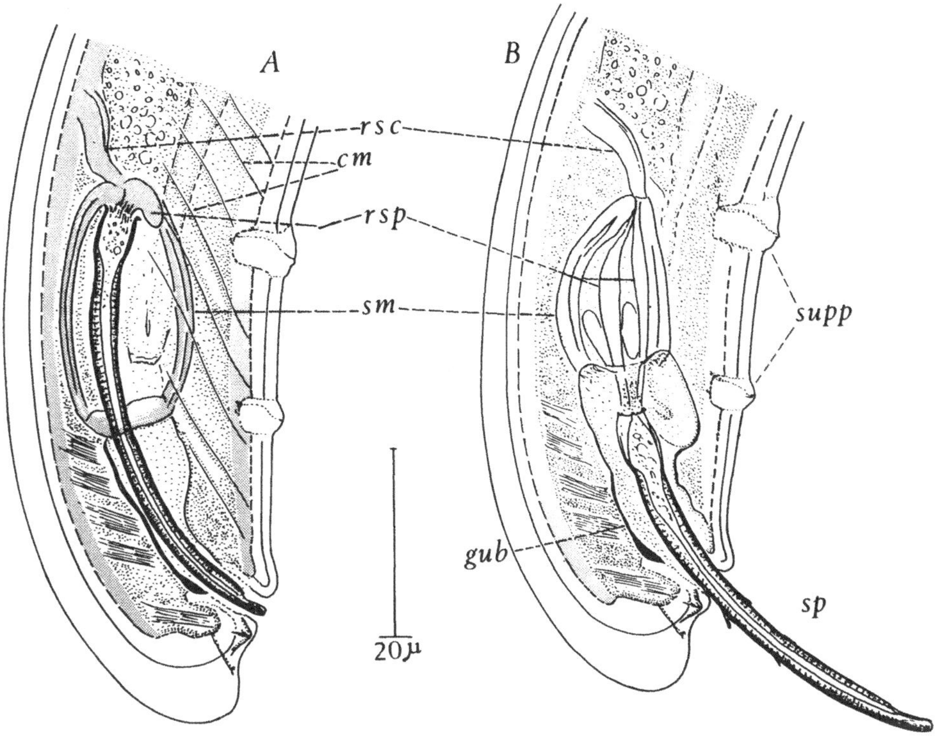

Fig. 10.2. Spicular muscles; in (A) retracted and (B) protruded spicule positions in *Trichodorus borneoensis*. cm = copulatory muscles; gub = gubernaculum; rsc = retractor muscles of spicular capsule; rsp = retractor muscles of spicule; sm = suspensor muscle of spicule; sp = spicule; supp = supplementary papillae. (After Siddiqi, 1974. Courtesy *Nematologica*.)

the body axis. The anus is a subterminal crescentic or oval aperture in the female, the cloaca being located more anteriad in the male.

Female genital system: The vulva is median to postmedian (V = 50–60%) in *Trichodorus* and *Paratrichodorus*, but far posterior in *Monotrichodorus* (V = 75–86%) and *Allotrichodorus* (V = 80–90%). The lips are non-protuberant and the vulval aperture takes the form of a small pore or a slit, either transverse or longitudinal, and often located in a shallow depression (Fig. 10.1, I–K). The vagina may be relatively short and with weakly developed, inconspicuous musculature, as in *Paratrichodorus* or extending further into

female; I. *Paratrichodorus* vulva, ventral view; J. *Trichodorus* vulva, ventral view showing lateral advulval pores; K. *Monotrichodorus vangundyi*, female posterior end, ventral view; L. *Paratrichodorus* porosus, ventral body pores near vulva; M. Female tail; N. *Trichodorus*, vaginal region; O. *Monotrichodorus*, vaginal region showing advulval lateral pore; P. *Trichodorus*, male posterior end; Q. *Allotrichodorus campanullatus*, posterior end, ventral view; R. *Paratrichodorus*, male posterior end; S. *Trichodorus lusitanicus* male spicule region; T. *Monotrichodorus monohystera*, spicule showing 'bristles'; U. Ventral view of female tail showing caudal pores. (After various authors. Not to scale.)

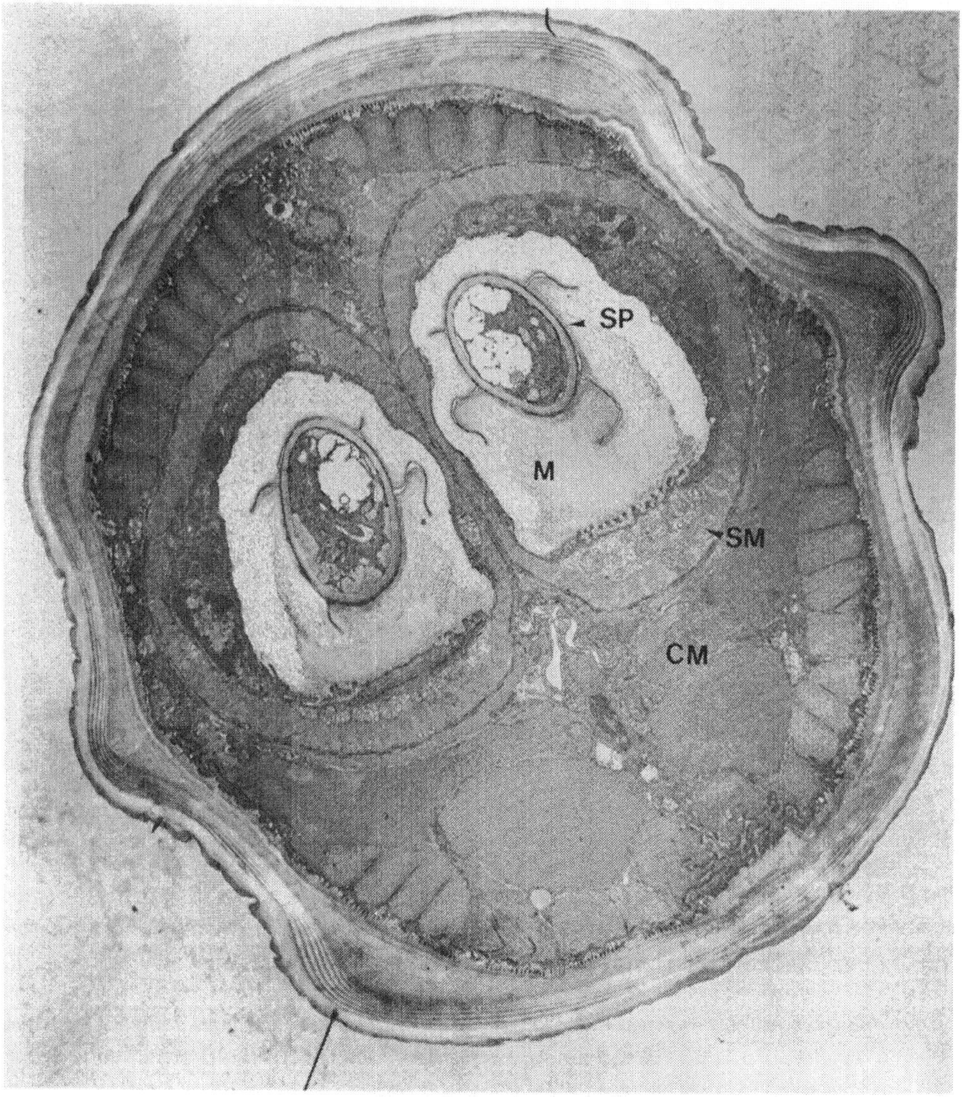

Fig. 10.3. Transverse section through spicule region of *Trichodorus primitivus*;: sp = spicule, sm = suspensor muscle of spicule, m = matrix, cm = copulatory muscle. (TEM courtesy W.M. Robertson and Scottish Crop Research Institute.)

the body with thicker walls and strong musculature as, for example, in *Trichodorus* (Fig. 10.1, N). Distally, at the junction of the vulva and vagina, there is a sclerotized ring which may be weak or strong in development and which, in lateral view, appears as two, variously shaped, sclerotized bodies, one anterior and one posterior to the vaginal lumen (Fig. 10.1, N, O).

The female genital tract is either amphididelphic with the telogonic ovary reflexed in those forms having a median vulva, or monoprodelphic, reflexed

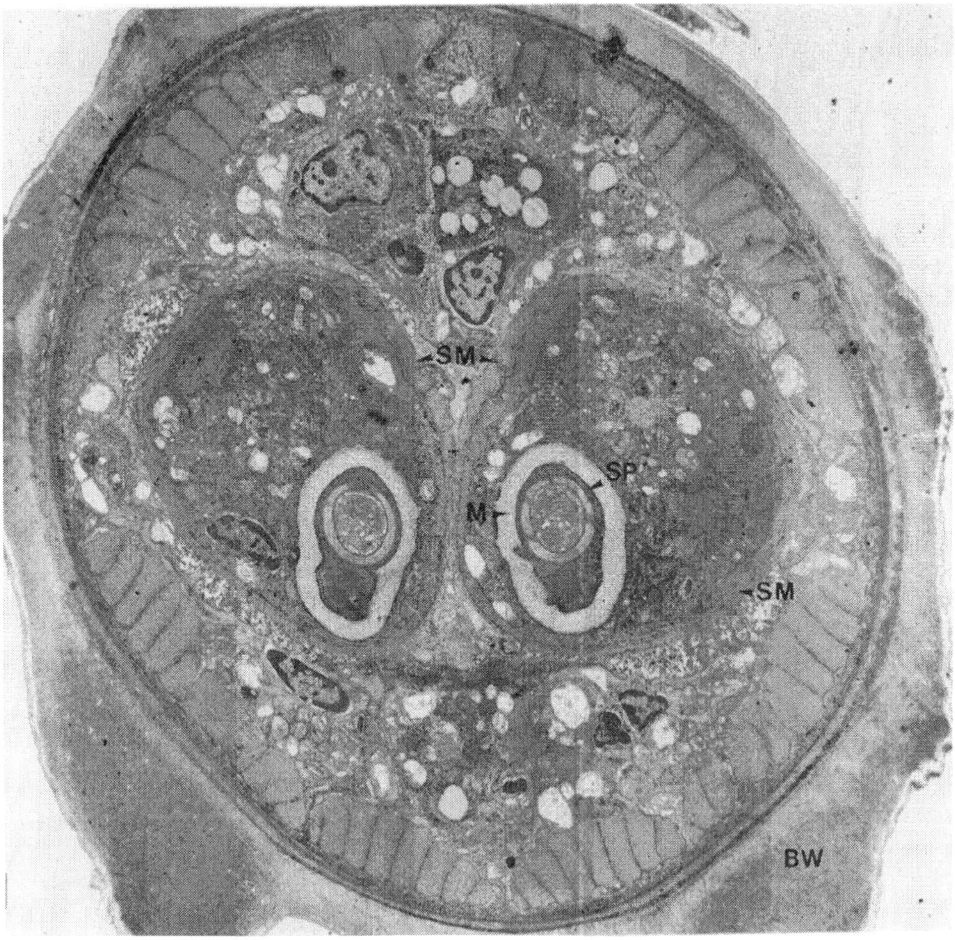

Fig. 10.4. Transverse section through spicule region of *Paratrichodorus pachydermus*;: sp = spicule, sm = suspensor muscle of spicule, m = matrix, bw = bursal wing. (TEM courtesy W.M. Robertson and Scottish Crop Research Institute.)

in those with a posterior vulva (Figs 10.1, E, F; 10.5). The latter group has a postuterine sac which may be short or well developed. An axial spermatheca is usually well developed in *Trichodorus*, but may be present or absent in the other genera. Sperm may be stored in the spermatheca, where present, but in some *Paratrichodorus* the uterus near to its junction with the oviduct may be utilized (Decraemer, 1989a). The uterus is a simple tube and the oviduct consists of two granular cells.

Male genital system: The male genital tract is monorchic and outstretched. The sperm are usually large and subcylindroid with sausage-shaped nuclei, but may be small and rounded or even filiform (Decraemer and Chaves, 1988;

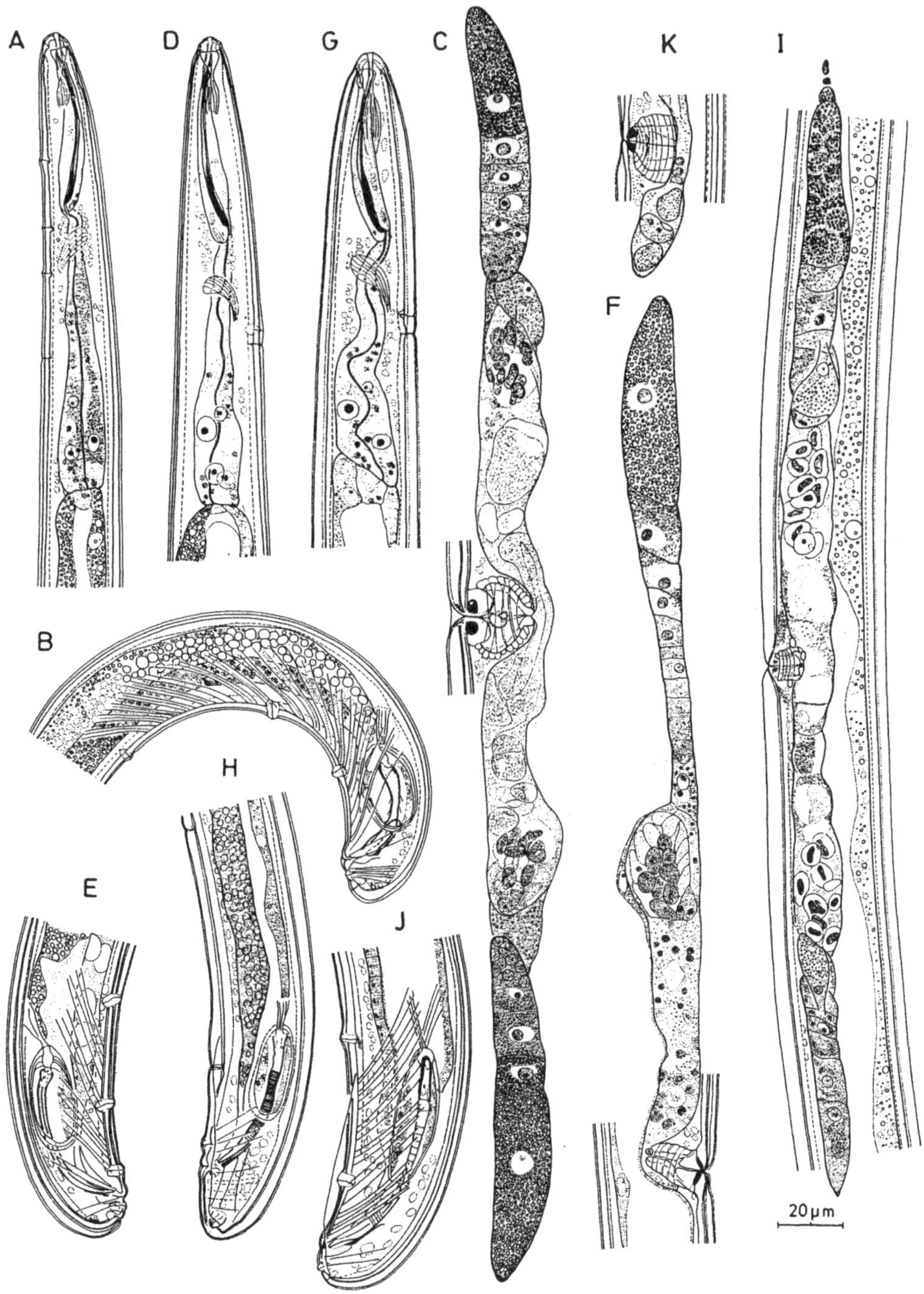

Fig. 10.5. The genera of the Trichodoridae. Anterior body region in male. A. *Trichodorus velatus* (paratype); D. *Monotrichodorus vangundyi*; G. *Paratrichodorus weischeri* (paratype.) Male posterior body region. B. *T. velatus* (paratype); E. *M. vangundyi*; H. *P. pachydermus*; J. *Allotrichodorus campanullatus*. Female reproductive system. C. *T. taylori* (holotype); F. *M. vangundyi*; K. *A. campanullatus*; I. *P. weischeri* (paratype.) (After Decraemer, 1991. Courtesy Marcel Dekker Inc.)

Decraemer, 1989a). Each spicule consists of an inner spongy core surrounded by a thick layer of cuticle which may bear a sheath of overlapping scales (Rodriguez-M. and Bell, 1978). The spicules are paired, separate and either ventrally arcuate, or modified therefrom, as in *Trichodorus*, or more or less straight with the distal tips curved ventrad (Fig. 10.1, P–T). The spicules may be plain, ornamented with transverse striations and/or setose in appearance, especially when protracted (Rodriguez-M. and Bell, 1978a; Decraemer, 1989a). Rarely, the spicule has a ventral flange, or velum, near the distal tip. There is a small, weakly delineated capitulum and, rarely, a rod-like capitular extension (some species of *Allotrichodorus* only). The dorsal wall of the cloaca is sclerotized to form a gubernaculum-like structure which is linear and non-protrusible (Fig. 10.1, R, S).

The spicule protractor system is highly unusual and was elegantly described and illustrated in some detail by Siddiqi (1974a). In essence, each spicule is suspended in a spicular pouch consisting of a spongy tissue or matrix surrounded by a capsule of longitudinal protractors (Figs 10.2, 10.3, 10.4). This muscular capsule is not attached directly to the spicule but, on contraction, squeezes the spicule out, presumably by the increased turgor pressure of the matrix. The spicules are each retracted by a pair of spicule retractor muscles which run from the capitulum to the capsule, the capsule itself being repositioned by a pair of muscles running from its proximal end to the subdorsal body walls.

Anterior to the cloaca is a series of oblique copulatory muscles whose anterior extent is negatively correlated with bursal development (Fig. 10.5). In *Trichodorus*, where a bursa is lacking, the copulatory muscles extend anteriorly for several spicule lengths and, on contraction, produce the typical J-shaped heat-relaxed form. Conversely, in *Paratrichodorus* the bursa is well developed, the muscles extend for less than a spicule length and, as a consequence, the heat-relaxed male dies more or less straight.

There is a series of one to four ventromedian copulatory supplements anterior to the cloaca, the disposition and anterior extent of which are taxonomically significant (Fig. 10.5, B, E, H, J). Posterior to the cloaca is one pair, or sometimes two pairs, of ventrosubmedian or subventral papillae. A bursa may be absent, as in *Trichodorus*, well developed as in most *Paratrichodorus* or merely represented by a flattening and thickening of the ventrosublateral cuticle of the cloacal region.

11 Bionomics

Trichodorids are polyphagous migratory ectoparasites of the roots of perennial and woody plants. Interest in the biology of trichodorids increased markedly in the sixties and seventies when they were shown to be capable of transmitting plant viruses (Harrison, 1961, 1966, 1967; van Hoof, 1962, 1964; Sänger, 1961; Sol *et al.*, 1960; Walkinshaw *et al.*, 1961).

Pitcher (1967), working from an underground laboratory fitted with observation panels, reported large aggregations of *Trichodorus viruliferus* feeding 1–3 mm behind the tip of apple roots. The nematodes fed in the zone of elongation and migrated with the root as it grew. Eventually root growth slowed and then ceased altogether and at this point the nematodes dispersed. Emerging lateral roots appear to be detected by the nematodes (Wyss, 1971a) and are also attacked, thus leading to the characteristic 'stubby root' symptoms.

Feeding on an individual cell by *T. similis* usually lasts for only a few minutes (Wyss, 1971a), but Rhoades (1965), studying *T. proximus*, recorded feed times of up to 170 minutes. According to Wyss (1971a), the contents of up to 15 cells may be ingested in an hour. As with longidorids, the feeding process can be divided into five phases, viz. exploration, penetration, salivation, ingestion and withdrawal. In the exploration phase the lips are rubbed over the cell wall until a satisfactory site is located. The lips are then pressed against the cell which is penetrated by rapid thrusts of the onchiostyle. Wyss (1971b, 1972) reported that *T. similis* penetrated with the onchiostyle tip always being directed at one site, but other workers reported a rasping action in *Paratrichodorus minor* (Rohde and Jenkins, 1957; Russell and Perry, 1966). Subsequent to penetration, the rate of onchiostyle thrusting drops and the cell cytoplasm streams towards the feeding site and accumulates. Several slow thrusts into the accumulated mass are followed by ingestion of large volumes of cytoplasm with each retraction of the onchiostyle. When nearly all the cytoplasm has been ingested, the nematode withdraws, leaving behind a feeding tube attached to the cell wall (Wyss, 1971a, b; 1972; 1975; 1974a).

Root pathology includes browning of epidermal tissues and cessation of root elongation. Cortical cells are only fed upon if made accessible by root cracking or collapse of epidermal tissues as a result of a mass attack. Cytoplasmic streaming to the feeding site occurs even after onchiostyle withdrawal. Cells can recover if the nematode abandons a site soon after penetration, thus allowing the possibility of virus transmission from infected vectors.

Although the direct feeding of trichodorids can cause severe damage to the host, e.g. 'Docking disorder' of sugarbeet caused by, amongst other factors, several species of *Trichodorus* and *Paratrichodorus* feeding on young seedlings and resulting in patchy growth (Whitehead and Hooper, 1970; Whitehead *et al.*, 1971), their major pest status stems from their known ability to transmit three tobraviruses: tobacco rattle virus (TRV); pea early browning virus (PEBV) and pepper ring spot virus (PRV). TRV is currently known from Europe, Japan and USA, PEBV only from Europe and PRV only from South America. Virus particles are selectively adsorbed on the lining of the oesophagus (Harrison *et al.*, 1974). Dissociation occurs as the nematode saliva is injected into the host , but the cell must not be killed or damaged for transmission to be a success (Harrison *et al.*, 1974). All juvenile stages can vector the viruses, but need to be reinfected after each moult as the virus retaining cuticle of the oesophagus is sloughed (Harrison *et al.*, 1974). Adult vectors can remain infective for long periods (Ayala and Allen, 1968; van Hoof, 1970). The literature on trichodorids and virus transmission is voluminous, but useful information and more extensive bibliographies can be found in: Harrison *et al.*, 1974; Hooper, 1973, 1976; Taylor, 1978; Harrison and Robinson, 1981, 1986; Lima, 1986; Brown *et al.*, 1989; Martelli and Taylor, 1990; Decraemer, 1991.

Trichodorids appear to be most widespread and abundant in light sandy soils (Winfield and Cooke, 1975) although this does not preclude their occurrence in heavier soils (e.g. Seinhorst, 1963; Cooper, 1971). They appear to be highly susceptible to mechanical damage (Bor and Kuiper, 1966) and as a result are usually found below the depth of cultivation (Cooke, 1973) with highest densities tending to occur at depths of 30–40 cm. Trichodorids are susceptible to desiccation, but can migrate vertically for considerable distances to avoid such stresses (Rössner, 1972), sometimes reaching depths of 250 cm (Rössner, 1969). Because of their sluggish behaviour they are best extracted using passive techniques such as sieving or Seinhorst's Two Flask Method (Seinhorst, 1955) in order to physically separate the nematodes from the soil bulk.

In general, *Trichodorus* species are predominantly distributed in cooler, temperate soils and *Paratrichodorus* species mostly in warmer, tropical and subtropical soils. *Monotrichodorus* and *Allotrichodorus* appear to be restricted to the Neotropical Region of South and Central America and probably represent a distinct evolutionary line (see Rashid *et al.* (1986b) for a more detailed discussion).

Reproduction is usually regarded as being amphimictic or parthenogenetic, but Sturhan (1989) reported syngonic hermaphroditism (automixis) in some species of *Paratrichodorus*. Decraemer and Chaves (1988) reported the presence of both sperm cells and spermatids in the uterus of *Paratrichodorus minor*, a predominantly monosexual species, but concluded that the former were non-fertile. Life cycle periods vary from 16–17 days at 30°C for *Paratrichodorus minor* (Rohde and Jenkins, 1957) to about 45 days at 15–20°C for *Trichodorus viruliferus* (Pitcher and McNamara, 1970). There are four juvenile stages, the first emerging from the egg.

12 Systematics and Taxonomy

The systematic relationships of the trichodorids and their closest relatives, the diphtherophorids, have not yet been satisfactorily resolved, the main bone of contention being whether they belong with the dorylaims or not. Thorne (1935b) elevated the subfamily Diphtherophorinae Micoletzky, 1922 to family rank and proposed that a new subfamily, the Trichodorinae, should embrace *Trichodorus* with Diphtherophorinae being reserved for *Diphtherophora* de Man, 1880 and *Tylolaimophorus* de Man, 1880. Thorne placed the Diphtherophoridae under the suborder Dorylaimina in the Dorylaimoidea but indicated (1939, 1961) that ordinal status might be eventually necessary for the group.

Siddiqi (1961) raised the Trichodorinae to family status, an elevation also proposed in the same year, but at a later date, by Clark (1961). Clark (1961) also proposed the superfamily Diphtherophoroidea. The following year Clark (1962) placed the Diphtherophoroidea under the suborder Alaimina within the Dorylaimida, a move followed by J.B. Goodey (1963) in his classic revision of T. Goodey's *Soil and Freshwater Nematodes*.

Subsequently, Coomans and Loof (1970) proposed the new suborder Diphtherophorina which they placed under Dorylaimida. Siddiqi (1974a) proposed the new rank of Trichodoroidea embracing the single family Trichodoridae. He retained the Trichodoroidea under Dorylaimida, but pointed out some of the unique characters of the group, such as the spicule musculature, possession of a single testis, presence of a gubernaculum, prominent excretory pore and the absence of a prerectum, features not shared by other dorylaims. On the basis of these morphological differences he eventually placed Diphtherophorina, embracing the trichodorids, in the order Triplonchida (Siddiqi, 1983b). Of the most recent authors, Decraemer (1991) still regards the suborder Diphtherophorina as belonging under Dorylaimida, whilst Jairajpuri and Ahmad (1992) accept Triplonchida as the appropriate repository for trichodorids.

I propose following the general scheme of Decraemer (1991), but accept, on morphological grounds, the view of Siddiqi (1983b) and Jairajpuri and Ahmad (1992) vis à vis the ordinal placement:

OUTLINE CLASSIFICATION

TRIPLONCHIDA Cobb, 1920
  Diphtherophorina Coomans and Loof, 1970
    Trichodoroidea Thorne, 1935 (Siddiqi, 1974)
      Trichodoridae Thorne, 1935 (Siddiqi, 1961)
        Trichodorinae Thorne, 1935
          Trichodorus Cobb, 1913
          Paratrichodorus Siddiqi, 1974
            syn. *P.* (*Paratrichodorus*) Siddiqi, 1974
              P. (*Atlantadorus*) Siddiqi, 1974
              Atlantadorus Siddiqi, 1974 (Siddiqi, 1980)
              P. (*Nanidorus*) Siddiqi, 1974
              Nanidorus Siddiqi, 1974 (Siddiqi, 1980)
          Monotrichodorus Andrássy, 1976
          Allotrichodorus Rodriguez-M., Sher and Siddiqi, 1978

Superfamily Trichodoroidea Thorne, 1935 (Siddiqi, 1974)

DIAGNOSIS: **Body** 0.5–1.5 mm long, **plump, females cigar-shaped, males straight or J-shaped on heat relaxation**. Cuticle thick, smooth; may swell abnormally on death. **Amphidial apertures wide, gaping ellipses; sensilla sac separated from the fovea only by a constriction. Onchiostyle distally solid, dorsally convex, attached to the dorsal wall of the pharynx**. Simple anterior guiding ring present. **Oesophagus consisting of a narrow anterior section expanding into a posterior bulboid section of spathulate or pyriform shape** containing five glands – one dorsal, two anterior ventrosublateral and two posterior ventrosublateral. **Distinct excretory pore present**, located either within the oesophageal region or slightly posterior. **Prerectum absent**. Vulva small, median to slightly postmedian or more posterior in position depending on whether the genital system is amphididelphic, reflexed, or monoprodelphic, reflexed. Uterus a simple tube; oviduct consisting of two cells. Spermatheca present or absent. **Female anus almost terminal. Male genital tract monorchic**, outstretched. Copulatory supplements present in a precloacal, ventromedian position. Spicules straight or curved with the **spicule protractor muscles forming a capsule around the proximal half of the retracted spicules**. Bursa present or absent. Tail very short and rounded.

TYPE GENUS: *Trichodorus* Cobb, 1913

TYPE FAMILY: Trichodoridae Thorne, 1935 (Siddiqi, 1961)

OTHER FAMILIES: Monotypic.

Family Trichodoridae Thorne, 1935 (Siddiqi, 1961)

Diagnosis: Trichodoroidea. With the characters of the superfamily.

Type genus: *Trichodorus* Cobb, 1913

Other genera:
Allotrichodorus Rodriguez-M., Sher and Siddiqi, 1978
Monotrichodorus Andrássy, 1976
Paratrichodorus Siddiqi, 1974
syn. *P.* (*Paratrichodorus*) Siddiqi, 1974
P. (*Atlantadorus*) Siddiqi, 1974
Atlantadorus Siddiqi, 1974 (Siddiqi, 1980)
P. (*Nanidorus*) Siddiqi, 1974
Nanidorus Siddiqi, 1974 (Siddiqi, 1980)

Keys to genera

Female

1. Female genital tract amphididelphic 2
– Female genital tract monoprodelphic 3
2. Vagina extending halfway into body, musculature well developed; lateral body pores within one body width of vulva; cuticle not abnormally swollen on fixation . . . . *Trichodorus*
– Vagina extending for up to a third of the corresponding body width; no lateral body pores within one body width of vulva; cuticle abnormally swollen on fixation . . . . *Paratrichodorus*
3. Lateral body pores present near vulva . . . . *Monotrichodorus*
– Lateral body pores absent . . . . *Allotrichodorus*

Male:

1. Diagonal copulatory muscles extending far anterior to head of retracted spicules. Tail region sharply curved (habitus J-shaped) on heat relaxation . . . . *Trichodorus*
– Diagonal copulatory muscles either within the range of the retracted spicules or only extending about one body width anterior to spicule head. Tail region almost straight and not sharply recurved on heat relaxation 2
2. Three ventromedian copulatory supplements more or less within the range of the retracted spicules, bursa present . . . . *Allotrichodorus*
– Only one or two ventromedian copulatory supplements within the range of the retracted spicules, bursa present or absent 3

3. Spicule protractor capsule weakly developed, bursa usually prominent, diagonal copulatory muscles within range of the retracted spicules. . . . . *Paratrichodorus*

– Spicule protractor capsule prominent, bursa absent, diagonal copulatory muscles extending just anterior to retracted spicules. . . . *Monotrichodorus*

GENUS *TRICHODORUS* COBB, 1913 (Fig. 12.1)

Diagnosis: Trichodoridae. Body plump, cylindrical with rounded ends. **Heat relaxed females die ventrally arcuate, the males J-shaped with the tail region more sharply curved ventrad. Cuticle not swelling strongly** on fixation. **One to four pairs of lateral body pores usually present. One pair of lateral body pores** (i.e. one pore on each side) **always situated within a body width of the vulva** (reportedly absent in two species), and usually located posteriorly. Onchiostyle dorsally convex with a simple, anterior, guiding ring. Oesophagus consisting of a narrow anterior section which expands posteriorly to form a spathulate bulb. **Bulb usually non-overlapping**, but in some species a ventral overlap develops whereas in others the intestine extends dorsally along the bulb to form an overlap. Posterior ventrosublateral nuclei located anterior to the oesophago-intestinal junction and with the dorsal nucleus usually at the same level. Vulva a median pore or a transverse or longitudinal slit. **Vagina extending into the body for about half the corresponding diameter. Vaginal musculature well developed and prominent and sclerotization usually strong.** Genital tract amphididelphic reflexed; spermatheca present, although weakly developed in a few species. Anus subterminal; tail rounded. Caudal pores paired. **Males usually with one to three ventromedian cervical papillae**; exceptionally absent or as many as four present. **Lateral cervical pores usually present**, the pair (one pore on each side) being between the onchiostyle base and the nerve ring. Male genital tract monorchic, outstretched. Sperm large, subcylindroid, with a large sausage-shaped or, rarely, a rounded nucleus. **Spicules more or less ventrally arcuate, never straight**; either smooth or with various ornamentations, bristles, etc. A ventral flange, or velum, is present in the eponymous *T. velatus*. Gubernaculum present. **Spicule suspensor muscles forming a prominent oval capsule** around the spicules. **Bursa absent** (but regarded as being present, although very small, in *T. cylindricus*). Three, sometimes four, ventromedian copulatory supplements; the first being within the range of the retracted spicules. **Oblique copulatory muscles extending for several body widths anterior to the retracted spicules**. Tail short, rounded, with one pair of ventrosublateral papillae and a pair of caudal pores.

Type species:

Trichodorus primitivus (de Man, 1880) Micoletzky, 1922
syn. *Dorylaimus primitivus* de Man, 1880
T. castellanensis Arias, Jiminez and Lopez, 1965

OTHER SPECIES:

T. aequalis Allen, 1957
T. aquitanensis Baujard, 1980
T. azorensis Almeida, De Waele, Santos and Sturhan, 1989
T. beirensis Almeida, De Waele, Santos and Sturhan, 1989
T. borai Rahman, Jairajpuri and Ahmad, 1985
T. borneoensis Hooper, 1962
T. californicus Allen, 1957
T. carlingi Bernard, 1992
T. cedarus Yokoo, 1964
  syn. *T. kurumeensis* Yokoo, 1966
    T. longistylus Yokoo, 1964
T. complexus Rahman, Jairajpuri and Ahmad, 1985
T. coomansi De Waele and Carbonell, 1983
T. cottieri Clark, 1963
T. cylindricus Hooper, 1962
T. dilatatus Rodriguez-M. and Bell, 1978
T. eburneus De Waele and Carbonell, 1983
T. elefjohnsoni Bernard, 1992
T. elegans Allen, 1957
T. hooperi Loof, 1973
T. intermedius Rodriguez-M. and Bell, 1978
T. lusitanicus Siddiqi, 1974
T. minzi De Waele and Cohn, 1992
T. nanjingensis Liu and Cheng, 1990
T. obscurus Allen, 1957
  syn. *T. primitivus apud* Thorne, 1939 and T. Goodey, 1951
T. orientalis De Waele and Hashim, 1984
T. pakistanensis Siddiqi, 1962
  syn. *T. litchi* Edward and Misra, 1970
T. parorientalis Decraemer and Kilian, 1992
T. paucisetosus Bernard, 1992
T. persicus De Waele and Sturhan, 1987
T. petrusalberti De Waele, 1988
T. philipi De Waele, Meyer and Van Mieghem, 1990
T. proximus Allen, 1957
T. rinae Vermeulen and Heyns, 1984
T. sanniae Vermeulen and Heyns, 1984
T. similis Seinhorst, 1963
T. sparsus Szczygiel, 1968
T. taylori De Waele, Mancini, Roca and Lamberti, 1982
T. tricaulatus Shishida, 1979
T. vandenbergae De Waele and Kilian, 1992
T. variopapillatus Hooper, 1972
T. velatus Hooper, 1972
T. viruliferus Hooper, 1963
T. yokooi Eroshenko and Teplyakov, 1975

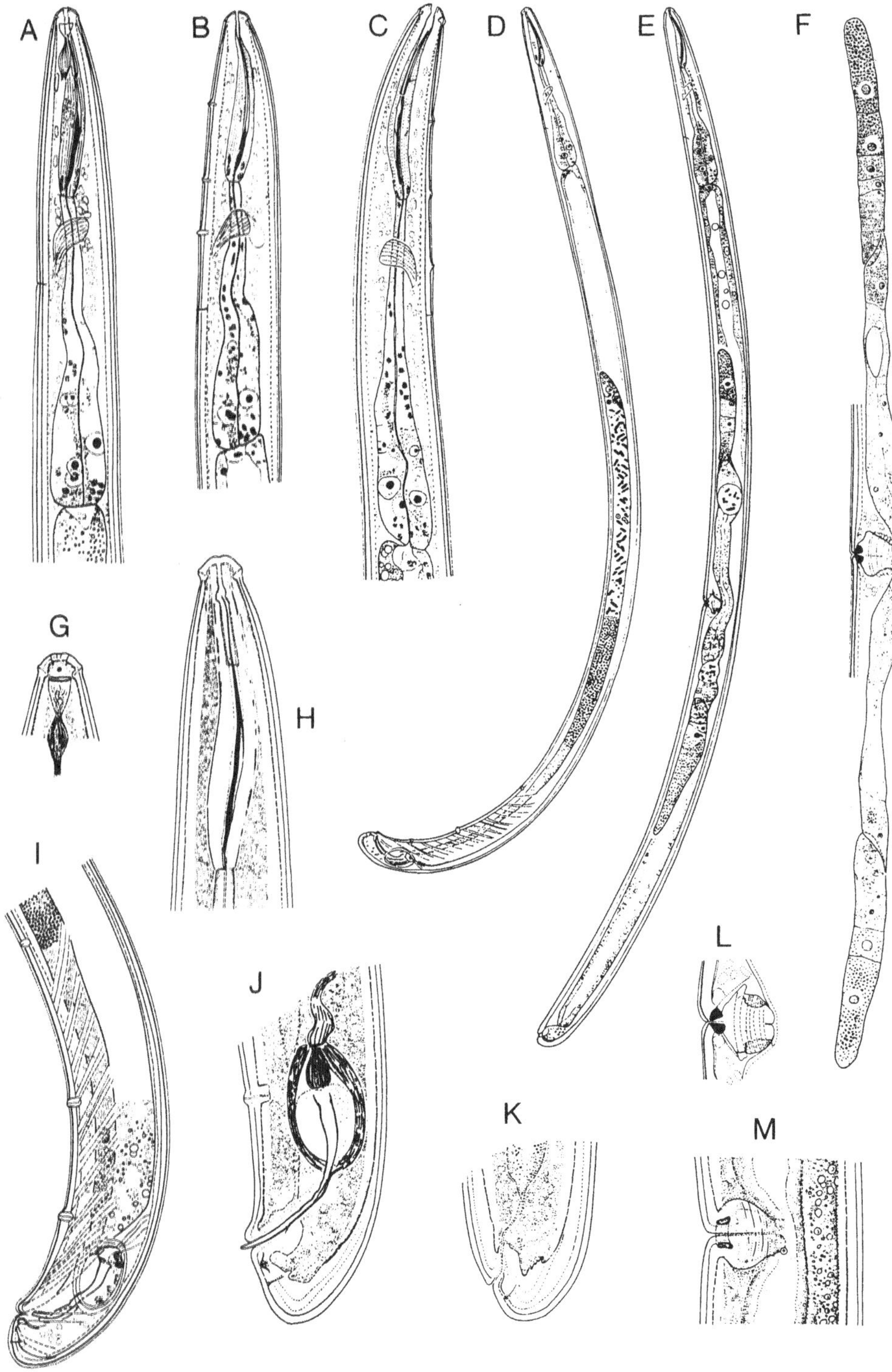

Fig. 12.1. General morphology of *Trichodorus*. A. *T. primitivus* female oesophagus; B. *T. primitivus* male oesophagus; C. *T. viruliferus* male oesophagus; D,E. entire male and female of *T. similis*; F. *T. similis* genital tract; G. Amphid; H. Onchiostyle; I. *T. viruliferus* male tail; J. *T. primitivus* male tail; K. Female tail; L. *T. similis* vagina;

Species inquirenda: *T. obtusus* Cobb, 1913

Bionomics: Migratory polyphagous ectoparasites of plant roots found mainly in sandy or sandy-loam soils in temperate regions. In addition to causing stubby root symptoms from their direct feeding activities, some species are known to be virus vectors and are of considerable economic importance in Europe and the USA.

Distribution: Almost worldwide (Africa, Asia, Australia, Europe, North America, Central America), but predominantly in the more temperate regions of Europe and North America. Some of the present distribution is probably due to nematodes being introduced with non-indigenous plants.

Etymology: Derived from the Greek *trichos* = a hair or bristle, and *dorys* = a spear.

Species descriptions

Trichodorus primitivus (de Man, 1880) Micoletzky, 1922 (Fig. 12.2)

syn. *Dorylaimus primitivus* de Man, 1880

Measurements:
Female (n = 6) (after Allen, 1957): L = 0.71–0.81 mm; a = 22–27; b = 4.9–5.5; c = ? (anus subterminal); $V = {}^{22-19}55\text{–}57^{19-25}$; onchiostyle = 45–52 μm.
Male (n = 6) (after Seinhorst, 1963): L = 0.66–0.87 mm; a = 24–29; b = 4.4–5.4; c = 40–58; spicules = 34–41 μm; onchiostyle = 50–55 μm.

Description: (modified from Hooper and Siddiqi, 1972)
Males and females usually occur in about equal numbers.
Female: Body only very slightly curved ventrally when killed by heat. Anterior region similar to male but without ventromedian papillae. Lateral body pores present; one slightly posterior to the vulva, one three body widths and the other several body widths anterior to the vulva. Caudal pores ventrally subterminal, tail bluntly rounded. Genital tracts amphididelphic, with reflexed ovaries; **spermatheca usually containing sperm. Refractive thickenings at the vulva are conspicuous and characteristically rod-shaped and appear at right angles to the body axis when viewed laterally**.

Male: Body rather stout, 'cigar-shaped', curved ventrally when killed by heat; tapers anteriorly to the slightly offset cephalic region. Amphidial apertures sub-

M. *T. viruliferus* vagina. (A-F, I, L, after Decraemer, 1991, courtesy Marcel Dekker Inc.; remainder after Hooper and Siddiqi, 1972. Not to scale.)

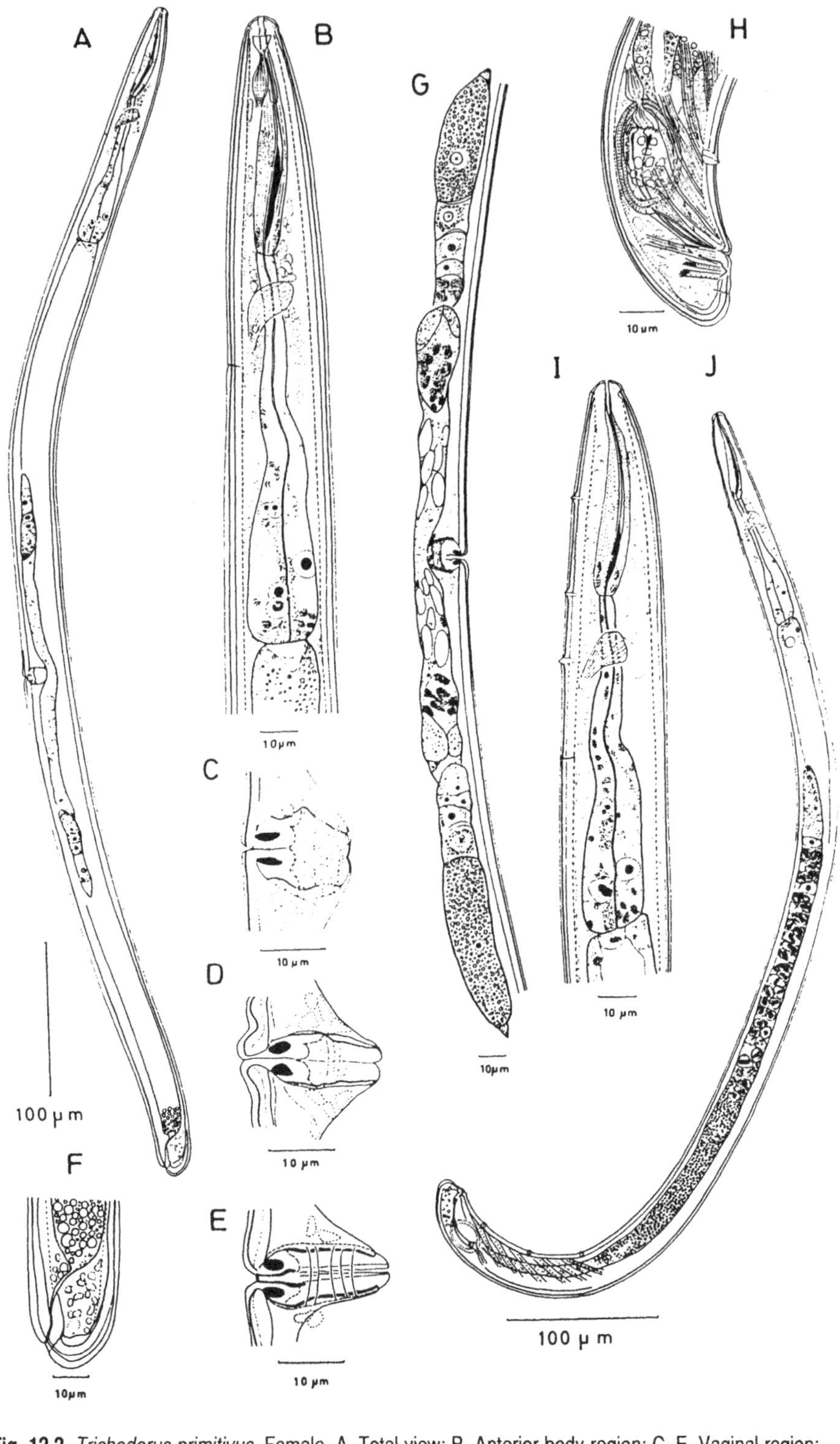

Fig. 12.2. *Trichodorus primitivus*. Female. A. Total view; B. Anterior body region; C–E. Vaginal region; F. Tail. G. Genital system. Male. H. Copulatory apparatus and tail; I. Anterior body region; J. Total view. (After Decraemer, 1991. Courtesy Marcel Dekker Inc.)

labial, ellipsoid; amphid pouches elongate, connected via a narrow tube to fibre-like bundles of sensillae. Onchiostyle narrow anteriorly, but gradually widens posteriorly where it sometimes has a tripartite appearance. It is surrounded by an indistinct guide ring about one third of the distance from the tip, the whole being enveloped in a dense sheath of muscles. Pharyngeal lumen ventral to onchiostyle, leading into the narrow anterior section of the oesophagus that gradually expands into a pyriform to spathulate **basal bulb** that **abuts the intestine**. The nerve ring surrounds the anterior part of the oesophagus with the excretory pore at the same level or just posterior. **Three ventromedian papillae are located anterior to the excretory pore**; one about opposite the middle of the onchiostyle; one towards the base of the onchiostyle and one just anterior to, or opposite the nerve ring. A pair of lateral cervical pores also present at the basal third of the onchiostyle. **Three ventromedian preanal copulatory supplements present at the tail end**; the first opposite the proximal end of the spicules, the second about one and three quarter body widths anterior to the first and the third, which is slightly less conspicuous, about two body widths anterior to the second. Paired postanal ventrosubmedian papillae and a pair of subventral caudal pores also present. Testis single. **Spicules paired, without transverse striations, about 4 μm wide at the proximal end and gradually narrowing to a shaft about 1 μm wide in the middle and continuing very narrow, and curved ventrally, to the distal end. Gubernaculum of characteristic shape and appears to lie between or even ventral to the spicules for much of its length.**

BIONOMICS: *T. primitivus* usually frequents porous sandy or sandy-loam soils, but is occasionally found in heavier soils. According to De Pelsmaeker *et al.* (1985), highest populations occurred at a depth of 20–40 cm under potato and *T. primitivus* was one of the most tolerant trichodorids to high acidity. Other workers have recorded *T. primitivus* as deep as 120 cm (Pietler, 1977). Reproduction is amphimictic and males are as numerous as females. The life cycle lasts about 45 days at 15–20°C (Pitcher and McNamara, 1970). De Pelsmaeker (1987b) reported, that under potato, maximal densities of *T. primitivus* occurred in spring, but Boag and Alphey (1988) stated that *T. primitivus* showed no obvious seasonal cycling, and had a K survival strategy.

In addition to damaging plant roots as a result of direct feeding, e.g. 'Docking disorder' of sugarbeet, *T. primitivus* vectors both tobacco rattle virus (TRV), causing spraing in potato and pea early browning virus (PEBV) to peas, lucerne, cucumber, tobacco, sugarbeet, etc. The species is polyphagous, hosts including such diverse plants as brassicas, clover, cucumber, lucerne, maize, potato, sugarbeet, tobacco, wheat and various trees such as *Juglans*, *Picea*, *Pinus*, *Prunus*, *Larix*, *Malus*, etc. (Christie *et al.*, 1950; Gibbs and Harrison, 1964; Hooper and Siddiqi, 1972; Boag, 1978; Noffsinger, 1984; Almeida *et al.*, 1989).

DISTRIBUTION: Recorded from northern and eastern Europe, where it is widespread and also from the USA and New Zealand.

TRICHODORUS SIMILIS SEINHORST, **1963** (Fig. 12.3)

MEASUREMENTS:
(After Seinhorst, 1963) From the Netherlands: Female (n = 10): L = 0.75–0.83 mm; a = 21–27; b = 5–6; c = ? (anus subterminal); V = 51–58; onchiostyle = 41–43 μm.
Male (n = 8): L = 0.76–0.87 mm; a = 24–30; b = 5.2–6.3; c = 63–79; onchiostyle = 38–42 μm; spicules = 36–40 μm.
(After Wyss, 1974) Female (n = 25): L = 0.75–0.90 (0.82) mm; a = 25–30 (27); b = 5.3–6.1 (5.6); c = 54–73 (62); V = 52–58 (56); onchiostyle = 38–43 (40) μm.
Male (n = 25): L = 0.74–0.91 (0.81) mm; a = 25–30 (27); b = 5.3–6.1 (5.6); c = 54–73 (62); onchiostyle = 39–42 (40) μm; spicules = 35–39 (37) μm.

DESCRIPTION: (modified from Wyss, 1974)
Males and females usually occur in about equal numbers.
Female: Body only slightly curved ventrally when heat relaxed. Basal bulb of oesophagus abuts onto intestine. Excretory pore usually about one onchiostyle-length behind the onchiostyle base. A pair of lateral body pores present close behind the vulva. In ventral view the vulva shows as a narrow transverse slit. **Refractive sclerotized thickenings at the vulva very conspicuous, approximately triangular in shape when seen laterally**. Vagina of characteristic shape in lateral view. Genital tracts amphididelphic, reflexed; spermatheca usually with sperm. Anus subterminal, in the form of a short, slightly curved transverse slit in ventral view; tail rounded with two terminal caudal pores.

Male: Posterior part of body curves ventrally when killed by heat. Onchiostyle characteristic. About 18 μm behind the oral aperture the lining of the pharyngeal lumen is attached dorsally and laterally to the onchiostyle, where it forms the so-called 'guide ring'. Excretory pore about one body width behind the nerve ring. **Three ventromedian cervical papillae located anterior to the excretory pore**; the first usually opposite the posterior third of the onchiostyle, the second about one third, and the third about three quarters of the onchiostyle length behind the base of the onchiostyle. The arrangement of these papillae in relation to the onchiostyle is variable; the first papilla may be opposite or even behind the onchiostyle base. A prominent lateral cervical pore is present on either side near the base of the onchiostyle. Basal bulb of oesophagus abuts onto intestine. Testis single, anteriorly outstretched. **Three ventromedian preanal copulatory supplements**; the first opposite or just above the proximal end of the spicules; the second about 1.5 body widths anterior to the first and the third, which is less conspicuous, slightly more than 1.5 body widths from the second. Paired postanal papillae and caudal pores also present. **Spicules without transverse striations, ventrally curved, proximal end distinctly cephalated, expanding in the mid-region nearly to the width of the proximal end, then narrowing towards the distal end. In dissected spicules, bristles or setae are clearly visible in the narrow 'neck' region. Spicule tips slightly bifid. Gubernaculum about 16 μm long, with a characteristic proximal hook and distal knob.**

Bionomics: *T. similis* is most abundant and widespread in sandy or sandy-loam soils. Highest densities occur at 20–30 cm (De Pelsmaeker *et al.*, 1985), rarely deeper, and the nematode shows a high tolerance to low pH soils. Reproduction is amphimictic with males being as numerous as females. De Pelsmaeker and Coomans (1986) reported that *T. similis* was most abundant in Belgium in acidic, sandy soil after a graminaceous crop. *T. similis* causes direct damage to root systems as a result of its feeding, but also transmits tobacco rattle virus (TRV) to potato and gladiolus. It is polyphagous, hosts including brassicas, carrot, clover, *Gladiolus*, lucerne, onion, peas, potato, sugarbeet, tobacco and trees such as *Juglans*, *Picea*, *Prunus* and *Picea*.

Distribution: *T. similis* appears to be restricted to the cooler regions of Europe (Belgium, France, Germany, Italy, Netherlands, Norway, Poland, Sweden, United Kingdom) with one record from the USA.

Trichodorus viruliferus Hooper, 1963 (Fig. 12.4)

Measurements:

(After Hooper, 1963) From England: Male (n = 20): L = 0.62–0.89 (0.77) mm; a = 23–33 (26); b = 4.6–7.3 (5.9); c = 47–71 (58); T = 57–73 (64); onchiostyle = 40–48 (45) μm; spicule length = 31–37 (34) μm; gubernaculum length = 13–18 (15) μm.

Female (n = 20): L = 0.64–0.86 (0.75) mm; a = 20–26 (24); b = 4.7–7.0 (5.4); c = ? (anus subterminal); V = $^{15-25}51-59^{15-20}$ ($^{19}56^{19}$); onchiostyle = 40–48 (46) μm.

Specimens from Ede, Holland: Male (n = 5): L = 0.66–0.72 (0.69) mm; a = 20.4–27.9 (22.8); b ≐ 5.3–5.8 (5.5); c = 47–55 (50); T = 58–70 (63); onchiostyle = 42–45 (44) μm; spicule length = 30–33 (31) μm; gubernaculum length = 13–15 (14) μm.

Female (n = 8): L = 0.63–0.70 (0.67) mm; a = 20.0–23.0 (21.9); b = 5.1–5.9 (5.4); c = ? (anus subterminal); V = $^{17\ 25}55-58^{17\ 24}$ ($^{20}56^{10}$); onchiostyle = 38–42 (41) μm.

Description: (modified from Hooper, 1976)

Male and females usually occur in about equal numbers.

Female: Body cylindrical, tapering towards anterior end, with slight ventral curvature when killed by heat. Excretory pore just over one onchiostyle length posterior to onchiostyle base and about opposite the middle of the basal oesophageal bulb. Anus subterminal; caudal pores almost terminal, tail terminus bluntly rounded. There is a single pore located laterally on each side of the body in the vulval region. Vulva median with amphididelphic, reflexed genital tracts; a **spermatheca present at junction of oviduct and uterus**. Vulval opening pore-shaped in ventral view. **Refractive thickening at junction of vulva and vagina appears as a ring in ventral view and as two small, rather indistinct oval shaped pieces in lateral view; these oval pieces are directed towards the vulva at an angle of about 45°** and their ends nearer the vulva appear slightly

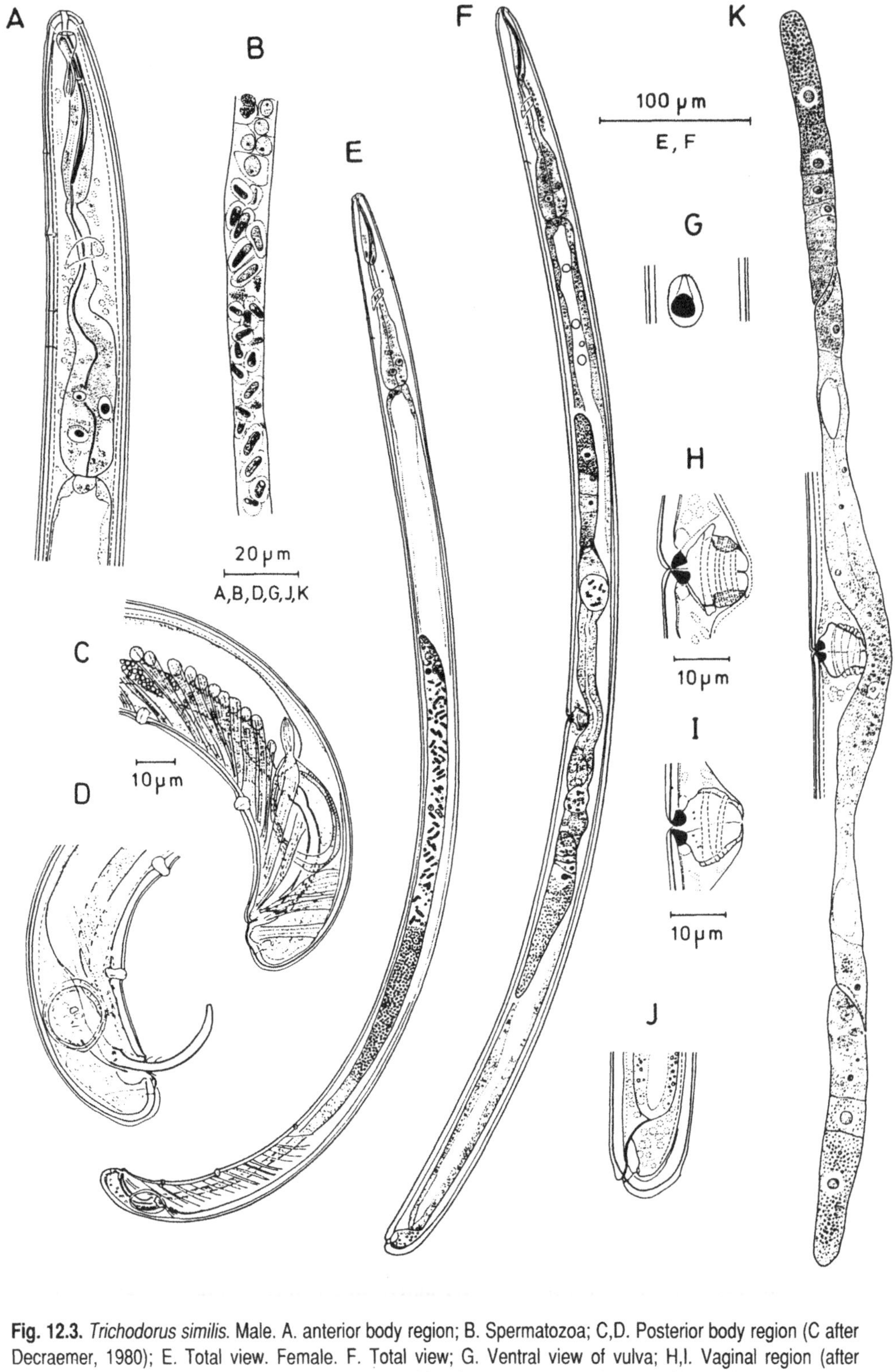

Fig. 12.3. *Trichodorus similis.* Male. A. anterior body region; B. Spermatozoa; C,D. Posterior body region (C after Decraemer, 1980); E. Total view. Female. F. Total view; G. Ventral view of vulva; H,I. Vaginal region (after Decraemer, 1980); J. Tail; K. Reproductive system. (After Decraemer, 1991. Courtesy Marcel Dekker Inc.)

more pointed and more distinct than the outer edges. **Vagina about as wide as long, of a characteristic rhomboid shape, the innermost half being surrounded by a prominent ring of muscles.**

Male: General appearance typical for the genus. Posterior part of body curved ventrally when killed by heat. Onchiostyle with a guide ring about one-third of its length from its tip. The anterior part of the oesophagus gradually expands into a spathulate basal bulb that **abuts the intestine, sometimes at an oblique angle and with a very slight ventrolateral overlap**. The nerve ring surrounds **the anterior part of the oesophagus and the excretory pore is usually opposite the middle of the basal oesophageal bulb. Three ventromedian cervical papillae are located anterior to the excretory pore**; they are usually almost equally spaced, one about a quarter of the onchiostyle length anterior to the onchiostyle base, one opposite the nerve ring and one two-thirds the onchiostyle length behind the onchiostyle base. A single lateral cervical pore is situated on each side just behind the onchiostyle base. **Three ventromedian preanal copulatory supplements are present at the tail end**; the first just anterior to the proximal end of the spicules; the second, 1–1.5 body widths anterior to the first; and the third, which is not quite so prominent, 2–3.5 body widths anterior to the first. Paired postanal ventrosubmedian papillae and a pair of subventral caudal pores also present. **Spicules paired, without transverse striations, curved ventrally; proximal part of spicule wide, gradually tapering to about half way along its length where there is a distinct and characteristic narrow bend**; behind this bend the shaft widens and then gradually tapers to the distal end. gubernaculum with its anterior two thirds spathulate lying between, and dorsal to, the spicule shafts; posteriorly there is a distinct dorsal keel.

Bionomics: *T. viruliferus* is most prevalent in sandy or sandy-loam soils at a depth of 10–20 cm (De Pelsmaeker *et al.*, 1985). It is one of the least tolerant trichodorids to low soil pH (De Pelsmaeker *et al.*, 1985). Reproduction is amphimictic with abundant males, the life cycle lasting about 45 days according to Pitcher and McNamara (1970). Maximal densities occurred in spring and autumn in Belgium (De Pelsmaeker, 1987b). The direct feeding activities result in stubby-root symptoms, but the species also transmits tobacco rattle virus (TRV), causing spraing in potato, and pea early browning virus (PEBV). *T. viruliferus* is polyphagous (see Hooper, 1976; Roca and Lamberti, 1984), attacking such crops as apple, barley, maize, pea, potato, rye, sugarbeet and wheat.

Distribution: Widespread in northern and eastern Europe and also recorded from the USA.

Control of trichodorids: Attempts at controlling trichodorids usually stem from their importance as virus vectors, rather than from the direct effects of their parasitism. As they are polyphagous, crop rotation is usually ineffective

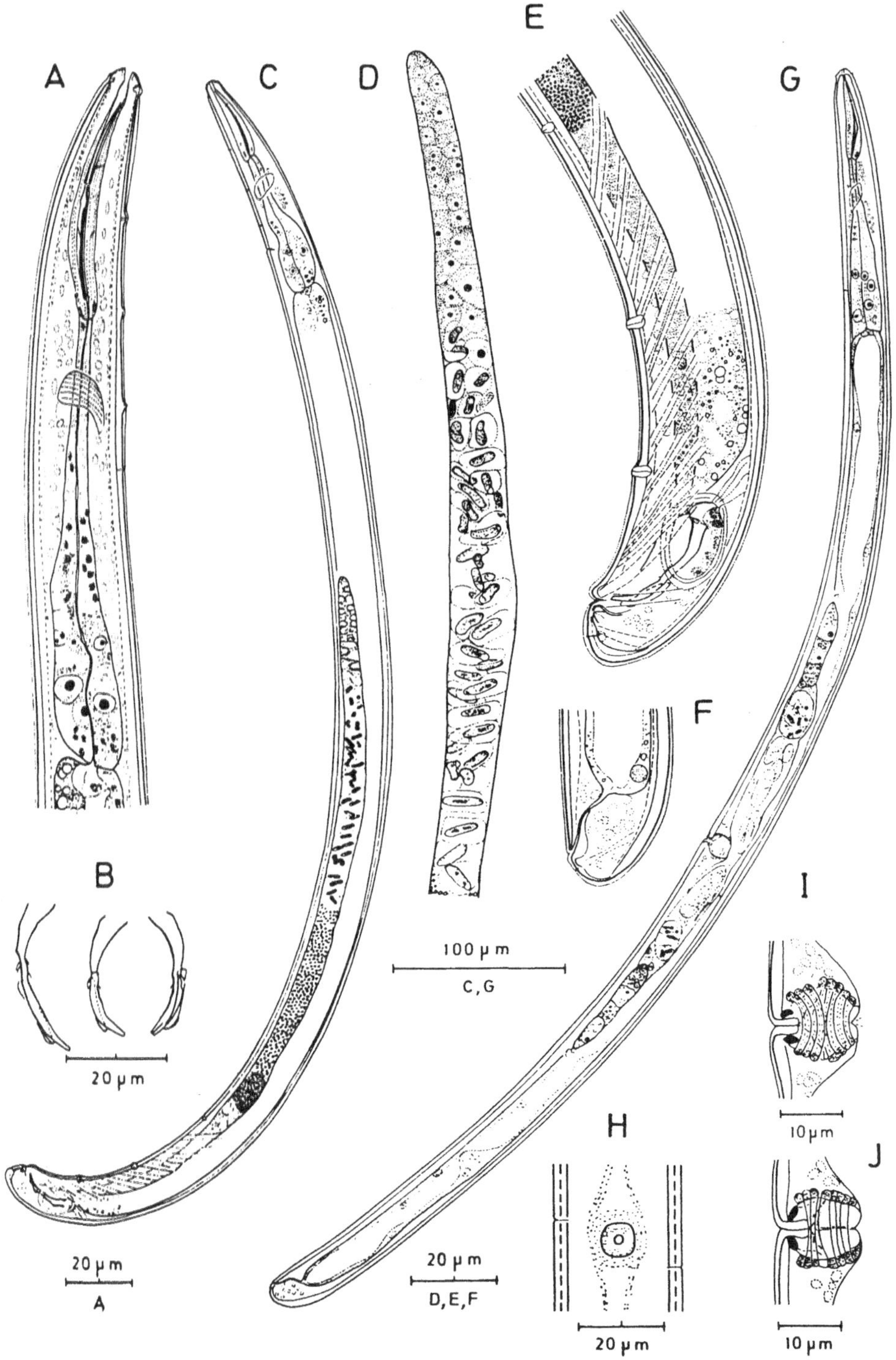

Fig. 12.4. *Trichodorus viruliferus.* Male. A. Anterior body region; B. Copulatory apparatus (after Decraemer, 1980); C. Total view; D. Testis; E. Posterior body region. Female. F. Tail; G. Total view; H. Ventral view of vulva (redrawn after Hooper, 1963); I,J. Vaginal region (after Decraemer, 1980). (After Decraemer, 1991. Courtesy Marcel Dekker Inc.)

or impractical, although certain crops have been shown to be beneficial in terms of reducing populations, e.g. groundnut and soyabean suppress *P. minor* (Johnson *et al.*, 1975) and *Crotalaria* is a poor host. Other non-chemical techniques such as tillage and fallowing have also been tried and found to be effective in reducing nematode numbers (e.g. Christie, 1959; Kuiper, 1977).

With chemical control, the majority of the earlier attempts used various volatile nematicidal compounds such as DBCP, D–D or EDB. Although effective, such products can be both difficult and dangerous to apply and some have now been banned in Europe and the USA. Other, non-volatile, nematicides such as aldicarb, carbofuran, fenamiphos and oxamyl, have now largely replaced the volatile compounds. Some of these nematicides appear to work not by reducing trichodorid populations, but by altering the behaviour and orientation abilities of the nematodes, thus affecting their ability to transmit viruses and thereby achieving the desired effect of reduced transmission. For more detail on chemical control the following papers provide useful information: Alphey *et al.*, 1975; Alphey, 1978; Rhoades, 1985; De Waele *et al.*, 1985; De Pelsmaeker, 1987a; Decraemer, 1991.

GENUS *PARATRICHODORUS* SIDDIQI, 1974 (Fig. 12.5)

syn. *P.* (*Paratrichodorus*) Siddiqi, 1974
P. (*Atlantadorus*) Siddiqi, 1974
P. (*Nanidorus*) Siddiqi, 1974
Atlantadorus Siddiqi, 1974 (Siddiqi, 1980)
Nanidorus Siddiqi, 1974 (Siddiqi, 1980)

DIAGNOSIS: Trichodoridae. Body plump, cigar-shaped, heat relaxed form more or less straight in both sexes. Cuticle swelling strongly after heat relaxation and/or acid fixation. **Lateral body pores** present in about half the species, but typically **not present within a body-width of the vulva** (reportedly present in four or five species). Onchiostyle dorsally convex. Oesophagus consisting of a narrow anterior section expanding posteriorly to form a **bulb** which **usually overlaps the intestine ventrally**. Posterior ventrosublateral nuclei located near to the oesophago-intestinal junction, the dorsal nucleus usually being near the oesophageal enlargement. An extension of the intestine along the dorsal side of the bulb may be present or absent. Vulva a minute median pore or small transverse or longitudinal slit. **Vagina extending into the body for up to one third the corresponding diameter, musculature weakly developed and inconspicuous; sclerotization poorly developed**. Genital tract amphididelphic, reflexed. Spermatheca present or absent. Anus subterminal, tail rounded. Males lacking or rare in about 40% of the nominal species. **Ventromedian cervical papillae absent** or, if present, consisting of a single papilla near, and anterior to, the excretory pore. **Lateral cervical pores absent** or, if present, usually represented by a single pair (one pore on each side) near the onchiostyle base or excretory pore. Testis single, outstretched. Sperm

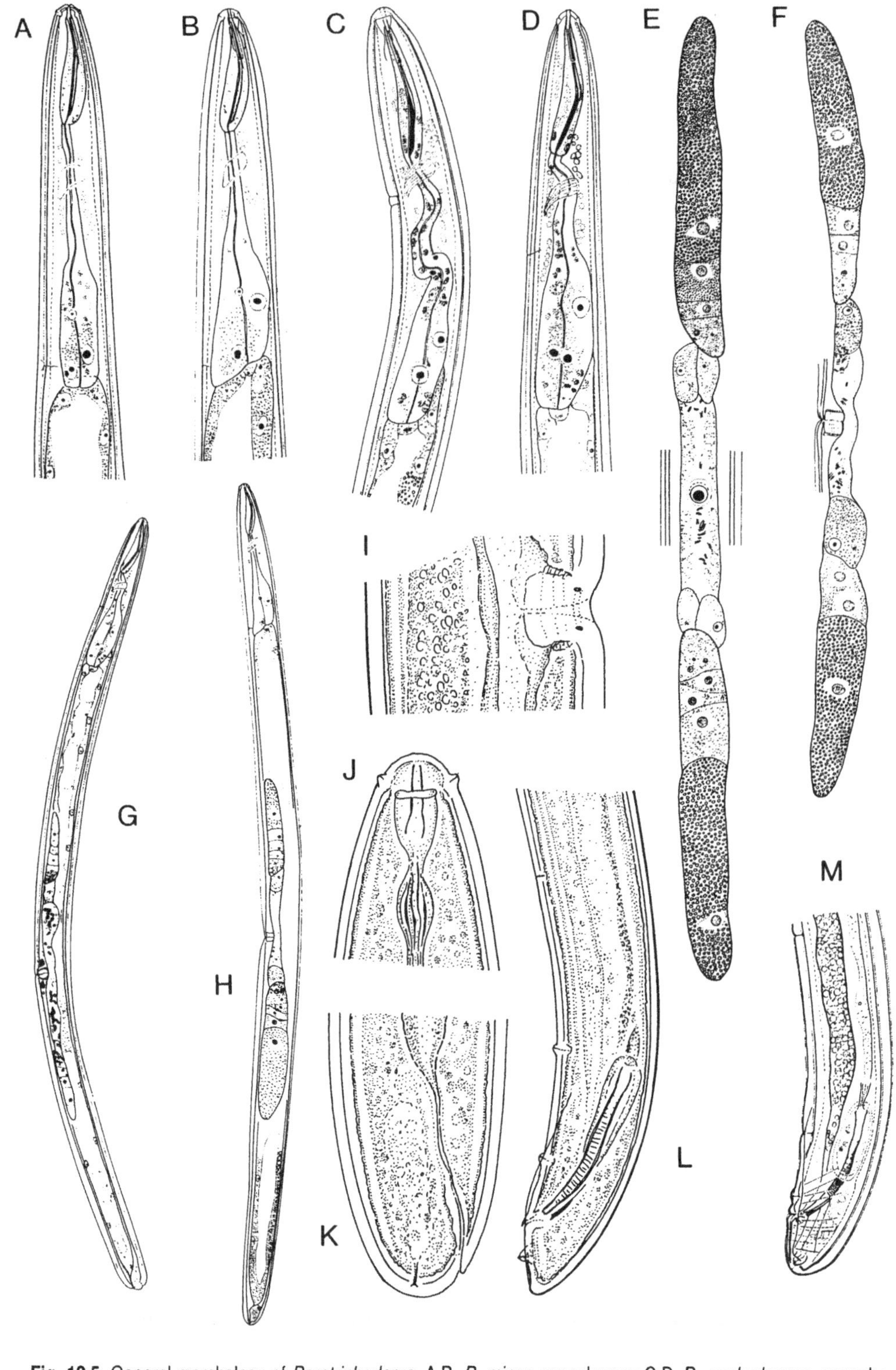

Fig. 12.5. General morphology of *Paratrichodorus*. A,B. *P. minor*, oesophagus; C,D. P. *pachydermus*, oesophagus; E,F. *P. minor*, genital tracts, ventral and lateral view; G. *P. pachydermus*, entire female; H. *P. minor*, entire

variously shaped (large subcylindroid; small and oval or rounded; filiform). **Spicules normally straight; transversely striated**, except at the extremities; gubernaculum present. **Spicule suspensor muscles forming an inconspicuous elongate-oval capsule. Bursa present**, but may be inconspicuous. Ventromedian copulatory supplements numbering one to three, exceptionally four; the first two supplements usually being within the range of the bursa. **Oblique copulatory muscles extending no further than the proximal tips of the retracted spicules**. Tail short, rounded, with one, sometimes two pairs of ventrosubmedian papillae. Caudal pores usually present, but absent in those species formerly included under *Nanidorus*.

Type species:

Paratrichodorus tunisiensis (Siddiqi, 1963) Siddiqi, 1974
  syn. *Trichodorus tunisiensis* Siddiqi, 1963
    Paratrichodorus (*Paratrichodorus*) *tunisiensis* (Siddiqi, 1963) Siddiqi, 1974

Other species:

P. acaudatus (Siddiqi, 1960) Siddiqi, 1974
  syn. *T. acaudatus* Siddiqi, 1960
    P. (*P.*) *acaudatus* (Siddiqi, 1960) Siddiqi, 1974
P. acutus (Bird, 1967) Siddiqi, 1974
  syn. *T. acutus* Bird, 1967
    Paratrichodorus (*Nanidorus*) *acutus* (Bird, 1967) Siddiqi, 1974
    Nanidorus acutus (Bird, 1967) Siddiqi, 1974
P. alleni (Andrássy, 1968) Siddiqi, 1974
  syn. *T. alleni* Andrássy, 1968
    P. (*P.*) *alleni* (Andrássy, 1968) Siddiqi, 1974
P. allius (Jensen, 1963) Siddiqi, 1974
  syn. *T. allius* Jensen, 1963
    P. (*P.*) *allius* (Jensen, 1963) Siddiqi, 1974
    P. tansaniensis Siddiqi, 1974
P. anemones (Loof, 1965) Siddiqi, 1974
  syn. *T. anemones* Loof, 1965
    Paratrichodorus (*Atlantadorus*) *anemones* (Loof, 1965) Siddiqi, 1974
    Atlantadorus anemones (Loof, 1965) Siddiqi, 1974
P. anthuriae Baujard and Germani, 1985
P. atlanticus (Allen, 1957) Siddiqi, 1974
  syn. *T. atlanticus* Allen, 1957
    P. (*A.*) *atlanticus* (Allen, 1957) Siddiqi, 1974
    A. atlanticus (Allen, 1957) Siddiqi, 1974

female; I. *P. paramirzai*, vaginal region; J. *P. paramirzai*, amphid; K. *P. paramirzai*, female tail; L. *P. paramirzai*, male posterior end; M. *P. pachydermus*, male posterior end. (A–H,M, after Decraemer, 1991; I–L, after Siddiqi, 1992. Not to scale. Courtesy Marcel Dekker Inc. and *Afro-Asian Journal of Nematology*.)

P. catharinae Vermeulen and Heyns, 1983
P. grandis Rodriguez-M. and Bell, 1978
P. hispanus Roca and Arias, 1986
P. lobatus (Colbran, 1965) Siddiqi, 1974
  syn. *T. lobatus* Colbran, 1965
    P. (*P.*) *lobatus* (Colbran, 1965) Siddiqi, 1974
    T. clarki Yeates, 1967
P. macrostylus Popovici, 1989
P. meyeri De Waele and Kilian, 1992
P. minor (Colbran, 1956) Siddiqi, 1974
  syn. *T. minor* Colbran, 1956
    P. (*N.*) *minor* (Colbran, 1956) Siddiqi, 1974
    N. minor (Colbran, 1956) Siddiqi, 1974
    T. christiei Allen, 1957
    P. christiei (Allen, 1957) Siddiqi, 1974
    P. (*N.*) *christiei* (Allen, 1957) Siddiqi, 1974
    N. christiei (Allen, 1957) Siddiqi, 1974
    P. obesus (Razjivin and Penton, 1975) Rodriguez-M. and Bell, 1978
    T. obesus Razjivin and Penton, 1975
    P. (*N.*) *obesus* (Razjivin and Penton, 1975) Rodriguez-M. and Bell, 1978
P. mirzai (Siddiqi, 1960) Siddiqi, 1974
  syn. *T. mirzai* Siddiqi, 1960
    P. (*P.*) *mirzai* (Siddiqi, 1960) Siddiqi, 1974
    T. musambi Edward and Misra, 1970
P. nanus (Allen, 1957) Siddiqi, 1974
  syn. *T. nanus* Allen, 1957
P. orrae Decraemer and Reay, 1991
P. pachydermus (Seinhorst, 1954) Siddiqi, 1974
  syn. *T. pachydermus* Seinhorst, 1954
    P. (*A.*) *pachydermus* (Seinhorst, 1954) Siddiqi, 1974
    A. pachydermus (Seinhorst, 1954) Siddiqi, 1974
P. paramirzai Siddiqi, 1992
P. paraporosus Khan, Jairajpuri and Ahmad, 1990
P. porosus (Allen, 1957) Siddiqi, 1974
  syn. *T. porosus* Allen, 1957
    P. (*A.*) *porosus* (Allen, 1957) Siddiqi, 1974
    A. porosus (Allen,1957) Siddiqi, 1974
    T. bucrius Lordello and Zamith, 1958
P. queenslandensis Decraemer and Reay, 1991
P. renifer Siddiqi, 1974
  syn. *P.* (*N.*) *renifer* Siddiqi, 1974
    N. renifer Siddiqi, 1974 (Siddiqi, 1980)
P. rhodesiensis (Siddiqi and Brown, 1965) Siddiqi, 1974
  syn. *T. rhodesiensis* Siddiqi and Brown, 1965
    P. (*P.*) *rhodesiensis* (Siddiqi and Brown, 1965) Siddiqi, 1974
P. sacchari Vermeulen and Heyns, 1983
P. teres (Hooper, 1962) Siddiqi, 1974

syn. *T. teres* Hooper, 1962
P. (P.) teres (Hooper, 1962) Siddiqi, 1974
T. flevensis Kuiper and Loof, 1962
P. weischeri Sturhan, 1985

Bionomics: Migratory polyphagous ectoparasites of plant roots. In addition to causing stubby root symptoms from their direct feeding activities, some species are known to be virus vectors, but information on the vast majority of species is lacking.

Distribution: Worldwide, but mainly in tropical and subtropical regions. The genus may well have been introduced relatively recently to Central and South America with non-indigenous plants (Rashid *et al.*, 1986b).

Etymology: Derived from the Greek *para* = beside or near, and *trichodorus*.

Species descriptions

Paratrichodorus minor (Colbran, 1956) Siddiqi, 1974 (Fig. 12.6)

syn. *Trichodorus* minor Colbran, 1956
Paratrichodorus (Nanidorus) minor (Colbran, 1956) Siddiqi, 1974
Nanidorus minor (Colbran, 1956) Siddiqi, 1974
Trichodorus christiei Allen, 1957
Paratrichodorus (Nanidorus) christiei (Allen, 1957) Siddiqi, 1974

Note: Allen (1957a), Hooper (1962) and Siddiqi (1962c) remarked on the similarities between *Paratrichodorus christiei* and *P. minor*. Siddiqi (1963) suggested that the two species could be separated by differences in spicule length of the very rarely occurring males and Siddiqi (1974a) suggested that the two species could be separated on onchiostyle length. However, Loof (1975), after studying various populations including paratypes of *P. christiei*, concluded that these differences were not as marked as first supposed and formally declared *P. christiei* a junior synonym of *P. minor*.

Measurements:
(After Colbran, 1956) From Queensland, Australia: Female: L = 0.55–0.70 mm; a = 19–25; b = 5.2–6.0; V = $^{20-25}52–55^{20-25}$; onchiostyle = 29–34 μm.
(After Siddiqi, 1962c) From India: Female (n = 15): L = 0.48–0.55 (0.5) mm; a = 17–23 (20); b = 4.3–5.8 (5.1); V = 52–57 (55); onchiostyle = 26–30 μm.
(After Mamiya, 1967) From Japan: Female (n = 30): L = 0.49–0.59 (0.53) mm; a = 18–28 (23); b = 5.0–6.2 (5.4); V = $^{14-25}53–64^{14-22}$ ($^{19}57^{18}$); onchiostyle = 30–33 (31) μm.
(After Siddiqi, 1963) From Nicaragua: Male: L = 0.6 mm; a = 25; b = 6; c = 40; T = 53; onchiostyle = 28 μm; spicules = 50 μm; gubernaculum = 17 μm.

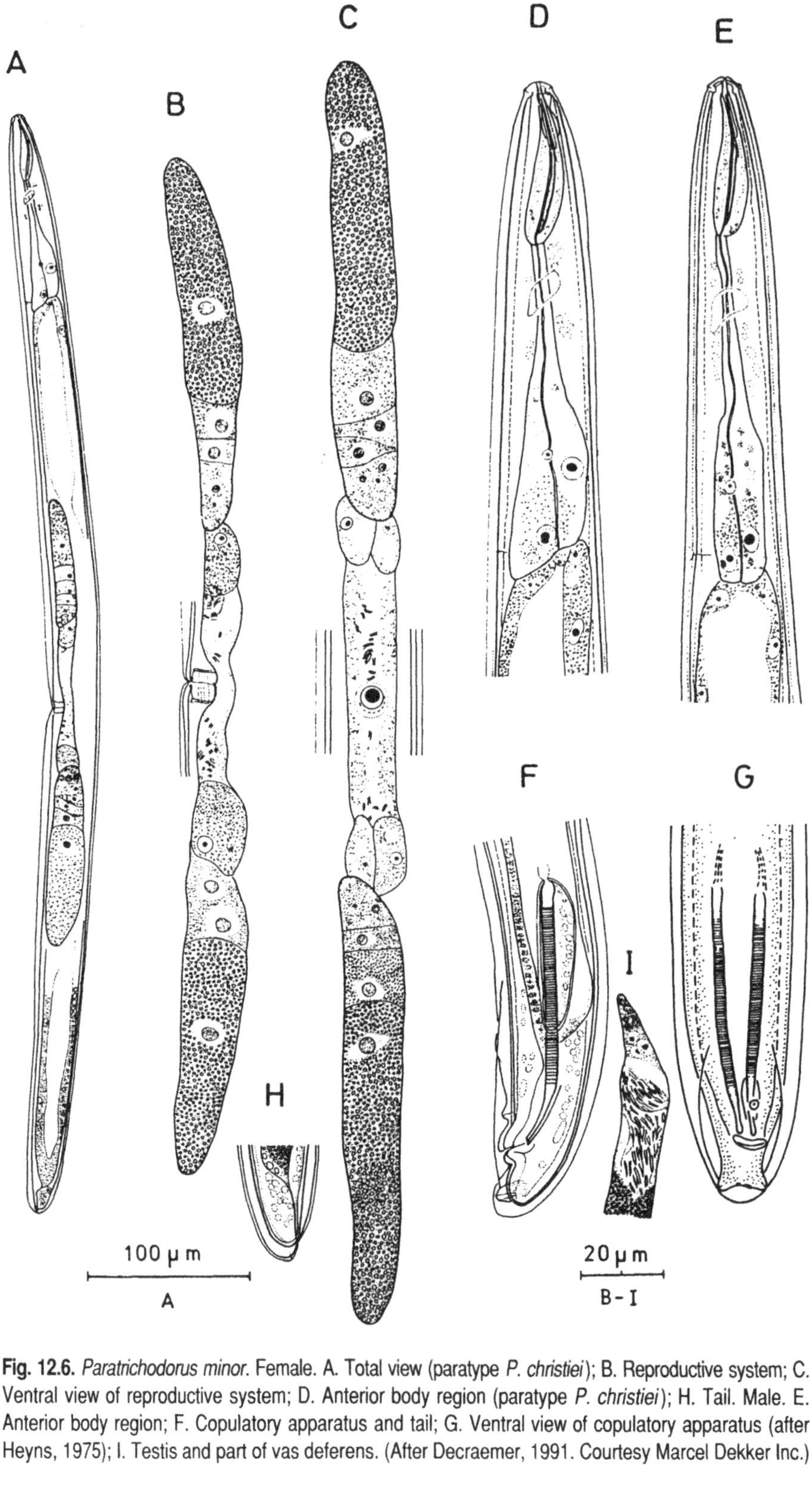

Fig. 12.6. *Paratrichodorus minor*. Female. A. Total view (paratype *P. christiei*); B. Reproductive system; C. Ventral view of reproductive system; D. Anterior body region (paratype *P. christiei*); H. Tail. Male. E. Anterior body region; F. Copulatory apparatus and tail; G. Ventral view of copulatory apparatus (after Heyns, 1975); I. Testis and part of vas deferens. (After Decraemer, 1991. Courtesy Marcel Dekker Inc.)

DIAGNOSIS: (Modified from Hooper, 1977)

Female: Body almost straight when heat relaxed, short, tapering slightly at both ends. Cephalic region rounded with slightly protruding papillae. Onchiostyle typical for the genus. The pharyngeal lumen, ventral to the onchiostyle, leads into the lumen of the narrow anterior part of the oesophagus which gradually expands into a spathulate to pyriform **basal bulb overlapping the intestine ventrally and subventrally**. Narrow part of oesophagus surrounded by a nerve ring. **Excretory pore usually opposite the base of the oesophagus. Lateral body pores and caudal pores absent**. Anus subterminal and the very short tail bluntly rounded. Genital tracts amphididelphic, reflexed; **no distinct spermatheca. Vulva a short transverse slit; vagina weakly developed, extending for only about one-third of the body width. Refractive thickenings at the vulva inconspicuous, almost rod-like and nearly parallel to the cuticle in lateral view**.

Male: Extremely rare. The oesophageal region of the male resembles the female described above. Testis single, outstretched. **Spicules large with slight ventral curvature towards the tip, marked with fine transverse striations. Gubernaculum with a small distal keel. A single, ventromedian papilla 9 μm anterior to the cloaca. A single, postanal, subventral papilla present on left side near tail terminus** [specimen apparently aberrant, as noted by Siddiqi (1963), in that the corresponding subventral papilla, usually found in trichodorid males, is missing from the right side]. **Bursa envelops the tail beginning at a level just anterior to the precloacal papilla**.

BIONOMICS: *P. minor* (= *P. christiei*) prefers sandy or sandy-loam soils with highest populations at a depth of 30 cm (Brodie, 1976). Reproduction is parthenogenetic or, according to Sturhan (1989), hermaphroditic. Males are exceedingly rare. The life cycle lasts 21–22 days at 22°C and 16–17 days at 30°C (Rohde and Jenkins, 1957). The direct feeding activities result in the typical stubby-root symptoms, the first record of which was by Christie and Perry (1951). *P. minor* has been recorded as transmitting tobacco rattle virus (TRV) in the USA and also pepper ring spot virus (PRV) in South America. It is extremely polyphagous, feeding on crops such as avocado, brassicas, lucerne, onion, potato, sugarbeet, sugarcane, sweet potato, tomato, etc.

DISTRIBUTION: Mainly recorded from the tropical and subtropical regions of the world where it is often almost ubiquitous (Africa, Asia, Central and South America, USA). According to Rashid *et al.* (1986b) *P. minor* is not indigenous to South and Central America, but has been introduced with crop plants, a suggestion borne out by its distribution and host associations. *P. minor* is occasionally recorded from Europe.

***PARATRICHODORUS PACHYDERMUS* (SEINHORST, 1954) SIDDIQI, 1974** (Fig. 12.7)

syn. *Trichodorus pachydermus* Seinhorst, 1954

Paratrichodorus (Atlantadorus) pachydermus (Seinhorst, 1954) Siddiqi, 1974
Atlantadorus pachydermus (Seinhorst, 1954) Siddiqi, 1974

MEASUREMENTS:
(After Seinhorst, 1954) Female: L = 0.73–1.02 mm; a = 23–32; b = 4.1–5.4; V = $^{18-25}51-63^{17-23}$.
Male: L = 0.61–0.99 mm; a = 24.5–26.3; b = 5.0–5.5; c = 35–45; T = 55–69; spicules = 40 μm; gubernaculum = 11 μm.
(After Allen, 1957a) Female (n = 14): L = 0.66–0.99 mm; a = 16–28; b = 4.5–7.4; c = ? (anus subterminal); V = $^{16-27}51-61^{13-29}$; onchiostyle = 44–54 μm.
Male (n = 7): L = 0.70–0.95 mm; a = 23–29; b = 5.5–6.4; c = 25–38; T = ? ; onchiostyle = 45–54 μm; spicules = 44–54 μm; gubernaculum = 12–16 μm.
(After Baujard, 1980) Female: L = 0.60–1.02 (0.83) mm; a = 13.9–23.3 (16.3); b = 4.4–7.7 (5.7); V = $^{21-36}55-61.5^{21-31}$ ($^{25}61^{24}$); onchiostyle = 48–58 (52.5) μm.
Male (n = 30): L = 0.67–0.97 (0.82) mm; a = 17.2–25.6 (21.8); b = 4.1–8.1 (5.8); c = 30–45.2 (35.8); T = 50–70 (62); onchiostyle = 48–60 (53.5) μm; spicules = 43–54 (50) μm; gubernaculum = 12–19 (14.5) μm.

DESCRIPTION: (Modified from De Waele *et al.*, 1985)
Female: Body tapering anteriorly: lip region only one third as wide as the oesophageal base. Cuticle smooth, in living specimens about 1 μm thick near the lip region, gradually thickening to about 2.5 μm at the posterior end of the body; cuticle strongly swollen when treated with fixative. Cephalic region rounded, with slightly protruding papillae. Amphidial aperture about one third of the head width; amphidial pouch rather long, connected via a narrow tube to bundles of fibre-like sensilla. Onchiostyle typical for the genus. Excretory pore at the level of the anterior part of the oesophageal bulb. Oesophageal bulb occupying about one third of the total oesophagus. Five oesophageal gland nuclei usually disinct. The posterior ventrosublateral gland nuclei lie in the posterior third of the oesophageal bulb while the smaller anterior pair is situated in the anterior third of the bulb. The position of the large dorsal gland nucleus varies. **In adults the intestine usually clearly overlaps the posterior part of the bulb dorsally and dorsolaterally**. However, Decraemer and De Waele (1981) noted that in juveniles the overlap is not always obvious and that an overlap of the intestine by the posterior ventrosublateral oesophageal glands is also possible. Nerve ring located in the anterior third of the anterior oesophagus. A pair of lateral cervical pores present at some distance anterior to the excretory pore. **Lateral body pores posterior to the vulva, two to five on each side**. One pair of subterminal caudal pores. Anus subterminal. Tail short, bluntly rounded. Female reproductive system amphididelphic, reflexed; **spermatheca absent**, uteri usually filled with sperm. The vulva, in ventral view, is pore-like with the walls showing radial folds surrounded by a sclerotized disc; **in lateral view the refractive thickenings at the vulva are small and rounded**.

Male: General appearance similar to female but posterior end usually slightly curved ventrally after fixation. A single rather indistinct ventromedian cervical papilla is present, situated immediately anterior to the excretory pore. Only one pair of lateral cervical body pores present and situated slightly anterior to the level of the excretory pore. Testis single and outstretched. Spicules showing a slight ventral curvature distally and transversely striated. Gubernaculum also slightly curved and slightly swollen distally. **Three ventromedian preanal copulatory supplements** present: the posterior one situated just anterior to the cloaca, the middle one near the beginning of the bursa and the anterior one further forward at about four times the distance between the posterior one and the middle one. **Distance between anus and middle preanal supplement equal to 75–94% of spicule length** according to Loof (1975) **and 53–83% of spicule length** according to Baujard (1980). Two pairs of large ventrosubmedian postcloacal papillae present, the first pair near the cloacal aperture, the second pair subterminal and near the caudal pores. Bursa present, beginning slightly posterior to the proximal end of the spicules and surrounding the tail tip.

BIONOMICS: *P. pachydermus* shows a preference for sandy or sandy-loam soils, but is not particularly tolerant of high acidity when compared to other trichodorids (De Pelsmaeker *et al.*, 1985). Its depth preference ranges from 15–70 cm (De Pelsmaeker *et al.*, 1985; De Waele *et al.*, 1985). De Pelsmaeker and Coomans (1986) found that a graminaceous crop markedly increased populations and De Waele *et al.* (1985) commented further on the influence of biotic factors on *P. pachydermus* and other trichodorids. De Pelsmaeker (1987b) reported peak populations in spring and autumn, although Boag and Alphey (1988) indicated no obvious seasonal cycling. Reproduction is amphimictic with abundant males. *P. pachydermus* causes direct root damage from its feeding activities, but also vectors tobacco rattle virus (TRV) and pea early browning virus (PEBV). It is polyphagous, feeding on clover, onion, potato, sugarbeet, tobacco, etc. (see De Waele *et al.*, 1985)

DISTRIBUTION: Widespread in temperate Europe and also found in Canada and the USA (Noffsinger, 1984).

PARATRICHODORUS POROSUS (ALLEN, 1957) SIDDIQI, 1974 (Fig. 12.8)

syn. *Trichodorus porosus* Allen, 1957
Trichodorus bucrius Lordello and Zamith, 1958

MEASUREMENTS:
(From Decraemer, 1991) Females: L = 0.42–0.80 mm; a = 13.6–25.4; b = 3.1–6.2; c = ? (anus subterminal); V = 51–59%; onchiostyle = 39–58 μm.
Males: L = 0.53–0.77 mm; a = 15–25; b = 4–6; c = 60–90; onchiostyle = 43–48 μm; spicules = 36–38 μm.

DESCRIPTION: (Modified from Decraemer, 1991)
Female: Body stout, cigar-shaped, dying almost straight on heat relaxation. Cuticle swelling stongly on fixation. Onchiostyle and oesophagus typical for

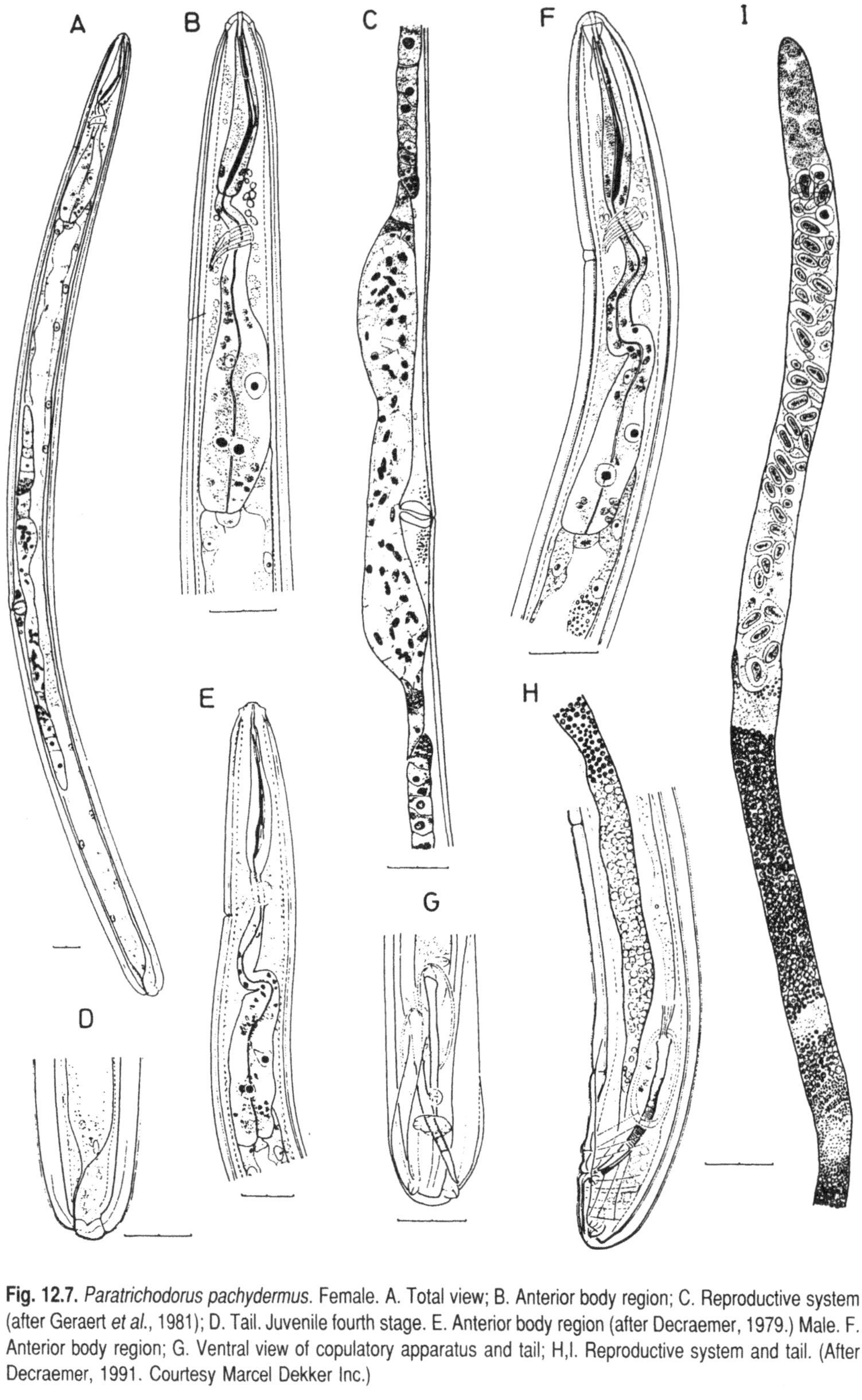

Fig. 12.7. *Paratrichodorus pachydermus*. Female. A. Total view; B. Anterior body region; C. Reproductive system (after Geraert *et al.*, 1981); D. Tail. Juvenile fourth stage. E. Anterior body region (after Decraemer, 1979.) Male. F. Anterior body region; G. Ventral view of copulatory apparatus and tail; H,I. Reproductive system and tail. (After Decraemer, 1991. Courtesy Marcel Dekker Inc.)

the genus, but with the **intestine often overlapping the oesophageal bulb for some distance dorsally**. Nerve ring located at the anterior extremity of the oesophagus, the excretory pore opening between the nerve ring and the oesophageal bulb. **Lateral body pores absent within one body width of the vulva. Vulva** median in position, **pore-like**. Vagina barrel-shaped, weakly developed and extending into the body for about one-third of the corresponding diameter. **Vaginal sclerotization weak and inconspicuous. Four ventromedian advulvar body pores present**, two anterior to, and two posterior to, the vulva. Some specimens, particularly from Japan, have six such pores, and up to ten have been recorded on a Californian specimen.

Male: **Rare**, known only from the type locality. Body cigar-shaped, dying almost straight on heat relaxation. **A single ventromedian cervical papilla is located slightly anterior to the excretory pore** with a pair of lateral cervical pores at about the same level. **Two ventromedian copulatory supplements present**, both within the range of the retracted spicules. One pair of postanal ventrosublateral papillae present. **Spicules straight, transversely striated. Gubernaculum** straight, orientated parallel to the spicules and **with the distal end thickened**. Bursa inconspicuous.

Bionomics: *P. porosus* is most widespread and abundant in sandy or sandy-loam soils, achieving its highest density at about 45 cm (Gomes Carneiro *et al.*, 1982). Reproduction is almost certainly parthenogenetic as males are rare. The direct feeding of *P. porosus* can cause serious damage, e.g. to the root systems of *Camellia*, maize, sorghum, etc. The species is extremely polyphagous, the host list ranging from brassicas, cotton, maize, peas, sugarcane and tomato to numerous broad-leaved and coniferous trees (see Mamiya, 1967; Ayala and Allen, 1968; Ayala *et al.*, 1970; Shishida, 1979 and Noffsinger, 1984).

Distribution: Widely distributed in tropical and subtropical areas (Africa, Asia, Australasia, South and Central America, USA). Rashid *et al.* (1986b) suggest that the species has been introduced into South America with crop plants and is therefore not indigenous.

Control: Principles are the same as mentioned under *Trichodorus* (see p.269).

GENUS *MONOTRICHODORUS* ANDRÁSSY, 1976 (Fig. 12.9)

Diagnosis: Trichodoridae. Body plump, cigar-shaped. **Heat relaxed females die straight, the males with the tail region curved ventrad. Cuticle not usually swollen abnormally when fixed. One pair of lateral body pores present near the vulva** (usually slightly anterior or at the same level). Onchiostyle dorsally convex with a simple, anterior, guide ring. Oesophagus consisting of a narrow anterior section expanding posteriorly to form a **non-overlapping bulb**. The intestine does not overlap the basal bulb dorsally. **Vulva posterior** at 75–86%, usually a transverse slit, but longitudinal in *M. sacchari* and *M.*

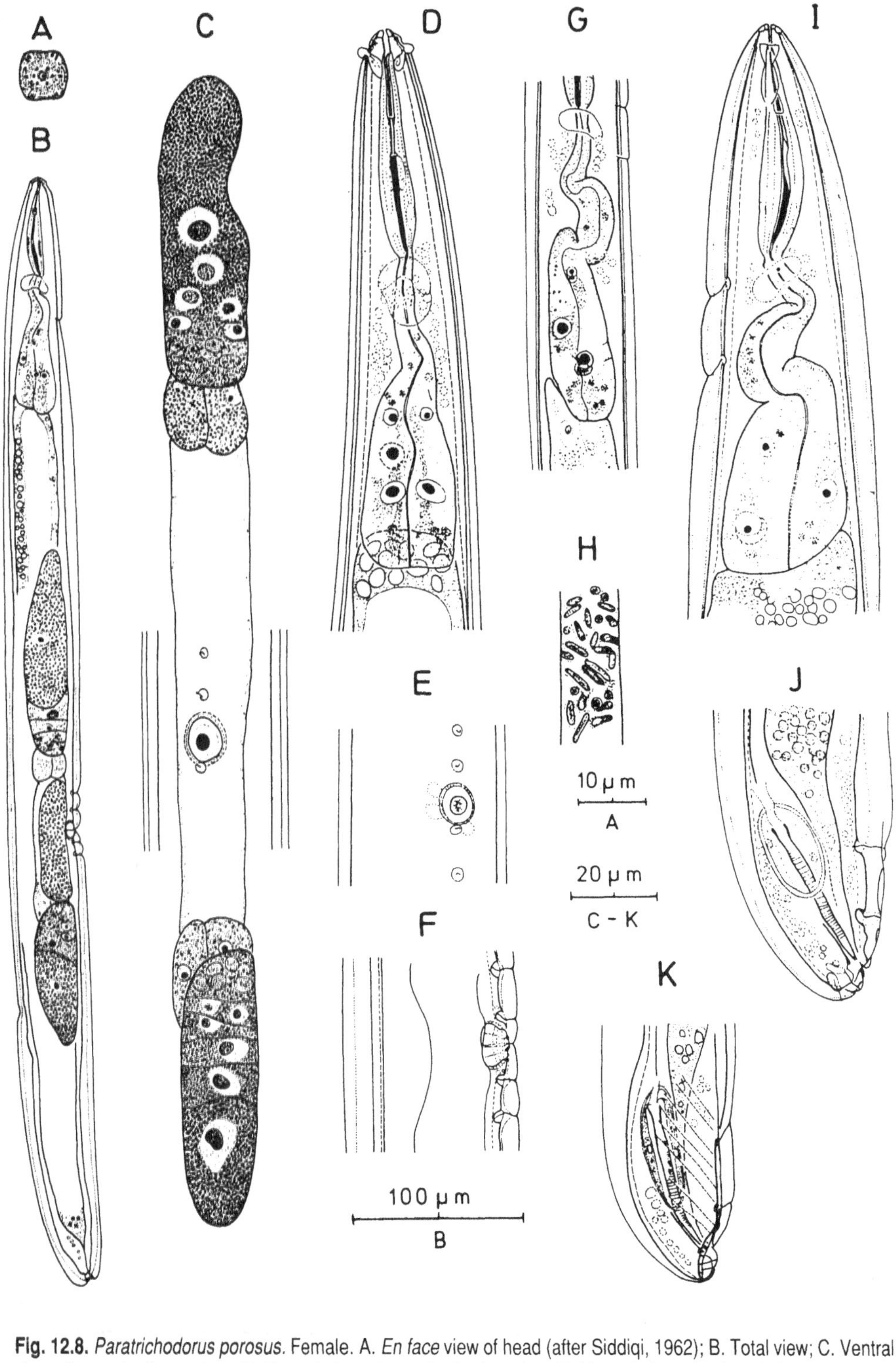

Fig. 12.8. *Paratrichodorus porosus*. Female. A. *En face* view of head (after Siddiqi, 1962); B. Total view; C. Ventral view of reproductive system; D. Ventral view of anterior body region; E. Ventral view of vulva; F. Vaginal region. Male. G. Oesophageal region (paratype); H. Spermatozoa (paratype); I. Anterior body region (paratype); J,K. Copulatory apparatus and tail (paratypes). (After Decraemer, 1991. Courtesy Marcel Dekker Inc.)

muliebris. Vagina anteriorly directed, extending into the body for half or more of the corresponding diameter; well developed musculature with prominent distal sclerotization. **Genital tract monoprodelphic**, reflexed. Spermatheca present. **Postuterine sac present**, very short to long. Anus subterminal; tail rounded. **A single caudal pore present** according to Siddiqi (1992), but Decraemer (1987) claimed that two pores were present, the orifi being very close together so as to give the impression of one. Males with one ventromedian cervical papilla just anterior to the excretory pore. One pair of lateral cervical pores near the nerve ring. Genital tract monorchic, outstretched. Spicules slender, ventrally curved, long; with bristles or striations; lacking a capitular extension. **Suspensor muscles forming a well developed oval capsule. Bursa usually absent**, but present, if very low in *M. sacchari*. Three ventromedian copulatory supplements present. Oblique copulatory muscles extending for less than a body width beyond the proximal tips of the retracted spicules. Tail short, rounded, with one pair of ventrosublateral papillae.

Type species:
Monotrichodorus monohystera (Allen, 1957) Andrássy, 1976
syn. *Trichodorus monohystera* Allen, 1957

Other species:
M. acuparvus Siddiqi, 1992
M. muliebris Andrássy, 1989
M. parvus Siddiqi, 1992
M. proporifer Siddiqi, 1992
M. sacchari Baujard and Germani, 1985
M. vangundyi Rodriguez-M., Sher and Siddiqi, 1978

Bionomics: Recorded from around the roots of various plants (grasses, banana, citrus, sugarcane, oilpalm, etc.) in the Neotropical Region. Several species have been reported from primary forest, thus supporting the contention that *Monotrichodorus* is endemic to South and Central America. Males have been described for six of the seven species (the exception being *M. muliebris*, the females of which, however, contained numerous sperm) and reproduction is presumably amphimictic. No details on their biology have been published.

Distribution: *Monotrichodorus* species are apparently restricted to the Neotropical Region. *M. monohystera* has been recorded from Venezuela by Allen (1957a) and Loof (1964), from Costa Rica, El Salvador and Peru by Rodriguez-M. *et al.* (1978) and from Colombia (Stanton *et al.*, 1989). Rodriguez-M. *et al.* (1978) reported *M. vangundyi* from Ecuador and Panama whilst Baujard and Germani (1985) described *M. sacchari* from French Guyana. *M. muliebris* was described from Ecuador by Andrássy (1989) and three new species were added to the genus by Siddiqi (1992): *M. acuparvus* and *M. parvus* from Colombia and *M. proporifer* from Para State, Brazil. This record from Brazil may represent a significant southeastward extension of the range towards the *Allotrichodorus* centre of distribution in Bahia State and could undermine the geographical support for the notion that monodelphy evolved twice in the region in the form of *Monotrichodorus* and *Allotrichodorus*. However, the geographic distributions are still allopatric, *Monotrichodorus* being widely recorded to the north and north west of the Amazon Basin and *Allotrichodorus* found only to the south east of the Basin and restricted to Bahia State.

Etymology: Derived from the Greek *monas* = single (referring to the female genital tract) and *trichodorus*.

GENUS *ALLOTRICHODORUS* RODRIGUEZ-M., SHER & SIDDIQI, 1978 (Fig. 12.10)

Diagnosis: Trichodoridae. Body plump, cigar-shaped. Heat relaxed males may have the tail straight or curved ventrad. Cuticle swells on fixation in some, but not all species. **Lateral body pores absent near the vulva** and elsewhere. Onchiostyle curved dorsally with a simple, anterior, guiding ring. Oesophagus consisting of a narrow anterior portion expanding posteriorly to form a non-overlapping, or slightly ventrally overlapping, bulb. Dorsal extension of intestine over bulb present or absent. **Vulva posterior, typically at 81–89%** (but at 60–65% in the dubiously placed *A. westindicus*) and in the form of a small transverse slit. **Vagina extending into the body for about half the corresponding diameter** (but shorter in *A. westindicus*) and with well developed musculature and prominent sclerotization. In some of the nominal species the vagina is directed anteriorly, but in others this is not the case. **Genital tract monoprodelphic**, reflexed. **Spermatheca present**. Postuterine sac present, small to well developed. Anus subterminal; tail rounded. **One pair of caudal pores**. Male usually lacking ventromedian cervical papillae, but one may be present near the excretory pore. One pair of lateral cervical pores situated at about the level of the nerve ring. Testis single, outstretched. **Spicules, long, slender, ventrally curved**. Bristles or striations may be present or absent. Capitular extension present in two species, including the type, but absent in the remainder. **Suspensor muscles forming a weakly developed elongate-oval capsule. Bursa present**; well developed or minute. Ventromedian copulatory supplements numbering three or four. Oblique copulatory muscles extending for less than a body width beyond the proximal tips of the retracted spicules. Tail short, rounded with one pair of more or less ventrosubmedian situated papillae.

Type species: *Allotrichodorus campanullatus* Rodriguez-M., Sher and Siddiqi, 1978

Other species:

A. brasiliensis Rashid, De Waele and Coomans, 1986
A. guttatus Rodriguez-M., Sher and Siddiqi, 1978
A. longispiculis Rashid, De Waele and Coomans, 1986

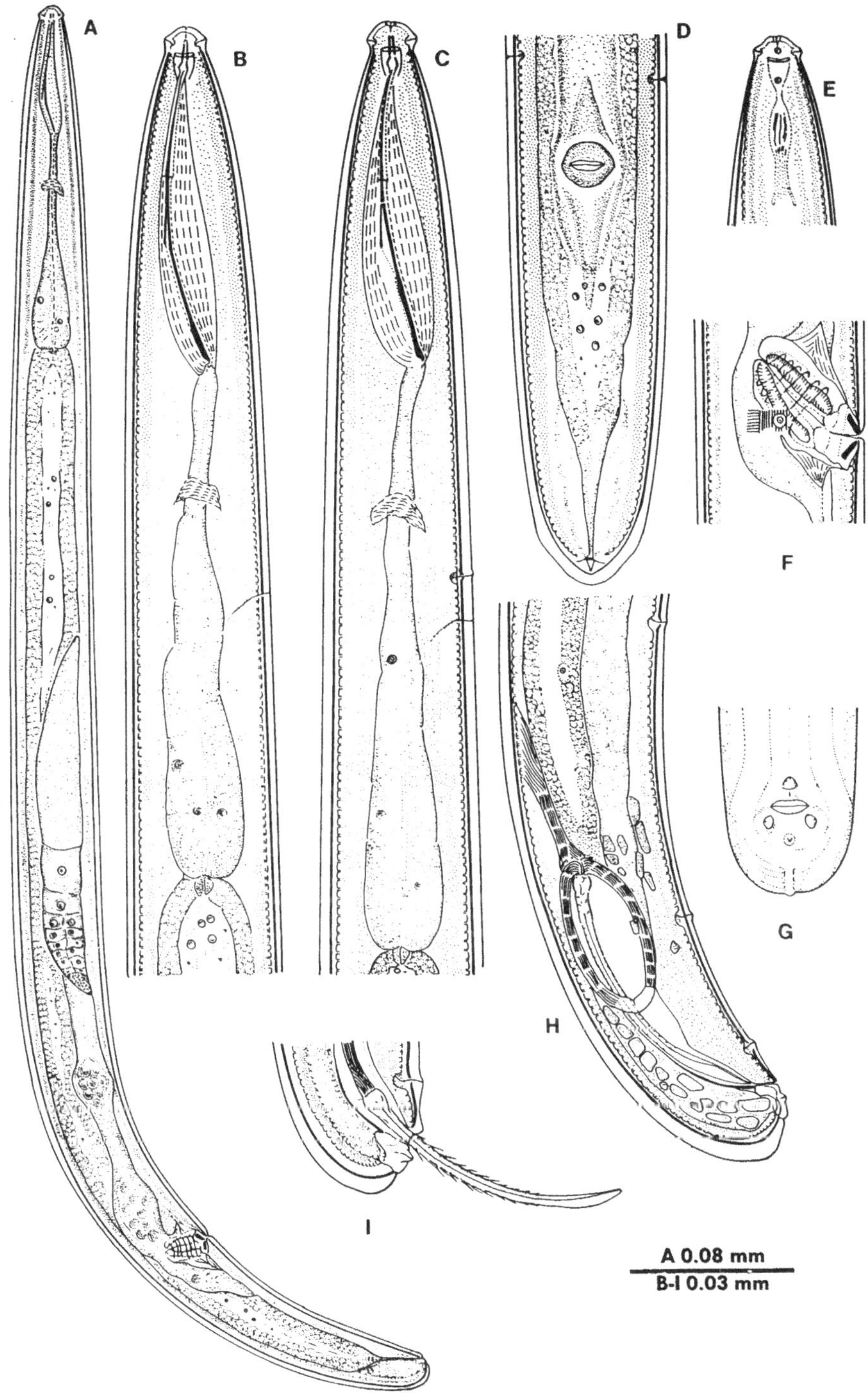

Fig. 12.9. *Monotrichodorus vangundyi*. A. Female; B,C; Oesophageal region of female and male; D. Female posterior end, ventral; E. Amphid; F. Vagina; G. Male tail, ventral; H. Male posterior end; I. Spicule. (After Rodriguez-M. *et al.*, 1978. Courtesy *Journal of Nematology*.)

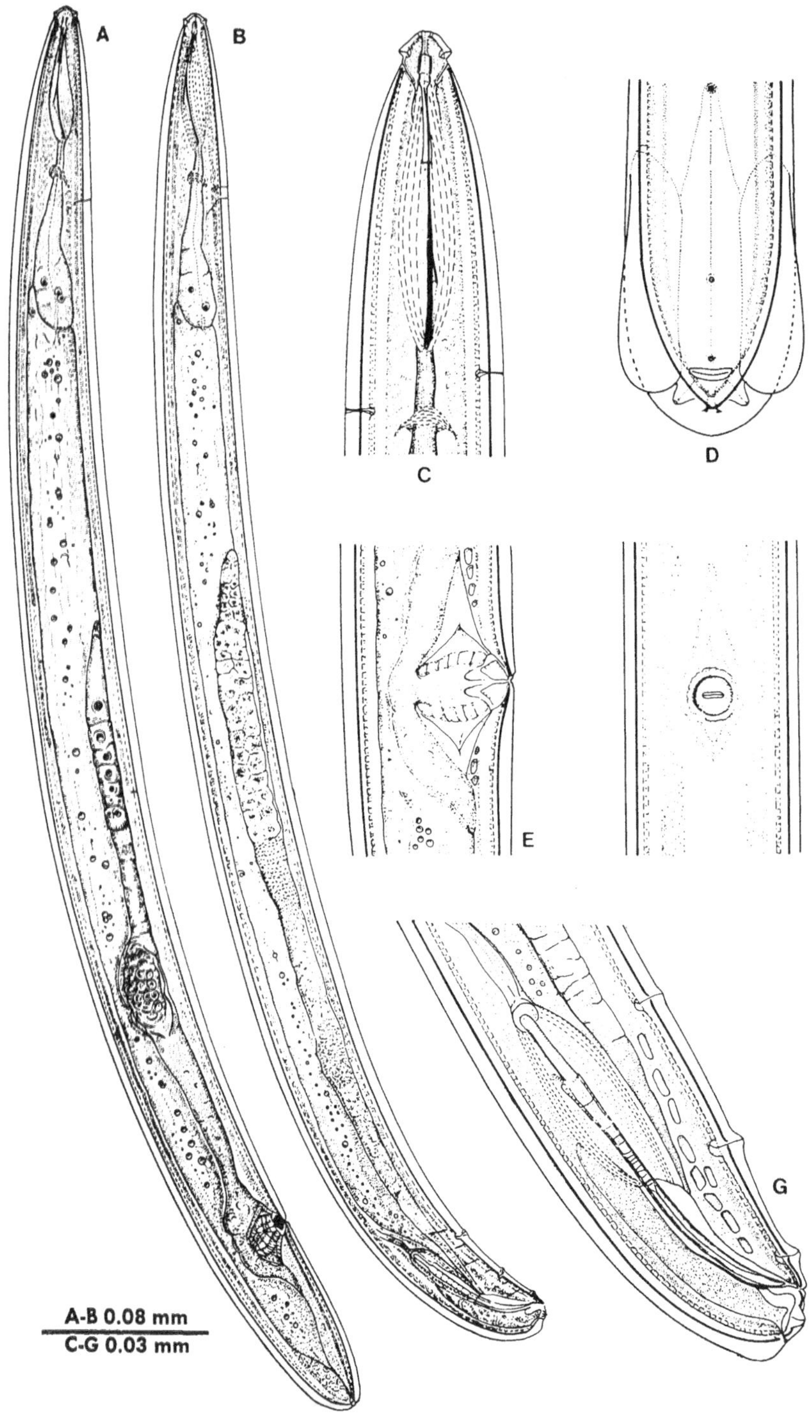

Fig. 12.10. *Allotrichodorus campanullatus.* A. Female; B. Male; C. Male anterior end, ventral; D. Male posterior end, ventral; E. Vulval region; F. Vulva, ventral; G. Male posterior end. (After Rodriguez-M. et al., 1978. Courtesy *Journal of Nematology.*)

A. loofi Rashid, De Waele and Coomans, 1986
A. sharmae Rashid, De Waele and Coomans, 1986

Species incertae sedis:
A. westindicus (Rodriguez-M., Sher and Siddiqi, 1978) Rashid, De Waele and Coomans, 1986
syn. *Paratrichodorus* (*Nanidorus*) *westindicus* Rodriguez-M., Sher and Siddiqi, 1978
Nanidorus westindicus Rodriguez-M., Sher and Siddiqi, 1978

Note: *Paratrichodorus westindicus* was placed in this genus by Rashid, De Waele and Coomans (1986), largely on the basis of its single functional genital tract. Decraemer (1991) also accepts this view. However, *westindicus* shares very little else with the other nominal members of *Allotrichodorus*: the excretory pore is located posterior to the oesophageal base as opposed to being near the nerve ring; the vulva is considerably more anterior (60–65% compared to 81–89%); the vaginal musculature is small and weakly developed; the female tail is conoid-subdigitate in form, relatively long and with the rectum at an oblique angle to the body axis (*Allotrichodorus* has the tail virtually nonexistent and the rectum parallel to the body axis as it leads to the subterminal anus); there are no lateral or caudal pores (present in *Allotrichodorus*); short body length (<0.5 mm as opposed to >0.5 mm); males are lacking in all three described populations as opposed to always present in the six other species of *Allotrichodorus* and the distribution is largely allopatric (one population described from Brazil, the other two from Martinique and Trinidad). I consider it to be closer to *Paratrichodorus* in general respects, but for the time being tentatively place it near *Allotrichodorus* on the basis of monodelphy, but regard it as *species incertae sedis*.

Bionomics: These species have been found around the roots of various plants, including cocoa, coffee, sugarcane and several tree species, in South America. A. *campanullatus* is the most common species and is somewhat variable in morphology (see Rashid *et al.*, 1986b). It occurred in heavy, loamy soil in Brazil in the rhizosphere of clove trees (*Eugenia caryophyllata*), rubber (*Hevea brasiliensis*), a wild palm and banana while *A. brasiliensis* and *A. sharmai* were found in light, sandy soils about the roots of cocoa (Rashid *et al.* 1986). No details have been published on their biology, but as males are known for all species, except for the dubiously placed *A. westindicus*, amphimixis would appear to be the usual mode of reproduction.

Distribution: Restricted to the Neotropical Region. The majority of species have been described solely from Bahia State, Brazil (Rodriguez-M. *et al.*, 1978; Rashid *et al.*, 1986b). *A. westindicus* was originally described from Trinidad (Rodriguez-M., *et al.*, 1978), but has subsequently been recorded from

Martinique (Baujard and Germani, 1985) and Brazil (Rashid *et al.* 1986b). As discussed above, it has little in common with the other Brazilian species apart from a single functional genital tract.

Etymology: Derived from the Greek *alloi* = of another sort, and *trichodorus*.

APPENDIX

(taxa not included in the text)

TAXA:

Aphelenchoides sinensis (Wu and Hoeppli, 1929) Andrássy, 1960
syn. *Aphelenchus parietinus* var. *sinensis* Wu and Hoeppli, 1929
Longidorus lignosus Chizhov, Subbotin, Romanenko and Kruchina, 1991
Longidorus silvae Roca, 1993
Seinura paraoxyura Mavlyanov, 1976
Trichodorus kilianae Decraemer and Marais, 1993
Trichodorus magnus Decraemer and Marais, 1993
Xiphinema ilyasi Ahmad and Baqri, 1987 (= *X. ensiculiferum* according to Loof and Luc, 1993)
Xiphinema riparium Chizhov, Subbotin, Romanenko and Kruchina, 1991 (=*riparia* emend. Loof and Luc, 1993)

REFERENCES:

Ahmad, N. and Baqri, Q.H. (1987) The species of the genus *Xiphinema* Cobb, 1913 (Xiphinematidae: Dorylaimida) from Malaysia. *Bulletin of the Zoological Survey of India* 8, 143–148.

Andrássy, I. (1960) Beiträge zur Kenntnis der freilebenden Nematoden Chinas. *Annales historico-naturales Musei nationalis hungarici, n.s.* 52, 201–216.

Chizhov, V.N., Subbotin, S.A., Romanenko, N.D. and Kruchina, S.N. (1991) [New nematode species of the genera *Xiphinema* and *Longidorus* (Nematoda, Longidoridae) from West Caucasus]. *Zoologicheskii Zhurnal* 70, 130–133 (in Russian).

Decraemer, W. and Marais, M. (1993) Two new species of *Trichodorus* from South Africa, with a note on *T. petrusalberti* (Nemata: Trichodoridae). *Fundamental and Applied Nematology*, 16, 273–282.

Loof, P.A.A. and Luc, M. (1993) A revised polytomous key for the identification of species of the genus *Xiphinema* Cobb, 1913 (Nematoda: Longidoridae) with exclusion of the *X. americanum*-group: Supplement 1. *Systematic Parasitology* 24, 185–189.

Mavlyanov, O.M. (1976) [Parasitic and soil nematodes of technical crops of south Uzbekistan]. *Izdatelstvo 'FAN' Tashkent, Uzbek SSR*, pp. 1–82 (in Russian).

Roca, F. (1993) *Longidorus silvae* n. sp. (Nematoda: Longidoridae) from Italy. *Fundamental and Applied Nematology*, 16, 211–214.

Wu, H.W. and Hoeppli, R.J.C. (1929) [Free-living nematodes from Fookien and Chekiang]. Teil III. *Archiv für Schiffs- und Tropenhygiene* 33, 35–43 (in German).

Bibliography*

Abad, P., Tares, S., Brugieur, N. and De Guiran, G. (1991) Characterization of the relationships in the pinewood nematode species complex (*Bursaphelenchus* spp.) using a heterologous unc-22 DNA probe from *Caenorhabditis elegans*. *Parasitology* 102, 303–308.

Aboul-Eid, H.Z. (1970) Systematic notes on *Longidorus* and *Paralongidorus*. *Nematologica* 16, 159–179.

Adilova, N.B. (1973) [The nematode fauna on maize and its rhizosphere; its dynamics in the Tashkent area. In: *The Question of Phytonematodes in the Uzbekistan. Book 3.*] Isdatel'stvo 'FAN', Tashkent, pp. 15–68 (in Russian).

Akhtar, S.A. (1962) *Metaphelenchus saccharae* n.sp. (Nematoda: Aphelenchoidea Fuchs, 1937) associated with the roots of sugarcane. *Nematologica* 7, 53–56.

Ali, S.M., Farooqui, M.N. and Suryawanshi, M.V. (1970) On a new species of *Paraphelenchus* (Micoletzky, 1922) Micoletzky, 1925 (Nematoda: Paraphelenchidae) from Marathwada, India. *Rivista di Parassitologia* 31, 139–142.

Alkemada, J.R.M. and Loof, P.A.A. (1989) Observations on the ontogeny of some *Xiphinema* species (Nematoda: Dorylaimida). *Mededelingen van de Faculteit Landbouwwetenschappen Gent* 54/3b, 1177–1186.

Allen, M.W. (1940) *Anomyctus xenurus*, a new genus and species of Tylenchoidea (Nematoda). *Proceedings of the Helminthological Society of Washington* 7, 96–98.

Allen, M.W. (1941) *Aphelenchoides megadorus*, a new species of Tylenchoidea (Nematoda). *Proceedings of the Helminthological Society of Washington* 8, 21–23.

Allen, M.W. (1952) Taxonomic status of the bud and leaf nematodes related to *Aphelenchoides fragariae* (Ritzema Bos, 1891). *Proceedings of the Helminthological Society of Washington* 19, 108–120.

Allen, M.W. (1957a) A review of the nematode genus *Trichodorus* with descriptions of ten new species. *Nematologica* 2, 32–62.

Allen, M.W. (1957b) Supplement: Dorylaimoidea described since the monographs on dorylaims and Dorylaimoidea were published in 1936 and 1939. In: G. Thorne, *A mono-*

* Wherever possible, the date following the author(s) refers to the *actual* year of publication and not necessarily to the chronological *volume* year of the journal.

graph of the nematodes of the superfamily Dorylaimoidea, 2nd edition. *Capita zoologica* 8, 262–263.

Allgén, C. (1933) Uber einige freilebende Nematoden aus dem Niederkongo. *Zoologischer Anzeiger* 103, 312–320.

Almeida, T., De Waele, D., Santos, S. and Sturhan, D. (1989) Species of *Trichodorus* (Nematoda: Trichodoridae) from Portugal. *Revue de Nématologie* 12, 219–233.

Alphey, T.J.W. (1978) Oxamyl sprays for the control of potato spraing disease caused by nematode-transmitted tobacco rattle virus. *Annals of Applied Biology* 88, 75–80.

Alphey, T.J.W. and Taylor, C. (1986) *European Atlas of the Longidoridae and Trichodoridae.* Scottish Crop Research Institute, Dundee, Scotland.

Alphey, T.J.W., Cooper, J.I. and Harrison, B.D. (1975) Systemic nematicides for the control of trichodorid nematodes and of potato spraing disease caused by tobacco rattle virus. *Plant Pathology* 24, 117–121.

Altherr, E. (1953) Nématodes du sol du Jura vandois et Français (1). *Bulletin de la Societé Vaudoise des sciences naturelles* 65, 429–460.

Altherr, E. (1958) Nématodes du bassin inférieur de la Weser et des dunes d'Héligoland. Espèces nouvelles ou incomplètement décrites. *Mémoires de la Société Vaudoise des sciences naturelles* 74, 45–63.

Altherr, E. (1963) Contribution à la connaissance de la faune des sables submergés en Lorraine. Nématodes. *Annales de Spéléologie* 18, 53–98.

Altherr, E. (1974) Nématodes de nappe phréatique du réseau fluvial de la Saale. (Thuringe) II. *Limnologica* 9, 81–132.

Altherr, E. (1976) La faune des eaux profondes interstitielles de la région de Wiesbaden. *Bulletin de la Societé Vaudoise des sciences naturelles* 73, 97–116.

Anderson, R.V. and Hooper, D.J. (1980) Diagnostic value of the vagina structure in the taxonomy of *Aphelenchus* Bastian, 1865 (Nematoda: Aphelenchidae) with a description of *A. (Anaphelenchus) isomerus* n.subgen., n.sp. *Canadian Journal of Zoology* 58, 924–928.

Anderson, R.V. and Laumond, C. (1990) *Noctuidonema guyanense* Remillet and Silvain, 1988: morphologie du spicule, redescription du mâle et diagnose amendée du genre *Noctuidonema*. *Revue de Nématologie* 13, 433–436.

Anderson, R.V. and Laumond, C. (1992) *Noctuidonema daptria,* n.sp. (Nematoda: Aphelenchoididae), an ectoparasite of the moth *Lesmone porcia* (Stoll). *Journal of Nematology* 24, 16–22.

Andrássy, I. (1957) *Aphelenchoides citri* n.sp., ein neuer Wurzelparasit der Zitrone. *Nematologica* 2, 238–241.

Andrássy, I. (1958) Erd-und Süsswassernematoden aus Bulgarien. *Acta Zoologica Academiae Scientiarum Hungaricae* 4, 1–88.

Andrássy, I. (1959) Neue und wenig bekannte Nematoden aus Jugoslawien. *Annales historico-naturales Musei nationalis Hungarici* 51, 259–275.

Andrássy, I. (1960) Nematoden aus dem Periphyton der Landungsmolen der Donau zeischen Budapest und Mohács. *Annales Universitatis scientiarum budapestinensis de Rolando Eötvös Nominatae. Sectio Biologica* 3, 3–21.

Andrássy, I. (1966) Erd-und Süsswasser-Nematoden aus Ghana klasse Secernentea (Phasmidia) *Annales Universitatis scientiarum budapestinensis de Rolando Eötvös* 8, 5–24.

Andrássy, I. (1968) The scientific results of the Hungarian soil zoological expedition to the Brazzaville-Congo. 31. Nematoden aus Grundwasser. *Annales Universitatis scientiarum budapestinensis* 9–10, 3–26.

Andrássy, I. (1970) Nematoden aus einigen Fluss-Systemen Südafrikas. *Opuscula Zoologica* 10, 179–219.

Andrássy, I. (1973) Ein Meeresrelikt und einige andere bemerkenswerte Nematodenarten aus Neuguinea. *Opuscula Zoologica* 12, 3–19.

Andrássy, I. (1976) *Evolution as a Basis for the Systematization of Nematodes.* Pitman Publishing Ltd., London, UK, 288pp.

Andrássy, I. (1986) *Paralongidorus rex* sp.n., a new nematode species from Hungary. *Allattani Kozlemenyek* 73, 115–118.

Andrássy, I. (1989) Six new nematode species from South America. *Acta zoologica hungarica* 35, 1–16.

Andrés, M.F. and Arias, M. (1988a) A new species of *Longidorus* (Nematoda: Longidoridae) associated with forest soils and notes on *L. congoensis* Aboul-Eid, 1970 and *L. intermedius* Kozlowska & Seinhorst, 1979, new records for Spain. *Nematologica* 33, 386–392.

Andrés, M.F. and Arias, M. (1988b) *Longidorus belloi* n.sp. (Nematoda: Longidoridae) from Spain. *Revue de Nématologie* 11, 415–421.

Andrés, M., Bleve-Zacheo, T. and Arias, M. (1989) The ultrastructure of cereal and leguminous root tips parasitized by *Longidorus belloi. Revue de Nématologie* 12, 365–374.

Anon. (1984) *Xiphinema americanum* Cobb, 1913. *Bulletin EPPO/OEPP Bulletin* 14, 67–71.

Arias Delgado, M., Jiménez Millán, F. and López Pedregal, J.M. (1965) Tres nuevas especies de nematodos posibles fito-parásitos en suelos españoles. *Publ. Inst. Biol. Apl.* 38, 47–58.

Arias, M., Andrés, M.F. and Navas, A. (1986) *Longidorus carpetanensis* sp.n. and *L. unedoi* sp.n. (Nematoda: Longidoridae) from Spain. *Revue de Nématologie* 9, 101–106.

Ashby, S.F. (1921) Some recent observations on red ring disease of the coco-nut. *Agricultural News* 20, 350–351.

Ayala, A. and Allen, M.W. (1968) Transmission of the Californian tobacco virus by three species of the nematode genus *Trichodorus. Journal of Agriculture of the University of Puerto Rico* 52, 101–125.

Ayala, A., Allen, M.W. and Noffsinger, E.M. (1970) Host range, biology, and factors affecting survival and reproduction of the stubby root nematode. *Journal of Agriculture of the University of Puerto Rico* 54, 341–369.

Bajaj, H.K. and Bhatti, D.S. (1982) Aphelenchid nematodes from Haryana. *Indian Journal of Nematology* 12, 258–262.

Bajaj, H.K. and Jairajpuri, M.S. (1977) Two new species of *Xiphinema* from India. *Nematologia Mediterranea* 4, 195–200.

Bajaj, H.K. and Jairajpuri, M.S. (1979) A review of the genus *Xiphinema* Cobb, 1913 with descriptions of species from India. *Records of the Zoological Survey of India* 75, 255–325.

Baker, A.D. (1962) *Check Lists of the Nematode Superfamilies Dorylaimoidea, Rhabditoidea, Tylenchoidea, and Aphelenchoidea.* E.J. Brill, Leiden, 261pp.

Bala, G. (1988) A review of plant-nematological research conducted in Trinidad and Tobago. *Nematropica* 18, 71–74.

Bally, W. and Reydon, G.A. (1931) De tegenwoordige stand van het vraagstuk van de wortelaaltjes in de koffiecultuur. *Arch. Koffiecult. Nederl. Indië* 5, 23–216.

Baranovskaya, I.A. (1958a) [Contribution to the knowledge of the genus *Paraphelenchus* (Micoletzky, 1922) Micoletzky, 1925 (Nematoda: Aphelenchidae)]. *Zoologicheskii Zhurnal* 37, 13–19 (in Russian).

Baranovskaya, I.A. (1958b) [New species of nematodes on cereals from vicinity of Moscow]. *Izdatel. Akad. Nauk SSDR., Moskva,* 55–58 (in Russian).

Baranovskaya, I.A. (1963) [Two new species of *Aphelenchoides* Fischer, 1894 (Nematoda, Aphelenchoididae)]. In: *[Helminths of man, animals and plants and their control: Papers on helminthology presented to Academician K.I. Skryabin on his 85th Birthday]*. Moscow: Izdatelstvo Akad. Nauk SSSR, pp.480–483 (in Russian).

Baranovskaya, I.A. (1981) [*Plant and Soil Nematodes (Aphelenchoididae and Seinuridae)*]. Moscow, USSR; 'Nauka', 234pp (in Russian).

Baranovskaya, I.A. (1984) [Nematodes of the genus *Paraphelenchus* (Micoletzky, 1922) Micoletzky, 1925]. In: Turlygina, E.S. (ed.), *Taksonomiya i Biologiya Fitogel'mintov*. pp.5–35 (in Russian).

Baranovskaya, I.A. and Haque, M.M. (1968) [*Aphelenchoides graminis* n.sp. (Nematoda: Aphelenchoididae)]. *Zoologicheskii Zhurnal* 47, 631–634 (in Russian).

Barker, K.R. (1963) Parasitism of tobacco callus and Kentucky bluegrass by *Aphelenchus avenae*. *Phytopathology* 53, 870.

Bastian, H.C. (1865) Monograph on the Anguillulidae, or free nematoids, marine, land, and freshwater; with descriptions of 100 new species. *Transactions of the Linnean Society of London* 25, 73–184.

Baujard, P. (1980a) *Trichodorus aquitanensis* n.sp. et donnés nouvelles sur *Paratrichodorus (Atlantadorus) pachydermus* (Seinhorst, 1954) Siddiqi, 1974 (Nematoda: Trichodoridae) *Revue de Nématologie* 3, 21–27.

Baujard, P. (1980b) Trois nouvelles espèces de *Bursaphelenchus* (Nematoda: Tylenchida) et remarques sur le genre. *Revue de Nématologie* 3, 167–177.

Baujard, P. (1981) Revue taxonomique du genre *Laimaphelenchus* Fuchs, 1937 et description de *Laimaphelenchus pini* n.sp. *Revue de Nématologie* 4, 75–92.

Baujard, P. (1984) Remarques sur la sous-famille des Ektaphelenchinae Paramonov, 1964 et proposition d'*Ektaphelenchoides* n.gen. (Nematoda: Aphelenchoididae). *Revue de Nématologie* 7, 147–171.

Baujard, P. (1985) Remarks on the genus *Laimaphelenchus* Fuchs, 1937. *Revue de Nématologie* 8, 186–187.

Baujard, P. (1988) Identification of aphelenchids. In: Fortuner, R. (ed.), *Nematode Identification and Expert System Technology*. Plenum Press, New York and London, pp. 153–156.

Baujard, P. (1989) Remarques sur les genres des sous-familles Bursaphelenchinae Paramonov, 1964 et Rhadinaphelenchinae Paramonov, 1964 (Nematoda: Aphelenchoididae). *Revue de Nématologie* 12, 323–324.

Baujard, P. and Germani, G. (1985) Description de *Monotrichodorus sacchari* n.sp., *Paratrichodorus anthurii* n.sp. et d'une population de *Paratrichodorus westindicus* Rodriguez-M, Sher and Siddiqi, 1978 (Nematoda: Trichodoridae). *Revue de Nématologie* 8, 35–39.

Baujard, P., Boulbria, A., Ham, R., Laumond, C. and Scotto La Massese, C. (1979) Premières données sur la nématofaune associée aux dépérissements du pin maritime dans l'Ouest de la France. *Annales des Sciences Forestières* 36, 331–339.

B'Chir, M.M. (1979) Description and biometrical, morphological and biological studies of *Aphelenchoides tuzeti* n.sp. (Nematoda: Aphelenchoidea) and discussion on the specific criteria applied to the genus Aphelenchoides. *Nematologica* 25, 22–31.

Bedford, G.O. (1974) Parasitism of the palm weevil *Rhynchophorus bilineatus* (Montrouzier) (Coleoptera: Curculionidae) by *Praecocilenchus rhaphidophorus* Poinar (Nematoda: Aphelenchoidea) in New Britain. *Journal of the Australian Entomological Society* 13, 155–156.

Bergeson, G.B., Athow, K.L., Laviolette, F.A. and Thomasine, Sister M. (1964) Vector relationships of tobacco ringspot virus (TRSV) and *Xiphinema americanum* and the importance of this vector in TRSV infection of soybean. *Phytopathology* 54, 723–728.

Bernard, E.C. (1992a) Terrestrial nematodes of Alaska 1. Trichodoridae (Nemata). *Journal of Nematology* 24, 67–77.

Bernard, E.C. (1992b) *Trichodorus elefjohnsoni* n.sp. (Nemata: Trichodoridae) from undisturbed Appalachian forest. *Journal of Nematology* 24, 78–83.

Bessarabova, L.M. (1966) [*Aphelenchoides conimucronatus* n.sp. from leguminous plants in the Moscow region]. *Zoologicheskii Zhurnal* 45, 1569–1570 (in Russian).

Bird, A.F. and Bird, J. (1991) *The Structure of Nematodes.* 2nd edition, Academic Press, San Diego, xviii + 316pp.

Bird, G.W. (1967) *Trichodorus acutus* n.sp. (Nematoda: Diphtherophoroidea) and a discussion of allometry. *Canadian Journal of Zoology* 45, 1201–1204.

Bleve-Zacheo, T., Zacheo, G., Melillo, M.T. and Lamberti, F. (1984) On the structure of the Z pseudo-organ in *Xiphinema diversicaudatum. International Journal for Parasitology* 14, 405–413.

Bleve-Zacheo, T., Zacheo, G. and Lamberti, F. (1985) Ultrastructure of the Z organ in *Xiphinema ifacolum. International Journal for Parasitology* 15, 453–461.

Boag, B. (1978) Nematodes in Scottish forest nurseries. *Annals of Applied Biology* 88, 279–286.

Boag, B. and Alphey, T.J.W. (1988) Influence of interspecific competition on the population dynamics of migratory plant-parasitic nematodes with r and K survival strategies. *Revue de Nématologie* 11, 321–326.

Boag, B., Brown, D.J.F. and Topham, P.B. (1987) Vertical and horizontal distribution of virus-vector nematodes and implications for sampling procedures. *Nematologica* 33, 83–96.

Boag, B., Crawford, J.W. and Neilson, R. (1991) The effect of potential climate changes on the geographical distribution of the plant-parasitic nematodes *Xiphinema* and *Longidorus* in Europe. *Nematologica* 37, 312–323.

Bolla, R.I., Nosser, C. and Tamura, H. (1989) Chemistry of response of pines to *Bursaphelenchus xylophilus*: Resin acids. *Japanese Journal of Nematology* 19, 1–6.

Bor, N.A. and Kuiper, K. (1966) Gevoeligheid van *Trichodorus teres* en *T. pachydermus* voor uitwendige invloeden. *Mededelingen Rijks. Landbouw wetenschappen Gent* 31, 609–616.

Bos, W.S. (1977) *Aphelenchoides arachidis* n.sp. (Nematoda: Aphelenchoidea), an endoparasite of the testa of groundnut in Nigeria. *Zeitschrift für Pflanzenkrankheiten (Pflanzenpathologie) und Pflanzenschutz*, 84, 95–99.

Bos, W.S. and Loof, P.A.A. (1985) Nigerian species of the genus *Xiphinema* Cobb, 1913 (Nematoda: Dorylaimida) Part I. *Nematologica* 30, 395–418.

Braithwaite, C.W.D. and Siddiqi, M.R. (1975) *Rhadinaphelenchus cocophilus*. In: *CIH Descriptions of Plant-parasitic Nematodes*. Set 5, No. 72, 4pp.

Bridge, J. and Hunt, D.J. (1985) *Aphelenchoides arachidis*. In: *CIH Descriptions of Plant-parasitic Nematodes*. Set 8, No. 116, 3pp.

Bridge, J., Luc, M. and Plowright, R.A. (1990) Nematode parasites of rice. In: Luc, M., Sikora, R.A. and Bridge, J. (eds), *Plant Parasitic Nematodes in Subtropical and Tropical Agriculture*, CAB International, Wallingford, UK, pp.69–108.

Brinkman, H., Loof, P.A.A. and Barbez, D. (1987) *Longidorus dunensis* n.sp. and *L. kuiperi* n.sp. from the sand dune region of the Netherlands (Nematoda: Longidoridae) *Revue de Nématologie* 10, 299–308.

Brodie, B.B. (1976) Vertical distribution of three nematode species in relation to certain soil properties. *Journal of Nematology* 8, 243–247.

Brown, D.J.F. (1989) Viruses transmitted by nematodes. *EPPO/OEPP Bulletin* 19, 453–461.

Brown, D.J.F. and Boag, B. (1977) *Longidorus attenuatus*. In: *CIH Descriptions of Plant-parasitic Nematodes*. Set 7, No. 101, 4pp.

Brown, D.J.F. and Taylor, C.E. (1987) Comments on the occurrence and geographical distribution of longidorid nematodes in Europe and the Mediterranean Region. *Nematologia Mediterranea* 15, 333–373.

Brown, D.J.F., Luc, M. and Saka, V.W. (1983) Two new species of *Xiphinema* Cobb, 1913 (Nematoda, Dorylaimida). *Bulletin du Muséum National d'Histoire Naturelle. Paris, 4ème Série, 5, Section A*: 521–529.

Brown, D.J.F., Ploeg, A.T. and Robinson, D.J. (1989) A review of reported associations between *Trichodorus* and *Paratrichodorus* species (Nematoda: Trichodoridae) and tobraviruses with a description of laboratory methods for examining virus transmission by trichodorids. *Revue de Nématologie* 12, 235–241.

Brown, D.J.F., Taylor, C.E., Choleva, B. and Romanenko, N.D. (1990) The occurrence of Longidoridae (Nematoda: Dorylaimida) in western USSR with further comments on longidorid nematodes in Europe and the Mediterranean basin. *Nematologia Mediterranea* 18, 199–207.

Brown, D.J.F., Faria, A., Lamberti, F., Halbrendt, J.M., Agostinelli, A. and Jones, A.T. (1993) A description of *Xiphinema madeirense* sp.n. and the occurrence and virus vector potential of *X. diversicaudatum* (Nematoda: Dorylaimida) from Santana, Madeira. *Nematologia Mediterranea* 20, 251–259.

Brown, R.H. (1968) *Xiphinema monohysterum* n.sp. (Nematoda: Dorylaimoidea) from Southern New South Wales. *Nematologica* 13, 633–637.

Bütschli, O. (1873) Beiträge zur Kenntnis der freilebenden Nematoden. *Nova Acta Acad. Nat. Curios* 36, 1–124.

Bütschli, O. (1874) Zur Kenntnis der freilebenden Nematoden insbesondere der des Kieler Hafens. *Abhandlungen hrsg. von der Senckenbergischen naturfofschenden Gesellschaft* 9, 237–292.

Calaway, W.T. and Tarjan, A.C. (1973) A compendium of the genus *Mononchoides* Rahm, 1928 (Diplogastrinae: Nematoda). *Journal of Nematology* 5, 107–116.

Carvalho, J.C. (1955) Plantas ornamentais parasitadas por espécies do gênero *Xiphinema*. *Revista Instituto Adolfo Lutz* 15, 180–185.

Carvalho, J.C. (1962) *Xiphinema itanhaense* n.sp. (Nematoda: Dorylaimoidea). *Arquivos Instituto Biológico Sao Paulo* 29, 223–225.

Carvalho, J.C. (1965) *Xiphinema paulistanum*, una nova espécie de Nematoide. *Arquivos Instituto Biológico Sao Paulo* 32, 77–79.

Chaturvedi, Y. and Khera, S. (1977) *Caballeroides olitorius*, gen. nov., (Nematoda: Aphelenchoididae) from the rhizosphere of jute. In: *Excerta Parasitológica en Memoria del Doctor Eduardo Caballero y Caballero*. University of Mexico, pp.419–422.

Chaturvedi, Y. and Khera, S. (1979) Studies on taxonomy, biology and ecology of nematodes associated with jute crop. *Technical Monograph, Zoological Survey of India, No. 2*, v + 105pp.

Chaves, E. and Coomans, A. (1984) Three new species of *Xiphidorus* from Argentina with comments on *Xiphinema sandellum* Heyns, 1966. *Revue de Nématologie* 7, 3–12.

Chaves, E. and Geraert, E. (1977) Observations sur quelques Mononchides et Dorylaimides du Cameroun et du Zaire (Vermes-Nematoda). *Revue de Zoologie Africaine* 91, 766–780.

Chawla, M.L. and Khan, E. (1979) Two nomenclatorial corrections. *Indian Journal of Nematology* 7, 100.

Chawla, M.L. and Samathanam, G.J. (1981) Two new species of the superfamily Longidoroidea (Dorylaimida: Nematoda) from India. *Indian Journal of Nematology* 10, 205–210.

Chawla, M.L., Bhamburkar, B.L., Khan, E. and Prasad, S.K. (1968) One new genus and seven new species of nematode from India. *Labdev Journal of Science and Technology, Series B* 6, 86–100.

Chin, D.A. and Estey, R.H. (1966) Studies on the pathogenicity of *Aphelenchus avenae* Bastian, 1865. *Phytoprotection* 47, 66–72.

Chitwood, B.G. (1958) The designation of official names for higher taxa of invertebrates. *Bulletin of Zoological Nomenclature* 15(25/28), 860–895.

Chizhov, V.N., Tiev, R.A. and Turkina, A. Yu. (1986) Dva novykh vida nematod iz roda *Xiphinema* (Nematoda: Longidoridae) severnogo Kavkaze. *Zoologicheskii Zhurnal* 65, 932–935.

Cho, M.R. and Robbins, R.T. (1990) Scanning electron microscopy of *Xiphinema, Longidorus* and *Californidorus* stylet morphology. *Journal of Nematology* 22, 162–169.

Christie, J.R. (1932) Recent observations on the strawberry dwarf nematode in Massachusetts. *Plant Disease Reporter* 16, 113–114.

Christie, J.R. (1939) Predaceous nematodes of the genus *Aphelenchoides* from Hawaii. *Journal of the Washington Academy of Science*, 29, 161–170.

Christie, J.R. (1942) A description of *Aphelenchoides besseyi* n.sp., the summer-dwarf nematode of strawberries, with comments on the identity of *Aphelenchoides subtenuis* (Cobb, 1926) and *Aphelenchoides hodsoni* Goodey, 1935. *Proceedings of the Helminthological Society of Washington* 9, 82–84.

Christie, J.R. (1959) *Plant Nematodes, their Bionomics and Control.* University of Florida, Gainsville, Florida, xi + 256pp.

Christie, J.R. and Arndt, C.H. (1933) Further notes on the nematodes associated with the soreshin of cotton. *Plant Disease Reporter* 17, 10–12.

Christie, J.R. and Perry, V.G. (1951) A root disease of plants caused by a nematode of the genus *Trichodorus. Science* 113, 491–493.

Christie, J.R., Perry, V.G. and Wilson, J.W. (1950) Control of nematodes injurious to vegetable crops. *Report, Florida Agriculture Experiment Station* 4950, 144–145.

Clark, S.A. and Shepherd, A.M. (1977) Structure of the spicules and caudal sensory equipment in the male of *Aphelenchoides blastophthorus* (Nematoda: Tylenchida, Aphelenchina). *Nematologica* 23, 103–111.

Clark, W.C. (1961) A revised classification of the Order Enoplida (Nematoda). *New Zealand Journal of Science* 4, 123–150.

Clark, W.C. (1962) The systematic position of the Alaimidae and the Diphtherophoroidea (Enoplida, Nematoda). *Nematologica* 7, 119–121.

Clark, W.C. (1963a) A new species of *Trichodorus* (Nematoda: Enoplida) from Westland, New Zealand. *New Zealand Journal of Science* 6, 414–417.

Clark, W.C. (1963b) A new species of Longidorus (Micol.) Dorylaimida, Nematoda. *New Zealand Journal of Science* 6, 607–611.

Cobb, N.A. (1888) Beiträge zur Anatomie und Ontogenie der Nematoden. *Jenaische Zeitschrift für Naturwissenschaft v. 23,* n.s. 16, 41–76.

Cobb, N.A. (1891) Strawberry-bunch. (A new disease caused by nematodes). *Agricultural Gazette of New South Wales* 2, 390–400.

Cobb, N.A. (1893a) Plant diseases and their remedies. Diseases of the sugar-cane. 3. Nematode worms found attacking sugar-cane. *Agricultural Gazette of New South Wales* 4, 808–833.

Cobb, N.A. (1893b) Nematodes, mostly Australian and Fijian. *Macleay Memorial Volume, Linnean Society of New South Wales*, pp.252–308.

Cobb, N.A. (1913) New nematode genera found inhabiting fresh water and non-brackish soils. *Journal of the Washington Academy of Sciences* 3, 432–444.

Cobb, N.A. (1919) A newly discovered nematode (*Aphelenchus cocophilus*, n.sp.), connected with a serious disease of the coco-nut palm. *West Indian Bulletin* 17, 203–210.

Cobb, N.A. (1920) One hundred new nemas. *Contributions to the Science of Nematology* 9, 217–343.

Cobb, N.A. (1923) Observations on nemas, salivary glands of the nemic genera *Tylenchus* and *Aphelenchus*. *Journal of Parasitology* 9, 236–239.

Cobb, N.A. (1926) Nemic diseases of narcissus. *Official Records, USDA* 5, 3.

Cobb, N.A. (1927) Note on a new nema, *Aphelenchus retusus*, with a proposed division of *Aphelenchus* into three subgenera. *Journal of Parasitology* 14, 57–58.

Cobbold, T.S. (1865) Report of experiments respecting the development and migrations of the entozoa. *Report of the British Association for the Advancement of Science*, 34th meeting, Bath, pp.111–120.

Cohn, E. (1969) The occurrence and distribution of species of *Xiphinema* and Longidorus in Israel. *Nematologica* 15, 179–192.

Cohn, E. (1970) Observations on the feeding and symptomatology of *Xiphinema* and *Longidorus* species on selected host roots. *Journal of Nematology* 2, 167–173.

Cohn, E. (1975) Relation between *Xiphinema* and *Longidorus* and their host plants. In: Lamberti, F., Taylor, C.E. and Seinhorst, J.W. (eds), *Nematode Vectors of Plant Viruses*. Plenum Press, London and New York, pp.365–386.

Cohn, E. (1977) *Xiphinema italiae*. In: *CIH Descriptions of Plant-parasitic Nematodes*. Set 7, No. 95, 3pp.

Cohn, E. and Mordechai, M. (1969) Investigations on the life cycles and host preference of some species of *Xiphinema* and *Longidorus* under controlled conditions. *Nematologica* 15, 295–302.

Cohn, E. and Orion, D. (1970) The pathological effect of representative *Xiphinema* and *Longidorus* species on selected host plants. *Nematologica* 16, 423–428.

Cohn, E. and Sher, S.A. (1972) A contribution to the taxonomy of the genus *Xiphinema* Cobb, 1913. *Journal of Nematology* 4, 36–65.

Colbran, R.C. (1956) Studies of plant and soil nematodes. 1. Two new species from Queensland. *Queensland Journal of Agricultural Science* 13, 123–126.

Colbran, R.C. (1965) Studies of plant and soil nematodes. 9. *Trichodorus lobatus* n.sp. (Nematoda: Trichodoridae), a stubby-root nematode associated with citrus and peach trees. *Queensland Journal of Agricultural and Animal Science* 22, 273–276.

Cooke, D.A. (1973) The effect of plant parasitic nematodes, rainfall and other factors on Docking disorder of sugar beet. *Plant Pathology* 22, 161–170.

Coomans, A. (1965) *Xiphinema basilgoodeyi* n.sp. with observations on its larval stages (Nematoda: Dorylaimida). *Nematologica* 10, 581–593.

Coomans, A. (1985) A phylogenetic approach to the classification of the Longidoridae (Nematoda: Dorylaimida). *Agriculture, Ecosystems and Environment* 12, 335–354.

Coomans, A. and Heyns, J. (1985) The genus *Xiphinema* in South Africa. X. Species of the X. *meridianum*-group (Nematoda: Dorylaimida). *Phytophylactica* 17, 183–194.

Coomans, A. and Loof, P.A.A. (1970) Morphology and taxonomy of Bathyodontina (Dorylaimida). *Nematologica* 16, 180–196.

Coomans, A., Rashid, F. and Luc, M. (1990) Observations on *Xiphinema vitis* Heyns, 1974, X. *elongatum* Schuurmans Stekhoven and Teunissen, 1938 and X. *fatikae fatikae* Bos and Loof, 1985, and description of X. *fatikae eburnense* subsp.n. (Nemata: Longidoridae) from Africa. *Revue de Nématologie* 13, 239–248.

Cooper, J.I. (1971) The distribution in Scotland of tobacco rattle virus and its nematode vectors in relation to soil type. *Plant Pathology* 20, 52–58.

Corbett, D.C.M. (1964) *Longidorus utriculoides* n.sp. (Nematoda: Dorylaimidae) from Nyasaland. *Nematologica* 10, 496–499.

Corbett, M.K. (1959) Diseases of the coconut palm. *Principes* 3, 83–86.

Cotten, J. (1974) Feeding behaviour and reproduction of *Xiphinema index* on some herbaceous test plants. *Nematologica* 19, 516–520.

Dalmasso, A. (1967) Description du mâle de *Longidorus laevicapitatus* Williams, J.R., 1959, Nematoda Dorylaimina, parasite de la canne à sucre aux Antilles. *Annales des Épiphyties* 18, 313–316.

Dalmasso, A. (1969) Etude anatomique et taxonomique des genres *Xiphinema, Longidorus* et *Paralongidorus* (Nematoda: Dorylaimidae). *Mémoires du Muséum National d'Histoire Naturelle. Paris, Série A. Zoologie* 61, 33–82.

Dalmasso, A. (1970) Influence directe de quelques facteurs écologiques sur l'activité biologique et la distribution des espèces françaises de la famille des Longidoridae. *Annales de Zoologie: Ecologie Animale* 2, 163–200.

Darekar, K.S. and Khan, E. (1983) *Xiphinema tugewai* sp.n. and *Longidorus paramirus* sp.n. (Longidoroidea: Nematoda) from Maharashtra, India. *Indian Journal of Nematology* 12(1982), 153–157.

Das, S.N. (1986) Transmission of plant viruses by nematode vectors. In: Swarup, G. and Dasgupta, D.R. (eds), *Plant Parasitic Nematodes of India, Problems and Progress*. pp.412–423.

Das, V.M. (1960) Studies on the nematode parasites of plants in Hyderabad (Andhra Pradesh, India). *Zeitschrift für Parasitenkunde* 19, 553–605.

Das, V.M. and Singh, S.D. (1968) *Paraphelenchus crenatus* n.sp. (Aphelenchidae) from Andhra Pradesh. *Journal of Helminthology* 42, 29–32.

Dassonville, A.F. and Heyns, J. (1984) Freshwater nematodes from South Africa. 7. New and known species collected in Skinnerspruit, Pretoria. *Phytophylactica* 16, 15–32.

Davis, R.A. and Jenkins, W.R. (1960) Nematodes associated with roses and the root injury caused by *Meloidogyne hapla* Chitwood, 1949, *Xiphinema diversicaudatum* (Micoletzky, 1927) Thorne, 1939 and *Helicotylenchus nannus* Steiner, 1945. *Bulletin of the Maryland Agricultural Experiment Station A-106*, 16pp.

Dean, C.G. (1979) Red ring disease of *Cocos nucifera* L. caused by *Rhadinaphelenchus cocophilus* (Cobb, 1919) Goodey, 1960. An annotated bibliography and review. *Technical Communication No. 47*, Commonwealth Agricultural Bureaux, Farnham Royal, UK, 70pp.

Dean, C.G. and Velis, M. (1976) Differences in the effects of red ring disease on coconut palms in Central America and the Caribbean and its control. *Oléagineaux* 31, 321–324.

Decker, H. (1962) Zur Biologie und Ökologie von *Aphelenchus avenae* Bastian. *Nematologica* 7, 9 (abstract).

Decker, H. (1972) Zur Systematik der Phytonematoden der Ordnung Tylenchida Thorne, 1949. In: *Probleme der Phytonematologie. Gross Lüsewitz, East Germany, June 2, 1972*, 51–75.

Decraemer, W. (1979) Differentiation of juvenile specimens to species level in Trichodoridae (Nematoda) from fields of seed-potatoes in Belgium. *Mededelingen van de Faculteit Landbouwwetenschappen, Rijksuniversiteit, Gent* 44, 273–287.

Decraemer, W. (1980) Systematics of the Trichodoridae (Nematoda) with keys to their species. *Revue de Nématologie* 3, 81–99.

Decraemer, W. (1987) Morphological comparison between the monodelphic genera *Monotrichodorus and Allotrichodorus* with some remarks on their relation to both didelphic genera of the Trichodoridae (Nematoda). *Nematologica* 32, 256–264.

Decraemer, W. (1988) Morphometric variability and value of the characters used for species identification in *Trichodorus* Cobb, 1913. *Bulletin van het Koninklijk Belgisch Instituut voor Natuurwetenschappen, Biologie* 58, 29–44.

Decraemer, W. (1989a) Identification of trichodorids. In: Fortuner, R. (ed.), *Nematode Identification and Expert System Technology*. Plenum Press, New York, pp.157–170.

Decraemer, W. (1989b) Morphologic variability and value of the characters used for species identification in *Paratrichodorus* Siddiqi, 1974 (Nematoda: Trichodoridae) *Nematologica* 35, 37–61.

Decraemer, W. (1991) Stubby root and virus vector nematodes: *Trichodorus, Paratrichodorus, Allotrichodorus* and *Monotrichodorus*. In: Nickle, W.R. (ed.), *Manual of Agricultural Helminthology*, Marcel Dekker, New York, pp.587–625.

Decraemer, W. and Chaves, E. (1988) Morphology of the reproductive system and structure of the genus *Paratrichodorus* with uncommon males (Nematoda: Diphtherophorina). *Revue de Nématologie* 11, 405–409.

Decraemer, W. and Kilian, S. (1992) Description of *Trichodorus parorientalis* n.sp. (Nemata: Diphtherophoroidea) from South Africa. *Fundamental and Applied Nematology* 15, 539–543.

Decraemer, W. and Reay, F. (1991) Trichodorid nematodes from Australia with description of two new species from native vegetation. *Australasian Plant Pathology* 20, 52–66.

Decraemer, W. and De Waele, D. (1981) Taxonomic value of the position of oesophageal gland nuclei and of oesophageal gland overlap in the Trichodoridae (Diphtherophorina). *Nematologica* 27, 82–94.

De Guiran, G. and Boulbria, A. (1986) Le nématode des pins. Caractéristiques de la souche française et risque d'introduction et d'extension de *Bursaphelenchus xylophilus* en Europe. *Bulletin OEPP/EPPO Bulletin* 16, 445–452.

De Guiran, G. and Brugieur, N. (1989) Hybridization and phylogeny of the pine wood nematode (*Bursaphelenchus* spp.). *Nematologica* 35, 321–330.

de Man, J.G. (1876) Onderzoekingen over vrij in de aarde levende Nematoden. *Tijdschrift der Nederlandsche dierkundige Vereeniging* 2, 78–196.

de Man, J.G. (1880) Die Einheimischen, frei in der reinen Erde und im süssen Wasser lebende Nematoden. Vorläufiger Bericht und descriptiv-systematischer Theil. *Tijdschrift der Nederlandsche dierkundige Vereeniging* 5, 1–104.

de Man, J.G. (1881) Ueber einige neue oder noch unvollständig bekannte Arten von frei in der reinen Erde lebenden Nematoden. (1. Supplement zu der Aufsatz auf S. 1 dieses Bandes.) *Tijdschrift der Nederlandsche dierkundige Vereeniging* 5, 138–143.

de Man, J.G. (1884) *Die frei in der reinen Erde und im süssen Wasser lebenden Nematoden der Niederländischen Fauna*. Brill, Leiden, 206pp.

de Man, J.G. (1895) Description of three species of Anguillulidae, observed in diseased pseudo-bulbs of tropical orchids. *Proceedings and Transactions of the Liverpool Biological Society* 9, 76–94.

de Man, J.G. (1907) Observations sur quelques espèces de nématodes terrestres libres de l'Ile de Walcheren. *Annales de la Société Royale Zoologique et Malacologique de Belgique* 41 (1906), 161–174.

de Man, J.G. (1921) Nouvelles recherches sur les nématodes libres terricoles de la Hollande. *Capita zoologica* 1, 3–62.

De Pelsmaeker, M. (1987a) Effect of Vydate 10 G on trichodorids, on tobacco rattle virus disease and on potato yields. *Revue de l'Agriculture* 40, 251–262.

De Pelsmaeker, M. (1987b) [*The importance and the control of virus vector nematodes and soil borne viruses in cultures of potatoes, hops and strawberries*]. Instituut tot Aanmoediging van het Wetenschappelijk Onderzoe, Brussels, Belgium, 93pp (in Dutch)

De Pelsmaeker, M. and Coomans, A. (1986) [Nematodes as vectors of viruses in potato fields]. *Parasitica* 42, 11–16 (in Dutch).

De Pelsmaeker, M. and Coomans, A. (1987) Nematodes in potato fields and the relation to some biotic and abiotic factors. *Mededelingen van de Faculteit Landbouwwetenschappen, Rijksuniversiteit Gent* 52, 561–569.

De Pelsmaeker, M., Calus, A. and Coomans, A. (1985) Vertical distribution of trichodorids and TRV. *Mededelingen van de Faculteit Landbouwwetenschappen Rijksuniversiteit Gent* 50, 769–775.

Devdariani, Ts.G. (1970) [*Tylaphelenchus georgiensis* n.sp. (Nematoda: Aphelenchoididae)]. *Soobschcheniya Akademii Nauk Gruzinskoi SSR* 58, 717–720. (in Russian).

Devdariani, Ts.G. (1971) [*Cryptaphelenchus bicoloris* n.sp. (Nematoda: Aphelenchoididae)]. *Soobschcheniya Akademii Nauk Gruzinskoi SSR* 63, 221–224. (in Georgian).

Devdariani, Ts.G. (1974) [New nematode species from *Monochamus sutor*]. *Soobschcheniya Akademii Nauk Gruzinskoi SSR* 76, 709–712 (in Russian).

De Waele, D. (1988) *Trichodorus petrusalberti* n.sp. (Nematoda: Trichodoridae) from rice with additional notes on the morphology of *T. sanniae* and *T. rinae*. *Journal of Nematology* 20, 85–90.

De Waele, D. and Carbonell, E. (1983) Two new species of *Trichodorus* (Nematoda: Diphtherophorina) from Africa. *Nematologica* 28, 387–397.

De Waele, D. and Cohn, E. (1992) Trichodoridae from Israel, with a description of *Trichodorus minzi* n.sp. *Fundamental and Applied Nematology* 15, 201–207.

De Waele, D. and Coomans, A. (1990) Occurrence and ecology of longidorid nematodes in Belgium. *Revue de Nématologie* 13, 197–202.

De Waele, D. and Hashim, Z. (1984) *Trichodorus orientalis* n.sp. (Nematoda: Trichodoridae) from Jordan and Iran. *Systematic Parasitology* 6, 63–67.

De Waele, D. and Kilian, S. (1992) *Trichodorus vandenbergae* n.sp. and *Paratrichodorus meyeri* n.sp. (Nematoda: Trichodoridae) from South Africa, with notes on *P. teres*, *P. catherinae* and *P. sacchari*. *Fundamental and Applied Nematology* 15, 397–406.

De Waele, D. and Sturhan, D. (1987) *Trichodorus persicus* n.sp. (Nematoda: Trichodoridae) from Iran. *Systematic Parasitology* 10, 79–83.

De Waele, D., Mancini, G., Roca, F. and Lamberti, F. (1982) *Trichodorus taylori* n.sp. (Nematoda: Dorylaimida) from Italy. *Nematologia Mediterranea* 10, 27–37.

De Waele, D., Alphey, T.J.W. and Barbez, D. (1985) *Paratrichodorus pachydermus*. In: *CIH Descriptions of Plant-Parasitic Nematodes*, Set 8, No. 112, 4pp.

De Waele, D., Meyer, A.J. and Van Mieghem, A.P. (1990) *Trichodorus philipi* n.sp. from South Africa, with notes on *Paratrichodorus lobatus* and *P. acutus*. *Journal of Nematology* 22, 200–206.

Dmitrenko, M.A. (1966) [Two new species of plant nematodes]. *Zoologicheskii Zhurnal* 45, 761–766 (in Russian).

Doncaster, C.C. and Seymour, M.K. (1974) Passive ingestion in a plant nematode, *Hexatylus viviparus* (Neotylenchidae: Tylenchida). *Nematologica* 20, 297–307.

Doucet, M.E. (1992) A new species of the genus *Laimaphelenchus* Fuchs, 1937 (Nemata: Aphelenchina). *Fundamental and Applied Nematology* 15, 1–6.

Drozdovsky, E.M. (1967) [Use of the characteristics of embryonal development in the classification of eelworms]. *Trudy gel'mintologicheskoi Laboratorii* 18, 22–29 (in Russian).

Ebsary, B.A. (1991) *Catalog of the Order Tylenchida (Nematoda)*. Agriculture Canada, iv + 196pp.

Ebsary, B.A., Potter, J.W. and Allen, W.R. (1984) Redescriptions and distribution of *Xiphinema rivesi* Dalmasso, 1969 and *Xiphinema americanum* Cobb, 1913 in Canada with a description of *Xiphinema occiduum* n.sp. (Nematoda: Longidoridae). *Canadian Journal of Zoology* 62, 1696–1702.

Ebsary, B.A., Vrain, T.C. and Graham, M.B. (1989) Two new species of *Xiphinema*

(Nematoda: Longidoridae) from British Columbia vineyards. *Canadian Journal of Zoology* 67, 801–804.

Edward, J.C. and Misra, S.L. (1969) Occurrence of some new species of Aphelenchoidea in the rhizosphere of certain field crops of Uttar Pradesh, India, with a note on an intersex. *Allahabad Farmer* 43, 1–6.

Edward, J.C. and Misra, S.L. (1970) Two new species of *Trichodorus* from Uttar Pradesh, India. *Allahabad Farmer* 44, 167–171.

Edward, J.C. and Sharma, N.N. (1982) *Xiphinema uasi* sp. nov. (Nematoda: Longidoroidea) from India. *Journal of Soil Biology and Ecology* 2, 87–89.

Edward, J.C., Misra, S.L. and Singh, G.R. (1964a) *Longidorus pisi* n.sp. (Nematoda: Dorylaimoidea) associated with the rhizosphere of *Pisum sativum* from Uttar Pradesh, India. *Japanese Journal of Applied Entomology and Zoology* 8, 310–312.

Edward, J.C., Misra, S.L. and Singh, G.R. (1964b) A new species of *Paralongidorus* (Nematoda: Dorylaimoidea) from Allahabad, Uttar Pradesh, India. *Japanese Journal of Applied Entomology and Zoology* 8, 313–316.

Elmiligy, I. and Geraert, E. (1972) *Laimaphelenchus deconincki* n.sp. (Nematoda: Tylenchida). *Biologische Jaarboek* 39, 145–149.

Eroshenko, A.S. (1966) [Three new nematode species of the genus *Paraphelenchus* (Micoletzky, 1922), Micoletzky, 1925 (Nematoda: Aphelenchidae)]. *Zoologicheskii Zhurnal* 45, 1873–1876 (in Russian).

Eroshenko, A.S. (1967a) [The nematode fauna of oats in the Primorsk territory]. *Materialy nauchoi Konferenskii Vsesoyuznogo Obshchestva Gel'mintologov, Year 1966, Part* 5, pp.159–160 (in Russian).

Eroshenko, A.S. (1967b) [Three new species of *Aphelenchoides* (Nematoda: Aphelenchoididae)]. *Zoologicheskii Zhurnal* 46, 617–620 (in Russian).

Eroshenko, A.S. (1968a) [Five new species of *Aphelenchoides* Fischer, 1894 (Nematoda: Aphelenchoididae)]. *Soobshch. dal'nevost. Fil. V. L. Komarova sib. Otdel. Akad. Nauk SSSR*, No.26, pp.58–66 (in Russian).

Eroshenko, A.S. (1968b) [Three new species of *Aphelenchoides* (Nematoda: Aphelenchoididae)]. In: [*Parasites of Animals and Plants*]. Moscow: 'Nauka', No.4, pp.224–228 (in Russian).

Eroshenko, A.S., and Teplyakov, A.A. (1975) [A new species of ectoparasitic nematode in *Trichodorus* from coniferous forest in the Far East]. *Parazitologiya* 9, 545–547 (in Russian).

Esser, R.P. (1966) *Xiphinema macrostylum* n.sp. (Nematoda: Longidoridae). *Proceedings of the Helminthological Society of Washington* 33, 162–165.

Esser, R.P. (1969) *Rhadinaphelenchus cocophilus*, a potential threat to Florida palms. *Nematology Circular, Division of Plant Industry, Florida Department of Agriculture*, No. 9, 2pp.

Esser, R.P. (1971) A compendium of the genus *Trichodorus* (Dorylaimoidea: Diphtherophoridae). *Proceedings – Soil and Crop Science Society of Florida* 31, 244–253.

Esser, R.P. and Meredith, J.A. (1987) Red ring nematode. *Nematology Circular, Division of Plant Industry, Florida Department of Agriculture and Consumer Services*, No. 141, 4pp.

Evans, A.A.F. and Fisher, J.M. (1970) Some factors affecting the number and size of nematodes in populations of *Aphelenchus avenae*. *Nematologica* 16, 295–304.

Eveleigh, E.S. and Allen, W.R. (1982) Description of *Longidorus diadecturus* n.sp. (Nematoda: Longidoridae), a vector of the peach rosette mosaic virus in peach orchards in southwestern Ontario, Canada. *Canadian Journal of Zoology* 60, 112–115.

Feng, Z.X. and Li, S.M. (1986) A new species of the genus *Aphelenchoides*. *Acta Zootaxonomica Sinica* 11, 245–249.

Filipjev, I.N. (1921) Die freilebenden Nematoden des Gouv. Petrograd. *Izdanie Detskoselo Entom. Stantsii*, s. C (7), 11pp.

Filipjev, I.N. (1931) Report on freshwater Nematoda. Mr. Omer-Cooper's investigation of the Abyssinian freshwaters (Hugh Scott Expedition). *Proceedings of the Zoological Society of London*, part 2, 429–443.

Filipjev, I.N. (1934a) [*Harmful and Useful Nematodes in Rural Economy*]. Moskva, Leningrad, 440pp (in Russian).

Filipjev, I N. (1934b) The classification of the free-living nematodes and their relation to parasitic nematodes. *Smithsonian Miscellaneous Collection* 89, 1–63.

Filipjev, I.N. and Schuurmans Stekhoven, J.H. (1941) *A Manual of Agricultural Helminthology*. E.J. Brill, Leiden, Holland.

Fischer, M. (1894) Über eine Clematis-krankheit. *Bericht aus dem Physiolischen Laboratorium des Landwirthschaftlichen, Instituts der Universitat Halle*, (11) 3, 1–11.

Fisher, J.M. (1964) *Dolichodorus adelaidensis* n.sp. and *Paralongidorus eucalypti* n.sp. from S. Australia. *Nematologica* 10, 464–470.

Fisher, J.M. and Evans, A.A.F. (1967) Penetration and feeding by *Aphelenchus avenae*. *Nematologica* 13, 425–428.

Fisher, J.M. and Raski, D.J. (1967) Feeding of *Xiphinema index* and *X. diversicaudatum*. *Proceedings of the Helminthological Society of Washington* 34, 68–72.

Flegg, J.J.M. (1968) Life-cycle studies of some *Xiphinema* and *Longidorus* species in south-eastern England. *Nematologica* 14, 197–210.

Fortuner, R. (1970) On the morphology of *Aphelenchoides besseyi* Christie, 1942 and *A. siddiqii* n.sp. (Nematoda: Aphelenchoidea). *Journal of Helminthology* 44, 141–152.

Fortuner, R. (1984) List and status of the genera and families of plant-parasitic nematodes. *Helminthological Abstracts, Series B* 53, 81–133.

Franklin, M.T. (1950) Two species of *Aphelenchoides* associated with strawberry bud disease in Britain. *Annals of Applied Biology* 37, 1–10

Franklin, M.T. (1952) A disease of *Scabiosa caucasica* caused by the nematode *Aphelenchoides blastophthorus* n.sp. *Annals of Applied Biology* 39, 54–60.

Franklin, M.T. (1955) A redescription of *Aphelenchoides parietinus* (Bastian, 1865) Steiner, 1932. *Journal of Helminthology* 29, 65–76.

Franklin, M.T. (1957) *Aphelenchoides composticola* n.sp. and *A. saprophilus* n.sp. from mushroom compost and rotting plant tissues. *Nematologica* 2, 306–313.

Franklin, M.T. and Hooper, D.J. (1962) *Bursaphelenchus fungivorus* n.sp. (Nematoda: Aphelenchoidea) from rotting gardenia buds infected with *Botrytis cinerea* Purs. ex. Fr. *Nematologica* 8, 136–142.

Franklin, M.T. and Siddiqi, M.R. (1963) *Aphelenchoides trivialis* n.sp. from South India. *Nematologica* 9, 15–18.

Franklin, M.T. and Siddiqi, M.R. (1972) *Aphelenchoides besseyi*. In: *CIH Descriptions of Plant-parasitic Nematodes*. Set 1, No. 4, 3pp.

Fuchs, A.G. (1915) Die Naturgeschichte der Nematoden und einiger anderen Parasiten. I. Des *Ips typographus* L. 2. Des *Hylobius abietis* L.. *Zoologische Jahrbücher, Abteilung für Systematik, Öekologie und Geographie der Tiere* 38, 109–222.

Fuchs, A.G. (1929) Die Parasiten einiger Rüssell- und Borkenkäfer. *Zeitschrift für Parasitenkunde* 2, 248–285.

Fuchs, A.G. (1930) Neue an Borken- und Rüsselkäfer gebundene Nematoden, halbparasitische und Wohnungseinmieter. *Zoologische Jahrbücher, Abteilung für Systematik, Öekologie und Geographie der Tiere* 59, 505–646.

Fuchs, A.G. (1931) *Seinura* gen. nov. *Zoologischer Anzeiger* 94, 226–228.

Fuchs, A.G. (1937) Neue parasitische und halbparasitische Nematoden bei Borkenkäfern und einige andere Nematoden. I. Teil. *Zoologische Jahrbücher, Abteilung für Systematik, Öekologie und Geographie der Tiere* 70, 291–380.

Ganguly, S., Patil, K.N. and Khan, E. (1981) Two new species of *Siddiqia* from India. *Indian Journal of Nematology* 10, 175–181.

Gasparrini, G. (1864) Sulla maturazione e la qualita dei fichi dei contorni di Napoli. *Atti acad. Pontaniana* 9, 99–118.

Geraert, E. (1962) De nematodenfauna in en om de wortels van *Musa paradisiaca normalis*. In: *Bijdragen tot de Kennis der Plantenparasitaire en der Vrijlevende Nematoden van Kongo. II.* Instituut voor Dierkunde, Laboratorium voor Systematiek, Rijksuniversiteit, Gent. pp. 1–73.

Geraert, E. (1966) The systematic position of the families Tylenchulidae and Criconematidae. *Nematologica* 12, 362–368.

Geraert, E., Grootaert, P. and Decraemer, W. (1981) Structure of the female reproductive system in some Dorylaimida and Enoplida (Nematoda). *Nematologica* 26, 255–271.

Gerber, K. and Giblin-Davis, R.M. (1990) Association of the red ring nematode and other nematode species with the palm weevil, *Rhynchophorus palmarum*. *Journal of Nematology* 22, 143–149.

Gerber, K., Giblin-Davis, R.M., Griffith, R. and Escobar-Goyes, J. (1989) Morphometric comparisons of geographic and host isolates of the red ring nematode, *Rhadinaphelenchus cocophilus*. *Nematropica* 19, 151–159.

Germani, G. (1989) Description of *Xiphinema guillaumeti* sp.n. (Nematoda: Longidoridae) and measurements on six other species of *Xiphinema* from Brazilian Amazonia. *Nematologica* 35, 69–75.

Germani, G. (1990) Description of *Dolichodorus pellegrini* sp.n. (Nematoda: Dolichodoridae) and *Xiphinema fagesi* sp.n. (Nematoda: Dorylaimidae) from New Caledonia. *Nematologica* 36, 73–80.

Gibbs, A.J. and Harrison, B.D. (1964) Nematode-transmitted viruses in sugar beet in East Anglia. *Plant Pathology* 13, 144–150.

Giblin, R.M. (1982) Biology and taxonomy of nematode associates of soil-dwelling bees. 131pp. PhD Thesis, University of California, Davis, USA.

Giblin, R.M. (1985) Association of *Bursaphelenchus* sp. (Nematoda: Aphelenchoididae) with nitidulid beetles (Coleoptera: Nitidulidae). *Revue de Nématologie* 8, 369–375.

Giblin, R.M. and Kaya, H.K. (1981) Association of the nematodes *Huntaphelenchoides* sp. (Aphelenchoididae) and *Acrostichus* sp. (Diplogasteridae) with the semisocial soil dwelling bee, *Halictus farinosus* (Halictidae: Hymenoptera). *Journal of Nematology* 13, 439–440.

Giblin, R.M. and Kaya, H.K. (1983a) *Bursaphelenchus seani* n.sp. (Nematoda: Aphelenchoididae) a phoretic associate of *Anthophora bomboides stanfordiana* Cockerell, 1904 (Hymenoptera: Anthophoridae). *Revue de Nématologie* 6, 39–50.

Giblin, R.M. and Kaya, H.K. (1983b) Host, temperature and media additive effects on the growth of *Bursaphelenchus seani*. *Revue de Nématologie* 7, 13–17.

Giblin, R.M. and Kaya, H.K. (1983c) Isolation of an unusual entaphelenchid (Tylenchida: Entaphelenchidae) from the hemocoel of *Platysoma punctigerum* Le Conte (Coleoptera: Histeridae). *Journal of Nematology* 15, 650–651.

Giblin, R.M. and Kaya, H.K. (1984) Associations of halictid bees with the nematodes *Aduncospiculum halicti* (Diplogasterida: Diplogasteroididae) and *Bursaphelenchus kevini* (Aphelenchida: Aphelenchoididae). *Journal of the Kansas Entomological Society* 57, 92–99.

Giblin, R.M., Kaya, H.K. and Brooks, R.W. (1981) Occurrence of *Huntaphelenchoides* sp. (Aphelenchoididae) and *Acrostichus* sp. (Diplogasteridae) in the reproductive tracts of soil nesting bees (Hymenoptera: Apoidea). *Nematologica* 27, 20–27.

Giblin, R.M., Powers, T.M. and Platzer, E.G. (1984a) *Bursaphelenchus* sp. and *Sheraphelenchus entomophagus* (Aphelenchoididae), phoretic associates of *Urophorus humeralis* (Coleoptera: Nitidulidae). *Proceedings of the 1st International Congress of Nematology, Guelph, Canada*, 1984, 31 (abstract).

Giblin, R.M., Swan, J.L. and Kaya, H.K. (1984b) *Bursaphelenchus kevini* n.sp. (Aphelenchida: Aphelenchoididae), an associate of bees in the genus *Halictus* (Hymenoptera: Halictidae). *Revue de Nématologie* 7, 177–187.

Giblin-Davis, R.M., Mundo-Ocampo, M., Baldwin, J.G., Gerber, K. and Griffith, R. (1989) Observations on the morphology of the red ring nematode, *Rhadinaphelenchus cocophilus* (Nemata: Aphelenchoididae). *Revue de Nématologie*, 12, 285–292.

Giudice, V.L. and Lanza, G. (1973) Presenza di *Sheraphelenchus entomophagus* su frutti di aranchio. *Annali dell'Istituto Sperimentale per l'Agrumicoltura* 6, 347–349.

Goffart, H. (1930) *Die Aphelenchiden der Kulturpflanzen*. Berlin, 105pp.

Gomes Carneiro, R.M.D., Carneiro, R.G. and Monteiro, A.R. (1982) Vertical distribution of four sugarcane nematodes in relation to soil factors. In: Lordello, L.G.E. (ed.), *Trabalhos Apresentados a VI reunio Brasileira de Nematologia, Fortaleza*. Publicacao No. 6, pp.117–132 (in Portuguese).

Goodey, J.B. (1958) *Paraphelenchus myceliophthorus* n.sp. (Nematoda: Aphelenchidae). *Nematologica*, 3, 1–5.

Goodey, J.B. (1960a) *Rhadinaphelenchus cocophilus* (Cobb, 1919) n.comb., the nematode associated with 'red-ring' disease of coconut. *Nematologica* 5, 98–102.

Goodey, J.B. (1960b) The classification of the Aphelenchoidea Fuchs, 1937. *Nematologica*, 5, 111–126.

Goodey, J.B. (1960c) Observations on the effects of the parasitic nematodes *Ditylenchus myceliophagus*, *Aphelenchoides composticola* and *Paraphelenchus myceliophthorus* on the growth and cropping of mushrooms. *Annals of Applied Biology* 48, 655–664.

Goodey, J.B. (1963) Speculation on the identity of the parts of the tylenchid spear. *Nematologica* 9, 468–470.

Goodey, J.B. (1963) see also under Goodey, T. (1963).

Goodey, J.B. and Hooper, D.J. (1965) A neotype of *Aphelenchus avenae* Bastian, 1865 and the rejection of *Metaphelenchus* Steiner, 1943. *Nematologica* 11, 55–65.

Goodey, J.B., Peacock, F.C. and Pitcher, R.S. (1960) redescription of *Xiphinema diversicaudatum* (Micoletzky, 1923 & 1927) Thorne, 1939 and observations on its larval stages. *Nematologica* 5, 127–135.

Goodey, T. (1923) A review of the plant parasitic members of the genus *Aphelenchus*. *Journal of Holminthology* 1, 113 156.

Goodey, T. (1927) On the nematode genus *Aphelenchus*. *Journal of Helminthology* 5, 203–220.

Goodey, T. (1928) The species of the genus *Aphelenchus*. *Journal of Helminthology* 6, 121–160.

Goodey, T. (1933) *Plant Parasitic Nematodes and the Diseases they Cause*. E.P. Dutton & Co. Inc., New York.

Goodey, T. (1935) *Aphelenchoides hodsoni* n.sp., a nematode affecting narcissus bulbs and leaves. *Journal of Helminthology* 13, 167–172.

Goodey, T. (1936) A new dorylaimid nematode, *Xiphinema radicicola* n.sp. *Journal of Helminthology* 14, 69–72.

Goodey, T. (1943) On the status of *Aphelenchus agricola* Maupas, 1900, with remarks on the genus *Paraphelenchus*. *Journal of Helminthology* 21, 20–21.

Goodey, T. (1951) *Soil and Freshwater Nematodes*. Methuen and Co., Ltd, London, 390pp.

Goodey, T. (1953) On two new species of nematodes associated with leaf-blotch in

Evodia roxburghiana, an Indian evergreen tree. *Thapar Commemoration Volume*. University of Lucknow, Lucknow, India, pp.95–102.

Goodey, T., revised by J.B. Goodey (1963) *Soil and Freshwater Nematodes*. Methuen and Co., London, 2nd ed., 544pp.

Grewal, P.S., Siddiqi, M.R. and Atkey, P.T. (1992) *Aphelenchoides richardsoni* sp. nov. and *Seinura paynei* sp. nov. from mushrooms in the British Isles and *S. obscura* sp. nov. from India (Nematoda: Aphelenchina). *Afro-Asian Journal of Nematology* 1, 204–211.

Griffith, R. (1968) The mechanism of transmission of the red ring nematode. *Journal of the Agricultural Society of Trinidad* 68, 437–457.

Griffith, R. (1987) Red ring disease of coconut palm. *Plant Disease* 71, 193–196.

Griffith, R. and Koshy, P.K. (1990) Nematode parasites of coconut and other palms. In: Luc, M., Sikora, R.A. and Bridge, J. (eds), *Plant Parasitic Nematodes in Subtropical and Tropical Agriculture*, CAB International, Wallingford, UK, pp.363–381.

Gritsenko, V.P. (1971) [*Ditylenchus tenuidens* n.sp. and *Aphelenchoides curiolis* n.sp. (Nematoda: Tylenchidae and Aphelenchoididae) from Kirgizia]. *Zoologicheskii Zhurnal* 50, 1402–1405 (in Russian).

Grootaert, P. and Coomans, A. (1981) The formation of the anterior feeding apparatus in dorylaims. *Nematologica* 26, 406–413.

Grube, A.E. (1849) Ueber einige Anguillulen und die Entwicklung von *Gordius aquaticus*. *Archiv für Naturgeschichte* 15, 358–375.

Gurando, E.V. (1979) [The nematodes of *Blastophagus minor* Hartig]. *Vestnik Zoologii* 4, 28–33. (in Russian).

Hagley, E.A.C. (1962) The palm weevil *Rhynchophorus palmarum* L., a probable vector of red ring disease of coconuts. *Nature* 193, 499.

Hagley, E.A.C. (1963) The role of the palm weevil *Rhynchophorus palmarum* as a vector of red ring disease of coconuts. I. Results of preliminary investigations. *Journal of Economic Entomology* 56, 375–380.

Halbrendt, J.M. and Brown, D.J.F. (1992) Morphometric evidence for three juvenile stages in some species of Xiphinema *americanum* sensu lato. *Journal of Nematology* 24, 305–309.

Hansen, E., Buecher, E.J. and Evans, A.A.F. (1970) Axenic culture of *Aphelenchus avenae*. *Nematologica* 16, 328–329.

Haque, M.M. (1966) [Description of *Seinura variobulbosa* n.sp. (Nematoda: Aphelenchoididae)]. *Zoologicheskii Zhurnal* 45, 293–295 (in Russian).

Haque, M.M. (1967a) [*Paraphelenchoides capsuloplanus* n.g., n.sp. (Nematoda: Aphelenchoididae)]. *Zoologicheskii Zhurnal* 46, 1251–1253 (in Russian).

Haque, M.M. (1967b) [Description of the new species belonging to the genus *Paraphelenchus*, Micol., 1922 (Nematoda: Paraphelenchidae)]. *Zoologicheskii Zhurnal* 46, 1842–1846 (in Russian).

Haque, M.M. (1968) [*Aphelenchoides echinocaudatus* n.sp. (Nematoda: Aphelenchoididae)]. *Zoologicheskii Zhurnal* 47, 287–289 (in Russian).

Harris, K.F. (1990) *Advances in Disease Vector Research*, No. 6, Springer, New York and London, xviii + 363pp.

Harrison, B.D. (1961) Soil-borne viruses – tobacco rattle virus. *Report, Rothamsted Experimental Station*, 1960, p.118.

Harrison, B.D. (1966) Further studies on a British form of pea early-browning virus. *Annals of Applied Biology* 57, 121–129.

Harrison, B.D. (1967) Pea early-browning virus (PEBV). *Report, Rothamsted Experimental Station* 1966, p.115.

Harrison, B.D. and Cadman, C.H. (1959) Role of a dagger nematode (*Xiphinema* sp.) in outbreaks of plant disease caused by arabis mosaic virus. *Nature* 184, 1624–1626.

Harrison, B.D. and Robinson, D.J. (1981) Tobraviruses. In: Kurstak, E.K. (ed.), *Handbook of Plant Virus Infections*. Elsevier/North-Holland Biomedical Press, Amsterdam, pp.515–540.

Harrison, B.D. and Robinson, D.J. (1986) Tobraviruses. In: van Regenmortel, M.H.V. and Fraenkel-Conrat, H. (eds), *The Plant Viruses. 2. The Rod-shaped Viruses*. Plenum Press, London and New York, pp.339–369.

Harrison, B.D. and Winslow, R.D. (1961) Laboratory and field studies on the relation of *Arabis* mosaic virus to its nematode vector *Xiphinema diversicaudatum* (Micoletzky). *Annals of Applied Biology* 49, 621–633.

Harrison, B.D., Robertson, W.M. and Taylor, C.E. (1974) Specificity of retention and transmission of viruses by nematodes. *Journal of Nematology* 6, 155–164.

Hechler, H.C. (1963) Description, developmental biology, and feeding habits of *Seinura tenuicaudata* (de Man) J.B. Goodey, 1960 (Nematoda: Aphelenchoididae), a nematode predator. *Proceedings of the Helminthological Society of Washington* 30, 182–195.

Hechler, H.C. and Taylor, D.P. (1965) Taxonomy of the genus *Seinura* (Nematoda: Aphelenchoididae), with descriptions of *S. celeris* n.sp. and *S. steineri* n.sp. *Proceedings of the Helminthological Society of Washington* 32, 205–219.

Hewitt, W.B., Raski, D.J. and Goheen, A.C. (1958) Nematode vector of soil-borne fanleaf virus of grapevines. *Phytopathology* 48, 586–595.

Heyns, J. (1962) A report on South African nematodes of the families Longidoridae, Belondiridae and Alaimidae (Nematoda: Dorylaimoidea), with descriptions of three new species. *Nematologica* 8, 15–20.

Heyns, J. (1964) *Aphelenchoides helicus* n.sp. and *Ditylenchus equalis* n.sp., two new soil-inhabiting nematodes. *South African Journal of Agricultural Science* 7, 147–150.

Heyns, J. (1965a) Four new species of the genus *Xiphinema* (Nematoda: Dorylaimoidea) from South Africa. *Nematologica* 11, 87–99.

Heyns, J. (1965b) New species of the genera *Paralongidorus* and *Longidorus* (Nematoda: Dorylaimoidea) from South Africa. *South African Journal of Agricultural Science* 8, 863–874.

Heyns, J. (1966a) Studies on South African *Xiphinema* species, with descriptions of two new species displaying sexual dimorphism of the tail. *Nematologica* 12, 369–384.

Heyns, J. (1966b) Further studies on South African Longidoridae (Nematoda) *South African Journal of Agricultural Science* 9, 927–944.

Heyns, J. (1966c) *Paralongidorus capensis* n.sp. and *Longidorus belondiroides* n.sp. with a note on *L. taniwha* Clark, 1963 (Nematoda: Longidoridae). *Nematologica* 12, 568–574.

Heyns, J. (1969) *Longidorus cohni* n.sp. a nematode parasite of alfalfa and Rhodes grass in Israel. *Israel Journal of Agricultural Research* 19, 179–183.

Heyns, J. (1971a) *A Guide to the Plant and Soil Nematodes of South Africa*. Balkema, Cape Town, 233pp.

Heyns, J. (1971b) Three *Xiphinema* species from the Western Cape province (Nematoda: Longidoridae). *Phytophylactica* 3, 107–114.

Heyns, J. (1972) Observations on South African *Paralongidorus* species (Nematoda: Longidoridae) with descriptions of two new species. *Phytophylactica* 4, 127–132.

Heyns, J. (1974) The genus *Xiphinema* in South Africa. II. X. *elongatum*-group (Nematoda: Dorylaimida). *Phytophylactica* 6, 249–260.

Heyns, J. (1975) *Paralongidorus maximus*. In: *CIH Descriptions of Plant Parasitic Nematodes*, Set 5, No. 75, 4pp.

Heyns, J. (1976) The genus *Xiphinema* in South Africa, III. Two new species of the X.

hallei-group, with additional information on X. *vanderlindei* Heyns, 1962 and X. *dimorphicaudatum* Heyns, 1966 (Nematoda: Dorylaimida). *Phytophylactica* 8, 67–78.

Heyns, J. (1977) The genus *Xiphinema* in South Africa. IV. X. *krugi* Lordello, 1955, X. *mediterraneum* Martelli and Lamberti, 1967 and a new species of the X. *hallei*-group (Nematoda: Dorylaimida). *Phytophylactica* 9, 109–114.

Heyns, J. (1979) The genus *Xiphinema* in South Africa. V. X. *zulu* Heyns, 1965 and related species in the X. *hallei*-group (Nematoda: Dorylaimida). *Phytophylactica* 11, 13–22.

Heyns, J. (1986) The genus *Xiphinema* in southern Africa. XII. Description of *Xiphinema loteni* n.sp. from Natal (Nematoda: Dorylaimida). *Phytophylactica* 18, 21–25.

Heyns, J. (1991) The genus *Xiphinema* in southern Africa. XXII. X. *miekeae* spec. nov., with notes on X. *clavatum* Heyns, 1965, X. *setariae* Luc, 1958 and X. *insigne* Loos, 1949 (Nematoda: Longidoridae). *Phytophylactica* 23, 207–211.

Heyns, J. and Coomans, A. (1983) Three *Xiphinema* species from Papua New Guinea and the Solomon Islands (Nematoda: Longidoridae). *Nematologica* 29, (1982), 1–10.

Heyns, J. and Coomans, A. (1989) *Paralongidorus cebensis* n.sp. from the Transkei (Nematoda: Longidoridae). *Phytophylactica* 21, 91–94.

Heyns, J. and Jacobs, P.J.F. (1989) Comparative ultrastructure of the cuticle in species of *Longidorus, Longidoroides, Paralongidorus* and *Xiphinema* (Nematoda: Longidoridae). *Phytophylactica* 21, 265–270.

Heyns, J. and Spaull, V. W. (1979) The genus *Xiphinema* in South Africa. VI. *Xiphinema umobae* n.sp. from sugarcane in Natal (Nematoda: Dorylaimida). *Phytophylactica* 11, 133–135.

Heyns, J. and Vermeulen, W.J.J. (1982) The genus *Xiphinema* in South Africa. VII. Two new species from Natal, with more information on *Xiphinema umobae* Heyns and Spaull, 1979 (Nematoda: Dorylaimida). *Phytophylactica* 14, 123–129.

Heyns, J., Coomans, A., Hutsebaut, M. and Swart, A. (1987) *Longidorus fursti* n.sp. from South Africa with a discussion of its relationships (Nematoda: Longidoridae). *Revue de Nématologie* 10, 381–385.

Heyns, J., Swart, A. and Chaves, E. (1989) A SEM study of the head region and stylet of *Xiphidorus balcarceanus* (Nematoda: Longidoridae). *Phytophylactica* 21, 121–124.

Hirling, W. (1979) Deutsche Namen pflanzenparasitärer Nematoden. *Gesunde Pflanzen* 30, 50–53.

Hirling, W. (1982) Eine Ubersicht über die Arten der Nematodengattungen *Laimaphelenchus* Fuchs, 1937 und *Ruidosaphelenchus* Laumond und Carle, 1971, mit Bestimmungstabellen und dre neue *Laimaphelenchus* – Arte (Nematoda: Tylenchida). *Zeitschrift für Pflanzenkrankheiten (Pflanzenpathologie) und Pflanzenschutz* 89, 30–42.

Hirling, W. (1986) *Laimaphelenchus penardi* (Nematoda: Tylenchida) von Typenfundort und Beitrage zur Eidonomie, Biologie und Verbreitung dieses Nematodan und dreir nahe verwandter Arten. *L. sylvaticus* und *L. praepenardi* n.sp. *Zoologische Beiträge* 29, 349–375.

Hofmänner, B. and Menzel, R. (1915) Die freilebenden Nematoden der Schweiz. *Revue Suisse de Zoologie* 23, 109–243.

Hooper, D.J. (1958) *Aphelenchoides dactylocercus* n.sp. and *A. sacchari* n.sp. (Nematoda: Aphelenchoidea). *Nematologica* 3, 229–235.

Hooper, D.J. (1961) A redescription of *Longidorus elongatus* (de Man, 1876) Thorne and Swanger, 1936 (Nematoda: Dorylaimoidea), and description of five new species of *Longidorus* from Great Britain. *Nematologica* 6, 237–257.

Hooper, D.J. (1962) Three new species of *Trichodorus* (Nematoda: Dorylaimoidea) and observations on *T. minor* Colbran, 1956. *Nematologica*, 7, 273–280.

Hooper, D.J. (1963) *Trichodorus viruliferus* n.sp. (Nematoda: Dorylaimida). *Nematologica* 9, 200–204.

Hooper, D.J. (1966) *Longidorus profundorum* n.sp. (Nematoda: Dorylaimidae). *Nematologica* 11, 489–495.

Hooper, D.J. (1972) Two new species of *Trichodorus* (Nematoda: Dorylaimida) from England. *Nematologica* 18, 59–65.

Hooper, D.J. (1973) *Longidorus elongatus*. In: *CIH Descriptions of Plant-parasitic Nematodes*. Set 2, No. 30, 4pp.

Hooper, D.J. (1974) *Aphelenchus avenae*. In: *CIH Descriptions of Plant-parasitic Nematodes*. Set 4, No. 50, 4pp.

Hooper, D.J. (1975) Morphology of trichodorid nematodes. In: Lamberti, F., Taylor, C.E. and Seinhorst, J.W. (eds), *Nematode Vectors of Plant Viruses*, Plenum Press, London and New York, pp.91–101.

Hooper, D.J. (1976) *Trichodorus viruliferus*. In: *CIH Descriptions of Plant-parasitic Nematodes*, Set 6, No. 86, 3pp.

Hooper, D.J. (1977) *Paratrichodorus (Nanidorus) minor*. In: *CIH Descriptions of Plant-parasitic Nematodes*, Set 7, No. 103.

Hooper, D.J. (1985) *Longidorus laevicapitatus*. In: *CIH Descriptions of Plant-parasitic Nematodes*. Set 8, No. 117, 3pp.

Hooper, D.J. and Clarke, S.A. (1980) Scanning electron micrographs of the head region of some species of Aphelenchoidea (Aphelenchina: Nematoda). *Nematologica* 26, 47–56.

Hooper, D.J. and Clark, S.A. (1985) *Paraphelenchus myceliophthorus*. In: *CIH Descriptions of Plant-parasitic Nematodes*. Set 8, No. 115, 3pp.

Hooper, D.J. and Cooke, D.A. (1967) Some observations on *Anomyctus xenurus* Allen, 1940. *Nematologica* 13, 320–321.

Hooper, D.J. and Cowland, J.A. (1986) Fungal hosts for the chrysanthemum nematode, *Aphelenchoides ritzemabosi*. *Plant Pathology* 35, 128–129.

Hooper, D.J. and Myers, R.F. (1971) *Aphelenchoides rutgersi* n.sp. (Nematoda: Aphelenchoidea), description and morphometrics, with observations on *A. dactylocercus* Hooper, 1958 and *A. cibolensis* Riffle, 1970. *Nematologica* 17, 295–302.

Hooper, D.J., and Siddiqi, M.R. (1972) *Trichodorus primitivus*. In: *CIH Descriptions of Plant-parasitic Nematodes*, Set 1, No. 15, 3pp.

Hooper, D.J. and Stone, A.R. (1975) Investigations on the morphology and function of the caudal tubercles of *Laimaphelenchus penardi*. *Rothamsted Experimental Station Report, 1975, Part 1*, 193.

Hooper, D.J., Kuiper, K. and Loof, P.A.A. (1961) Observations on the identity of *Trichodorus teres* Hooper, 1962 and *T. flevensis* Kuiper and Loof, 1962. *Nematologica* 9, 646.

Hopper, B.E. and Cairns, E.J. (1959) *Taxonomic Keys to Plant, Soil and Aquatic Nematodes*. Mimeograph. Sth. Reg. Nematode Project, Ala. Polyt. Inst. Auburn, Ala., USA, 176pp.

Horner, C.H. and Jensen, H.J. (1954) Nematodes associated with mints in Oregon. *Plant Disease Reporter*, 38, 39–41.

Huang, C. S., Uesugi, C. H. and Raski, D. J. (1987) *Xiphinema clavicaudatum* sp. n. and X. *fluminense* sp. n. (Nemata: Longidoridae) from Brazil. *Revue de Nématologie* 10, 23–27.

Hunt, D.J. (1980a) *Acugutturus parasiticus* n.g., n.sp., a remarkable ectoparasitic aphelenchoid nematode from *Periplaneta americana* (L.) with proposal of Acugutturinae n.subf. *Systematic Parasitology* 1, 167–170.

Hunt, D.J. (1980b) *Xiphinema machoni* n.sp. (Nematoda: Longidoridae) from St. Lucia, West Indies. *Revue de Nématologie* 3, 71–74.

Hunt, D.J. (1982) Two new species of *Xiphinema* (Nematoda: Dorylaimida) from Colombia. *Systematic Parasitology* 4, 89–92.

Hunt, D.J. and Hague, N.G.M. (1974) A redescription of *Parasitaphelenchus oldhami* Rühm, 1956 (Nematoda: Aphelenchoididae) a parasite of two elm bark beetles: *Scolytus scolytus* and *S. multistriatus*, together with some notes on its biology. *Nematologica* 20, 174–180.

Hunt, D.J. and Hague, N.G.M. (1976) The bionomics of *Cryptaphelenchoides scolyti* n.comb., syn. *Ektaphelenchus scolyti* Rühm, 1956, (Nematoda: Aphelenchoididae) a nematode associate of *Scolytus scolytus* (Coleoptera: Scolytidae). *Nematologica* 22, 212–216.

Hunt, D.J. and Rahman, L. (1991) Two new species of *Paralongidorus* (Nematoda: Dorylaimida) from deepwater rice in Bangladesh. *Afro-Asian Journal of Nematology* 1, 83–85.

Hunt, D.J. and Siddiqi, M.R. (1977) *Longidorus edmundsi* n.sp. (Dorylaimida: Longidoridae) from seagrape in Windward Islands. *Nematropica* 7, 32–35.

Hunt, D.J. and Towle, A. (1979) Feeding studies on *Xiphinema vulgare* Tarjan, 1964 (Nematoda: Longidoridae). *Revue de Nématologie* 2, 37–40.

Hurd, P.H., Smith, R.F. and Durham, J.W. (1962) The fossiliferous amber of Chiapas, Mexico. *Ciencia*, 21, 107–113.

Husain, S.I. and Khan, A.M. (1965) *Seinura nagini* n.sp. (Nematoda: Aphelenchoididae) from north India. *Proceedings of the Helminthological Society of Washington* 32, 179–181.

Husain, S.I. and Khan, A.M. (1967) On the status of the genera of the superfamily Aphelenchoidea (Fuchs, 1937) Thorne, 1949 with the descriptions of six new species of nematodes from India. *Proceedings of the Helminthological Society of Washington* 34, 167–174.

Hutsebaut, M. (1989) The genus *Xiphinema* in South Africa. XXI. Description of *X. pongolense* sp.n. and a parthenogenetic population of *X. clavatum* Heyns, 1965 (Nematoda: Dorylaimida). *Phytophylactica* 21, 343–352.

Hutsebaut, M., Heyns, J. and Coomans, A. (1988) The genus *Xiphinema* in South Africa, XVI. New species related to *X. meridianum* (Nematoda: Dorylaimida). *Nematologica* 33 (1987), 127–148.

Hutsebaut, M., Heyns, J. and Coomans, A. (1989a) The genus *Xiphinema* in southern Africa. XVIII. A study of the *X. zulu* and *X. mampara* species complex (Nematoda: Dorylaimida). *Nematologica* 34, 373–400.

Hutsebaut, M., Heyns, J. and Coomans, A. (1989b) The genus *Xiphinema* in southern Africa. XIX. Description of two new and one known species (Nematoda: Dorylaimida). *Phytophylactica* 21, 1–11.

Hutsebaut, M., Heyns, J. and Coomans, A. (1989c) The genus *Xiphinema* in southern Africa. XX. Two species related to *X. flagellicaudatum* Luc, 1961 (Nematoda: Dorylaimida). *Phytophylactica* 21, 113–120.

Hutsebaut, M., Heyns, J. and Coomans, A. (1990) Electron microscopic study of *Xiphidorus balcarceanus* Chaves & Coomans, 1984 (Longidoridae: Dorylaimida). *Phytophylactica* 22, 63–68.

Ikeda, T. (1984) Integrated pest management of Japanese pine wilt disease. *European Journal of Forest Pathology* 14, 398–414.

Imamura, S. (1931) Nematodes in the paddy field, with notes on their population before and after irrigation. *Journal of the College of Agriculture, Imperial University of Tokyo* 11, 193–240.

Ishibashi, N. and Choi, D.-R. (1991) Biological control of soil pests by mixed application of entomopathogenic and fungivorous nematodes. *Journal of Nematology* 23, 175–181.

Jacobs, P.J.F. and Heyns, J. (1982a) *Siddiqia* species from sugar cane in Natal (Nematoda: Longidoridae). *Phytophylactica* 14, 169–178.

Jacobs, P.J.F. and Heyns, J. (1982b) *Longidoroides* species from sugar cane in Natal (Nematoda: Longidoridae). *Phytophylactica* 14, 179–193.

Jacobs, P.J.F. and Heyns, J. (1982c) *Longidorus* species from sugar cane in Natal (Nematoda: Longidoridae). *Phytophylactica* 14, 195–204.

Jacobs, P.J.F. and Heyns, J. (1987a) New and known species of *Longidoroides* and *Paralongidorus* from South Africa (Nematoda: Longidoridae). Phytophylactica 19, 7–14.

Jacobs, P.J.F. and Heyns, J. (1987b) Eight new and two known species of *Longidorus* from South Africa (Nematoda: Longidoridae). *Phytophylactica* 19, 15–33.

Jacobs, P.J.F. and Heyns, J. (1989) A new name for *Longidorus conicaudatus* Jacobs & Heyns 1987, homonym of *Longidorus conicaudatus* Khan, 1987. *Nematologica* 34, 116.

Jacobs, P.J.F., Heyns, J., Coomans, A. and Loots, G.C. (1987) Studies on the ultrastructure of the pharyngeal region in Longidoridae (Nematoda). *Nematologica* 33, 199–212.

Jain, V.K. and Singh, S.P. (1984) Two new species of *Aphelenchoides* (Nematoda: Aphelenchoididae) from India. *Bulletin of Entomology* 25, 135–142.

Jairajpuri, M.S. (1968) Some studies on the morphology and biology of *Aphelenchus avenae* Bastian, 1865 (Nematoda: Aphelenchoidea). *Labdev Journal of Science and Technology* 6B, 27–34.

Jairajpuri, M.S. (1982a) The systematic position of *Californidorus* Robbins & Weiner, 1978 (Nematoda: Dorylaimida). *Systematic Parasitology* 4, 135–137.

Jairajpuri, M.S. (1982b) Three new species of nematodes from sugarcane fields in Nigeria. *Revue de Nématologie* 5, 241–246.

Jairajpuri, M.S. and Ahmad, W. (1992) *Dorylaimida. Free-living, Predaceous and Plant-parasitic Nematodes.* Oxford and IBH Publishing Co., New Delhi, xv + 458pp.

Jairajpuri, M.S. and Lamberti, F. (1981) *Xiphinema bajaji* nom. nov. for *Xiphinema luci* Bajaj and Jairajpuri, 1979 (Nematoda: Longidoridae). *Nematologia Mediterranea* 8 (1980), 211.

Jansson, H.B. (1986) Some possible fossil nematophagous fungi. *Transactions of the British Mycological Society* 87, 471–474.

Javed, R. (1983) Two new species of the subfamily Longidoroidea (Dorylaimida: Nematoda) from Maharashtra. *Indian Journal of Nematology* 13, 26–31.

Jensen, H.J. (1963) *Trichodorus allius*, a new species of stubby-root nematode from Oregon (Nemata: Dorylaimoidea). *Proceedings of the Helminthological Society of Washington* 30, 157–159.

Jha, A. and Posnette, A.F. (1959) Transmission of a virus to strawberry plants by a nematode. *Nature* 184, 962–963.

Johnson, A.W., Dowler, C.C. and Hauser, E.W. (1975) Crop rotation and herbicide effects on population densities of plant-parasitic nematodes. *Journal of Nematology* 7, 158–168.

Joubert, A. P., Kruger, J. C. de W. and Heyns, J. (1988) The genus *Xiphinema* in South Africa. XVII. A study of the X. *imitator* group with description of five new species (Nematoda: Dorylaimida). *Phytophylactica* 19, (1987), 419–432.

Kaiser, W.J., Bock, K.R., Guthrie, E.J. and Meredith, G. (1978) Occurrence of tomato black ring virus in potato cultivar Anett in Kenya. *Plant Disease Reporter* 62, 1088–1092.

Kakuliya, G.A. (1963a) [*Cryptaphelenchus quadridens* n.sp. from *Pityogenes quadridens* Hart.]. *Soobschcheniya Akademii Nauk Gruzinskoi SSR* 31, 433–438 (in Russian).

Kakuliya, G.A. (1963b) [*Tylaphelenchus paramonovi* n.sp.]. *Soobschcheniya Akademii Nauk Gruzinskoi SSR* 32, 649–654 (in Russian).

Kakuliya, G.A. (1966) [Nematode fauna of bark beetles from the Borzhomi-Bakurinsk Pass]. *Parazitologischeskii Sbornik, Tbilisi* 1, 48–64 (in Georgian).

Kakuliya, G.A. (1967) [New nematode genus, *Devibursaphelenchus* n.g. (Nematoda: Aphelenchoididae)]. *Soobschcheniya Akademii Nauk Gruzinskoi SSR* 47(2), 439–443 (in Russian).

Kakuliya, G.A. and Lazarevskaja, S.L. (1966) [*Ektaphelenchus piniperdae* nov. sp. (Tylenchida, Aphelenchoididae) – new nematode from *Blastophagus piniperda*]. *Proc. helminth. Lab. Acad. Sci. USSR* 15, 84–85 (in Russian).

Kakuliya, G.A., Devdariani, T.G. and Maglakelidze, K.L. (1980) [Nematodes of Cerambycidae, parasites of trees in Eastern Georgia]. In: *IX Konf. Ukr. Parazitol. Obshch. Tezisy Dokl. Chast'2. Kiev, USSR; 'Naukova Kumka'*, pp.109–110 (in Russian).

Kankina, V.K. and Ivanova, T.S. (1986) [A new species of nematode from the genus *Longidorus* Micoletzky, 1922 in southern Tajikistan]. *Izvestiya Akademii Nauk Tadzhikskoi SSR, Biologicheskie Nauk. No.* 2, 60–62 (in Russian).

Kankina, V.K. and Metlitskaya, K.V. (1983) [Parasitic nematodes of the family Longidoridae (Thorne, 1935) Meyl, 1961 on currant and raspberry]. *Izvestiya Akademii Nauk Tadzhikskoi SSR, Biologicheskie Nauk.* No. 4 (93), 50–52 (in Russian).

Karimova, S.M. (1957a) [Eelworms of crops on the left bank of the Amu-Darya basin]. In: Zemlyanskaya, A.I., Tikhonova, L. V. and Karimova, I.S., (eds), [*Eelworms of Agricultural Crops in the Uzbek SSR*] Tashkent, Akademiya Nauk Uzbekskol SSR, pp.133–208 (in Russian).

Karimova, S.M. (1957b) [Plant nematodes on the left bank of the lower Amu-Darya.]. In: *Parazitich. kruglye chervi-nematody Uzb., Tashkent* (in Russian).

Kazachenko, I.P. (1980) [New species of nematodes (Tylenchida) and a description of the male *Teratocephalus sigillarius* (Rhabditida) from forests in the Far East]. *Zoologicheskii Zhurnal* 59, 810–817 (in Russian).

Khan, E. (1964) *Longidorus afzali* n.sp. and *Xiphinema arcum* n.sp. (Nematoda: Longidoridae) from India. *Nematologica* 10, 313–318.

Khan, E. (1982) *Inagreius gloriosus* gen.n., sp.n. and descriptions of three new species of *Xiphinema* Cobb, 1913 along with report on X. *radicicola* T. Goodey, 1936 and X. *elongatum* Schuurmans Stekhoven and Teunissen, 1938 (Nematoda: Longidoroidea) from India. *Indian Journal of Nematology* 11, (1981), 189–204.

Khan, E. (1987a) Two new species of *Inagrius* [sic. *Inagreius*] Khan, 1982 (Nematoda: Longidoroidea) from India. *Indian Journal of Nematology* 16, 163–167.

Khan, E. (1987b) One new genus and four new species in the superfamily Longidoroidea (Nematoda). *Indian Journal of Nematology* 16, 185–193.

Khan, E., Seshadri, A.R., Weischer, B. and Mathen, K. (1971) Five new nematode species associated with coconut in Kerala, India. *Indian Journal of Nematology* 1, 116–127.

Khan, E., Chawla, M.L. and Prasad, S.K. (1972a) *Paralongidorus similis* sp.nov. (Nematoda: Longidoridae) from Poona, India. *Bulletin of Entomology* 12, 52–54.

Khan, E., Chawla, M.L. and Seshadri, A.R. (1972b) Longidorus mirus sp. nov. (Nematoda: Longidoridae) from soil around the roots of maize from Delhi, India. *Bulletin of Entomology* 12, 113–117.

Khan, E., Saha, M. and Seshadri, A.R. (1972c) Plant parasitic nematodes from Kumaon Hills, India. I. Two new species of *Paralongidorus* (Nematoda: Longidoridae). *Nematologica* 18, 38–43.

Khan, E., Chawla, M. L. and Saha, M. (1978) Comments on the classification of the Longidoroidea (Nematoda) with description of three new species. *Indian Journal of Nematology* 6, 47–62.

Khan, E., Singh, M. and Singh, L.K. (1981) *Siddiqia seclipsi* sp. n. (Nematoda: Longidoroidea) from India with a key to the species of *Siddiqia* Khan, Chawla and Saha, 1978. *Indian Journal of Nematology* 10, 211–215.

Khan, F.A. and Khan, A.M. (1972) Studies on distribution and population of *Longidorus brevicaudatus, Xiphinema basiri* and *X. americanum* in Uttar Pradesh and Rajasthan with description of *Longidorus psidia* n.sp. (Nematoda: Dorylaimoidea). *Indian Phytopathology* 25, 269–274.

Khan, M.A. (1960) Description of two nematodes, *Ektaphelenchus macrostylus* n.sp., and *Laimaphelenchus ulmi* n.sp., with a key to the species of *Laimaphelenchus*. *Canadian Journal of Zoology* 38, 91–97.

Khan, S. H. and Ahmad, S. (1975) Longidoroidea (Thorne, 1935) n.rank (Nematoda: Dorylaimina) with the description of *Xiphinema neoamericanum* n.sp. from India and proposal of a new name for *X. americanum* sensu Carvalho (1956) non Cobb, 1913. *Nematologia Mediterranea* 3, 23–27.

Khan, S.H. and Ahmad, S. (1977) *Xiphinema inaequale* nom.nov. (syn, *X. neoamericanum* Khan et Ahmad, 1975). *Nematologia Mediterranea* 5, 93.

Khan, T.H., Jairajpuri, M.S. and Ahmad, W. (1990) Descriptions of new and known species of dorylaim nematodes. *Nematologica* 35, 419–437.

Khera, S. (1970) Nematodes from the banks of still and running waters. VIII. Order Tylenchida. *Proceedings of the Zoological Society, Calcutta* 23, 53–65.

Kinn, D.N. (1986) Survival of *Bursaphelenchus xylophilus* in wood chips. *Bulletin OEPP/EPPO Bulletin* 16, 461–464.

Kirjanova, E.S. (1951) [Nematody pochvy khlipkovogo polya i tseliny v Golodnoi stepi (Uzbekistan)]. *Trudy Zoologicheskogo Instituta Akademii Nauk SSSR* 9, 625–657.

Klingler, J. (1970) The reaction of *Aphelenchoides fragariae* to slit-like micro-openings and to stomatal diffusion gases. *Nematologica* 16, 417–422.

Kolodge, C., Radewald, J.D. and Shibuya, F. (1987) Revised host range and studies on the life cycle of *Longidorus africanus*. *Journal of Nematology* 19, 77–81.

Konicek, D.E. and Jensen, H.J. (1961) *Longidorus menthasolanus*, a new plant parasite from Oregon (Nematoda: Dorylaimoidea). *Proceedings of the Helminthological Society of Washington* 28, 216–218.

Korenchenko, E.A. (1980) [New species of nematodes from the family Aphelenchoididae, parasites of trunk pests of *Larix dahurica*]. *Zoologicheskii Zhurnal* 59, 1768–1780 (in Russian).

Korenchenko, E.A. (1987) [*Cryptaphelenchus diversispicularis* n.sp. (Tylenchida, Aphelenchoididae) – a new nematode of the bark beetle, *Ips subelongatus* (Coleoptera: Ipidae)]. *Parazitologiya* 21, 73–78 (in Russian).

Körner, H. (1954) Die Nematodenfauna des vergehenden Holzes und ihre Beziehungen zu den Insekten. *Zoologische Jahrbücher, Abteilung für Systematik, Öekologie und Geographie der Tiere* 82, 245–353.

Kozlowska, J. and Seinhorst, J.W. (1979) *Longidorus elongatus* and closely related species in the Netherlands and lower Saxony (Germany), with the description of two new species, *L. cylindricaudatus* and *L. intermedius* (Nematoda: Dorylaimida) *Nematologica* 25, 42–53.

Krall, E.L. (1959) [New and little known tylenchs (Nematoda: Tylenchida) including a description of gynandromorphism in the genus *Aphelenchoides*]. *Eesti NSV Teaduste Akadeemia Toimetised. Biologiline Seer* 8, 190–198 (in Russian).

Kruger, J.C. de W. (1988) The uterine differentiation in *Xiphinema*. *Phytophylactica* 20, 233–251.

Kruger, J. C. de W. and Heyns, J. (1985) The genus *Xiphinema* in South Africa. IX. Description of *Xiphinema xenovariabile* n.sp. from Transvaal (Nematoda: Dorylaimida). *Phytophylactica* 16, 279–281.

Kruger, J. C. de W. and Heyns, J. (1985) The genus *Xiphinema* in southern Africa. XI. Description of *Xiphinema stockeri* n.sp. from Botswana (Nematoda: Dorylaimida). *South African Journal of Zoology* 20, 204–208.

Kruger, J. C. de W. and Heyns, J. (1986) The genus *Xiphinema* in South Africa. XIII. Description of *X. coomansi* and *X. magaliesmontanum* n.sp. from Transvaal (Nematoda: Dorylaimida). *Phytophylactica* 18, 71–79.

Kruger, J. C. de W. and Heyns, J. (1987) The genus *Xiphinema* in southern Africa. XIV. Description of two new species from Bourke's Luck, Eastern Transvaal and a redescription of *X. rarum* Heyns, 1979 (Nematoda: Dorylaimida). *Revue de Nématologie*, 10, 269–287.

Kruger, J.C. de W. and Heyns, J. (1990) Notes on the female reproductive system of *Xiphidorus* Monteiro, 1976 (Nematoda: Longidoridae). *Nematologia Mediterranea* 18, 47–51.

Kruger, J.C. de W., Kilian, S.M. and Heyns, J. (1992) The genus *Xiphinema* Cobb, 1913 in southern Africa. XXIII. *X. parasimplex* spec. nov., and notes on *X. machoni* Hunt, 1980 and *X. heynsi* Siddiqi, 1979 (Nematoda: Longidoridae). *Phytophylactica* 24 271–278.

Kuiper, K. (1977) Introductie en vestiging van planteparasitaire aaltjes in nieuwe polders, in het bijzonder van *Trichodorus teres*. Thesis, Wageningen School of Agriculture, No. 77–4, 140pp.

Kuiper, K. and Loof, P.A.A. (1962) *Trichodorus flevensis* n.sp. (Nematoda: Enoplida), a plant nematode from new polder soil. *Verslagen en Mededelingen, PlantenZiekten kundige Dienst Wageningen* 136, 193–200.

Kulinich, O.A. (1984) [*Aphelenchoides parasexlineatus* n.sp. (Nematoda: Aphelenchoididae) from conifer seedlings]. *Zoologicheskii Zhurnal* 63, 1570–1572 (in Russian).

Kumar, P. (1982) *Aphelenchoides loofi* sp. n. and *Aphelenchus siddiqii* sp. n. (Nematoda: Aphelenchoidea) from Lucknow, India. *Journal of Advanced Zoology* 3, 151–156.

Kumari, R.V. and Reddy, Y.N. (1984) Studies on the association of a new nematode species *Schistonchus hispida* sp. n. (Aphelenchoidea, Nickle 1971) and wasp. *Proceedings of the Indian Academy of Parasitology* 5, 21–25.

Kurashvili, B.E., Kakuliya, G.A. and Devdariani, Ts.G. (1980) [*Parasitic Nematodes of Bark Beetles of the Georgian SSR*]. Zool. Inst. Acad. Sci., Tbilisi, 170pp (in Russian).

Lal, A. and Mathur, V.K. (1986) Nematodes in the rhizosphere of date palm with description of *Longidoroides bikanerensis* sp.n. (Nematoda: Dorylaimoidea). *Date Palm Journal* 5, 208–214.

Lall, B.S. and Das, K.K. (1959) A preliminary note on the root knot nematodes (*Meloidogyne* spp.) affecting vegetable crops in Bihar. *Science and Culture* 25, 76–77.

Lamberti, F. (1969) Morphological variations and geographical distribution of *Longidorus africanus* Merny (Nematoda: Longidoridae). *Phytopathologia Mediterranea* 8, 137–141.

Lamberti, F. (1970a) A new character for species identification in the genus *Longidorus* (Micol.) Filipjev. *Proceedings of the Xth International Nematology Symposium, Pescara*, 1970, 18–19.

Lamberti, F. (1970b) *Longidorus africanus* Merny, un grave parassita di Barbabietola da zucchero e Lattuga e largamente diffuso nel Bacino del Mediterraneo. *Phytopathologia Mediterranea* 9, 176–179.

Lamberti, F. (1975) Laboratoio di nematologia agraria applicata ai vegetali, Bari. Attività scientifica svolta nel 1974. *Ricerca Scientifica (1975)* 45, 58–66 (in Italian).

Lamberti, F. and Bleve-Zacheo, T. (1977) Two new species of *Longidorus* (Nematoda: Longidoridae) from Italy. *Nematologia Mediterranea*, 5, 73–83.

Lamberti, F. and Bleve-Zacheo, T. (1979) Studies on *Xiphinema americanum* sensu *lato* with descriptions of fifteen new species (Nematoda: Longidoridae). *Nematologia Mediterranea* 7, 51–106.

Lamberti, F. and Carone, M. (1991) A dichotomous key for the identification of species of *Xiphinema* (Nematoda: Dorylaimida) within the X. *americanum*-group. *Nematologia Mediterranea* 19, 341–348.

Lamberti, F. and Golden, A.M. (1986) On the identity of *Xiphinema americanum* sensu lato in the nematode collection of Gerald Thorne with description of X. *thornei* sp.n. *Nematologia Mediterranea* 14, 163–171.

Lamberti, F. and Tarjan, A.C. (1974) *Xiphinema costaricense* n.sp. (Longidoridae, Nematoda), a new species of dagger nematode from Costa Rica. *Nematologia Mediterranea* 2, 1–11.

Lamberti, F., Bleve-Zacheo, T. and Arias, M. (1982) The Longidoridae of the Maltese Islands with the description of *Longidorus magnus* sp.n. and *Xiphinema melitense* sp.n. *Nematologia Mediterranea* 10, 183–200.

Lamberti, F., Choleva, B. and Agostinelli, A. (1983) Longidoridae from Bulgaria (Nematoda: Dorylaimida) with description of three new species of *Longidorus* and two new species of *Xiphinema*. *Nematologia Mediterranea* 11, 49–72.

Lamberti, F., Roca, F. and Chinappen, M. (1985) *Paralongidorus buchae*, a new Longidoridae species associated with declining chilli in Mauritius. *Nematologia Mediterranea* 13, 213–216.

Lamberti, F., Roca, F., Agostinelli, A. and Bleve-Zacheo, T. (1986) *Xiphinema barense* sp.n. (Nematoda: Dorylaimida) from Italy. *Nematologia Mediterranea* 14, 101–106.

Lamberti, F., Roca, F. and Agostinelli, A. (1989) *Xiphinema macroacanthum* (Nematoda: Dorylaimida) a new species from southern Italy closely resembling X. *ingens* Luc et Dalmasso. *Nematologia Mediterranea* 17, 115–119.

Lamberti, F., Castillo, P., Gomez-Barcina, A. and Agostinelli, A. (1992) Descriptions of six new species of *Xiphinema* (Nematoda, Dorylaimida) from the Mediterranean region. *Nematologia Mediterranea* 20, 125–139.

Lamberti, F., Cicancio, A., Agostinelli, A. and Coiro, M.I. (1992) Relationship between *Xiphinema brevicolle* and X. *diffusum* with a redescription of X. *brevicolle* and descriptions of three new species of *Xiphinema* (Nematoda: Dorylaimida). *Nematologia Mediterranea* 19, 311–326.

Lamberti, F., Coiro, M.I. and Agostinelli, A. (1992) *Longidorus aethesinus* sp.n. (Nematoda: Dorylaimida) from Northern Italy. *Nematologia Mediterranea* 19, 163–167.

Laumond, C. and Carle, P. (1971) Nématodes associés et parasites de *Blastophagus destruens* Woll. (Col. Scolytidae). *Entomophaga* 10, 51–60.

Lazarevskaya, S.L. (1961a) [New species from *Pissodes pini* L.]. *Trudi Gelmintologicheskoi Laboratorii Akademiya Nauk SSSR* 11, 144–152 (in Russian).

Lazarevskaya, S.L. (1961b) [On the nematode fauna of *Acanthocinus aedilis* (Cerambycidae)]. *Helminthologia* 3, 212–220 (in Russian).

Lazarevskaya, S.L. (1963) [Nematode fauna of *Orthotomicus laricis* and *O. proximus* (Coleoptera: Ipidae)]. *Helminthologia* 4, 254–265 (in Russian).

Lieutier, F. (1984) Ovarian and fat body protein concentrations in *Ips sexdentatus* (Coleoptera: Scolytidae) parasitised by nematodes. *Journal of Invertebrate Pathology* 43, 21–31.

Lieutier, F. (1985) Influence des nematodes parasites sur l'hemolymphe des adultes d'*Ips sexdentatus* Boern. (Coleoptera: Scolytidae). *Annales de Parasitologie Humaine et Comparée* 60, 195–204.

Lieutier, F. and Laumond, C. (1978) Nématodes parasites et associés à *Ips sexdentatus et Ips typographus* (Coleoptera, Scolytidae) en région parisienne. *Nematologica* 24, 184–200.

Lieutier, F., Jastrabsky, M. and Bonnafe, P. (1985) Modifications du spectre proteique de l'hemolymphe et des ovaires d'*Ips sexdentatus* (Coleoptera: Scolytidae) sous l'action des nematodes parasites. *Annales de Parasitologie Humaine et Comparée* 60, 715–725.

Lima, A.J. (1986) [Virus transmission to plants from nematodes]. *Nematologia-Brasileira* 10, 33–57 (in Portuguese).

Linde, W.J. Van der. (1938) A contribution to the study of nematodes. *Entomology Memoir, Union of South Africa* 2, 1–40.

Linford, M.B. (1937) The feeding of some hollow-stylet nematodes. *Proceedings of the Helminthological Society of Washington* 4, 41–46.

Linford, M.B. and Oliveira, J.M. (1937) The feeding of hollow-spear nematodes on other nematodes. *Science* 85, 295–297.

Linit, M.J. (1988) Nematode vector relationships in the pine wilt disease system. *Journal of Nematology* 20, 227–235.

Linit, M.J. (1990) Transmission of pinewood nematode through feeding wounds of *Monochamus carolinensis* (Coleoptera: Cerambycidae). *Journal of Nematology* 22, 231–236.

Lisetskaya, L.F. (1971) [*Aphelenchoides menthae* n.sp. (Nematoda: Aphelenchoididae)]. In: *Parazity zhivotnykh i rastenii. Kishinev, RIO Akademii Nauk Moldavskoi SSR* No.6, pp.123–126 (in Russian).

Liu, R. and Cheng, H. (1990) Occurrence of trichodorid species (Nemata: Trichodoridae) in China. *Journal of Nanjing Agricultural University*, 13, 50–54.

Loof, P.A.A. (1964) Free-living and plant parasitic nematodes from Venezuela. *Nematologica* 10, 201–300.

Loof, P.A.A. (1965) *Trichodorus anemones* n.sp. with a note on *T. teres* Hooper, 1962 (Nematoda: Enoplida). *Verslagen van den PlantenZiektenkundigen Dienst te Wageningen* 142, 132–136.

Loof, P.A.A. (1973) Taxonomy of the *Trichodorus-aequalis* complex (Diphtherophorina). *Nematologica* 19, 49–62.

Loof, P.A.A. (1975) Taxonomy of Trichodoridae. In: Lamberti, F., Taylor, C.E. and Seinhorst, J.W. (eds) *Nematode Vectors of Plant Viruses*. Plenum Press, London, and New York, pp.103–126.

Loof, P.A.A. (1983) Two new species of Longidoridae (Dorylaimida) from Saudi Arabia. *Nematologica* 28 (1982), 307–317.

Loof, P.A.A. and Luc, M. (1990) A revised polytomous key for the identification of species of the genus *Xiphinema* Cobb, 1913 (Nematoda: Longidoridae) with exclusion of the *X. americanum* group. *Systematic Parasitology* 16, 35–66.

Loof, P.A.A. and Maas, P.W.T. (1972) The genus *Xiphinema* (Dorylaimida) in Surinam. *Nematologica* 18, 92–119.

Loof, P.A.A. and Sharma, R.D. (1979) Plant parasitic nematodes from Bahia State, Brazil, The genus *Xiphinema* Cobb, 1913 (Dorylaimoidea). *Nematologica* 25, 111–127.

Loof, P.A.A. and Yassin, A.M. (1971) Three new plant-parasitic nematodes from Sudan, with notes on *Xiphinema basiri* Siddiqi, 1959. *Nematologica* 16 (1970), 537–546.

Loos, C.A. (1949) Notes on free-living and plant-parasitic nematodes from Ceylon, 5. *Journal of the Zoological Society of India* 1, 23–29.

Lordello, L.G.E. (1951a) *Xiphinema brasiliense* nova especie de nematoide do Brasil, parasita de *Solanum tuberosum* L. *Bragantia* 11, 87–90.

Lordello, L.G.E. (1951b) *Xiphinema campinense*, nova especie (Nematoda: Dorylaimidae). *Bragantia* 11, 313–316.

Lordello, L.G.E. (1955) *Xiphinema krugi* n.sp. (Nematoda: Dorylaimidae) from Brazil with a key to the species of *Xiphinema*. *Proceedings of the Helminthological Society of Washington* 22, 16–21.

Lordello, L.G.E. (1958) Parasitismo de *Aphelenchus avenae* em raizes de cantalupe (Nematoda: Aphelenchidae). *Revista Brasileira de Biologia* 18, 33–36.

Lordello, L.G.E. and Da Costa, C.P. (1961) A new nematode parasite of coffee roots in Brazil. *Revista Brazileira de Biologia* 21, 363–366.

Lordello, L.G.E. and Zamith, A.P.L. (1958) Nota sôbre o gênero *Trichodorus* Cobb, 1913 com descricao de *Trichodorus bucrius* n.sp. (Nematoda: Dorylaimoidea). *Ann. Acad. bras. Sci.* 30, 103–105.

Luc, M. (1958) *Xiphinema* de l'ouest africain: description de cinq nouvelles espèces (Nematoda: Dorylaimidae). *Nematologica* 3, 57–72.

Luc, M. (1961) *Xiphinema* de l'ouest africain (Nematoda: Dorylaimidae) Deuxième note. *Nematologica* 6, 107–122.

Luc, M. (1973) Redescription de *Xiphinema hallei* Luc, 1958 et description de six nouvelles espèces de *Xiphinema* Cobb, 1913 (Nematoda: Dorylaimoidea). *Cahiers de l'Office de la Recherche Scientifique et Technique Outre-Mer, série Biologie* No. 21, 45–65.

Luc, M. (1975) Trois nouvelles espèces de *Xiphinema* à queue arrondie-mucronée (Nematoda: Longidoridae). *Cahiers de l'Office de la Recherche Scientifique et Technique Outre-Mer, série Biologie* 10, 293–302.

Luc, M. (1981a) Observations on some *Xiphinema* species with the female anterior genital branch reduced or absent (Nematoda: Longidoridae). *Revue de Nématologie* 4, 157–167.

Luc, M. (1981b) Redescription of *Xiphinema yapoense* and description of male and juveniles of *X. douceti* (Nematoda: Longidoridae). *Nematologica* 27, 199–206.

Luc, M. and Aubert, V. (1985) *Xiphinema conurum* Siddiqi, 1964, a valid species (Nematoda: Longidoridae). *Revue de Nématologie* 8, 9–13.

Luc, M. and Baujard, P. (1983) *Xiphinema stenocephalum* n.sp. (Nematoda: Longidoridae) *Bulletin du Muséum National d'Histoire Naturelle de Paris, 4ème Série, 5 Section A*, 515–519.

Luc, M. and Coomans, A. (1988) *Xiphinema smoliki* n.sp. and *Longidorus pawneensis* n.sp. (Nemata: Longidoridae) from Colorado, USA. *Revue de Nématologie* 11, 137–142.

Luc, M. and Dalmasso, A. (1964) Trois nouveaux *Xiphinema* associés à la vigne (Nematoda: Dorylaimidae). *Nematologica* 9 (1963), 531–541.

Luc, M. and Dalmasso, A. (1975a) A 'lattice' for the identification of species of *Xiphinema* Cobb, 1913. In: Lamberti, F., Taylor, C. E. and Seinhorst, J. W. (eds) *Nematode Vectors of Plant Viruses*. Plenum Press, London and New York, pp.53–70.

Luc, M. and Dalmasso, A. (1975b) Considerations on the genus *Xiphinema* Cobb, 1913 (Nematoda: Longidoridae) and a 'lattice' for the identification of species. *Cahiers de l'Office de la Recherche Scientifique et Technique Outre-Mer, Biologie* 10, 303–327.

Luc, M. and Doucet, M.E. (1984) Description of *Xiphidorus achalae* n.sp. and proposal for a classification of longidorids (Nematoda: Dorylaimoidea). *Revue de Nématologie* 7, 103–112.

Luc, M. and Hunt, D.J. (1978) Redescription of *Xiphinema longicaudatum* Luc, 1961 and observations on *Xiphinema krugi* Lordello, 1955 (Nematoda: Longidoridae). *Nematologica* 24, 1–18.

Luc, M. and Kostadinov, A. (1982) *Xiphinema algeriense* n.sp. (Nematoda: Longidoridae). *Bulletin du Muséum National d'Histoire Naturelle de Paris 4ème Série, 3 Section A* (1981), 777–781.

Luc, M. and Southey, J.F. (1980) Study on biometrical variability in *Xiphinema insigne* Loos, 1949 and *X. elongatum* Schuurmans Stekhoven and Teunissen, 1938; description of X. *savanicola* n.sp. (Nematoda: Longidoridae) and comments on the thelytokous species. *Revue de Nématologie* 3, 243–269.

Luc, M. and Williams, J.R. (1978) *Xiphinema guirani* n.sp. and X. *silvaticum* n.sp. (Nematoda: Longidoridae). *Revue de Nématologie* 1, 87–97.

Luc, M., Lima, M.B., Weischer, B. and Flegg, J.J.M. (1964) *Xiphinema vuittenezi* n.sp. (Nematoda: Dorylaimidae). *Nematologica* 10, 151–163.

Luc, M., Brown, D.J.F. and Cohn, E. (1982) *Xiphinema israeliae* n.sp. (Nematoda: Dorylaimoidea). *Revue de Nématologie* 5, 233–239.

Luc, M., Loof, P.A.A. and Brown, D.J.F. (1985) On the systematics of eleven *Xiphinema* species (Nematoda: Longidoridae) described from India. *Revue de Nématologie* 7 (1984), 399–405.

Luc, M., Loof, P.A.A. and Coomans, A. (1986) Descriptions of *Xiphinema thorneanum* n.sp. and observations on some species of the genus (Nemata: Longidoridae). *Revue de Nématologie* 9, 337–346.

Luc, M., Maggenti, A.R., Fortuner, R., Raski, D.J. and Geraert, E. (1987) A reappraisal of Tylenchina (Nemata) 1. For a new approach to the taxonomy of Tylenchina. *Revue de Nématologie* 10, 127–134.

Macara, A.M. (1970) *Xiphinema amarantum* sp. nov. (Nematoda: Dorylaimidae). *Revista Ibérica de Parasitologia* 30, 649–658.

Macara, A.M. (1986) Two new species of *Longidorus* (Nematoda: Longidoridae) associated with forest plants in Portugal. *Nematologica* 31, 410–423.

McBeth, C.W. (1937) Observations on a predaceous nematode. *Proceedings of the Helminthological Society of Washington* 4, 18.

McElroy, F.D. and van Gundy, S.D. (1968) Observations on the feeding processes of *Hemicycliophora arenaria*. *Phytopathology* 58, 1558–1565.

McGuire, J.M., Kim, J.S. and Dou Thit, L.B. (1970) Tobacco ringspot in the nematode *Xiphinema americanum*. *Virology* 42, 212–216.

McLeod, R.W. and Khair, G.T. (1971) *Xiphinema australiae* n.sp., its host range, observations on X. *radicicola* Goodey, 1936 and X. *monohysterum* Brown, 1968 and a key to monodelphic *Xiphinema* spp. (Nematoda: Longidoridae). *Nematologica* 17, 58–68.

Maggenti, A.R. (1979) The role of cuticular strata nomenclature in the systematics of Nemata. *Journal of Nematology* 11, 94–98.

Maggenti, A.R. (1981) *General Nematology*. Springer-Verlag, New York, 372pp.

Maggenti, A.R. (1982) Nemata. In: Parker, S.P. (ed.), *Synopsis and Classification of Living Organisms*, McGraw-Hill, New York, pp.879–929.

Maggenti, A.R. (1983) Nematode higher classification as influenced by species and family concepts. In: Stone, A.R., Platt, H.M. and Khalil, L.F. (eds), *Concepts in Nematode Systematics*. The Systematics Association Special Volume No. 22, Academic Press, London and New York, pp.25–40.

Maggenti, A.R. (1991a) General nematode morphology. In: Nickle, W.R. (ed.), *Manual of Agricultural Nematology*. Marcel Decker, New York, pp.3–46.

Maggenti, A.R. (1991b) Nemata, Higher classification. In: Nickle, W.R. (ed.), *Manual of Agricultural Nematology*. Marcel Decker, New York, pp.147–187.

Maggenti, A.R., Luc, M., Raski, D.J., Fortuner, R. and Geraert, E. (1987) A reappraisal of Tylenchina (Nemata) 2. Classification of the suborder Tylenchina (Nemata: Diplogasteria). *Revue de Nématologie* 10, 135–142.

Magnusson, C. (1986) Potential for establishment of *Bursaphelenchus xylophilus* and the pine wilt disease under Nordic conditions. *Bulletin OEPP/EPPO Bulletin* 16, 465–471.

Malek, R.B. and Appleby, J.E. (1984) Epidemiology of pine wilt in Illinois. *Plant Disease* 68, 180–186.

Mali, V.R. and Hooper, D.J. (1974) Observations on *Longidorus euonymus* n.sp. and *Xiphinema vuittenezi* Luc *et al.*, 1964 (Nematoda: Dorylaimida) associated with spindle trees infected with euonymus mosaic virus in Czechoslovakia. *Nematologica* 19, 459–467.

Mamiya, Y. (1967) Descriptive notes on three species of *Trichodorus* (Dorylaimida: Trichodoridae) from forest nurseries in Japan. *Applied Entomology and Zoology* 2, 61–68.

Mamiya, Y. (1975) The life history of the pine wood nematode, *Bursaphelenchus lignicolus*. *Japanese Journal of Nematology* 5, 16–25.

Mamiya, Y. (1983) Pathology of pine wilt disease caused by *Bursaphelenchus xylophilus*. *Annual Review of Phytopathology* 21, 201–220.

Mamiya, Y. and Enda, N. (1979) *Bursaphelenchus mucronatus* n.sp. (Nematoda: Aphelenchoididae) from pinewood and its biology and pathogenicity to pine trees. *Nematologica* 25, 353–361.

Mamiya, Y. and Kiyohara, T. (1972) Description of *Bursaphelenchus lignicolus* n.sp. (Nematoda: Aphelenchoididae) from pine wood and histopathology of nematode-infested trees. *Nematologica* 18, 120–124.

Mankau, R. and Mankau, S.K. (1963) The role of mycophagous nematodes in the soil. I. The relationships of *Aphelenchus avenae* to phytopathogenic soil fungi. In: Doeksen, J. and Drift, J. van der, (eds), *Soil Organisms. Proceedings of Colloquium on Soil Fauna, Soil Microflora and their Relationships*, Oosterbeek, The Netherlands, pp.271–280.

Martelli, G.P. (1975) Some features of nematode-borne viruses and their relationship with the host plants. In: Lamberti, F., Taylor, C.E. and Seinhorst, J.W. (eds), *Nematode Vectors of Plant Viruses*, Plenum Press, London and New York, pp.223–252.

Martelli, G.P. and Lamberti, F. (1967) Le specie di *Xiphinema* Cobb, 1913 trovate in Italia a commenti sulla presenza di *Xiphinema americanum* Cobb. *Phytopathologia Mediterranea* 6, 65–85.

Martelli, G.P. and Taylor, C.E. (1990) Distribution of viruses and their nematode vectors. In: Harris, K.F. (ed.), *Advances in Disease Vector Research*, vol. 6, Springer, New York and London, pp.151–189.

Martelli, G.P., Cohn, E. and Dalmasso, A. (1966) A redescription of *Xiphinema italiae* Meyl, 1953 and its relationship to *Xiphinema arenarium* Luc and Dalmasso, 1963 and *Xiphinema conurum* Siddiqi, 1964. *Nematologica* 12, 183–194.

Marti, O.G., Rogers, C.E., Silvain, J.F. Simmons, A.M. (1990) Pathological effects of an ectoparasitic nematode *Noctuidonema guyanense* (Nematoda: Aphelenchoididae) on adults of the fall armyworm (Lepidoptera: Noctuidae). *Review of Agricultural Entomology* 83, 956–960.

Maslen, N.R. (1979) Six new nematode species from the maritime Antarctic. *Nematologica* 25, 288–308.

Massey, C.L. (1956) Nematode parasites and associates of the Englemann spruce beetle (*Dendroctonus engelmanni* Hopk). *Proceedings of the Helminthological Society of Washington* 23, 14–24.

Massey, C.L. (1960) Nematode parasites and associates of the California five-spined engraver, *Ips confusus* (Lec.). *Proceedings of the Helminthological Society of Washington* 27, 14–22.

Massey, C.L. (1964a) The nematode parasites and associates of the fir engraver beetle, *Scolytus ventralis* LeConte, in New Mexico. *Journal of Insect Pathology* 6, 133–155.

Massey, C.L. (1964b) Two new species of the nematode genus *Ektaphelenchus* (Nematoda: Aphelenchoididae), parasites of bark beetles in the southwestern United States. *Proceedings of the Helminthological Society of Washington* 31, 37–40.

Massey, C.L. (1966) The nematode parasites and associates of Dendroctonus *adjunctus* (Coleoptera: Scolytidae) in New Mexico. *Annals of the Entomological Society of America* 59, 424–440.

Massey, C.L. (1971a) Nematode associates of several species of *Pissodes* (Coleoptera: Curculionidae) in the United States. *Annals of the Entomological Society of America* 64, 162–169.

Massey, C.L. (1971b) *Omemeea maxbassiensis* n.gen., n.sp. (Nematoda: Aphelenchoididae) from galleries of the bark beetle *Lepersinus californicus* Sw. (Coleoptera: Scolytidae) in North Dakota. *Journal of Nematology* 3, 289–291.

Massey, C.L. (1974) *Biology and Taxonomy of Nematode Parasites and Associates of Bark Beetles in the United States.* Agriculture Handbook No. 446, USDA, Forest Service, Washington, v + 233pp.

Massey, C.L. and Hinds, T.E. (1970) Nematodes from aspen cankers in Colorado and New Mexico. *Canadian Journal of Zoology* 48, 97–108.

Maupas, E.F. (1900) Modes et formes de reproduction des nématodes. *Archives de Zoologie expérimentale et générale* 3, 463–624.

Merny, G. (1966) Nematode d'Afrique tropicale, un nouveau *Paratylenchus* (Criconematidae), deux nouveaux *Longidorus* et observations sur *Longidorus laevicapitatus* Williams, 1959 (Dorylaimidae). *Nematologica* 12, 385–395.

Merzheevskaja, O.I. (1951) Novye vidy nematod. *Sbornik Nauchnykh Trudov Akademii Nauk Belorussk. SSR, Inst. Biol.* 2, 112–120.

Meyl, A. H. (1953) Beiträge zur Kenntnis der Nematodenfauna vulkanisch erhitzter Biotope. I. Mitt., Die terrikolen Nematoden im Bereich von Fumarolen auf der Insel Ischia. *Zeitschrift für Morphologie und Oekologie der Tiere* 42, 67–116.

Meyl, A.H. (1954) Die bisher in Italien gefundenen freilebenden Erd- und Süsswasser-Nematoden. *Archivio Zoologico Italiano* 39, 161–264.

Meyl, A.H. (1961) Die freilebenden Erd- und Süsswassernematoden (Fadenwurmer) In: *Die Tierwelt Mitteleuropas.* Quelle and Meyer, Leipzig, 164pp.

Micoletzky, H. (1913) Die freilebenden Süsswassernematoden der Ost-Alpen. 1. Teil der vorlaüfige Mitteilung, Die freilebenden Süsswassernematoden des Lunzer Seengebietes. *Sitzungsb. K. Akad. Wissensch., Wien, Math.-Naturw.-Kl., Abt. 1* 122, 111–122.

Micoletzky, H. (1922) Die freilebenden Erd-Nematoden mit besonderer Berucksichtigung der Steiermark und der Bukowina, zugleich mit einer Revision sämtlicher nicht mariner, freilebender Nematoden in Form von Genus - Beschreibunger und Bestimmungsschlüsseln. *Archiv für Naturgeschichte,* Berlin 87 (1921), 1–650.

Micoletzky, H. (1923) Freilebende Nematoden der Wolga mit besonderer Berücksichtigung der Umgebung von Saratow. *Raboty Volzhskoi Biologicheskoi Stantsii* 7, 3–29.

Micoletzky, H. (1925) Die Freilebenden Süsswasser- und Moornematoden Dänemarks. Nebst. Anhang, Ueber Amöbosporidien und andere Parasiten bei freilebenden Nematoden. *K. Danske Vidensk. Selsk. Skr. Naturv. og. Math. Afd.* 8, R. 10, 57–310.

Micoletzky, H. (1927) Neue und seltene freilebende Nematoden aus dem Wolgagebiet (Kama). *Zoologischer Anzeiger* 73, 113–123.

Minagawa, N. (1992) *Aphelenchoides involutus* n.sp. (Tylenchida: Aphelenchoididae) from Ishigaki Is., Japan. *Japanese Journal of Nematology* 22, 21–25.

Minton, N.A. and Baujard, P. (1990) Nematode parasites of peanut. In: Luc, M., Sikora, R.A. and Bridge, J. (eds), *Plant Parasitic Nematodes in Subtropical and Tropical Agriculture,* CAB International, Wallingford, UK, pp.285–320.

Monteiro, A. R. (1976) *Xiphidorus yepesara* n.gen. n.sp. (Nemata: Longidoridae) from Brazil. *Nematologia Mediterranea* 4, 1–5.

Monteiro, A.R., Lordello, L.G.E. and Nakasono, K. (1981) *Xiphidorus parthenus* n.sp. (Nemata: Longidoridae) from Brazil. *Revista de Agricultura* 56, 93–97.

Morin, J.P., Lucchini, F., De Araujo, J.C.A., Ferreira, J.M.S. and Fraga, L.S. (1986) *Rhynchophorus palmarum* control using traps made of oil palm cubes. *Oléagineaux* 41, 57–62.

Muchina, T.I. (1988) [Two new species of nematodes (Paraphelenchidae) with a description of anomalies in their reproductive system]. *Zoologicheskii Zhurnal* 67, 1240–1245 (in Russian).

Naeem, S.A. (1967) *Some Studies on the Biology and Taxonomy of the Genus* Paraphelenchus. Thesis for Diploma of Imperial College, London, 104pp.

Nasira, K., Firoza, K. and Maqbool, M.A. (1992) Description of *Xiphinema karachiense* sp.n. and morphometric data on *Enchodelus macrodorus* (de Man, 1880), Thorne, 1939 (Nematoda: Dorylaimida) from Pakistan. *Fundamental and Applied Nematology* 15, 421–426.

Navas, A. and Arias, M. (1986) On the distribution and ecology of *Xiphinema index* and *X. italiae* in Spain. *Nematologia Mediterranea* 14, 207–215.

Nesterov, P.I. (1973) [New species of plant nematodes from the arctic tundra of the USSR (Yamal Peninsula)]. *Izvestiya Akademii Nauk, Moldavia SSR. Ser. Biol. Khim. Nauk* 4, 68–70 (in Russian).

Nesterov, P.I. (1985) [Two new species of plant nematodes-*Merlinius salechardicus* n.sp. (Tylenchida: Tylenchorhynchidae) and *Aphelenchoides bimucronatus* n.sp. (Tylenchida: Aphelenchoididae) from the arctic zone of the Soviet Union]. *Ekologiya i prakticheskoe znachenie zoo- i fitoparaziticheskikh organizmov*, pp.55–58 (in Russian).

Nesterov, P.I. and Lisetskaya, L.F. (1965) [Nematode fauna of some soils in the Moldavian SSR]. In: *Parasites of animals and plants*. Kishinev, 'Kartya Moldovenyaske' 1, 3–16 (in Russian).

Nickle, W.R. (1970a) Description of Entaphelenchidae fam.n., *Roveaphelenchus jonesi* gen.n., sp.n., and *Sheraphelenchus entomophagus* gen.n., sp.n. (Nematoda: Aphelenchoidea). *Proceedings of the Helminthological Society of Washington* 37, 105–109.

Nickle, W.R. (1970b) A taxonomic review of the genera of the Aphelenchoidea (Fuchs, 1937) Thorne, 1949 (Nematoda: Tylenchida). *Journal of Nematology* 2, 375–392.

Nickle, W.R. (1971) Description of a plesiotype male for *Anomyctus xenurus* Allen, 1940 (Nematoda: Aphelenchoididae). *Proceedings of the Helminthological Society of Washington* 38, 135–136.

Nickle, W.R. and Hooper, D.J. (1991) The Aphelenchina: bud, leaf and insect nematodes. In: Nickle, W.R. (ed.), *Manual of Agricultural Nematology*. Marcel Decker, New York, pp.465–507.

Nickle, W.R., Golden, A.M., Mamiya, Y. and Wergin, W.P. (1981) On the taxonomy and morphology of the pine wood nematode, *Bursaphelenchus xylophilus* (Steiner & Buhrer, 1934) Nickle, 1970. *Journal of Nematology* 13, 385–392.

Noack, F. (1898) Die Pfahlwurzelfäule des Kaffees, eine Nematoden-Krankheit. *Zeitschrift für Pflanzenkrankheiten* 8, 137–142.

Noffsinger, E.M. (1984) Adenophorea. In: *Distribution of Plant-parasitic Nematode Species in North America*. Society of Nematologists, Hyattsville, Maryland, USA, pp.20–29.

Norton, D.C. and Hoffmann, J.K. (1975) *Longidorus breviannulatus* n.sp. (Nematoda: Longidoridae) associated with stunted corn in Iowa. *Journal of Nematology* 7, 168–171.

Norton, D.C., Donald, P., Kimpinski, J., Meyers, R.F., Noel, G.R., Noffsinger, E.M., Robbins, R.T., Schmitt, D.C., Sosa-Moss, C. and Vrain, T.C. (1984) *Distribution of*

Plant-parasitic Nematodes in North America. Society of Nematologists, Hyattsville, Maryland, USA, pp.1–19.

Nowell, W. (1923) Diseases of crop plants in the Lesser Antilles. *Circ. W. Indian Comm. (Lond.)*, 383pp.

OEPP/EPPO (1990) Specific quarantine requirements. *EPPO Technical Documents*, No. 1008.

Oerley, L. (1880) Az anguillulidák magánrajza. A kir. m. természettudom. társulat által a bugatdijjal jutalmazott palyamü. *Természet rajzi Füzetek* 4, 16–150.

Oostenbrink, M. (1963) Symptomen en aaltjespopulaties in 'red ring'–zieke cocos–en olie-palmen. *Mededelingen van de Landbouwhogeschool en de Opzoekingsstations van de Staat te Gent* 28, 663–671.

Orton Williams, K.J. (1984) *Xiphinema bacaniboia* n.sp. (Nematoda: Dorylaimida) from Fiji. *Systematic Parasitology* 6, 207–211.

Paesler, F. (1946) Beitrag zur Kenntnis der im Dünger lebenden Nematoden. *Osterreichische Zoologische Zeitschrift* 1, 87–128.

Paesler, F. (1957) Beschreibung einiger Nematoden aus Champignonbeeten. *Nematologica* 2, 314–328.

Paesler, F. (1959) Beitrag zur Nematodenfauna des Siebengebirges und des Rodderberges. *Decheniana*, Bonn, No.7, pp.69–89.

Paramonov, A.A. (1953) [A revision of the superfamily Aphelenchoidea Fuchs, 1937 (Nematoda: Tylenchata)]. *Papers on helminthology presented to Academician K.I. Skryabin for his 75th birthday*, Moskva, Izdat'stvo Akad. Nauk. SSSR, pp.488–496 (in Russian).

Paramonov, A.A. (1962) [*Principles of Phytonematology. Vol. I. The Origin of Nematodes. Ecological and Morphological Characteristics of Phytonematodes. General Principles of Taxonomy.*] Izdatelsko 'Nauka', Moscow, 480pp (in Russian).

Paramonov, A.A. (1964) [Fundamentals of phytohelminthology. Vol II. Taxonomy of phytonematodes]. *Doklady Akademii Nauk SSSR, Helminthol. Lab.*, Moscow, 466pp (in Russian).

Paramonov, A.A. (1970) [*Principles of Phytonematology. Vol. III. Taxonomy of Nematodes of the Superfamily Tylenchoidea*]. Moscow, Izdatelstvo 'Nauka,' 253pp (in Russian).

Pearse, A.A. (1936) *Zoological names. A list of phyla, classes and orders prepared for Section F, American Association for Advancement of Science*. Durham, N.C., USA, 24pp.

Pearse, A.A. (1942) *Introduction to Parasitology*. Springfield, Baltimore, USA, 357 pp.

Phelps, R.J. and Mokry, J.E. (1976) *Culicoides* species (Diptera: Ceratopogonidae) as potential vectors of filariasis in Rhodesia. *Journal of the Entomological Society of Southern Africa* 39, 201–206.

Phukan, P.N. and Sanwal, K.C. (1982) Taxonomic studies on six species of *Xiphinema* from Assam. *Journal of Research of the Assam Agricultural University* 3, 76–83.

Phukan, P.N. and Sanwal, K.C. (1983) *Siddiqia indicus* sp.n. (Nematoda: Longidoroidea) from Assam, India. *Indian Journal of Nematology* 12, 188–191.

Pietler, K. (1977) Zur populationsdynamik virusübertragender Nematoden der Gattungen *Longidorus*, *Xiphinema* and *Trichodorus* unter Berücksichtigung der Standortverhältnisse und des Witterungverlaufes. *Biol. Gesellsch. Deutsch. Demokr. Rep. Rostock*, 70–80.

Pitcher, R.S. (1967) The host-parasite relations and ecology of *Trichodorus viruliferus* on apple roots, as observed from an underground laboratory. *Nematologica* 13, 547–557.

Pitcher, R.S. and McNamara, D.G. (1970) The effect of nutrition and season of year on the reproduction of *Trichodorus viruliferus*. *Nematologica* 16, 99–106.

Pitcher, R.S., Siddiqi, M.R. and Brown, D.J.F. (1974) *Xiphinema diversicaudatum*. In: *CIH Descriptions of Plant-parasitic Nematodes*, Set 4, No. 60. 4pp.

Poinar, G.O. (1969) *Praecocilenchus rhaphidophorus* n.gen., n.sp. (Nematoda: Aphelenchoidea) parasitizing *Rhynchophorus bilineatus* (Montrouzier) (Coleoptera, Curculionidae) in New Britain. *Journal of Nematology* 1, 227–231.

Poinar, G.O. (1974) Insect immunity to parasitic nematodes. In: Cooper, E.L. (ed.), *Contempory Topics in Immunobiology*. Volume 4, Plenum Press, New York, pp.167–178.

Poinar, G.O. (1977a) *CIH Key to the Groups and Genera of Nematode Parasites of Invertebrates*. Commonwealth Agricultural Bureaux, Farnham Royal, UK , iv + 43pp.

Poinar, G.O. (1977b) Fossil nematodes from Mexican amber. *Nematologica* 23, 232–238

Popovici, I. (1989) A new species of *Paratrichodorus* (Nematoda: Diphtherophorina) from Romania. *Nematologica* 35, 455–460.

Prabha, M.J. (1973) Two species of *Longidorus* from Marathwada Region. *Nematologica* 19, 62–68.

Prota, U., Lamberti, F., Bleve-Zacheo, T. and Martelli, G.P. (1971) I Longidoridae (Nematoda: Dorylaimoidea) dei vigneti sardi. *Redia* 52, 601–618.

Quraishi, M.A. and Das, V.M. (1984) Two new species of the genus *Xiphinema* from grape vineyards of Hyderabad City (Andhra Pradesh). *Indian Journal of Nematology* 14, 18–21.

Radewald, J.D. and Raski, D.J. (1962) Studies on the host range and pathogenicity of *Xiphinema index*. *Phytopathology* 52, 748–749.

Rahm, G.F. (1925) Beitrag zur Kenntnis der Moostierwelt der preussischen Rheinlande. I. Systematisch beschreibender Teil. *Archiv für Naturgeschichte* (1924), Abt. A 90, 153–214.

Rahm, G.F. (1938) Freilebende und saprophytische Nematoden aus Insel Hainan (Mit besonderer Berüchsichtigung der bischer bekannt gewordenen Nematoden Nordchinas und Japans). *Annotationes Zoologicae Japonenses* 17, 646–667.

Rahman, M.F., Jairajpuri, M.S. and Ahmad, I. (1985) Two new species of the genus *Trichodorus* Cobb, 1913 (Nematoda: Dorylaimida) from India. *Revue de Nématologie* 8, 103–107.

Raina, R. (1966) *Longidorus reneyii* sp.nov. (Nematoda: Longidoridae) from Srinagar, Kashmir. *Indian Journal of Entomology* 28, 438–441.

Rao, P.N. and Reddy, Y.N. (1980) Description of a new nematode *Praecocilenchus ferruginophorus* n.sp. , from weevil pests (Coleoptera) of coconut palms in South India. *Rivista di Parassitologia* 41, 93–98.

Rashid, F., Coomans, A. and Sharma, R.D. (1986a) Longidoridae (Nematoda: Dorylaimida) from Bahia State, Brazil. *Nematologia Mediterranea* 14, 235–250.

Rashid, F., De Waele, D. and Coomans, A. (1986b) Trichodoridae (Nematoda) from Brazil. *Nematologica* 31, 289–320.

Rashid, F., Geraert, E. and Sharma, R.D. (1986c) *Seinura*, *Aphelenchoides* and *Aphelenchus* from Brazil (Nematoda: Aphelenchina). *Biologisch Jaarboek Dodonaea* 54, 30–45.

Raski, D.J. and Valenzuela, A. (1988) Descriptions of four new species of Criconematidae, male of *Ogma terrestris* (Tylenchida: Criconematoidea) and *Tylaphelenchus yamani* sp.n. (Aphelenchina: Aphelenchoididae). *Nematologica* 33, 149–166.

Raski, D.J., Jones, N.O. and Roggen, D.R. (1969) On the morphology and ultrastructure of the esophageal region of *Trichodorus allius* Jensen. *Proceedings of the Helminthological Society of Washington* 36, 106–118.

Raski, D.J., Maggenti, A.R. and Jones, N.O. (1973) Location of grapevine fanleaf and yellow mosaic virus particles in *Xiphinema index*. *Journal of Nematology* 5, 208–211.

Razjivin, A.A. and Penton, G. (1975) [A new species of the genus *Trichodorus* (Nematoda) from the rhizosphere of the sugar cane in Cuba]. *Zoologicheskii Zhurnal* 54, 1082–1083 (in Russian).

Razjivin, A.A., O'Relly, J.P. and Millan, J.P. (1973) Nuevas especies de Nematodos (Nematoda: Dorylaimidae y Hoplolaimidae) parasitos de la cana de azucar en Cuba. *Poeyana* 108, 1–12.

Reddy, Y.N. and Rao, P.N. (1986) *Schistonchus racemosa* sp.n., a nematode parasite of wasp (*Ceratosolen* sp.) associated with the fig, *Ficus racemosa* L. *Indian Journal of Nematology* 16, 135–137.

Remillet, M. and Silvain, J.F. (1988) *Noctuidonema guyanense* n.g., n.sp. (Nematoda: Aphelenchoididae) ectoparasite of noctuids of the genus *Spodoptera* (Lepidoptera: Noctuidae). *Revue de Nématologie* 11, 21–24.

Renubala, K., Gambhir, R.K. and Dhanachand, C. (1991) Three known and two new species of *Xiphinema* (Dorylaimida) from Manipur State. *Current Nematology* 2, 55–58.

Rhoades, H.L. (1965) Parasitism and pathogenicity of Trichodorus proximus to St. Augustine grass. *Plant Disease Reporter* 49, 259–262.

Rhoades, H.L. (1985) Comparison of fenamiphos and *Arthrobotrys amerospora* for controlling plant nematodes in central Florida. *Nematropica* 15, 1–7.

Riffle, J.W. (1970) *Aphelenchoides cibolensis* (Nematoda: Aphelenchoididae), a new mycophagous nematode species. *Proceedings of the Helminthological Society of Washington* 37, 78–80.

Riga, E., Beckenbach, K. and Webster, J.M. (1992) Taxonomic relationships of *Bursaphelenchus xylophilus* and *B. mucronatus* based on interspecific and intraspecific cross-hybridization and DNA analysis. *Fundamental and Applied Nematology* 15, 391–396.

Ritzema Bos, J. (1890) De bloemkoolziekte der aardbeien, veroorzaakt door *Aphelenchus fragariae* nov. spec. (Voorloopige mededeeling). *Maandblad Natuurwetensch.* 16, 107–117.

Ritzema Bos, J. (1891) Zwei neue Nematodenkrankheiten der Erdbeerpflanze. (Vorläufige Mitteilung.) *Zeitschrift für Pflanzenkrankheiten* 1, 1–16.

Ritzema Bos, J. (1892) *Aphelenchus olesistus* nov. spec., een nematode, schadelijk aan begonia's en varens (Voorloopige mededeeling). *Maandblad Natuurwetensch.* 17, 101–112.

Ritzema Bos, J. (1893) Neue Nematodenkrankheiten bei Topfpflanzen. *Zeitschrift für Pflanzenkrankheiten* 3, 69–82.

Robbins, R.T. (1978) Descriptions of females (emended), a male, and juveniles of *Paralongidorus microlaimus* (Nematoda: Longidoridae). *Journal of Nematology* 10, 28–34.

Robbins, R.T. (1985) Descriptions of three new species of *Californidorus* from Arkansas and the male of *C. pinguicaudatus* (Nematoda: Dorylaimida). *Revue de Nématologie* 8, 215–227.

Robbins, R.T. and Brown, D.J.F. (1991) Comments on the taxonomy, occurrence and distribution of Longidoridae (Nematoda) in North America. *Nematologica* 37, 395–419.

Robbins, R.T. and Weiner, A.C. (1978) *Californidorus pinguicaudatus* n.gen., n.sp. from California (Nematoda: Longidoridae). *Revue de Nématologie* 1, 189–195.

Robertson, W.M. (1975) A possible gustatory organ associated with the odontophore in *Longidorus leptocephalus* and *Xiphinema diversicaudatum*. *Nematologica* 21, 443–448.

Robertson, W.M. and Taylor, C.E. (1975a) The structure and musculature of the feeding apparatus in *Longidorus* and *Xiphinema*. In: Lamberti, F., Taylor, C.E. and Seinhorst, J.W. (eds), *Nematode Vectors of Plant Viruses.* Plenum Press, London and New York, pp.179–194.

Robertson, W.M. and Taylor, C.E. (1975b) The ultrastructure of the feeding apparatus in trichodorids. In: Lamberti, F., Taylor, C.E. and Seinhorst, J.W. (eds), *Nematode Vectors of Plant Viruses.* Plenum Press, London and New York, pp.195–202.

Robertson, W.M. and Wyss, U. (1979) Observations on the ultrastructure and function of the dorsal oesophageal gland cell in *Xiphinema index. Nematologica* 25, 391–396.

Robertson, W.M. and Wyss, U. (1983) Feeding processes of virus-transmitting nematodes. In: Harris, K.F. (ed.), *Current Topics in Vector Research, vol. 1.* Praeger Scientific, New York, pp.271–295.

Roca, F. (1991) *Xiphinema coronatum* sp.n. (Nematoda: Longidoridae) from Italy. *Revue de Nématologie* 14, 597–601.

Roca, F. and Agostinelli, A. (1986) *Xiphinema porosum* sp.n. (Nematoda: Dorylaimida) from Italy. *Nematologia Mediterranea* 14, 257–261.

Roca, F. and Arias, M. (1986) A new *Paratrichodorus* species (Nematoda: Trichodoridae) from Spain. *Nematologia Mediterranea* 14, 181–185.

Roca, F. and D'Errico, F.P. (1987) *Longidorus major*, a new Longidoridae species from Italy. *Nematologia Mediterranea*, 15, 59–63.

Roca, F. and Lamberti, F. (1981) *Longidorus fasciatus* sp.n. from Greece and Italy. *Nematologia Mediterranea* 9, 175–179.

Roca, F. and Lamberti, F. (1984) Trichodorids (Nematoda) from Italy. *Nematologia Mediterranea* 12, 95–118.

Roca, F. and Lamberti, F. (1988) *Xiphinema aequum* sp.n. (Nematoda: Dorylaimida) from Italy, with description of the male of *Longidorus eridanicus. Nematologia Mediterranea* 16, 87–91.

Roca, F. and Pereira, M.J. (1991) *Xiphinema abrantium* sp.n. (Nematoda: Longidoridae) from Portugal. *Revue de Nématologie* 14, 485–490.

Roca, F. and Pereira, M.J. (1992) *Xiphinema belmontense* sp.n. (Nematoda: Longidoridae) from Portugal. *Fundamental and Applied Nematology* 15, 251–255.

Roca, F., Lamberti, F. and Agostinelli, A. (1984) Three new species of *Longidorus* (Nematoda: Dorylaimida) from Italy. *Nematologia Mediterranea* 12, 187–200.

Roca, F., Lamberti, F., Agostinelli, A. and Vinciguerra, M.T. (1986) *Longidorus aetnaeus*, a new Longidoridae species from Italy. *Nematologia Mediterranea* 14, 1–5.

Roca, F., Lamberti, F. and Agostinelli, A. (1988a) *Xiphinema fortuitum*, a new longidorid nematode from Italy. *Nematologia Mediterranea* 15, 219–223.

Roca, F., Pereira, M.J. and Lamberti, F. (1988b) Two new *Xiphinema* species from Portugal (Nematoda: Dorylaimida). *Nematologia Mediterranea* 15 (1987), 323–332.

Roca, F., Lamberti, F., Santos, M.S.N. de A. and Abrantes, I.M. de O. (1989a) *Xiphinema diversum* and *X. exile* (Nematoda: Dorylaimida), two new species from Portugal. *Nematologia Mediterranea* 16, 213–219.

Roca, F., Pereira, M.J. and Lamberti, F. (1989b) *Longidorus alvegus* sp.n. (Nematoda: Dorylaimida) from Portugal. *Nematologia Mediterranea* 17, 1–4.

Rodriguez-M., R., and Bell, A.H. (1978a) External morphology of the spicules of some Trichodoridae. *Journal of Nematology* 10, 127–132.

Rodriguez-M., R., and Bell, A.H. (1978b) Three new species of Trichodoridae (Nematoda: Diphtherophorina) with observations on the vulva in *Paratrichodorus. Journal of Nematology* 10, 132–141.

Rodriguez-M., R., Sher, S.A. and Siddiqi, M.R. (1978) Systematics of the monodelphic species of Trichodoridae (Nematoda: Diphtherophorina) with descriptions of a new genus and four new species. *Journal of Nematology* 10, 141–152.

Rogers, C.E., Marti, O.G., Simmons, A.M. and Silvain, J.F. (1990a) Host range of *Noctuidonema guyanense* (Nematoda: Aphelenchoidide), an ectoparasite of moths in French Guiana. *Review of Agricultural Entomology* 19, 795–798.

Rogers, C.E., Simmons, A.M. and Marti, O.G. (1990b) Parasitism of Lepidoptera adults by *Noctuidonema guyanense* Remillet and Silvain (Nematoda: Aphelenchoididae) in southeastern United States. *Journal of Agricultural Entomology* 7, 241–245.

Roggen, D.R. (1975) Functional morphology of Dorylaimida. In: Lamberti, F., Taylor, C.E., Seinhorst, J.W. (eds), *Nematode Vectors of Plant Viruses*. London, UK and New York, USA; Plenum Press, pp.129–137.

Roggen, D.R. (1979) Functional morphology of the nematode pharynx. II. Spherical bulbs. *Nematologica* 25, 127–135.

Rohde, R.A. and Jenkins, W.R. (1957) Host range of a species of *Trichodorus* and its host-parasite relationships on tomato. *Phytopathology* 47, 295–298.

Romanenko, N.D. (1981) Obnaryzhenye novogo vida nematod *Xiphinema paramonovi* n.sp. (Nematoda: Longidoridae) na territorii Sovjetskogo Soyuza. Tezisy Dokladov, Pervaya Konferentsiya po Nematodam Rastenii, Nasekomykh, Pochvy i Vod, pp.68–69.

Romaniko, V.I. (1966) [Two new species of plant nematodes from wheat]. *Zoologicheskii Zhurnal* 45, 929–931 (in Russian).

Romaniko, V.I. (1968) [A new nematode species of the genus *Paraphelenchus* (Micoletzky, 1922) Micoletzky, 1925]. *Trudy Biologicheskogo Chelyabinskogo Gosudarstvennogo Pedagogicheskogo Instituta* pp.40–42 (in Russian).

Rössner, J. (1969) Populationsentwicklung pflanzenparasitarer Nematoden unter Koniferen. *Mitteilungen der Biologischen Bundesanstalt für Land- u. Forstwirtschaft* 136, 50–58.

Rössner, J. (1972) Einfluss der austrocknung des bodens auf wandernde wurzelnematoden. *Nematologica* 17, 127–144.

Roy, T.K. and Gupta, A.N. (1974) Comments on the taxonomic position of some species of the genus *Xiphinema* Cobb, 1913 (Nematoda: Dorylaimoidea) with the creation of a new subgenus. *Acta Morphologica Neerlando-Scandinavica* 12, 345–354.

Rühm, W. (1954) Einige neue, ipidenspezifische Nematodenarten. *Zoologischer Anzeiger* 153, 221–242.

Rühm, W. (1955) Über einige an holzbrütende Ipiden gebundene Nematodenarten. *Zoologischer Anzeiger* 155, 70–83.

Rühm, W. (1956) Die Nematoden der Ipiden. *Parasitologische Schriftenreihe* 6, 1–435.

Rühm, W. (1957) *Aphelenchoides sinodendroni* n.sp. und *Ectaphelenchus zwölferi* n.sp., zwei neue, mit *Sinodendron cylindricum* L. vergesellschafttete Nematodenarten. *Zoologischer Anzeiger* 158, 72–82.

Rühm, W. (1960) Ein Beitrag zur Nomenklatur und Systematik einige mit Scolytiden vergesellschafteter Nematodenarten. *Zoologischer Anzeiger* 164, 201–213.

Rühm, W. (1964) Ein Beitrag zur Vergesellschaftung zwischen Nematoden und Insekten [*Pelodera bakeri* n.sp. (Nematoda: Rhabditoidea, Rhabditidae) eine mit *Calvertius tuberosus* Perm. et. Germ. (Coleoptera, Curculionidae, Hylobiinae) vergesellschaftete Nematodenart an *Araucaria araucana* (Mol.) Koch]. *Zoologischer Anzeiger* 173, 212–220.

Rühm, W. (1965) Brutbiologie und Morphologie einer Scolytidenart als Voraussetzung einer neuartigen Spezialisierung zweier Nematodenarten. *Zeitschrift für Angewandte Entomologie* 55, 264–275.

Rühm, W. and Chararas, C. (1957) Description, biologie et histologie de quatre espèces nouvelles de nematodes parasites de *Dryocoetes hectographus* Reit. (Col. Scolytidae). *Entomophaga* 2, 253–269.

Russell, C.C. and Perry, V.G. (1966) Parasitic habit of *Trichodorus christiei* on wheat. *Phytopathology* 56, 357–358.
Rutherford, T.A. and Webster, J.M. (1987) Distribution of pine wilt disease with respect to temperature in North America, Japan and Europe. *Canadian Journal of Forest Research* 17, 1050–1059.
Rutherford, T.A., Mamiya, Y. and Webster, J.M. (1990) Nematode induced pine wilt disease, factors influencing its occurrence and distribution. *Forest Science* 36, 145–155.
Saltukoglu, M.E., Geraert, E. and Coomans, A. (1976) Some Tylenchida from the Istanbul-area (Turkey). *Nematologia Mediterranea* 4, 139–153.
Samibaeva, K.X. (1966) [New species of nematodes in tobacco and adjacent soil in Urgutskom Province of the Samarkand territory]. *Samarkand Govt. Univ. Scient. Work* 156, 121–128 (in Russian).
Sänger, H.L. (1961) Untersuchungen über schwer übertragbare Formen des Rattlevirus. *Proc. 4th Conf. Potato Virus. Dis.*, Braunschweig, Germany, 1960, pp.22–28.
Sanwal, K.C. (1961a) A key to the species of the nematode genus *Aphelenchoides* Fischer, 1894. *Canadian Journal of Zoology* 39, 143–148.
Sanwal, K.C. (1961b) *Aphelenchoides subparietinus* n.sp. (*Nematoda: Aphelenchoididae*) from diseased lily bulbs. *Canadian Journal of Zoology* 39, 573–577.
Sanwal, K.C. (1965) Two new species of the genus *Aphelenchoides* Fischer, 1894 (Nematoda: Aphelenchoididae) from the Canadian Arctic. *Canadian Journal of Zoology* 43, 933–940.
Sato, H., Sakuyama, T. and Kobayashi, M. (1987) Transmission of *Bursaphelenchus xylophilus* (Steiner et Buhrer) Nickle (Nematoda: Aphelenchoididae) by *Monochamus saltuarius* (Gebler) (Coleoptera, Cerambycidae). *Journal of the Japanese Forestry Society* 69, 492–496.
Sauer, M.R. (1970) In: *Report, CSIRO Australia, Division of Horticultural Research* (1967–1969), p.98.
Saunders, J.N. and Norris, D.M. (1961) Nematode parasites and associates of the smaller European elm bark beetle, *Scolytus multistriatus* (Marsham). *Annals of the Entomological Society of America* 54, 792–798.
Saxena, P.K., Chabbra, H.K. and Joshi, R. (1973) *Xiphinema neoamericanum* sp.n. (Nematoda: Longidoridae). *Zoologischer Anzeiger* 191, 130–132.
Schindler, A.F. (1957) Parasitism and pathogenicity of *Xiphinema diversicaudatum*, an ectoparasitic nematode. *Nematologica* 2, 25–31.
Schneider, G. (1927) Dritter Beitrag zur Kenntnis der Brackwassernematoden Finlands. *Acta Societatis pro fauna et flora Fennica* 56, 1–24.
Schneider, W. (1935) Voyage de Ch. Alluaud et P.A. Chappuis en Afrique occidentale francaise (Déc. 1930–Avril 1931) VI, Freilebende Nematoden. *Archiv für Hydrobiologie* 28, 1–20.
Schuurmans Stekhoven, J.H. (1951) Nématodes saprozoaires et libres du Congo Belge. *Mémoires de l'Institut Royal des Sciences Naturelles de Belgique, 2me Série* 39, 1–79.
Schuurmans Stekhoven, J.H. and Teunissen, R.J.H. (1938) Nématodes libres terrestres. *Exploration du Parc national Albert, Mission de Witte* 22, 1–229.
Schwartz, M. (1911) Die Aphelenchen der Veilchengallen und Blattflecken an Farnen und Chrysanthemum. *Arb. kaiserl. biol. Anstalt Land. Forstwirtsch.* 8, 303–334.
Scognamiglio, A. (1974) *Aprutides guidettii* n.sp. (Nematoda: Aphelenchoididae). *Bolletino del Laboratorio di entomologia agraria 'Filippo Silvestri'* 31, 17–21.
Scognamiglio, A., Talame', M. and s'Jacob, J.J. (1970) *Aprutides martuccii* (Nematoda: Aphelenchoididae) n.g., n.sp. *Bolletino del Laboratorio di entomologia agraria 'Filippo Silvestri'* 28, 3–11.

Seinhorst, J.W. (1954) On *Trichodorus pachydermus* n.sp. (Nematoda: Enoplida). *Journal of Helminthology* 28, 111–114.

Seinhorst, J.W. (1955) Een eenvoudige methode voor het afschieden van aaltjes uit grond. *Tijdschriftover Plantenziekten* 61, 188–190.

Seinhorst, J.W. (1963) A redescription of the male of *Trichodorus primitivus* (de Man), and the description of a new species *T. similis*. *Nematologica* 9, 125–130.

Seth, A. and Sharma, N.K. (1986) Five new species of genus *Aphelenchoides* (Nematoda: Aphelenchida) infesting mushroom in northern India. *Indian Journal of Nematology* 16, 205–215.

Sharma, N.N. and Edward, J.C. (1985) Two new species of *Paralongidorus* (Nematoda: Longidoridae) from Maharashtra, India. *Journal of Soil Biology and Ecology* 5, 14–19.

Sharma, R.K. and Saxena, V. (1981) Two new species of the genus *Xiphinema* Cobb, 1913 (Dorylaimida: Nematoda). *Indian Journal of Parasitology* 5, 95–99.

Sharma, S.B. and Siddiqi, M.R. (1990) *Paralongidorus bullatus* n.sp. from groundnut soils in Niger and comments on *Xiphinema parasetariae* Luc. *Journal of Nematology* 22, 579–584.

Shavrov, G.N. (1967a) [New species of *Aphelenchus* Bastian, 1865 (Nematoda: Aphelenchidae)]. *Zoologicheskii Zhurnal* 46, 278–279 (in Russian).

Shavrov, G.N. (1967b) [Three new species of *Aphelenchoides* Fischer, 1894 (Nematoda: Aphelenchoididae)]. *Zoologicheskii Zhurnal* 46, 762–764 (in Russian).

Shavrov, G.N. (1968) [New species of *Paraphelenchus* (Micoletzky, 1922) Micoletzky, 1925 (Nematoda: Aphelenchinae)]. *Soobshch. dal'nevost. Fil. V. L. Komarova sib. Otdel. Akad. Nauk SSR* 26, 135–136 (in Russian).

Shepherd, A.M. and Clark, S.A. (1976a) Spermatogenesis and the ultrastructure of sperm and of the male reproductive tract of *Aphelenchoides blastophthorus* (Nematoda: Tylenchida, Aphelenchina). *Nematologica* 22, 1–9.

Shepherd, A.M. and Clark, S.A. (1976b) Structure of the anterior alimentary tract of the passively feeding nematode *Hexatylus viviparus* (Neotylenchidae: Tylenchida). *Nematologica* 22, 332–342.

Shepherd, A.M., Clark, S.A. and Hooper, D.J. (1980) Structure of the anterior alimentary tract of *Aphelenchoides blastophthorus* (Nematoda: Tylenchida, Aphelenchina). *Nematologica* 26, 313–357.

Shishida, Y. (1979) Studies on nematodes parasitic on woody plants. 1. Family Trichodoridae (Thorne, 1933) Clark, 1961. *Japanese Journal of Nematology* 9, 28–44.

Shishida, Y. (1983) Studies on nematodes parasitic on woody plants. 2. Genus *Xiphinema* Cobb, 1913. *Japanese Journal of Nematology* 12, 1–14.

Siddiqi, M.R. (1959) Studies on *Xiphinema* spp. (Nematoda: Dorylaimoidea) from Aligarh (North India), with comments on the genus Longidorus Micoletzky, 1922. *Proceedings of the Helminthological Society of Washington* 26, 151–163.

Siddiqi, M.R. (1960) Two new species of the genus *Trichodorus* (Nematoda: Dorylaimoidea) from India. *Proceedings of the Helminthological Society of Washington* 27, 22–27.

Siddiqi, M.R. (1961) On *Xiphinema opisthohysterum* n.sp. and *X. pratense* Loos, 1949, two dorylaimid nematodes attacking fruit trees in India. *Zeitschrift für Parasitenkunde* 20, 457–465.

Siddiqi, M.R. (1962a) Studies on the genus *Longidorus* Micoletzky, 1922 (Nematoda: Dorylaimoidea), with descriptions of three new species. *Proceedings of the Helminthological Society of Washington* 29, 177–188.

Siddiqi, M.R. (1962b) Longidorus *tarjani* n.sp. found around oak roots in Florida. *Nematologica* 8, 152–156.

Siddiqi, M.R. (1962c) *Trichodorus pakistanensis* n.sp. (Nematoda: Trichodoridae) with observations on *T. porosus* Allen, 1957, *T. mirzai* Siddiqi, 1960 and *T. minor* Colbran, 1956, from India. *Nematologica* 8, 193–200.

Siddiqi, M.R. (1963) *Trichodorus* spp. (Nematoda: Trichodoridae) from Tunisia and Nicaragua. *Nematologica* 9, 69–75.

Siddiqi, M.R. (1964a) Three new species of *Dorylaimoides* Thorne and Swanger, 1936, with a description of *Xiphinema orbum* n.sp. (Nematoda: Dorylaimoidea). *Nematologica* 9, 626–634.

Siddiqi, M.R. (1964b) *Xiphinema conurum* n.sp. and *Paralongidorus microlaimus* n.sp. with a key to the species of *Paralongidorus* (Nematoda: Longidoridae). *Proceedings of the Helminthological Society of Washington* 31, 133–137.

Siddiqi, M.R. (1965) *Longidorus nirulai* n.sp. a parasite of potato plants in Shillong, India, with a key to species of *Longidorus* (Nematoda: Dorylaimoidea). *Proceedings of the Helminthological Society of Washington* 32, 95–99.

Siddiqi, M.R. (1973) *Xiphinema americanum*. In: *CIH Descriptions of Plant-parasitic Nematodes*. Set 2, No. 29, 4pp.

Siddiqi, M.R. (1974a) Systematics of the genus *Trichodorus* Cobb, 1913 (Nematoda: Dorylaimida), with descriptions of three new species. *Nematologica* 19, 259–278.

Siddiqi, M.R. (1974b) *Xiphinema index*. In: *CIH Descriptions of Plant-parasitic Nematodes*. Set 3, No. 45, 4pp.

Siddiqi, M.R. (1974c) *Aphelenchoides ritzemabosi*. In: *CIH Descriptions of Plant-parasitic Nematodes*. Set 3, No. 32, 4pp.

Siddiqi, M.R. (1975) *Aphelenchoides fragariae*. In: *CIH Descriptions of Plant-parasitic Nematodes*. Set 5, No. 74, 4pp.

Siddiqi, M.R. (1979) Four new species of *Xiphinema* Cobb, 1913 (Nematoda: Dorylaimida) from Africa. *Revue de Nématologie* 2, 51–64.

Siddiqi, M.R. (1980a) On the generic status of *Atlantadorus* Siddiqi, 1974 and *Nanidorus* Siddiqi, 1974 (Nematoda: Trichodoridae). *Systematic Parasitology* 1, 151–152.

Siddiqi, M.R. (1980b) The origin and phylogeny of the nematode orders Tylenchida Thorne, 1949 and Aphelenchida n.ord. *Helminthological Abstracts, Series B* 49, 143–170.

Siddiqi, M.R. (1983a) Evolution of plant parasitism in nematodes. In: Stone, A.R., Platt, H.M. and Khalil, L.F. (eds), *Concepts in Nematode Systematics*. The Systematics Association Special Volume No. 22, Academic Press, London and New York, pp.113–129.

Siddiqi, M.R. (1983b) Phylogenetic relationships of the soil orders Dorylaimida, Mononchida, Triplonchida and Alaimida, with a revised classification of the subclass Enoplia. *Pakistan Journal of Nematology* 1, 79–110.

Siddiqi, M.R. (1992) Studies on plant-parasitic nematode genus *Monotrichodorus*, with descriptions of three new species from South America and *Paratrichodorus paramirzai* sp.n. from India. *Afro-Asian Journal of Nematology* 1, 180–191.

Siddiqi, M.R. and Brown, K.F. (1965) *Trichodorus rhodesiensis* and *Amphidelus trichurus*, two new nematode species from cultivated soils of Africa. *Proceedings of the Helminthological Society of Washington* 32, 239–242.

Siddiqi, M.R. and Franklin, M. (1967) *Aphelenchoides goodeyi* n.sp. (Nematoda: Aphelenchoidea), a mycophagous nematode from South India. *Nematologica* 13, 125–130.

Siddiqi, M.R. and Hawksworth, D.L. (1982) Nematodes associated with galls on *Cladonia glauca*, including two new species. *Lichenologist* 14, 175–184.

Siddiqi, M.R. and Husain, Z. (1965) *Paralongidorus beryllus* n.sp. (Nematoda: Dorylaimoidea) from India. *Proceedings of the Helminthological Society of Washington* 32, 243–245.

Siddiqi, M.R. and Lenne, J.M. (1990) *Xiphinema llanosum* and *Trophurus vultus*, two new plant nematode species from pasture soils in Colombia. *Journal of Nematology* 22, 262–267.

Siddiqi, M.R., Hooper, D.J. and Khan, E. (1963) A new nematode genus *Paralongidorus* (Nematoda: Dorylaimoidea) with description of two new species and observations on *Paralongidorus citri* (Siddiqi, 1959) n.comb. *Nematologica* 9, 7–14.

Siddiqi, M.R., Husain, S.I. and Khan, A.M. (1967) *Seinura propora* n.sp. and *Aphelenchoides aligarhiensis* n.sp. (Nematoda: Aphelenchoididae) from North India. *Nematologica* 13, 287–290.

Siddiqui, I.A. (1976) *Aphelenchoides bicaudatus*. In: *CIH Descriptions of Plant-parasitic Nematodes*. Set 6, No. 84, 3pp.

Siddiqui, I.A. and Taylor, D.P. (1967) Redescription of *Aphelenchoides bicaudatus* (Imamura, 1931) Filipjev & Schuurmans Stekhoven, 1941 (Nematoda: Aphelenchoididae), with a description of the previously undescribed male. *Nematologica* 13, 581–585.

Siddiqui, I.A. and Taylor, D.P. (1969) Feeding mechanisms of *Aphelenchoides bicaudatus* on three fungi and an alga. *Nematologica* 15, 503–509.

Simmons, A.M. and Rogers, C.E. (1990a) Distribution and prevalence of an ectoparasitic nematode, *Noctuidonema guyanense*, on moths of the fall armyworm (Lepidoptera: Noctuidae) in the tropical Americas. *Journal of Entomological Science* 25, 510–518.

Simmons, A.M. and Rogers, C.E. (1990b) Temperature and humidity effects on *Noctuidonema* (Nematoda: Aphelenchoididae), an ectoparasite of adult *Spodoptera frugiperda* (Lepidoptera: Noctuidae), and transfer success during host mating. *Annals of the Entomological Society of America* 83, 1084–1087.

Simmons, A.M. and Rogers, C.E. (1991) Dispersal and seasonal occurrence of *Noctuidonema guyanense*, an ectoparasitic nematode of adult fall armyworm (Lepidoptera: Noctuidae), in the United States. *Journal of Entomological Science* 26, 136–148.

Singh, S.D. (1967) On two new species of the genus *Aphelenchoides* Fischer, 1894 (Nematoda: Aphelenchoididae) from North India. *Journal of Helminthology* 41, 63–70.

Singh, S.P. (1969) A new species of the genus *Aphelenchoides* Fischer, 1894 (Nematoda: Aphelenchoididae) from rootlets of radish (*Raphanus sativus*) in Lucknow. *Journal of Helminthology* 43, 193–196.

Singh, S.P. (1976) *Xiphinema codiaei* n.sp. (Nematoda: Dorylaimoidea) from around the roots of garden croton. *Indian Journal of Zootomy* 17, 17–19.

Singh, S.P. (1977a) *Aphelenchoides wallacei* sp.n. and *Aphelenchoides jonesi* sp.n. (Nematoda: Aphelenchoididae) from inside the roots of papaya and egg plant. *Indian Journal of Nematology* 5, 207–213.

Singh, S.P. (1977b) Two plant parasitic nematodes of the genus *Seinura* Fuchs, 1931 (Nematoda: Aphelenchoididae) from Lucknow. *Records of the Zoological Survey of India* 72, 325–331.

Singh, S.P. and Agrawal, N. (1976) Description of a new species of the genus *Aphelenchus* Bastian (Nematoda: Aphelenchidae) from zamikand (*Amorphophallus campanulatus*). *Indian Journal of Zoology* 15, 193–195.

Skarbilovich, T.S. (1947) [Revision of the systematics of the nematode family Anguillulinidae Baylis and Daubney, 1926]. *Doklady Akademii Nauk SSR* 57, 307–308 (in Russian).

Skarbilovich, T.S. (1957) [Parasitic nematodes of clover and maize on collective farms of the Moscow, Voronezh, Lipetsk and Smolensk regions]. *Byulletin Nauchno - Tekhnicheskoi Informatsii Vsesoyuznogo Instituta Gel'mintologii Akad. K. I. Skryabina*, No. 4, 71–74 (in Russian).

Skarbilovich, T.S. and Potekhina, L.F. (1959) [Observations on nematodes in *Panax gin-*

seng]. *Proceedings of the All-Union Institute of Helminthology* 6, 411–414 (in Russian).

Skrjabin, K.I., Shikhobalova, N.P., Sobolev, A.A., Paramonov, A.A. and Sudarikov, V.E. (1954) [Descriptive catalogue of parasitic nematodes. Camallanata, Rhabditata, Tylenchata, Trichocephalata, Dioctophymata and the distribution of parasitic nematodes in their hosts]. In: Skrjabin, K.I. (ed.), *Opredelitel' Paraziticheskikh Nematod 4. Izdatel. Akad. Nauk SSSR* Moskva; 927pp (in Russian).

Slankis, A. (1967) [*Aphelenchoides macromucrons* n.sp. (Tylenchida) from *Ips typographus* L.]. *Materialy Nauchnoi Konferentsii Vsesoyuznogo Obshchestra Gel'mintologov, Year 1966*, Part 5, pp.279–282 (in Russian).

Sol, H.H., van Heuven, J.C. and Seinhorst, J.W. (1960) Transmission of rattle virus and *Atropa belladonna* mosaic virus by nematodes. *Tijdschrift over Plantenziekten*, 66, 228–231.

Southey, J.F. and Luc, M. (1974) Redescription of *Xiphinema ensiculiferum* (Cobb, 1893) Thorne, 1937 and description of *Xiphinema loosi* n.sp. and *Xiphinema hygrophilum* n.sp. (Nematoda: Dorylaimoidea). *Nematologica* 19 (1973), 293–307.

Spaull, V.W. and Heyns, J. (1991) Distribution of longidorid nematodes in sugar cane fields in South Africa. *Phytophylactica* 23, 167–172.

Stanton, J.M., Siddiqi, M.R. and Lenne, J.M. (1989) Plant-parasitic nematodes associated with tropical pastures in Colombia. *Nematropica* 19, 169–175.

Stegarescu, O.P. (1980) K sistematike nematod semeistva Longidoridae Thorne, 1935 (Nematoda: Dorylaimoidea) In: *Vozbuditeli Parazitarnykh Zabolevanii. Kishinev, USSR, 'Shtiintsa'*, 135–140.

Steiner, G. (1914a) Freilebende nematoden aus der Schweiz. 1. Teil einer vorlaüfigen Mitteilung. *Archiv für Hydrobiologie und Planktonkunde* 9, 259–276.

Steiner, G. (1914b) Freilebende nematoden aus der Schweiz. 2. Teil einer vorlaüfigen Mitteilung. *Archiv für Hydrobiologie und Planktonkunde* 9, 420–438.

Steiner, G. (1920) Freilebende Süsswassernematoden aus peruanischen Hochgebirgsseen (Huaron 5140 m. ü. M. und Naticocha 5140 m. ü. M.). *Revue Suisse de Zoologie* 28, 11–44.

Steiner, G. (1926) Parasitic nemas on peanuts in South Africa. *Zentralblatt für Bakteriologie, Parasitenkunde, Infektionskrankheiten und Hygiene. Abteilung* 2, 67, 351–365.

Steiner, G. (1931) On the status of the nemic genera *Aphelenchus* Bastian, *Pathoaphelenchus* Cobb, *Paraphelenchus* Micoletzky, *Parasitaphelenchus* Fuchs, *Isonchus* Cobb, and *Seinura* Fuchs. *Journal of the Washington Academy of Science* 21, 468–475.

Steiner, G. (1932a) Annotations on the nomenclature of some plant parasitic nematodes. *Journal of the Washington Academy of Science* 22, 517–518.

Steiner, G. (1932b) Notes on nemic diseases. *Plant Disease Reporter* 16, 2–3.

Steiner, G. (1934a) Observations on nematodes parasitic in tubers of the cinnamon-vine (*Dioscorea batatas*). *Proceedings of the Helminthological Society of Washington* 1, 15–17.

Steiner, G. (1934b) A new species of the nematode genus *Aphelenchoides* living in sugar cane. *Journal of the Washington Academy of Science* 24, 141–143.

Steiner, G. (1935) Opuscula miscellanea nematologica, II. *Proceedings of the Helminthological Society of Washington* 2, 104–110.

Steiner, G. (1936a) Opuscula miscellanea nematologica, IV. *Proceedings of the Helminthological Society of Washington* 3, 74–80.

Steiner, G. (1936b) The status of the nematode *Aphelenchus avenae* Bastian, 1865, as a plant parasite. *Phytopathology* 26, 294–295.

Steiner, G. (1937) Opuscula miscellanea nematologica, VI. *Proceedings of the Helminthological Society of Washington* 4, 48–52.

Steiner, G. (1941) Nematodes parasitic on and associated with roots of marigolds (*Tagetes* hybrids). *Proceedings of the Biological Society of Washington* 54, 31–34.

Steiner, G. (1942) Opuscula miscellanea nematologica IX. *Proceedings of the Helminthological Society of Washington* 9, 32–38.

Steiner, G. (1943) New nematodes associated with a disease of the papaya in Chile. *Boln. Dep. Sanid. veg. Santiago, Chile* 3, 95–116.

Steiner, G. (1949) Plant nematodes the grower should know. *Proceedings of the Soil Science Society of Florida* 4B, 72–117.

Steiner, G. and Buhrer, E.M. (1932a) Miscellaneous notes on nemic diseases. *Plant Disease Reporter* 16, 137.

Steiner, G. and Buhrer, E.M. (1932b) The nonspecificity of the brown-ring symptoms in narcissus attacked by nematodes. *Phytopathology* 22, 927–928.

Steiner, G. and Buhrer, E.M. (1933) The nematode *Tylenchus similis*, Cobb as a parasite of the tea plant (*Thea sinensis*, Linn.), its sexual dimorphism, and its nemic associates in the same host. *Zeitschrift für Parasitenkunde*, 5, 412–420.

Steiner, G. and Buhrer, E.M. (1934) *Aphelenchoides xylophilus*, n.sp. a nematode associated with blue-stain and other fungi in timber. *Journal of Agricultural Research*, 48, 949–951.

Stewart, F.H. (1921) The anatomy and biology of the parasitic Aphelenchi. *Parasitology* 13, 160–179.

Stirling, G.R. and McCulloch, J.S. (1984) *Paralongidorus australis* n.sp. (Nematoda: Longidoridae), causing poor root growth of rice in Australia. *Nematologica* 30, 387–394.

Stirling, G.R. and Shannon, E.L. (1986) Needle nematodes. A potential problem for the Burdekin rice industry. *Queensland Agricultural Journal* 112, 258–261.

Stirling, G.R., Vawdrey, L.L. and Shannon, E.L. (1989) Options for controlling needle nematode (*Paralongidorus australis*) and preventing damage to rice in northern Queensland. *Australian Journal of Experimental Agriculture* 29, 223–232.

Stocker, G. and Kruger, J.C. de W. (1988) The genus *Xiphinema* in South Africa. XV. Redescription of *X. mluci* Heyns, 1976 and description of three closely related new species (Nematoda: Dorylaimida). *Phytophylactica* 19 (1987), 405–418.

Stoyanov, D. (1964) [A contribution to the nematode fauna of the grapevine]. *Rastitelna Zashtita* 12, 16–24 (in Bulgarian).

Strümpel, H. (1967) Beobachtungen zur Lebensweise von *Aphelenchoides fragariae* in Lorraine-Begonien. *Nematologica* 13, 67–72.

Sturhan, D. (1963) Beitrag zur Systematik der Gattung *Longidorus*. *Nematologica* 9, 131–142.

Sturhan, D. (1978) Zwei neue *Xiphinema*-Arten aus Deutschland (Nematoda: Dorylaimida). *Nematologica* 24, 19–28.

Sturhan, D. (1983) Descriptions of two new *Xiphinema* species from Portugal, with notes on *X. pachtaicum* and *X. opisthohysterum* (Nematoda: Longidoridae). *Nematologica* 29 (1982), 270–283.

Sturhan, D. (1985a) Ein neuer Phytonematode aus Deutschland: *Paratrichodorus weischeri* spec. nov. (Nematoda: Trichodoridae). *Mitteilungen aus der Biolischen Bundesanstalt für Land-und Forstwirtschaft* 226, 31–45.

Sturhan, D. (1985b) Untersuchungen über den *Xiphinema coxi*-Komplex (Nematoda: Longidoridae). *Nematologica* 30 (1984), 305–323.

Sturhan, D. (1989) Hermaphroditism in *Paratrichodorus* species (Nemata: Dorylaimida). *Revue de Nématologie* 12, 273–276.

Sturhan, D. and Argo, D. (1983) Studies on *Longidorus closelongatus* Stoyanov and *L.*

cohni Heyns, with description of *L. proximus* sp.nov. (Nematoda: Dorylaimida). *Revue de Nématologie* 6, 57–64.

Sturhan, D. and Barooti, Sh. (1983) *Longidorus iranicus* n.sp. (Nematoda: Dorylaimida) *Systematic Parasitology* 5, 21–24.

Sturhan, D. and Weischer, B. (1964) *Longidorus vineacola* n.sp. *Nematologica* 10, 335–341.

Sudakova, I.M. (1958) [On the phytonematode fauna of the Chuvash ASSR]. *Zoologicheskii Zhurnal* 37, 134–139 (in Russian).

Sukul, N.C. (1972) *Paralongidorus droseri* sp.nov. (Nematoda: Longidoridae) associated with an insectivorous plant from West Bengal. *Bulletin of Entomology* 12 (1971), 85–88.

Suryawanshi, M.V. (1971) Studies on Aphelenchoidea (Nematoda) from Marathwada, India, with descriptions of three new species and a discussion on the validity of *Aphelenchus radicicolus* (Cobb, 1913) Steiner, 1931. *Nematologica* 17, 417–427.

Sutherland, J.R. (1969) Feeding of *Xiphinema bakeri*. *Phytopathology* 59, 1963–1965.

Swart, A. and Heyns, J. (1987a) Comparative morphology of the head regions of some longidorid nematodes from South Africa using the SEM. *Phytophylactica* 19, 99–101.

Swart, A. and Heyns, J. (1987b) Morphological study of longidorid nematode stylets using the SEM. *Phytophylactica* 19, 103–106.

Szczygiel, A. (1968) *Trichodorus sparsus* sp.n. (Nematoda: Trichodoridae). *Bulletin de l'Académie Polonaise des Sciences* 16, 695–698.

Tandon, R.S. and Singh, S.P. (1970) On two new nematodes (Aphelenchoidea) from tobacco roots in India. *Journal of Helminthology* 44, 323–328.

Tandon, R.S. and Singh, S.P. (1973) Two plant parasites of two different families of nematodes parasitising lady finger (*Abelmoschus esculentus*) at Lucknow. *Zoologischer Anzeiger* 191, 139–150.

Tandon, R.S. and Singh, S.P. (1974) Nematode parasites of the common sponge gourd, *Luffa cylindrica* from Lucknow. *Geobios* 1, 24–27.

Tarjan, A.C. (1956) Known and suspected plant-parasitic nematodes of Rhode Island. II. *Xiphinema americanum* with notes on *Tylencholaimus brevicaudatus* n.comb. *Proceedings of the Helminthological Society of Washington* 23, 88–92.

Tarjan, A.C. (1958) Note on the selection of types for certain nematode species. *Nematologica* 3, 79–80.

Tarjan, A.C. (1964) Two new American dagger nematodes (*Xiphinema*: Dorylaimidae) associated with citrus, with comments on the variability of *X. bakeri* Williams, 1961. *Proceedings of the Helminthological Society of Washington* 31, 65–76.

Tarjan, A.C. and Aragon, C.B. (1982) An analysis of the genus *Bursaphelenchus* Fuchs, 1937. *Nematropica* 12, 121–144.

Tarjan, A.C. and Hopper, B.E. (1974) *Nomenclatorial Compilation of Plant and Soil Nematodes*. Society of Nematologists, 419pp.

Taylor, A.M. (1917) Blackcurrant eelworm. *Journal of Agricultural Science* 8, 247–275.

Taylor, C.E. (1978) Plant-parasitic Dorylaimida: Biology and virus transmission. In: J.E. Southey (ed.), *Plant Nematology*, HMSO, London, pp.232–243.

Taylor, C.E. and Brown. D.J.F. (1976) The geographical distribution of *Xiphinema* and *Longidorus* nematodes in the British Isles and Ireland. *Annals of Applied Biology* 84, 383–402.

Taylor, C.E. and Raski, D.J. (1964) On transmission of grape fanleaf by *Xiphinema index*. *Nematologica* 10, 489–495.

Taylor, C.E. and Robertson, W.M. (1969) The location of raspberry ringspot and tomato black ring viruses in the nematode vector, *Longidorus elongatus* (de Man). *Annals of Applied Biology*, 64, 233–237.

Taylor, C.E. and Robertson, W.M. (1970a) Sites of virus retention in the alimentary tract of the nematode vectors, *Xiphinema diversicaudatum* (Micol.) and *X. index* Thorne and Allen. *Annals of Applied Biology* 66, 375–380.

Taylor, C.E. and Robertson, W.M. (1970b) Location of tobacco rattle virus in the nematode vector, *Trichodorus pachydermus* Seinhorst. *Journal of General Virology* 6, 179–182.

Taylor, C.E. and Robertson, W.M. (1971) Ultrastructure of the guide ring and guiding sheath in *Xiphinema* and *Longidorus*. *Nematologica* 17, 303–307.

Taylor, C.E. and Robertson, W.M. (1975) Acquisition, retention and transmission of viruses by nematodes. In: Lamberti, F., Taylor, C.E. and Seinhorst, J.W. (eds), *Nematode Vectors of Plant Viruses*. Plenum Press, London and New York, pp.253–276.

Taylor, D.P. and Pillai, J.K. (1967) *Paraphelenchus acontioides* n.sp. (Nematoda: Paraphelenchidae), a mycophagous nematode from Illinois, with observations on its feeding habits and a key to the species of *Paraphelenchus*. *Proceedings of the Helminthological Society of Washington* 34, 51–54.

Thomas, P.R. (1969) Population development of *Longidorus elongatus* on strawberry in Scotland with observations on *Xiphinema diversicaudatum* on raspberry. *Nematologica* 15, 582–590.

Thong, C.H.S. and Webster, J.M. (1983) Nematode parasites and associates of *Dendroctonus* spp. and *Trypodendron lineatum* (Coleoptera: Scolytidae) with a description of *Bursaphelenchus varicauda* n.sp. *Journal of Nematology* 15, 312–318.

Thorne, G. (1929) Nematodes from the summit of Long's Peak, Colorado. *Transactions of the American Microscopical Society* 48, 181–195.

Thorne, G. (1934) The classification of the higher groups of dorylaims. *Proceedings of the Helminthological Society of Washington* 1, 19.

Thorne, G. (1935a) Nemic parasites and associates of the mountain pine beetle (*Dendroctonus monticolae*) in Utah. *Journal of Agricultural Research* 51, 131–144.

Thorne, G. (1935b) Notes on free-living and plant-parasitic nematodes II. *Proceedings of the Helminthological Society of Washington* 2, 96–98.

Thorne, G. (1937) Notes on free-living and plant-parasitic nematodes. III. *Proceedings of the Helminthological Society of Washington* 4, 16–18.

Thorne, G. (1939) A monograph of the nematodes of the superfamily Dorylaimoidea. *Capita zoologica* 8, 1–261.

Thorne, G. (1949) On the classification of the Tylenchida, new order (Nematoda: Phasmidia). *Proceedings of the Helminthological Society of Washington* 16, 37–73.

Thorne, G. (1961) *Principles of Nematology*. McGraw-Hill Book Co., Inc., New York, xiv + 553pp.

Thorne, G. (1974) Nematodes of the Northern Great Plains. Part II, Dorylaimoidea in part (Nemata: Adenophorea). *Technical Bulletin of the South Dakota State University Agricultural Experiment Station* 41, 1–120.

Thorne, G. and Allen, M.W. (1950) *Paratylenchus hamatus* n.sp. and *Xiphinema index* n.sp., two new nematodes associated with fig roots, with a note on *Paratylenchus anceps* Cobb. *Proceedings of the Helminthological Society of Washington* 17, 27–35.

Thorne, G. and Malek, R.B. (1968) Nematodes of the Northern Great Plains. Part I. Tylenchida (Nemata: Secernentea). *Technical Bulletin of the South Dakota Agricultural Experiment Station* 31, 1–111.

Thorne, G. and Swanger, H.H. (1936) A monograph of the nematode genera *Dorylaimus* Dujardin, *Aporcelaimus* n.g., *Dorylaimoides* n.g. and *Pungentus* n.g. *Capita zoologica* 6, 1–223.

Tikyani, M.G., Khera, S. and Bhatnagar, G.C., (1970) *Aphelenchoides jodhpurensis* n.sp. from soil of great millet from Rajasthan, India. *Zoologischer Anzeiger* 184, 239–241.

Timm, R.W. (1960) *Paraseinura* (Nematoda: Aphelenchoididae), a new genus from East Pakistan. *Nematologica* 5, 171–174.

Timm, R.W. and Franklin, M.T. (1969) Two marine species of *Aphelenchoides*. *Nematologica* 15, 370–375.

Tomalak, M., Michalski, J. and Grocholski, J. (1984) The influence of nematodes on the structure of genitalia of *Tomicus piniperda* (Coleoptera: Scolytidae). *Journal of Invertebrate Pathology* 43, 358–362.

Towle, A. and Doncaster, C.C. (1978) Feeding of *Longidorus caespiticola* on rye-grass, *Lolium perenne*. *Nematologica* 24, 277–285.

Triantaphyllou, A.C. (1971) Genetics and cytology. In: Zuckerman, B.M., Mai, W.F. and Rohde, R.A. (eds), *Plant Parasitic Nematodes*, Vol. II. Academic Press Inc., New York, pp.1–34.

Trudgill, D.L. (1976) Observations on the feeding of *Xiphinema diversicaudatum*. *Nematologica* 22, 417–423.

Truskova, G.M. (1973) [Two new species of *Aphelenchoides* Fischer, 1894, (Nematoda: Aphelenchoididae)]. *Parazitologiya* 7, 188–189 (in Russian).

Truskova, G.M. and Eroshenko, A.S. (1977) [The nematode fauna of herbaceous and ligneous plants in the pine plantations of the Primor'yal]. *Trudy Biol. Inst. Nov. Ser.* 47, 35–49 (in Russian).

Tulaganov, A.T. (1937) Nematoden der Tomate und des sie umgebenden Bodens. *Zoologischer Anzeiger* 118, 283–285.

Tulaganov, A.T. (1938) [The fauna of nematodes of cotton and surrounding soil in Katta-Kurgan district of the Uzbek SSR]. *Trudy Uzbekskogo Gosudarstvennogo Universiteta* 12, 1–25 (in Russian).

Tulaganov, A.T. (1949) [Plant parasitic and soil nematodes in the Uzbekistan]. *Inst. bot. zool. AN Uzb. SSR. Tashkent* 1–227 (in Russian).

Uesugi, C.H., Huang, C.S. and Cares, J.E. (1985) *Xiphidorus amazonensis* n.sp. (Nematoda: Longidoridae) from the Brazilian Amazon Basin. *Journal of Nematology* 17, 310–313.

Vanha, J.J. (1893) Neue Rübennematoden, ihre Schädlichkeit und Verbreitung. *Zeitschrift Zuckerindustrie Böhmen* 17, 281–298.

van Hoof, H.A. (1962) *Trichodorus pachydermus* and *T. teres*, vectors of the early browning virus of peas. *Tijdschrift Plantenziekten* 68, 391–396.

van Hoof, H.A. (1964) Serial transmission of rattle virus by a single male of *Trichodorus pachydermus* Seinhorst. *Nematologica* 10, 141–144.

van Hoof, H.A. (1970) Some observations on retention of tobacco rattle virus in nematodes. *Netherlands Journal of Plant Pathology* 76, 329–330.

Verma, R.S. (1973) Two new species in the subfamily Longidorinae (Nematoda) from Uttar Pradesh, India, with a key to species of *Paralongidorus* Siddiqi *et al*., 1963. *Zoologischer Anzeiger* 190, 170–174.

Vermeulen, W.J.J., and Heyns, J. (1983) Studies on Trichodoridae (Nematoda: Dorylaimida) from South Africa. *Phytophylactica* 15, 17–34.

Vermeulen, W.J.J., and Heyns, J. (1984) Further studies on southern African Trichodoridae (Nematoda: Dorylaimida). *Phytophylactica* 16, 301–305.

Vinciguerra, M. (1988) Identification of dorylaims, particularly longidorids and related genera. In: Fortuner, R. (ed.), *Nematode Identification and Expert System Technology*. Plenum Publishing Corporation, New York, pp.171–175.

von Linstow, O.F.B. (1876) Helminthologische Beobachtungen. *Archiv für Naturgeschichte, Berlin, 42 J.* 1, 1–18.

von Linstow, O.F.B. (1879) Helminthologische Studien. *Archiv für Naturgeschichte, Berlin, 45 J.* 1, 165–188.

Vovlas, N., Inserra, R.N. and Greco, N. (1992) *Schistonchus caprifici* parasitizing caprifig (*Ficus carica sylvestris*) florets and the relationship with its fig wasp (*Blastophaga psenes*) vector. *Nematologica* 38, 215–226.

Wachek, F. (1955) Die entoparasitischen Tylenchiden. *Parasitologische Schriftenreihe* 3, 1–119.

Walkinshaw, C.H., Griffin, G.D. and Larson, R.H. (1961) *Trichodorus christiei* as a vector of potato corky ring-spot (tobacco rattle) virus. *Phytopathology* 51, 806–808.

Warwick, D.R.N. and Bezerra, A.P.T. (1992) Possible root transmission of the red ring nematode (*Rhadinaphelenchus cocophilus*) to coconut palms. *Plant Disease*, 76, 809–811.

Wasilewska, L. (1969) *Aphelenchoides dubius* n.sp. from Poland (Nematoda: Aphelenchoidea). *Bulletin de l'Académie Polonaise der Sciences. Classe II. Série des Sciences Biologiques* 17, 455–458.

Webster, J.M., Anderson, R.V., Baillie, D.Z., Beckenbach, K., Curran, J. and Rutherford, T.A. (1990) DNA probes for differentiating isolates of the pine wood nematode species complex. *Revue de Nématologie* 13, 255–263.

Weischer, B. (1975) Ecology of *Xiphinema* and *Longidorus*. In: Lamberti, F., Taylor, C.E. and Seinhorst, J.W. (eds), *Nematode Vectors of Plant Viruses*. Plenum Press, London and New York, pp.291–307.

Weischer, B. and Wyss, U. (1976) Feeding behaviour and pathogenicity of *Xiphinema index* on grapevine roots. *Nematologica* 22, 319–325.

Whitehead, A.G., and Hooper, D.J. (1970) Needle nematodes (*Longidorus* spp.) and stubby-root nematodes (*Trichodorus* spp.) harmful to sugar beet and other field crops in England. *Annals of Applied Biology* 65, 339–350.

Whitehead, A.G., Dunning, R.A. and Cooke, D.A. (1971) Docking disorder and root ectoparasitic nematodes of sugar beet. *Report of Rothamsted Experimental Station*, 1970, pp.219–236.

Williams, J.R. (1959) Studies on the nematode soil fauna of sugar cane fields in Mauritius. 3. Dorylaimidae (Dorylaimoidea, Enoplida). *Occasional Papers of the Mauritius Sugar Cane Industry Research Institute* 3, 1–28.

Williams, T.D. (1961) *Xiphinema bakeri* n.sp. (Nematoda: Longidoridae) from the Fraser River Valley, British Columbia, Canada. *Canadian Journal of Zoology* 39, 407–412.

Winfield, A.L., and Cooke, D.A. (1975) The ecology of *Trichodorus*. In: Lamberti, F., Taylor, C.E. and Seinhorst, J.W. (eds), *Nematode vectors of Plant Viruses*. Plenum Press, London, pp.309–341.

Wingfield, M.J. (1983) Transmission of pine wood nematode to cut timber and girdled trees. *Plant Disease* 67, 35–37.

Wood, F.H. (1975) Biology of *Seinura demani* (Nematoda: Aphelenchoididae). *Nematologica* 20, 347–353.

Wright, K.A., Carter, R.F. and Robertson, W.M. (1983) The musculature of the anterior feeding apparatus of *Xiphinema* species (Nematoda: Dorylaimida). *Nematologica* 21, 49–64.

Wyss, U. (1970a) Untersuchungen zur Populationsdynamik von *Longidorus elongatus*. *Nematologica* 16, 74–84.

Wyss, U. (1970b) Zur Bestimmung von *Trichodorus* Arten. *Zeitschrift für Pflanzenkrankheiten und Pflanzenschutz* 9, 506–514.

Wyss, U.(1971a) Saugverhalten und Pathogenität von *Trichodorus* spp. in steriler Agarkultur. *Nematologica* 17, 501–507.

Wyss, U (1971b) Der Mechanismus der Nahrungsaufnahme bei *Trichodorus similis*. *Nematologica* 17, 508–518.

Wyss, U. (1972) *Trichodorus similis* (Nematoda) Saugen an Wurzeln von Sämlingen (Rübsen). *Film E 1763 der Enc. Cin. Göttingen.*

Wyss, U. (1974a) *Trichodorus similis* (Nematoda) Reaktion der Protoplasten von Wurzelhaaren (*Nicotiana tabacum*) auf den Saugvorgang. *Film E 2045 der Enc. Cin. Göttingen.*

Wyss, U. (1974b) *Longidorus elongatus* (Nematoda) Embryonalentwicklung. *Film E 2046 der Enc. Cin. Göttingen.*

Wyss, U. (1974c) *Trichodorus similis.* In: *CIH Descriptions of Plant-parasitic Nematodes,* Set 4, No.59, 4pp.

Wyss, U. (1975) Feeding of *Trichodorus, Longidorus* and *Xiphinema.* In: Lamberti, F., Taylor, C.E. and Seinhorst, J.W. (eds), *Nematode Vectors of Plant Viruses.* Plenum Press, London and New York, pp.203–221.

Wyss, U. (1978a) Feeding phases of *Xiphinema index* and associated processes in the feeding apparatus. *Nematologica* 23, 463–470.

Wyss, U. (1978b) Root and cell response to feeding by *Xiphinema index. Nematologica* 24, 159–166.

Wyss, U., Robertson, W.M. and Trudgill, D.L. (1988) Oesophageal bulb function of *Xiphinema index* and associated root cell responses, assessed by video-enhanced contrast light microscopy. *Revue de Nématologie* 11, 253–261.

Xu, J. and Cheng, H. (1990) A new species of *Longidorus* (Nematoda: Longidoridae). *Journal of Nanjing Agricultural University* 14 , 38–42.

Xu, J. and Cheng, H. (1992) *Longidorus litchii* n.sp. and *L. henanus* n.sp. (Nemata: Longidoridae) from China. *Fundamental and Applied Nematology* 15, 517–523.

Xu, J. and Hooper, D.J. (1990) Observations on some species of *Longidorus* (Nematoda: Longidoridae) from Jiangsu Province, China, with a description of *Longidorus jiangsuensis* n.sp. *Revue de Nématologie* 13, 323–330.

Yagita, H. (1975) The life-history and biology of the needle nematode, *Longidorus martini* Merny. I. Morphological measurements and developmental stages. *Japanese Journal of Nematology* 5, 10–15.

Yang, B.J. (1985) The identification of nematodes on pine trees. *Scientia silvae Sinica* 21, 305–309.

Yeates, G.W. (1967) Studies on nematodes from dune sands. 4. Diphtherophoroidea. *New Zealand Journal of Science* 10, 322–329.

Yeates, G.W., Van Etteger, M. and Hooper, D.J. (1992) *Longidorus orongorongensis* n.sp. (Nematoda: Dorylaimida) from subsoil of conifer/broadleaved forest. *New Zealand Journal of Zoology* 19, 25–31.

Yin, K., Fang, Y. and Tarjan, A.C. (1988) A key to species in the genus *Bursaphelenchus* with a description of *Bursaphelenchus hunanensis* sp.n. (Nematoda: Aphelenchoididae) found in pinewood in Hunan Province, China. *Proceedings of the Helminthological Society of Washington* 55, 1–11.

Yin, K.C. and Feng, Z.X. (1981) [The investigation of plant nematodes]. *Acta Phytophylactica Sinica* 8, 111–126 (in Chinese)

Yokobori, M. (1986) Catastrophic withering of pine forests by pine wilt disease in Ibaraki Prefecture and its environmental factors. *Bulletin of the Ibaraki Prefectural Forest Experiment Station, Japan,* No. 15, 26pp.

Yokoo, T. (1948) *Aphelenchoides oryzae* Yokoo n.sp., a nematode parasitic to rice plant. *Annals of the Phytopathology Society of Japan* 13, 40–43.

Yokoo, T. (1964a) On the stubby root nematodes from the Western Japan. *Bulletin of the Faculty of Agriculture, Saga University* 20, 57–62.

Yokoo, T. (1964b) On a new species of *Aphelenchoides* (Aphelenchoididae: Nematoda), parasite of bulb of lily, from Japan. *Bulletin of the Faculty of Agriculture, Saga University* 20, 67–69.

Yokoo, T. (1966) On a new stubby root nematode (*Trichodorus kurumeensis* n.sp.) from Kyushu, Japan. *Bulletin of the Faculty of Agriculture, Saga University* 23, 1–6.

Zeidan, A.B. and Geraert, E. (1992) *Aphelenchoides*, *Aphelenchus* and *Paraphelenchus* from Sudan with the description of two new species. *Nematologica* 31, 420–438.

Zhang, B.C. and Huang, Y.C. (1990) A list of important plant diseases in China. *Review of Plant Pathology* 69, 97–118.

Zimmermann, A.W.P. (1898) De Nematoden der koffiewortels. Deel I. *Mededeelingen uit's Lands plantentuin* 27, 1–64.

Zullini, A. (1973) Some soil and freshwater nematodes from Chiapas (Mexico). *Accademia Nazionale dei Lincei Quaderno* 171, 55–96.

Index of Nematode Taxa

Printed and bound by CPI Group (UK) Ltd, Croydon, CR0 4YY
21/06/2026
14904329-0001